SECOND EDITION

Linear Algebra

Stephen H. Friedberg
Arnold J. Insel
Lawrence E. Spence
Illinois State University

Prentice Hall, Englewood Cliffs, New Jersey 07632

Library of Congress Cataloging-in-Publication Data

FRIEDBERG, STEPHEN H.
Linear algebra/Stephen H. Friedberg, Arnold J. Insel, Lawrence E. Spence.—2nd ed.
p. cm.
Includes indexes.
ISBN 0-13-537102-3
1. Algebras, Linear. I. Insel, Arnold J. II. Spence, Lawrence E. III. Title.
QA184.F75 1989 88-28568
512'.5—dc19 CIP

Editorial/production supervision
and interior design: *Kathleen M. Lafferty*
Cover design: *Wanda Lubelska*
Manufacturing buyer: *Paula Massenaro*

Printed in the United States of America

10 9 8 7 6 5 4 3 2 1

ISBN 0-13-537102-3

Prentice-Hall International (UK) Limited, *London*
Prentice-Hall of Australia Pty. Limited, *Sydney*
Prentice-Hall Canada Inc., *Toronto*
Prentice-Hall Hispanoamericana, S.A., *Mexico*
Prentice-Hall of India Private Limited, *New Delhi*
Prentice-Hall of Japan, Inc., *Tokyo*
Simon & Schuster Asia Pte. Ltd., *Singapore*
Editora Prentice-Hall do Brasil, Ltda., *Rio de Janeiro*

To our families

Contents

*Sections denoted by an asterisk are optional.

Preface

The language and concepts of matrix theory and, more generally, of linear algebra have come into widespread usage in the social and natural sciences, computer science, and statistics. In addition, linear algebra continues to be of great importance in modern treatments of geometry and analysis.

The primary purpose of *Linear Algebra, Second Edition*, is to present a careful treatment of the principal topics of linear algebra and to illustrate the power of the subject through a variety of applications. Although the only formal prerequisite for the book is a one-year course in calculus, the material in Chapters 6 and 7 requires the mathematical sophistication of typical college juniors and seniors (who may or may not have had some previous exposure to the subject).

The book is organized to permit a number of different courses (ranging from three to six semester hours in length) to be taught from it. The core material (vector spaces, linear transformations and matrices, systems of linear equations, determinants, and diagonalization) is found in Chapters 1 through 5. The remaining chapters, treating inner product spaces and canonical forms, are completely independent and may be studied in any order. In addition, throughout the book are a variety of applications to such areas as differential equations, economics, geometry, and physics. These applications are not central to the mathematical development, however, and may be excluded at the discretion of the instructor.

We have attempted to make it possible for many of the important topics of linear algebra to be covered in a one-semester course. This goal has led us to develop the major topics with fewer unnecessary preliminaries than in a traditional approach. (Our treatment of the Jordan canonical form, for instance, does not require any theory of polynomials.) The resulting economy permits us to cover most of the book (omitting many of the optional sections and a detailed discussion of determinants) in a one-semester four-hour course for students who have had some prior exposure to linear algebra.

Chapter 1 of the book presents the basic theory of finite-dimensional vector spaces—subspaces, linear combinations, linear dependence and independence, bases, and dimension. The chapter concludes with an optional section in which we prove the existence of a basis in infinite-dimensional vector spaces.

Linear transformations and their relationship to matrices are the subject of Chapter 2. We discuss there the null space and range of a linear transformation, matrix representations of a transformation, isomorphisms, and change of coordinates. Optional sections on dual spaces and homogeneous linear differential equations end the chapter.

The applications of vector space theory and linear transformations to systems of linear equations are found in Chapter 3. We have chosen to defer this important subject so that it can be presented as a consequence of the preceding material. This approach allows the familiar topic of linear systems to illuminate the abstract theory and permits us to avoid messy matrix computations in the presentation of Chapters 1 and 2. There will be occasional examples in these chapters, however, where we shall want to solve systems of linear equations. (Of course, these examples will not be a part of the theoretical development.) The necessary background is contained in Section 1.4.

Determinants, the subject of Chapter 4, are of much less importance than they once were. In a short course we prefer to treat determinants lightly so that more time may be devoted to the material in Chapters 5 through 7. Consequently we have presented two alternatives in Chapter 4—a complete development of the theory (Sections 4.1 through 4.3) and a summary of the important facts that are needed for the remaining chapters (Section 4.4).

Chapter 5 discusses eigenvalues, eigenvectors, and diagonalization. One of the most important applications of this material occurs in computing matrix limits. We have therefore included an optional section on matrix limits and Markov chains in this chapter even though the most general statement of some of the results requires a knowledge of the Jordan canonical form. Section 5.4 contains material on invariant subspaces and the Cayley–Hamilton theorem.

Inner product spaces are the subject of Chapter 6. The basic mathematical theory (inner products; the Gram–Schmidt process; orthogonal complements; adjoint transformations; normal, self-adjoint, orthogonal, and unitary operators; orthogonal projections; and the spectral theorem) is contained in Sections 6.1 through 6.7. Sections 6.8, 6.9, and 6.10 contain diverse applications of the rich inner product space structure.

Canonical forms are treated in Chapter 7. Sections 7.1 and 7.2 develop the Jordan form, Section 7.3 presents the minimal polynomial, and Section 7.4 discusses the rational form.

There are five appendices. The first four, which discuss sets, functions, fields, and complex numbers, respectively, are intended to review basic ideas used throughout the book. Appendix E on polynomials is used primarily in Chapters 5 and 7, especially in Section 7.4. We prefer not to discuss the appendices independently but rather to refer to them as the need arises.

The following diagram illustrates the dependencies among the various chapters.

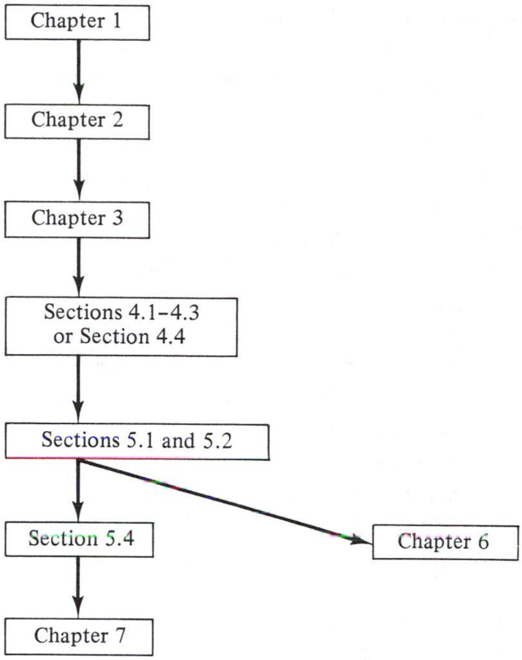

One final word is required about our notation. Sections denoted by an asterisk (*) are optional and may be omitted as the instructor sees fit. An exercise denoted by the dagger symbol (†) is not optional, however—we use this symbol to identify an exercise that will be cited at some later point of the text.

Special thanks are due to Kathleen M. Lafferty for her cooperation and fine work during the production process.

DIFFERENCES BETWEEN THE FIRST AND SECOND EDITIONS

In the 10 years that have elapsed since the first edition, we have discovered more efficient ways to present a number of key topics. Most of the material that relied on direct sums in the first edition has been rewritten, and the results about direct sums are contained in optional subsections or relegated to the exercises. Also, a number of exercises and examples have been added to reflect the changes made in the existing topics and the additional new material.

In Chapter 1 many proofs are more readable without the dependence on direct sums. Section 1.3 no longer contains any results about direct sums except in the exercises. Section 1.4 now employs Gaussian elimination.

In Chapter 2 the proof of the dimension theorem in Section 2.1 no longer uses direct sums. In Section 2.3 the proof of Theorem 2.15, and in Section 2.4 the proof of Theorem 2.21, have been improved. Also, exercises on quotient spaces have been added to Section 2.4. The material on the change of coordinate matrix in Section 2.5 has been rewritten. In particular, Theorem 2.24 is less general, but more understandable; the general case of Theorem 2.24 is included in the exercises.

In Chapter 3 systems of equations are now solved by the more efficient Gaussian elimination method, rather than by the Gauss–Jordan technique used in the first edition.

In Chapter 4 a considerably simplified proof of Cramer's rule is presented that no longer depends on the classical adjoint. In fact, the classical adjoint now appears as a simple exercise that exploits the new development of Cramer's rule.

In Chapter 5 the proofs of Theorems 5.10 and 5.12 in Section 5.2 have been simplified. Theorem 5.14 characterizes diagonalizability without the earlier dependence on direct sums. Direct sums are now treated as an optional subsection. Gerschgorin's disk theorem is now contained in Section 5.3, and Section 5.4 combines the Cayley–Hamilton theorem and invariant subspaces. The minimal polynomial is now found in Section 7.3.

Chapters 6 and 7 in the first edition have been interchanged in the second edition. This alteration not only reflects the authors' tastes, but those of a number of our readers.

In Chapter 6 the material in Section 6.2 has been rewritten to improve the clarity and flow of the topics. The results needed for least squares now appear in Section 6.2 and no longer rely on direct sums. Least squares and minimal solutions are introduced earlier and are now contained in Section 6.3. In Section 6.4 the addition of Schur's theorem allows a completely new and improved approach to normal and self-adjoint operators which no longer relies on results concerning T^*-invariant subspaces. These results are now exercises. Section 6.5 has been rewritten to take advantage of the new approach in Section 6.4. A new subsection of Section 6.5 treats rigid motions in the plane. Sylvester's law of inertia is now included with the material on bilinear forms in Section 6.6. For reasons of continuity, the optional sections on the special theory of relativity, conditioning, and the geometry of orthogonal operators are now located at the end of the chapter.

In Chapter 7 the proofs in Section 7.1 have been improved. In Section 7.2 the proof of the Jordan canonical form has been simplified and no longer uses direct sums, which are dealt with in a subsection. Section 7.3 now treats the minimal polynomial. In Section 7.4 the proof of the rational canonical form has been simplified and no longer uses either direct sums or quotient spaces.

Stephen H. Friedberg
Arnold J. Insel
Lawrence E. Spence

1

Vector Spaces

1.1 INTRODUCTION

Many familiar physical notions, such as forces, velocities,[†] and accelerations, involve both a magnitude (the amount of the force, velocity, or acceleration) and a direction. Any such entity involving both magnitude and direction is called a vector. Vectors are represented by arrows in which the length of the arrow denotes the magnitude of the vector and the direction of the arrow represents the direction of the vector. In most physical situations involving vectors, only the magnitude and direction of the vector are significant; consequently, we regard vectors with the same magnitude and direction as being equal irrespective of their positions. In this section the geometry of vectors is discussed. This geometry is derived from physical experiments that test the manner in which two vectors interact.

Familiar situations suggest that when two vectors act simultaneously at a point, the magnitude of the resultant vector (the vector obtained by adding the two original vectors) need not be the sum of the magnitudes of the original two. For example, a swimmer swimming upstream at the rate of 2 miles per hour against a current of 1 mile per hour does not progress at the rate of 3 miles per hour. For in this instance the motions of the swimmer and current oppose each other, and the rate of progress of the swimmer is only 1 mile per hour upstream. If, however, the swimmer is moving downstream (with the current), then his or her rate of progress is 3 miles per hour downstream.

Experiments show that vectors add according to the following parallelogram law (see Figure 1.1).

Parallelogram Law for Vector Addition. *The sum of two vectors* x *and* y *that act at the same point* P *is the vector in the parallelogram having* x *and* y *as adjacent sides that is represented by the diagonal beginning at* P.

[†]The word *velocity* is being used here in its scientific sense—as an entity having both magnitude and direction. The magnitude of a velocity (without regard for the direction of motion) is called its *speed*.

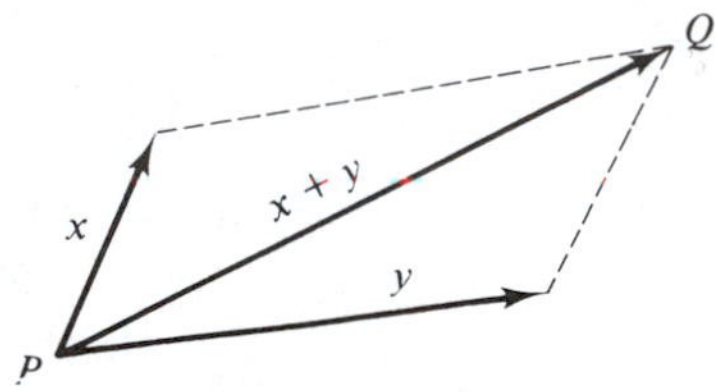

Figure 1.1

Since opposite sides of a parallelogram are parallel and of equal length, the endpoint Q of the arrow representing $x + y$ can also be obtained by allowing x to act at P and then allowing y to act at the endpoint of x. Similarly, the endpoint of the vector $x + y$ can be obtained by first permitting y to act at P and then allowing x to act at the endpoint of y. Thus two vectors x and y that both act at a point P may be added "tail-to-head"; that is, either x or y may be applied at P and a vector having the same magnitude and direction as the other may be applied to the endpoint of the first. If this is done, the endpoint of the second vector is the endpoint of $x + y$.

The addition of vectors can be described algebraically with the use of analytic geometry. In the plane containing x and y, introduce a coordinate system with P at the origin. Let (a_1, a_2) denote the endpoint of x and (b_1, b_2) denote the endpoint of y. Then as Figure 1.2(a) shows, the endpoint Q of $x + y$ is $(a_1 + b_1, a_2 + b_2)$. Henceforth, when a reference is made to the coordinates of the endpoint of a vector, the vector should be assumed to emanate from the origin. Moreover, since a vector beginning at the origin is completely determined by its endpoint, we will sometimes refer to *the point* x rather than *the endpoint of the vector* x if x is a vector emanating from the origin.

Besides the operation of vector addition there is another natural operation that can be performed on vectors—the length of a vector may be magnified or contracted without changing the direction of the vector. This operation, called scalar multiplication, consists of multiplying a vector by a real number. If the

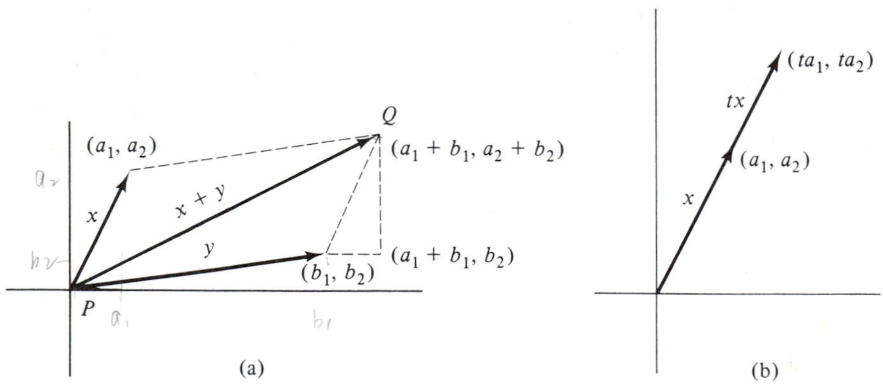

Figure 1.2

vector x is represented by an arrow, then for any real number $t \geqslant 0$ the vector tx will be represented by an arrow having the same direction as the arrow representing x but having length t times the length of the arrow representing x. If $t < 0$, the vector tx will be represented by an arrow having the opposite direction as x and having length $|t|$ times the length of the arrow representing x. Two nonzero vectors x and y are called *parallel* if $y = tx$ for some nonzero real number t. (Thus nonzero vectors having the same direction or opposite directions are parallel.)

To describe scalar multiplication algebraically, again introduce a coordinate system into a plane containing the vector x so that x emanates from the origin. If the endpoint of x has coordinates (a_1, a_2), then the coordinates of the endpoint of tx are easily shown to be (ta_1, ta_2) [see Figure 1.2(b)].

The algebraic descriptions of vector addition and scalar multiplication for vectors in a plane yield the following properties for arbitrary vectors x, y, and z and arbitrary real numbers a and b:

1. $x + y = y + x$.
2. $(x + y) + z = x + (y + z)$.
3. There exists a vector denoted 0 such that $x + 0 = x$ for each vector x.
4. For each vector x there is a vector y such that $x + y = 0$.
5. $1x = x$.
6. $(ab)x = a(bx)$.
7. $a(x + y) = ax + ay$.
8. $(a + b)x = ax + bx$.

Arguments similar to those given above show that these eight properties, as well as the geometric interpretations of vector addition and scalar multiplication, are true also for vectors acting in space rather than in a plane. We will use these results to write equations of lines and planes in space.

Consider first the equation of a line in space that passes through two distinct points P and Q. Let O denote the origin of a coordinate system in space, and let u and v denote the vectors that begin at O and end at P and Q, respectively. If w denotes the vector beginning at P and ending at Q, then "tail-to-head" addition shows that $u + w = v$, and hence $w = v - u$, where $-u$ denotes the vector $(-1)u$. (See Figure 1.3, in which quadrilateral $OPQR$ is a parallelogram.) Since a scalar multiple of w is parallel to w but possibly of a different length than w, any point on the line joining P and Q may be obtained as the endpoint of a vector that begins at P and has the form tw for some real number t. Conversely, the endpoint of every vector of the form tw that begins at P lies on the line joining P and Q. Thus an equation of the line through P and Q is $x = u + tw = u + t(v - u)$, where t is a real number and x denotes an arbitrary point on the line. Notice also that the endpoint R of the vector $v - u$ in Figure 1.3 has coordinates equal to the difference of the coordinates of Q and P.

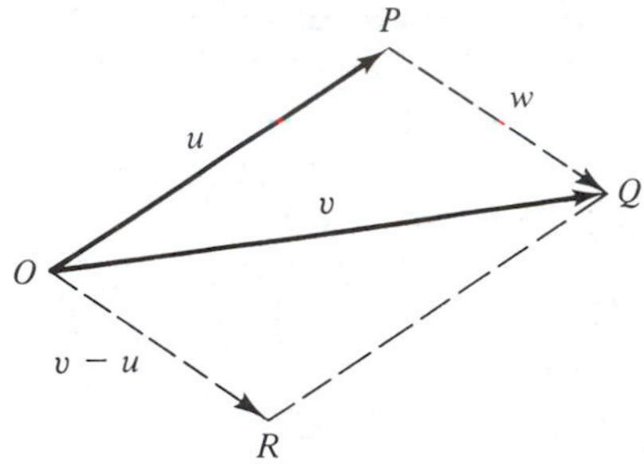

Figure 1.3

Example 1

We will find the equation of the line through the points P and Q having coordinates $(-2, 0, 1)$ and $(4, 5, 3)$, respectively. The endpoint R of the vector emanating from the origin and having the same direction as the vector beginning at P and terminating at Q has coordinates $(4, 5, 3) - (-2, 0, 1) = (6, 5, 2)$. Hence the desired equation is

$$x = (-2, 0, 1) + t(6, 5, 2). \quad \blacksquare$$

Now let P, Q, and R denote any three noncollinear points in space. These points determine a unique plane, whose equation can be found by use of our previous observations about vectors. Let u and v denote the vectors beginning at P and ending at Q and R, respectively. Observe that any point in the plane containing P, Q, and R is the endpoint S of a vector x beginning at P and having the form $t_1u + t_2v$ for some real numbers t_1 and t_2. The endpoint of t_1u will be the point of intersection of the line through P and Q with the line through S parallel to the line through P and R (see Figure 1.4). A similar procedure will locate the endpoint of t_2v. Moreover, for any real numbers t_1 and t_2, $t_1u + t_2v$ is a vector lying in the plane containing P, Q, and R. It follows that an equation of the plane containing P, Q, and R is

$$x = P + t_1u + t_2v,$$

where t_1 and t_2 are arbitrary real numbers and x denotes an arbitrary point in the plane.

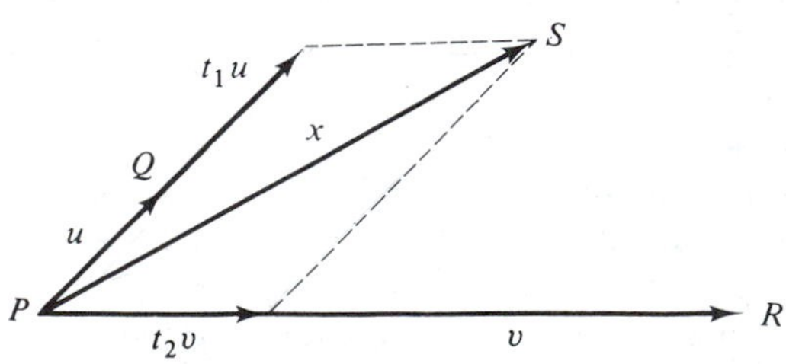

Figure 1.4

Example 2

Let P, Q, and R be the points having coordinates $(1, 0, 2)$, $(-3, -2, 4)$, and $(1, 8, -5)$, respectively. The endpoint of the vector emanating from the origin and having the same length and direction as the vector beginning at P and terminating at Q is

$$(-3, -2, 4) - (1, 0, 2) = (-4, -2, 2).$$

Similarly, the endpoint of the vector emanating from the origin and having the same length and direction as the vector beginning at P and terminating at R is $(1, 8, -5) - (1, 0, 2) = (0, 8, -7)$. Hence the equation of the plane containing the three given points is

$$x = (1, 0, 2) + t_1(-4, -2, 2) + t_2(0, 8, -7). \quad ∎$$

Any mathematical structure possessing the eight properties on page 3 is called a "vector space." In the next section we formally define a vector space and consider many examples of vector spaces other than the ones mentioned above.

EXERCISES

1. Determine if the vectors emanating from the origin and terminating at the following pairs of points are parallel.
(a) $(3, 1, 2)$ and $(6, 4, 2)$
(b) $(-3, 1, 7)$ and $(9, -3, -21)$
(c) $(5, -6, 7)$ and $(-5, 6, -7)$
(d) $(2, 0, -5)$ and $(5, 0, -2)$

2. Find the equations of the lines through the following pairs of points in space.
(a) $(3, -2, 4)$ and $(-5, 7, 1)$
(b) $(2, 4, 0)$ and $(-3, -6, 0)$
(c) $(3, 7, 2)$ and $(3, 7, -8)$
(d) $(-2, -1, 5)$ and $(3, 9, 7)$

3. Find the equations of the planes containing the following points in space.
(a) $(2, -5, -1)$, $(0, 4, 6)$, and $(-3, 7, 1)$
(b) $(3, -6, 7)$, $(-2, 0, -4)$, and $(5, -9, -2)$
(c) $(-8, 2, 0)$, $(1, 3, 0)$, and $(6, -5, 0)$
(d) $(1, 1, 1)$, $(5, 5, 5)$, and $(-6, 4, 2)$

4. What are the coordinates of the vector 0 in the Euclidean plane that satisfies condition 3 on page 3? Prove that this choice of coordinates does satisfy condition 3.

5. Prove that if the vector x emanates from the origin of the Euclidean plane and terminates at the point with coordinates (a_1, a_2), then the vector tx that emanates from the origin terminates at the point with coordinates (ta_1, ta_2).

6. Show that the midpoint of the line segment joining the points (a, b) and (c, d) is $((a + c)/2, (b + d)/2)$.
7. Prove that the diagonals of a parallelogram bisect each other.

1.2 VECTOR SPACES

Because such diverse entities as the forces acting in a plane and the polynomials with real number coefficients both permit natural definitions of addition and scalar multiplication that possess properties 1 through 8 on page 3, it is natural to abstract these properties in the following definition.

Definition. *A* vector space (*or* linear space) V over a field† F *consists of a set on which two operations (called* addition *and* scalar multiplication, *respectively) are defined so that for each pair of elements* x, y *in* V *there is a unique element* $\mathsf{x} + \mathsf{y}$ *in* V, *and for each element* a *in* F *and each element* x *in* V *there is a unique element* ax *in* V, *such that the following conditions hold:*

(VS 1) *For all* x, y *in* V, $\mathsf{x} + \mathsf{y} = \mathsf{y} + \mathsf{x}$ (*commutativity of addition*).

(VS 2) *For all* $\mathsf{x}, \mathsf{y}, \mathsf{z}$ *in* V, $(\mathsf{x} + \mathsf{y}) + \mathsf{z} = \mathsf{x} + (\mathsf{y} + \mathsf{z})$ (*associativity of addition*).

(VS 3) *There exists an element in* V *denoted by* 0 *such that* $\mathsf{x} + 0 = \mathsf{x}$ *for each* x *in* V.

(VS 4) *For each element* x *in* V *there exists an element* y *in* V *such that* $\mathsf{x} + \mathsf{y} = 0$.

(VS 5) *For each element* x *in* V, $1\mathsf{x} = \mathsf{x}$.

(VS 6) *For each pair* a, b *of elements in* F *and each element* x *in* V, $(\mathsf{ab})\mathsf{x} = \mathsf{a}(\mathsf{bx})$.

(VS 7) *For each element* a *in* F *and each pair of elements* x, y *in* V, $\mathsf{a}(\mathsf{x} + \mathsf{y}) = \mathsf{ax} + \mathsf{ay}$.

(VS 8) *For each pair* a, b *of elements in* F *and each element* x *in* V, $(\mathsf{a} + \mathsf{b})\mathsf{x} = \mathsf{ax} + \mathsf{bx}$.

The elements $\mathsf{x} + \mathsf{y}$ *and* ax *are called the* sum *of* x *and* y *and the* product *of* a *and* x, *respectively.*

The elements of the field F are called *scalars* and the elements of the vector space V are called *vectors*. The reader should not confuse this use of the word "vector" with the physical entity discussed in Section 1.1; the word "vector" is now being used to describe any element of a vector space.

A vector space will frequently be discussed in the text without explicitly mentioning its field of scalars. The reader is cautioned to remember, however,

†See Appendix C. With few exceptions, however, the reader may interpret the word "field" to mean "field of real numbers" (which we denote by R) or "field of complex numbers" (which we denote by C).

that *every vector space will be regarded as a vector space over a given field, which will be denoted by F.*

In the remainder of this section we introduce several important examples of vector spaces that will be studied throughout the text. Observe that in describing a vector space it is necessary to specify not only the vectors but also the operations of addition and scalar multiplication.

An object of the form $(a_1, \ldots, a_n)$, where the entries a_i are elements of a field F, is called an *n-tuple* with entries from F. Two n-tuples $(a_1, \ldots, a_n)$ and $(b_1, \ldots, b_n)$ are defined to be equal if and only if $a_i = b_i$ for $i = 1, 2, \ldots, n$.

Example 1

The set of all n-tuples with entries from a field F forms a vector space, which we denote by F^n, under the operations of coordinatewise addition and multiplication; that is, if $x = (a_1, \ldots, a_n) \in \mathsf{F}^n$, $y = (b_1, \ldots, b_n) \in \mathsf{F}^n$, and $c \in F$, then

$$x + y = (a_1 + b_1, \ldots, a_n + b_n) \quad \text{and} \quad cx = (ca_1, \ldots, ca_n).$$

For example, in R^4,

$$(3, \ -2, \ 0, \ 5) + (-1, \ 1, \ 4, \ 2) = (2, \ -1, \ 4, \ 7)$$

and

$$-5(1, \ -2, \ 0, \ 3) = (-5, \ 10, \ 0, \ -15).$$

Elements of F^n will often be written as *column vectors:*

$$\begin{pmatrix} a_1 \\ \vdots \\ a_n \end{pmatrix}$$

rather than as *row vectors* $(a_1, \ldots, a_n)$. Since a 1-tuple with entry from F may be regarded as an element of F, we will write F rather than F^1 for the vector space of 1-tuples from F. ∎

An $m \times n$ *matrix* with entries from a field F is a rectangular array of the form

$$\begin{pmatrix} a_{11} & a_{12} & \cdots & a_{1n} \\ a_{21} & a_{22} & \cdots & a_{2n} \\ \vdots & & & \\ a_{m1} & a_{m2} & \cdots & a_{mn} \end{pmatrix},$$

where each entry $a_{ij} (1 \leqslant i \leqslant m, \ 1 \leqslant j \leqslant n)$ is an element of F. The entries $a_{i1}, a_{i2}, \ldots, a_{in}$ of the matrix above compose the *ith row* of the matrix and will often be regarded as a row vector in F^n, whereas the entries $a_{1j}, a_{2j}, \ldots, a_{mj}$ compose the *jth column* of the matrix and will often be regarded as a column vector in F^m. The $m \times n$ matrix having each entry equal to zero is called the *zero matrix* and is denoted by O.

In this book we denote matrices by capital italic letters (e.g., A, B, and C), and we denote the entry of a matrix A that lies in row i and column j by A_{ij}. In addition, if the number of rows and columns of a matrix are equal, the matrix is called *square*.

Two $m \times n$ matrices A and B are defined to be equal if and only if all their corresponding entries are equal, that is, if and only if $A_{ij} = B_{ij}$ for $1 \leqslant i \leqslant m$ and $1 \leqslant j \leqslant n$.

Example 2

The set of all $m \times n$ matrices with entries from a field F is a vector space, which we denote by $\mathsf{M}_{m \times n}(F)$, under the following operations of addition and scalar multiplication: For $A, B \in \mathsf{M}_{m \times n}(F)$ and $c \in F$,

$$(A + B)_{ij} = A_{ij} + B_{ij} \quad \text{and} \quad (cA)_{ij} = cA_{ij}.$$

For instance,

$$\begin{pmatrix} 2 & 0 & -1 \\ 1 & -3 & 4 \end{pmatrix} + \begin{pmatrix} -5 & -2 & 6 \\ 3 & 4 & -1 \end{pmatrix} = \begin{pmatrix} -3 & -2 & 5 \\ 4 & 1 & 3 \end{pmatrix}$$

and

$$-3\begin{pmatrix} 1 & 0 & -2 \\ -3 & 2 & 3 \end{pmatrix} = \begin{pmatrix} -3 & 0 & 6 \\ 9 & -6 & -9 \end{pmatrix}$$

in $\mathsf{M}_{2 \times 3}(R)$. ∎

Example 3

Let S be any nonempty set and F be any field, and let $\mathcal{F}(S, F)$ denote the set of all functions from S into F. Two elements f and g in $\mathcal{F}(S, F)$ are defined to be equal if $f(s) = g(s)$ for each $s \in S$. The set $\mathcal{F}(S, F)$ is a vector space under the operations of addition and scalar multiplication defined for $f, g \in \mathcal{F}(S, F)$ and $c \in F$ by

$$(f + g)(s) = f(s) + g(s) \quad \text{and} \quad (cf)(s) = c[f(s)]$$

for each $s \in S$. Note that these are the familiar operations of addition and scalar multiplication for functions as used in algebra and calculus. ∎

A *polynomial* with coefficients from a field F is an expression of the form

$$f(x) = a_n x^n + a_{n-1} x^{n-1} + \cdots + a_1 x + a_0,$$

where n is a nonnegative integer and $a_n, \ldots, a_0$ are elements of F. If $f(x) = 0$, that is, if $a_n = \cdots = a_0 = 0$, then $f(x)$ is called the *zero polynomial* and the degree of $f(x)$ is said to be -1; otherwise, the *degree* of a polynomial is defined to be the largest exponent of x that appears in the representation

$$f(x) = a_n x^n + a_{n-1} x^{n-1} + \cdots + a_0$$

with a nonzero coefficient. Note that the polynomials of degree zero are of the form $f(x) = c$ for some nonzero scalar c. Two polynomials $f(x)$ and $g(x)$ are equal if and only if they have the same degree and their coefficients of like powers of x are equal.

When F is a field containing an infinite number of elements, we will usually regard a polynomial with coefficients from F as a function from F into F. In this case the value of the function

$$f(x) = a_n x^n + a_{n-1} x^{n-1} + \cdots + a_0$$

at $c \in F$ is the scalar

$$f(c) = a_n c^n + a_{n-1} c^{n-1} + \cdots + a_0.$$

Here either of the notations f or $f(x)$ will be used for the polynomial function

$$f(x) = a_n x^n + a_{n-1} x^{n-1} + \cdots + a_0.$$

Example 4

The set of all polynomials with coefficients from a field F is a vector space, which we denote by $\mathsf{P}(F)$, under the following operations: For

$$f(x) = a_n x^n + a_{n-1} x^{n-1} + \cdots + a_0$$

and

$$g(x) = b_n x^n + b_{n-1} x^{n-1} + \cdots + b_0$$

in $\mathsf{P}(F)$ and $c \in F$,

$$(f + g)(x) = (a_n + b_n)x^n + (a_{n-1} + b_{n-1})x^{n-1} + \cdots + (a_0 + b_0)$$

and

$$(cf)(x) = ca_n x^n + ca_{n-1} x^{n-1} + \cdots + ca_0. \quad \blacksquare$$

We will see in Exercise 21 of Section 2.4 that the vector space defined in the example below is essentially the same as $\mathsf{P}(F)$.

Example 5

Let F be any field. A *sequence* in F is a function σ from the positive integers into F. As usual, the sequence σ such that $\sigma(n) = a_n$ will be denoted by $\{a_n\}$. Let V consist of all sequences $\{a_n\}$ in F that have only a finite number of nonzero terms a_n. If $\{a_n\}$ and $\{b_n\}$ are sequences in V and $t \in F$, then $\{a_n\} + \{b_n\}$ is that sequence $\{c_n\}$ in V such that $c_n = a_n + b_n$ $(n = 1, 2, \ldots)$ and $t\{a_n\}$ is that sequence $\{d_n\}$ in V such that $d_n = ta_n$ $(n = 1, 2, \ldots)$. Under these operations V is a vector space. $\blacksquare$

Our next two examples contain sets on which an addition and scalar multiplication are defined, but which are not vector spaces.

Example 6

Let $S = \{(a_1, a_2): a_1, a_2 \in R\}$. For $(a_1, a_2), (b_1, b_2) \in S$ and $c \in R$, define

$$(a_1, a_2) + (b_1, b_2) = (a_1 + b_1, a_2 - b_2) \quad \text{and} \quad c(a_1, a_2) = (ca_1, ca_2).$$

Since (VS 1), (VS 2), and (VS 8) all fail to hold, S is not a vector space under these operations. ∎

Example 7

Let S be as in Example 6. For $(a_1, a_2), (b_1, b_2) \in S$ and $c \in R$, define

$$(a_1, a_2) + (b_1, b_2) = (a_1 + b_1, 0) \quad \text{and} \quad c(a_1, a_2) = (ca_1, 0).$$

Then S under these operations is not a vector space because (VS 3) [hence (VS 4)] and (VS 5) fail. ∎

We conclude this section with a few of the elementary consequences of the definition of a vector space.

Proposition 1.1 (Cancellation Law for Vector Addition). *If* x, y, *and* z *are elements of a vector space* V *such that* $x + z = y + z$, *then* $x = y$.

Proof. There exists an element v in V such that $z + v = 0$ (VS 4). Thus

$$\begin{aligned} x = x + 0 &= x + (z + v) = (x + z) + v \\ &= (y + z) + v = y + (z + v) = y + 0 = y \end{aligned}$$

by (VS 2) and (VS 3). ∎

Corollary 1. *The vector* 0 *described in* (*VS 3*) *is unique.*

Proof. Exercise. ∎

Corollary 2. *The vector* y *described in* (*VS 4*) *is unique.*

Proof. Exercise. ∎

The vector 0 in (VS 3) is called the *zero vector* of V, and the vector y in (VS 4) (that is, the unique vector such that $x + y = 0$) is called the *additive inverse* of x and is denoted by $-x$.

The following result contains some of the elementary properties of scalar multiplication.

Proposition 1.2. *In any vector space* V *the following statements are true:*
(a) $0x = 0$ *for each* $x \in V$.
(b) $(-a)x = -(ax) = a(-x)$ *for each* $a \in F$ *and each* $x \in V$.
(c) $a0 = 0$ *for each* $a \in F$.

Proof. (a) By (VS 8), (VS 1), and (VS 3) it follows that

$$0x + 0x = (0 + 0)x = 0x = 0 + 0x.$$

Hence $0x = 0$ by Proposition 1.1.

(b) The element $-(ax)$ is the unique element of V such that $ax + [-(ax)] = 0$. Hence if $ax + (-a)x = 0$, Corollary 2 of Proposition 1.1 implies that $(-a)x = -(ax)$. But by (VS 8),

$$ax + (-a)x = [a + (-a)]x = 0x = 0$$

by (a). Thus $(-a)x = -(ax)$.

The proof of (c) is similar to the proof of (a). ∎

EXERCISES

1. Label the following statements as being true or false.

(a) Every vector space contains a zero vector.
(b) A vector space may have more than one zero vector.
(c) In any vector space $ax = bx$ implies that $a = b$.
(d) In any vector space $ax = ay$ implies that $x = y$.
(e) An element of F^n may be regarded as an element of $M_{n \times 1}(F)$.
(f) An $m \times n$ matrix has m columns and n rows.
(g) In $P(F)$ only polynomials of the same degree may be added.
(h) If f and g are polynomials of degree n, then $f + g$ is a polynomial of degree n.
(i) If f is a polynomial of degree n and c is a nonzero scalar, then cf is a polynomial of degree n.
(j) A nonzero element of F may be considered to be an element of $P(F)$ having degree zero.
(k) Two functions in $\mathscr{F}(S, F)$ are equal if and only if they have the same values at each element of S.

2. Write the zero vector of $M_{3 \times 4}(F)$.

3. If

$$M = \begin{pmatrix} 1 & 2 & 3 \\ 4 & 5 & 6 \end{pmatrix},$$

what are M_{13}, M_{21}, and M_{22}?

4. Perform the operations indicated.

(a) $\begin{pmatrix} 2 & 5 & -3 \\ 1 & 0 & 7 \end{pmatrix} + \begin{pmatrix} 4 & -2 & 5 \\ -5 & 3 & 2 \end{pmatrix}$

(b) $\begin{pmatrix} -6 & 4 \\ 3 & -2 \\ 1 & 8 \end{pmatrix} + \begin{pmatrix} 7 & -5 \\ 0 & -3 \\ 2 & 0 \end{pmatrix}$

(c) $4\begin{pmatrix} 2 & 5 & -3 \\ 1 & 0 & 7 \end{pmatrix}$

(d) $-5\begin{pmatrix} -6 & 4 \\ 3 & -2 \\ 1 & 8 \end{pmatrix}$

(e) $(2x^4 - 7x^3 + 4x + 3) + (8x^3 + 2x^2 - 6x + 7)$

(f) $(-3x^3 + 7x^2 + 8x - 6) + (2x^3 - 8x + 10)$

(g) $5(2x^7 - 6x^4 + 8x^2 - 3x)$

(h) $3(x^5 - 2x^3 + 4x + 2)$

Exercises 5 and 6 show why the definitions of matrix addition and scalar multiplication (as defined in Example 2) are the appropriate ones.

5. Richard Gard (Effects of Beaver on Trout in Sagehen Creek, California, *J. Wildlife Management*, **25**, 221–242) reports the following number of trout having crossed beaver dams in Sagehen Creek:

Upstream Crossings

	Fall	Spring	Summer
Brook trout	8	3	1
Rainbow trout	3	0	0
Brown trout	3	0	0

Downstream Crossings

	Fall	Spring	Summer
Brook trout	9	1	4
Rainbow trout	3	0	0
Brown trout	1	1	0

Record the upstream and downstream crossings as data in two 3×3 matrices and verify that the sum of these matrices gives the total number of crossings (both upstream and downstream) categorized by trout species and season.

6. At the end of May a furniture store had the following inventory:

	Early American	Spanish	Mediterranean	Danish
Living room suites	4	2	1	3
Bedroom suites	5	1	1	4
Dining room suites	3	1	2	6

Record these data as a 3×4 matrix M. To prepare for its June sale, the store decided to double its inventory on each of the items above. Assuming that none of the present stock is sold until the additional furniture arrives, verify that the inventory on hand after the order is filled is described by the matrix $2M$. If the inventory at the end of June is described by the matrix

$$A = \begin{pmatrix} 5 & 3 & 1 & 2 \\ 6 & 2 & 1 & 5 \\ 1 & 0 & 3 & 3 \end{pmatrix},$$

interpret $2M - A$. How many suites were sold during the June sale?

7. Let $S = \{0, 1\}$ and $F = R$, the field of real numbers. In $\mathscr{F}(S, R)$, show that $f = g$ and $f + g = h$, where $f(x) = 2x + 1$, $g(x) = 1 + 4x - 2x^2$, and $h(x) = 5^x + 1$.

8. In any vector space V, show that $(a + b)(x + y) = ax + ay + bx + by$ for any $x, y \in$ V and any $a, b \in F$.

9. Prove Corollaries 1 and 2 to Proposition 1.1 and Proposition 1.2(c).

10. Let V denote the set of all differentiable real-valued functions defined on the real line. Prove that V is a vector space under the operations of addition and scalar multiplication defined in Example 3.

11. Let V $= \{0\}$ consist of a single vector 0, and define $0 + 0 = 0$ and $c0 = 0$ for each c in F. Prove that V is a vector space over F. (V is called the *zero vector space.*)

12. A real-valued function defined on the real line is called an *even function* if $f(-x) = f(x)$ for each real number x. Prove that the set of even functions defined on the real line with the operations of addition and scalar multiplication defined in Example 3 is a vector space.

13. Let V denote the set of ordered pairs of real numbers. If (a_1, a_2) and (b_1, b_2) are elements of V and c is an element of F, define

$$(a_1, a_2) + (b_1, b_2) = (a_1 + b_1, a_2 b_2) \quad \text{and} \quad c(a_1, a_2) = (ca_1, a_2).$$

Is V a vector space under these operations? Justify your answer.

14. Let $\mathsf{V} = \{(a_1, \ldots, a_n) \colon a_i \in C \text{ for } i = 1, 2, \ldots, n\}$. Is V a vector space over the field of real numbers with the operations of coordinatewise addition and multiplication?

15. Let $\mathsf{V} = \{(a_1, \ldots, a_n) \colon a_i \in R \text{ for } i = 1, 2, \ldots, n\}$. Is V a vector space over the field of complex numbers with the operations of coordinatewise addition and multiplication?

16. Let V denote the set of all $m \times n$ matrices with real number entries, and let F be the field of rational numbers. Is V a vector space over F under the usual definitions of matrix addition and scalar multiplication?

17. Let $\mathsf{V} = \{(a_1, a_2) \colon a_1, a_2 \in R\}$. For $(a_1, a_2), (b_1, b_2) \in \mathsf{V}$ and $c \in R$, define

$$(a_1, a_2) + (b_1, b_2) = (a_1 + b_1, a_2 + b_2)$$

and

$$c(a_1, a_2) = \begin{cases} (0, 0) & \text{if } c = 0 \\ \left(ca_1, \dfrac{a_2}{c}\right) & \text{if } c \neq 0. \end{cases}$$

Is V a vector space under these operations? Justify your answer.

18. Let $\mathsf{V} = \{(a_1, a_2) \colon a_1, a_2 \in C\}$. For $(a_1, a_2), (b_1, b_2) \in \mathsf{V}$ and $c \in C$, define

$$(a_1, a_2) + (b_1, b_2) = (a_1 + 2b_1, a_2 + 3b_2) \quad \text{and} \quad c(a_1, a_2) = (ca_1, ca_2).$$

Is V a vector space under these operations? Justify your answer.

19. Let $\mathsf{V} = \{(a_1, a_2) \colon a_1, a_2 \in F\}$, where F is an arbitrary field. Define addition of elements of V coordinatewise, and for $c \in F$ and $(a_1, a_2) \in \mathsf{V}$, define

$$c(a_1, a_2) = (a_1, 0).$$

Is V a vector space under these operations? Justify your answer.

20. How many elements are there in the vector space $\mathsf{M}_{m \times n}(Z_2)$?

1.3 SUBSPACES

In the study of any algebraic structure it is of interest to examine subsets that possess the same structure as the set under consideration. The appropriate notion of substructure for vector spaces is introduced in this section.

Definition. *A subset* W *of a vector space* V *over a field* F *is called a* subspace *of* V *if* W *is a vector space over* F *under the operations of addition and scalar multiplication defined on* V.

In any vector space V, note that V and $\{0\}$ are subspaces. The latter is called the *zero subspace* of V.

Fortunately, it is not necessary to verify all the vector space conditions in order to prove that a subset W of a vector space V is in fact a subspace. Since conditions (VS 1), (VS 2), (VS 5), (VS 6), (VS 7), and (VS 8) are known to hold for elements of V, these conditions automatically hold for elements of a subset of V. Thus a subset W of V is a subspace of V if and only if the following four conditions hold:

1. $x + y \in$ W whenever $x \in$ W and $y \in$ W.
2. $ax \in$ W whenever $a \in F$ and $x \in$ W.
3. The zero vector of V belongs to W.
4. The additive inverse of each element of W belongs to W.

Actually, condition 4 is redundant, as the following theorem shows.

Theorem 1.3. *Let* V *be a vector space and* W *a subset of* V. *Then* W *is a subspace of* V *if and only if the following three conditions hold for the operations defined in* V:

(a) $0 \in$ W.
(b) $x + y \in$ W *whenever* $x \in$ W *and* $y \in$ W.
(c) $ax \in$ W *whenever* $a \in F$ *and* $x \in$ W.

Proof. If W is a subspace of V, then W is a vector space under the operations of addition and scalar multiplication defined on V. Hence conditions (b) and (c) hold, and there exists an element $0' \in$ W such that $x + 0' = x$ for each $x \in$ W. But also $x + 0 = x$, and thus $0' = 0$ by Proposition 1.1. So condition (a) holds.

Conversely, if conditions (a), (b), and (c) hold, the discussion preceding this theorem shows that W is a subspace of V if the additive inverse of each element of W belongs to W. But if $x \in$ W, then $(-1)x$ belongs to W by condition (c), and $-x = (-1)x$ by Proposition 1.2. Hence W is a subspace of V. ∎

The theorem above provides a simple method for determining whether or not a given subset of a vector space is a subspace. Normally, it is this result that is used to prove that a subset is, in fact, a subspace.

The *transpose* M^t of an $m \times n$ matrix M is the $n \times m$ matrix obtained from M by interchanging the rows with the columns; that is, $(M^t)_{ij} = M_{ji}$. For example,

$$\begin{pmatrix} 1 & -2 & 3 \\ 0 & 5 & -1 \end{pmatrix}^t = \begin{pmatrix} 1 & 0 \\ -2 & 5 \\ 3 & -1 \end{pmatrix}.$$

A *symmetric matrix* is a matrix M such that $M^t = M$. Clearly, a symmetric matrix must be square. The set W of all symmetric matrices in $\mathsf{M}_{n \times n}(F)$ is a subspace of $\mathsf{M}_{n \times n}(F)$ since the conditions of Theorem 1.3 hold:

(a) The zero matrix is equal to its transpose and hence belongs to W.

It is easily proved that for any matrices A and B and any scalars a and b, $(aA + bB)^t = aA^t + bB^t$ (see Exercise 3). Using this fact, we can easily establish conditions (b) and (c) of Theorem 1.3 as follows.

(b) If $A \in \mathsf{W}$ and $B \in \mathsf{W}$, then $A = A^t$ and $B = B^t$. Thus $(A + B)^t = A^t + B^t = A + B$, so that $A + B \in \mathsf{W}$.
(c) If $A \in \mathsf{W}$, then $A^t = A$. So for any $a \in F$, $(aA)^t = aA^t = aA$. Thus $aA \in \mathsf{W}$.

The following examples provide further illustrations of the concept of a subspace. The first three are particularly important.

Example 1

Let M be an $n \times n$ matrix. The (*main*) *diagonal* of M consists of the entries M_{11}, $M_{22}, \ldots, M_{nn}$. An $n \times n$ matrix D is called a *diagonal matrix* if each entry not on the diagonal of D is zero, that is, if $D_{ij} = 0$ whenever $i \neq j$. The set of all diagonal matrices in $\mathsf{M}_{n \times n}(F)$ is a subspace of $\mathsf{M}_{n \times n}(F)$. ∎

Example 2

Let n be a nonnegative integer, and let $\mathsf{P}_n(F)$ consist of all polynomials in $\mathsf{P}(F)$ having degree less than or equal to n. (Notice that the zero polynomial is an element of $\mathsf{P}_n(F)$ since its degree is -1.) Then $\mathsf{P}_n(F)$ is a subspace of $\mathsf{P}(F)$. ∎

Example 3

The set $\mathsf{C}(R)$ consisting of all the continuous real-valued functions defined on R is a subspace of $\mathscr{F}(R, R)$, where $\mathscr{F}(R, R)$ is as defined in Example 3 of Section 1.2. ∎

Example 4

The *trace* of an $n \times n$ matrix M, denoted $\operatorname{tr}(M)$, is the sum of all the entries of M lying on the diagonal; that is,

$$\operatorname{tr}(M) = M_{11} + M_{22} + \cdots + M_{nn}.$$

The set of $n \times n$ matrices having trace equal to zero is a subspace of $\mathsf{M}_{n \times n}(F)$ (see Exercise 6). ∎

Example 5

The set of matrices in $\mathsf{M}_{m \times n}(F)$ having nonnegative entries is not a subspace of $\mathsf{M}_{m \times n}(F)$ because condition (c) of Theorem 1.3 does not hold. ∎

The next theorem provides a method of forming a new subspace from other subspaces.

Theorem 1.4. *Any intersection of subspaces of a vector space* V *is a subspace of* V.

Proof. Let $\mathscr{C}$ be a collection of subspaces of V, and let W denote the intersection of all the subspaces in $\mathscr{C}$. Since every subspace contains the zero vector, $0 \in$ W. Let $a \in F$ and x, y be elements of W; then x and y are elements of each subspace in $\mathscr{C}$. Hence $x + y$ and ax are elements of each subspace in $\mathscr{C}$ (because the sum of vectors in a subspace and the product of a scalar and a vector from the subspace both belong to that subspace). Thus $x + y \in$ W and $ax \in$ W, so that W is a subspace by Theorem 1.3. ∎

Having shown that the intersection of subspaces is a subspace, it is natural to consider the question of whether or not the union of subspaces is a subspace. It is easily seen that the union of subspaces must satisfy conditions (a) and (c) of Theorem 1.3 but that condition (b) need not hold. In fact, it can be readily shown (see Exercise 18) that the union of two subspaces is a subspace if and only if one of the subspaces is a subset of the other. It is natural, however, to expect that there should be a method of combining two subspaces W_1 and W_2 to obtain a larger subspace (that is, one that contains both W_1 and W_2). As we have suggested above, the key to finding such a subspace is condition (b) of Theorem 1.3. We explore this idea in Exercise 21.

EXERCISES

1. Label the following statements as being true or false.
   (a) If V is a vector space and W is a subset of V that is a vector space, then W is a subspace of V.
   (b) The empty set is a subspace of every vector space.
   (c) If V is a vector space other than the zero vector space $\{0\}$, then V contains a subspace W such that W $\neq$ V.
   (d) The intersection of any two subsets of V is a subspace of V.
   (e) An $n \times n$ diagonal matrix can never have more than n nonzero entries.
   (f) The trace of a square matrix is the product of its entries on the diagonal.
2. Determine the transpose of each of the following matrices. In addition, if the matrix is square, compute its trace.

   (a) $\begin{pmatrix} -4 & 2 \\ 5 & -1 \end{pmatrix}$ **(b)** $\begin{pmatrix} 0 & 8 & -6 \\ 3 & 4 & 7 \end{pmatrix}$

(c) $\begin{pmatrix} -3 & 9 \\ 0 & -2 \\ 6 & 1 \end{pmatrix}$

(d) $\begin{pmatrix} 10 & 0 & -8 \\ 2 & -4 & 3 \\ -5 & 7 & 6 \end{pmatrix}$

(e) $(1, -1, 3, 5)$

(f) $\begin{pmatrix} -2 & 5 & 1 & 4 \\ 7 & 0 & 1 & -6 \end{pmatrix}$

(g) $\begin{pmatrix} 5 \\ 6 \\ 7 \end{pmatrix}$

(h) $\begin{pmatrix} -4 & 0 & 6 \\ 0 & 1 & -3 \\ 6 & -3 & 5 \end{pmatrix}$

3. Prove that $(aA + bB)^t = aA^t + bB^t$ for any $A, B \in \mathsf{M}_{m \times n}(F)$ and any $a, b \in F$.

4. Prove that $(A^t)^t = A$ for each $A \in \mathsf{M}_{m \times n}(F)$.

5. Prove that $A + A^t$ is symmetric for any square matrix A.

6. Prove that $\mathrm{tr}(aA + bB) = a\,\mathrm{tr}(A) + b\,\mathrm{tr}(B)$ for any $A, B \in \mathsf{M}_{n \times n}(F)$.

7. Prove that diagonal matrices are symmetric matrices.

8. Determine if the following sets are subspaces of R^3 under the operations of addition and scalar multiplication defined on R^3. Justify your answers.

(a) $\mathsf{W}_1 = \{(a_1, a_2, a_3) \in \mathsf{R}^3\colon a_1 = 3a_2 \text{ and } a_3 = -a_2\}$
(b) $\mathsf{W}_2 = \{(a_1, a_2, a_3) \in \mathsf{R}^3\colon a_1 = a_3 + 2\}$
(c) $\mathsf{W}_3 = \{(a_1, a_2, a_3) \in \mathsf{R}^3\colon 2a_1 - 7a_2 + a_3 = 0\}$
(d) $\mathsf{W}_4 = \{(a_1, a_2, a_3) \in \mathsf{R}^3\colon a_1 - 4a_2 - a_3 = 0\}$
(e) $\mathsf{W}_5 = \{(a_1, a_2, a_3) \in \mathsf{R}^3\colon a_1 + 2a_2 - 3a_3 = 1\}$
(f) $\mathsf{W}_6 = \{(a_1, a_2, a_3) \in \mathsf{R}^3\colon 5a_1^2 - 3a_2^2 + 6a_3^2 = 0\}$

9. Let W_1, W_3, and W_4 be as in Exercise 8. Describe $\mathsf{W}_1 \cap \mathsf{W}_3$, $\mathsf{W}_1 \cap \mathsf{W}_4$, and $\mathsf{W}_3 \cap \mathsf{W}_4$ and observe that each is a subspace of R^3.

10. Verify that $\mathsf{W}_1 = \{(a_1, \ldots, a_n) \in \mathsf{F}^n\colon a_1 + \cdots + a_n = 0\}$ is a subspace of F^n but that $\mathsf{W}_2 = \{(a_1, \ldots, a_n) \in \mathsf{F}^n\colon a_1 + \cdots + a_n = 1\}$ is not.

11. Is the set $\mathsf{W} = \{f \in \mathsf{P}(F)\colon f = 0 \text{ or } f \text{ has degree } n\}$ a subspace of $\mathsf{P}(F)$ if $n \geqslant 1$? Justify your answer.

12. An $m \times n$ matrix A is called *upper triangular* if all entries lying below the diagonal are zero, that is, if $A_{ij} = 0$ whenever $i > j$. Verify that the upper triangular matrices form a subspace of $\mathsf{M}_{m \times n}(F)$.

13. Verify that for any $s_0 \in S$, $\mathsf{W} = \{f \in \mathcal{F}(S, F)\colon f(s_0) = 0\}$ is a subspace of $\mathcal{F}(S, F)$.

14. Is the set of all differentiable real-valued functions defined on R a subspace of $\mathsf{C}(R)$? Justify your answer.

15. Let $\mathsf{C}^n(R)$ denote the set of all real-valued functions defined on the real line that have a continuous nth derivative (and hence continuous derivatives of orders $1, 2, \ldots, n$). Verify that $\mathsf{C}^n(R)$ is a subspace of $\mathcal{F}(R, R)$.

16. Prove that a subset W of a vector space V is a subspace of V if and only if $W \neq \varnothing$, $ax \in W$, and $x + y \in W$ whenever $a \in F$ and $x, y \in W$.

17. Prove that a subset W of a vector space V is a subspace of V if and only if $0 \in W$ and $ax + y \in W$ whenever $a \in F$ and $x, y \in W$.

18. Let W_1 and W_2 be subspaces of a vector space V. Prove that $W_1 \cup W_2$ is a subspace of V if and only if $W_1 \subseteq W_2$ or $W_2 \subseteq W_1$.

19. Let F_1 and F_2 be fields. A function $g \in \mathcal{F}(F_1, F_2)$ is called an *even function* if $g(-x) = g(x)$ for each $x \in F_1$ and is called an *odd function* if $g(-x) = -g(x)$ for each $x \in F_1$. Prove that the set of all even functions in $\mathcal{F}(F_1, F_2)$ and the set of all odd functions in $\mathcal{F}(F_1, F_2)$ are subspaces of $\mathcal{F}(F_1, F_2)$.

20.† Prove that if W is a subspace of V and $x_1, \ldots, x_n$ are elements of W, then $a_1x_1 + \cdots + a_nx_n$ is an element of W for any scalars $a_1, \ldots, a_n$ in F.

Definition. *If S_1 and S_2 are nonempty subsets of a vector space V, then the* sum *of S_1 and S_2, denoted $S_1 + S_2$, is the set $\{x + y : x \in S_1$ and $y \in S_2\}$.*

21. Prove that if W_1 and W_2 are subspaces of a vector space V, then $W_1 + W_2$ is a subspace of V that contains both W_1 and W_2. Prove also that $W_1 + W_2$ is the smallest subspace that contains both W_1 and W_2.

Definition. *A vector space V is said to be the* direct sum *of W_1 and W_2, denoted by $V = W_1 \oplus W_2$, if W_1 and W_2 are subspaces of V such that $W_1 \cap W_2 = \{0\}$ and $W_1 + W_2 = V$.*

22. Show that F^n is the direct sum of the subspaces

$$W_1 = \{(a_1, \ldots, a_n) \in F^n : a_n = 0\}$$

and

$$W_2 = \{(a_1, \ldots, a_n) \in F^n : a_1 = \cdots = a_{n-1} = 0\}.$$

23. Let W_1 denote the set of all polynomials f in $P(F)$ such that $f(x) = 0$ or, in the representation

$$f(x) = a_nx^n + a_{n-1}x^{n-1} + \cdots + a_0,$$

the coefficients $a_0, a_2, a_4, \ldots$ of all even powers of x equal zero. Likewise, let W_2 denote the set of all polynomials g in $P(F)$ such that $g(x) = 0$ or, in the representation

$$g(x) = b_mx^m + b_{m-1}x^{m-1} + \cdots + b_0,$$

the coefficients $b_1, b_3, b_5, \ldots$ of all odd powers of x equal zero. Prove that $P(F) = W_1 \oplus W_2$.

†Exercises denoted by † will be referenced in other sections of the book.

24. Let $\mathsf{W}_1 = \{A \in \mathsf{M}_{m \times n}(F)\colon A_{ij} = 0$ whenever $i > j\}$ and $\mathsf{W}_2 = \{A \in \mathsf{M}_{m \times n}(F)\colon A_{ij} = 0$ whenever $i \leqslant j\}$. (W_1 is the set of upper triangular matrices defined in Exercise 12.) Show that $\mathsf{M}_{m \times n}(F) = \mathsf{W}_1 \oplus \mathsf{W}_2$.

25. Let V denote the vector space consisting of all upper triangular $n \times n$ matrices (as defined in Exercise 12), and let W_1 denote the subspace of V consisting of all diagonal matrices. Show that $\mathsf{V} = \mathsf{W}_1 \oplus \mathsf{W}_2$, where $\mathsf{W}_2 = \{A \in \mathsf{V}\colon A_{ij} = 0$ whenever $i \geqslant j\}$.

26. A matrix M is called *skew-symmetric* if $M^t = -M$. Clearly, a skew-symmetric matrix is square. Prove that the set of all skew-symmetric $n \times n$ matrices is a subspace W_1 of $\mathsf{M}_{n \times n}(R)$. Let W_2 be the subspace of $\mathsf{M}_{n \times n}(R)$ consisting of the symmetric $n \times n$ matrices. Prove that $\mathsf{M}_{n \times n}(R) = \mathsf{W}_1 \oplus \mathsf{W}_2$.

27. Let $\mathsf{W}_1 = \{A \in \mathsf{M}_{n \times n}(F)\colon A_{ij} = 0$ whenever $i \leqslant j\}$, and let W_2 denote the set of symmetric $n \times n$ matrices. Both W_1 and W_2 are subspaces of $\mathsf{M}_{n \times n}(F)$. Prove that $\mathsf{M}_{n \times n}(F) = \mathsf{W}_1 \oplus \mathsf{W}_2$. Compare Exercises 26 and 27.

28. Let W_1 and W_2 be subspaces of a vector space V. Prove that V is the direct sum of W_1 and W_2 if and only if each element in V can be *uniquely* written as $x_1 + x_2$, where $x_1 \in \mathsf{W}_1$ and $x_2 \in \mathsf{W}_2$.

29. Let W be a subspace of a vector space V over a field F. For any $v \in \mathsf{V}$ the set $\{v\} + \mathsf{W} = \{v + w\colon w \in \mathsf{W}\}$ is called the *coset of* W *containing* v. It is customary to denote this coset by $v + \mathsf{W}$ rather than $\{v\} + \mathsf{W}$. Prove the following:

(a) $v + \mathsf{W}$ is a subspace of V if and only if $v \in \mathsf{W}$.

(b) $v_1 + \mathsf{W} = v_2 + \mathsf{W}$ if and only if $v_1 - v_2 \in \mathsf{W}$.

Addition and scalar multiplication by elements of F can be defined in the collection $S = \{v + \mathsf{W}\colon v \in \mathsf{V}\}$ of all cosets of W as follows:

$$(v_1 + \mathsf{W}) + (v_2 + \mathsf{W}) = (v_1 + v_2) + \mathsf{W}$$

for all $v_1, v_2 \in \mathsf{V}$ and

$$a(v + \mathsf{W}) = av + \mathsf{W}$$

for all $v \in \mathsf{V}$ and $a \in F$.

(c) Prove that the operations above are well-defined; i.e., show that if $v_1 + \mathsf{W} = v_1' + \mathsf{W}$ and $v_2 + \mathsf{W} = v_2' + \mathsf{W}$, then

$$(v_1 + \mathsf{W}) + (v_2 + \mathsf{W}) = (v_1' + \mathsf{W}) + (v_2' + \mathsf{W})$$

and

$$a(v_1 + \mathsf{W}) = a(v_1' + \mathsf{W})$$

for all $a \in F$.

(d) Prove that the set S is a vector space under the operations defined above. This vector space is called the *quotient space of* V *modulo* W and is denoted by V/W.

1.4 LINEAR COMBINATIONS AND SYSTEMS OF LINEAR EQUATIONS

In Section 1.1 it was shown that the equation of the plane through three noncollinear points P, Q, and R in space is $x = P + t_1u + t_2v$, where u and v denote the vectors beginning at P and ending at Q and R, respectively, and t_1 and t_2 denote arbitrary real numbers. An important special case occurs when P is the origin. In this case the equation of the plane simplifies to $x = t_1u + t_2v$, and the set of all points in this plane is a subspace of R^3. (This will be proved as Theorem 1.5.) Expressions of the form $t_1u + t_2v$, where t_1 and t_2 are scalars and u and v are vectors, play a central role in the theory of vector spaces. The appropriate generalization of such expressions is presented in the following definition.

Definition. *Let* V *be a vector space and* S *a nonempty subset of* V. *A vector* x *in* V *is said to be a* linear combination of elements of S *if there exist a finite number of elements* $y_1, \ldots, y_n$ *in* S *and scalars* $a_1, \ldots, a_n$ *in* F *such that* $x = a_1y_1 + \cdots + a_ny_n$. *In this situation it is also customary to say that* x *is a* linear combination of $y_1, \ldots, y_n$.

Observe that in any vector space V, $0x = 0$ for each $x \in \mathsf{V}$. Thus the zero vector is a linear combination of any nonempty subset of V.

Example 1

Table 1.1 shows the vitamin content of 100 grams of 12 foods with respect to vitamins A, B_1 (thiamine), B_2 (riboflavin), niacin, and C (ascorbic acid).

We will record the vitamin content of 100 grams of each food as a column vector in R^5—for example, the vitamin vector for apple butter is

$$\begin{pmatrix} 0.00 \\ 0.01 \\ 0.02 \\ 0.20 \\ 2.00 \end{pmatrix}.$$

Considering the vitamin vectors for cupcake, coconut custard pie, brown rice, soy sauce, and wild rice, we see that

$$\begin{pmatrix} 0.00 \\ 0.05 \\ 0.06 \\ 0.30 \\ 0.00 \end{pmatrix} + \begin{pmatrix} 0.00 \\ 0.02 \\ 0.02 \\ 0.40 \\ 0.00 \end{pmatrix} + \begin{pmatrix} 0.00 \\ 0.34 \\ 0.05 \\ 4.70 \\ 0.00 \end{pmatrix} + 2\begin{pmatrix} 0.00 \\ 0.02 \\ 0.25 \\ 0.40 \\ 0.00 \end{pmatrix} = \begin{pmatrix} 0.00 \\ 0.45 \\ 0.63 \\ 6.20 \\ 0.00 \end{pmatrix}.$$

TABLE 1.1 Vitamin Content of 100 Grams of Certain Foods

	A (units)	B_1 (mg)	B_2 (mg)	Niacin (mg)	C (mg)
Apple butter	0	0.01	0.02	0.2	2
Raw, unpared apples (freshly harvested)	90	0.03	0.02	0.1	4
Chocolate-coated candy with coconut center	0	0.02	0.07	0.2	0
Clams (meat only)	100	0.10	0.18	1.3	10
Cupcake from mix (dry form)	0	0.05	0.06	0.3	0
Cooked farina (unenriched)	(0)[a]	0.01	0.01	0.1	(0)
Jams and preserves	10	0.01	0.03	0.2	2
Coconut custard pie (baked from mix)	0	0.02	0.02	0.4	0
Raw brown rice	(0)	0.34	0.05	4.7	(0)
Soy sauce	0	0.02	0.25	0.4	0
Cooked spaghetti (unenriched)	0	0.01	0.01	0.3	0
Raw wild rice	(0)	0.45	0.63	6.2	(0)

Source: Bernice K. Watt and Annabel L. Merrill, *Composition of Foods* (Agriculture Handbook Number 8), Consumer and Food Economics Research Division, U.S. Department of Agriculture, Washington, D.C., 1963.

[a]Zeros in parentheses indicate that the amount of a vitamin present is either none or too small to measure.

Thus the vitamin vector for raw wild rice is a linear combination of the vitamin vectors for cupcake, coconut custard pie, raw brown rice, and soy sauce. So 100 grams of cupcake, 100 grams of coconut custard pie, 100 grams of raw brown rice, and 200 grams of soy sauce provide exactly the same amounts of the five vitamins as 100 grams of raw wild rice. Similarly, since

$$2\begin{pmatrix}0.00\\0.01\\0.02\\0.20\\2.00\end{pmatrix}+\begin{pmatrix}90.00\\0.03\\0.02\\0.10\\4.00\end{pmatrix}+\begin{pmatrix}0.00\\0.02\\0.07\\0.20\\0.00\end{pmatrix}+\begin{pmatrix}0.00\\0.01\\0.01\\0.10\\0.00\end{pmatrix}+\begin{pmatrix}10.00\\0.01\\0.03\\0.20\\2.00\end{pmatrix}+\begin{pmatrix}0.00\\0.01\\0.01\\0.30\\0.00\end{pmatrix}=\begin{pmatrix}100.00\\0.10\\0.18\\1.30\\10.00\end{pmatrix},$$

200 grams of apple butter, 100 grams of apples, 100 grams of chocolate candy, 100 grams of farina, 100 grams of jam, and 100 grams of spaghetti provide exactly the same amounts of the five vitamins as 100 grams of clams. ∎

Throughout Chapters 1 and 2 we will encounter many different situations in which it is necessary to determine whether or not a vector can be expressed as a linear combination of other vectors, and if so, how. This question reduces to the problem of solving a system of linear equations. To illustrate this important technique, we will determine if the vector (2, 6, 8) can be expressed as a linear combination of

$$y_1 = (1, 2, 1), \qquad y_2 = (-2, -4, -2), \qquad y_3 = (0, 2, 3),$$
$$y_4 = (2, 0, -3), \quad \text{and} \quad y_5 = (-3, 8, 16).$$

Thus we must determine if there are scalars a_1, a_2, a_3, a_4, and a_5 such that

$$\begin{aligned}(2, 6, 8) &= a_1 y_1 + a_2 y_2 + a_3 y_3 + a_4 y_4 + a_5 y_5 \\ &= a_1(1, 2, 1) + a_2(-2, -4, -2) + a_3(0, 2, 3) \\ &\qquad\qquad + a_4(2, 0, -3) + a_5(-3, 8, 16) \\ &= (a_1 - 2a_2 + 2a_4 - 3a_5, 2a_1 - 4a_2 + 2a_3 + 8a_5, \\ &\qquad\qquad a_1 - 2a_2 + 3a_3 - 3a_4 + 16a_5).\end{aligned}$$

Hence (2, 6, 8) can be expressed as a linear combination of y_1, y_2, y_3, y_4, and y_5 if and only if there is a 5-tuple of scalars $(a_1, a_2, a_3, a_4, a_5)$ satisfying the system of linear equations

$$\begin{cases} a_1 - 2a_2 + 2a_4 - 3a_5 = 2 \\ 2a_1 - 4a_2 + 2a_3 + 8a_5 = 6 \\ a_1 - 2a_2 + 3a_3 - 3a_4 + 16a_5 = 8 \end{cases} \tag{1}$$

obtained by equating the corresponding coordinates in the preceding equation.

To solve system (1), we will replace the system by another system with the same solutions but which is easier to solve. The procedure to be used expresses some of the unknowns in terms of others by eliminating certain unknowns from all the equations except one. To begin, we eliminate a_1 from every equation except the first by adding -2 times the first equation to the second and -1 times the first equation to the third. The result is the following new system:

$$\begin{cases} a_1 - 2a_2 + 2a_4 - 3a_5 = 2 \\ 2a_3 - 4a_4 + 14a_5 = 2 \\ 3a_3 - 5a_4 + 19a_5 = 6. \end{cases} \tag{2}$$

In this case it happened that while eliminating a_1 from every equation except the first, we also eliminated a_2 from every equation except the first. This need not happen in general. We now want to solve the second equation in system (2) for a_3, and then eliminate a_3 from the third equation. To do this, we first multiply the second equation by $\frac{1}{2}$, which produces

$$\begin{cases} a_1 - 2a_2 + 2a_4 - 3a_5 = 2 \\ a_3 - 2a_4 + 7a_5 = 1 \\ 3a_3 - 5a_4 + 19a_5 = 6. \end{cases}$$

Next we add -3 times the second equation to the third, obtaining

$$\begin{cases} a_1 - 2a_2 + \phantom{a_3 - {}} 2a_4 - 3a_5 = 2 \\ \qquad\qquad a_3 - 2a_4 + 7a_5 = 1 \\ \qquad\qquad\qquad\quad a_4 - 2a_5 = 3. \end{cases} \tag{3}$$

We continue by eliminating a_4 from every equation of (3) except the third. This yields

$$\begin{cases} a_1 - 2a_2 + \qquad\qquad a_5 = -4 \\ \qquad\qquad a_3 + \quad 3a_5 = 7 \\ \qquad\qquad\quad a_4 - 2a_5 = 3. \end{cases} \tag{4}$$

System (4) is a system of the desired form: It is easy to solve for the first unknown present in each of the equations (a_1, a_3, and a_4) in terms of the other unknowns (a_2 and a_5). Rewriting system (4) in this form, we find

$$\begin{cases} a_1 = 2a_2 - a_5 - 4 \\ a_3 = \qquad - 3a_5 + 7 \\ a_4 = \qquad\quad 2a_5 + 3. \end{cases}$$

Thus for any choice of the scalars a_2 and a_5, a vector of the form

$$(a_1, a_2, a_3, a_4, a_5) = (2a_2 - a_5 - 4,\ a_2,\ -3a_5 + 7,\ 2a_5 + 3,\ a_5)$$

is a solution to system (1). In particular, the vector $(-4, 0, 7, 3, 0)$ obtained by setting $a_2 = 0$ and $a_5 = 0$ is a solution to (1). Therefore,

$$(2, 6, 8) = -4y_1 + 0y_2 + 7y_3 + 3y_4 + 0y_5,$$

so that $(2, 6, 8)$ is a linear combination of y_1, y_2, y_3, y_4, and y_5.

The procedure illustrated above uses three types of operations to simplify the original system:

1. Interchanging the order of any two equations in the system
2. Multiplying any equation by a *nonzero* constant
3. Adding any constant multiple of an equation to another equation

In Section 3.4 we will prove that these operations do not change the set of solutions to the original system. Note that we employed these operations to obtain a system of equations that had the following properties:

1. The first nonzero coefficient in each equation is one.
2. If an unknown is the first unknown with a nonzero coefficient in some equation, then that unknown occurs with a zero coefficient in each of the other equations.

3. The first unknown with a nonzero coefficient in any equation has a larger subscript than the first unknown with a nonzero coefficient in any preceding equation.

To help clarify the meaning of these properties, note that none of the following systems meets these requirements.

$$\begin{cases} x_1 + 3x_2 + x_4 = 7 \\ 2x_3 - 5x_4 = -1 \end{cases} \tag{5}$$

$$\begin{cases} x_1 - 2x_2 + 3x_3 + x_5 = -5 \\ x_3 - 2x_5 = 9 \\ x_4 + 3x_5 = 6 \end{cases} \tag{6}$$

$$\begin{cases} x_1 - 2x_3 + x_5 = 1 \\ x_4 - 6x_5 = 0 \\ x_2 + 5x_3 - 3x_5 = 2 \end{cases} \tag{7}$$

Specifically, the system in (5) does not satisfy condition 1 because the first nonzero coefficient in the second equation is 2; the system in (6) does not satisfy condition 2 because x_3, the first unknown with a nonzero coefficient in the second equation, occurs with a nonzero coefficient in the first equation; and the system in (7) does not satisfy condition 3 because x_2, the first unknown with a nonzero coefficient in the third equation, does not have a larger subscript than x_4, the first unknown with a nonzero coefficient in the second equation.

Once a system with properties 1, 2, and 3 has been obtained, it is easy to solve for some of the unknowns in terms of the others (as in the example above). *If, however, in the course of using operations* 1, 2, *and* 3 *a system containing an equation of the form* $0 = c$, *where* c *is nonzero, is obtained, then the original system has no solutions* (see Example 2).

We will return to the study of systems of linear equations in Chapter 3. We discuss there the theoretical basis for this method of solving systems of linear equations and further simplify the procedure by use of matrices.

Example 2

We will show that

$$2x^3 - 2x^2 + 12x - 6$$

is a linear combination of

$$x^3 - 2x^2 - 5x - 3 \quad \text{and} \quad 3x^3 - 5x^2 - 4x - 9$$

in $\mathsf{P}_3(R)$, but that

$$3x^3 - 2x^2 + 7x + 8$$

is not such a linear combination. In the first case we wish to find scalars a and b such that

$$2x^3 - 2x^2 + 12x - 6 = a(x^3 - 2x^2 - 5x - 3) + b(3x^3 - 5x^2 - 4x - 9)$$
$$= (a + 3b)x^3 + (-2a - 5b)x^2 + (-5a - 4b)x + (-3a - 9b).$$

Thus we are led to the following system of linear equations:

$$\begin{cases} a + 3b = 2 \\ -2a - 5b = -2 \\ -5a - 4b = 12 \\ -3a - 9b = -6. \end{cases}$$

Adding appropriate multiples of the first equation to the others in order to eliminate a, we find

$$\begin{cases} a + 3b = 2 \\ b = 2 \\ 11b = 22 \\ 0 = 0. \end{cases}$$

Now adding the appropriate multiples of the second equation to the others yields

$$\begin{cases} a = -4 \\ b = 2 \\ 0 = 0 \\ 0 = 0. \end{cases}$$

Hence

$$2x^3 - 2x^2 + 12x - 6 = -4(x^3 - 2x^2 - 5x - 3) + 2(3x^3 - 5x^2 - 4x - 9).$$

In the second case we wish to show that there are no scalars a and b for which

$$3x^3 - 2x^2 + 7x + 8 = a(x^3 - 2x^2 - 5x - 3) + b(3x^3 - 5x^2 - 4x - 9).$$

As above, we obtain a system of linear equations

$$\begin{cases} a + 3b = 3 \\ -2a - 5b = -2 \\ -5a - 4b = 7 \\ -3a - 9b = 8. \end{cases} \tag{8}$$

Eliminating a as before yields

$$\begin{cases} a + 3b = 3 \\ b = 4 \\ 11b = 22 \\ 0 = 17. \end{cases}$$

But the presence of the inconsistent equation $0 = 17$ indicates that system (8) has no solutions. Hence $3x^3 - 2x^2 + 7x + 8$ is not a linear combination of $x^3 - 2x^2 - 5x - 3$ and $3x^3 - 5x^2 - 4x - 9$. ∎

The set of linear combinations of the elements of a nonempty subset of a vector space provides another example of a subspace, as the following result shows.

Theorem 1.5. *If* S *is a nonempty subset of a vector space* V, *then the set* W *consisting of all linear combinations of elements of* S *is a subspace of* V. *Moreover,* W *is the smallest subspace of* V *containing* S *in the sense that* W *is a subset of any subspace of* V *that contains* S.

Proof. First, we will use Theorem 1.3 to prove that W is a subspace of V. Since $S \neq \varnothing$, $0 \in$ W. If y and z are elements of W, then y and z are linear combinations of elements of S. So there exist elements $x_1, \ldots, x_n$ and $w_1, \ldots, w_m$ in S such that $y = a_1x_1 + \cdots + a_nx_n$ and $z = b_1w_1 + \cdots + b_mw_m$ for some choice of scalars $a_1, \ldots, a_n$ and $b_1, \ldots, b_m$. Now

$$y + z = a_1x_1 + \cdots + a_nx_n + b_1w_1 + \cdots + b_mw_m$$

and

$$cy = ca_1x_1 + \cdots + ca_nx_n$$

are linear combinations of elements of S; so $y + z$ and cy are elements of W for any scalar c. Thus W is a subspace of V.

Now let W′ denote any subspace of V that contains S. If y is an element of W, then y is a linear combination of elements of S; say $y = a_1x_1 + \cdots + a_nx_n$, where $a_1, \ldots, a_n \in F$ and $x_1, \ldots, x_n \in S$. Because $S \subseteq$ W′, $x_1, \ldots, x_n \in$ W′. Therefore, $y = a_1x_1 + \cdots + a_nx_n$ is an element of W′ by Exercise 20 of Section 1.3. Since y, an arbitrary element of W, belongs to W′, W ⊆ W′. This completes the proof. ∎

Definition. *The subspace* W *described in Theorem 1.5 is called the* span of S *(or the* subspace generated by the elements of S) *and is denoted* span(S). *For convenience we define* span(∅) = {0}.

Observe that Theorem 1.5 shows that x is a linear combination of elements of S if and only if x is an element of span(S). For instance, in R^3, it is easily seen that span({(1, 0, 0), (0, 1, 0)}) is the xy-plane.

Definition. *A subset* S *of a vector space* V generates (*or* spans) V *if* span(S) = V. *In this situation we also say that the* elements of S generate (*or* span) V.

Example 3

The vectors (1, 1, 0), (1, 0, 1), and (0, 1, 1) generate R^3 since an arbitrary element (a_1, a_2, a_3) of R^3 is a linear combination of the three given vectors; in fact, the scalars r, s, and t for which

$$r(1, 1, 0) + s(1, 0, 1) + t(0, 1, 1) = (a_1, a_2, a_3)$$

are

$$r = \tfrac{1}{2}(a_1 + a_2 - a_3), \qquad s = \tfrac{1}{2}(a_1 - a_2 + a_3), \qquad \text{and} \qquad t = \tfrac{1}{2}(-a_1 + a_2 + a_3). \quad \blacksquare$$

Example 4

The polynomials $x^2 + 3x - 2$, $2x^2 + 5x - 3$, and $-x^2 - 4x + 4$ generate $\mathsf{P}_2(R)$ since each of the three given polynomials belongs to $\mathsf{P}_2(R)$ and each polynomial $ax^2 + bx + c$ in $\mathsf{P}_2(R)$ is a linear combination of these three; namely,

$$(-8a + 5b + 3c)(x^2 + 3x - 2) + (4a - 2b - c)(2x^2 + 5x - 3) + (-a + b + c)(-x^2 - 4x + 4) = ax^2 + bx + c. \quad \blacksquare$$

Example 5

The matrices

$$\begin{pmatrix} 1 & 1 \\ 1 & 0 \end{pmatrix}, \begin{pmatrix} 1 & 1 \\ 0 & 1 \end{pmatrix}, \begin{pmatrix} 1 & 0 \\ 1 & 1 \end{pmatrix}, \quad \text{and} \quad \begin{pmatrix} 0 & 1 \\ 1 & 1 \end{pmatrix}$$

generate $\mathsf{M}_{2\times 2}(R)$ since an arbitrary element of $\mathsf{M}_{2\times 2}(R)$

$$\begin{pmatrix} a_{11} & a_{12} \\ a_{21} & a_{22} \end{pmatrix},$$

can be expressed as a linear combination of the four given matrices as follows:

$$\left(\tfrac{1}{3}a_{11} + \tfrac{1}{3}a_{12} + \tfrac{1}{3}a_{21} - \tfrac{2}{3}a_{22}\right)\begin{pmatrix} 1 & 1 \\ 1 & 0 \end{pmatrix}$$

$$+ \left(\tfrac{1}{3}a_{11} + \tfrac{1}{3}a_{12} - \tfrac{2}{3}a_{21} + \tfrac{1}{3}a_{22}\right)\begin{pmatrix} 1 & 1 \\ 0 & 1 \end{pmatrix}$$

$$+\left(\tfrac{1}{3}a_{11}-\tfrac{2}{3}a_{12}+\tfrac{1}{3}a_{21}+\tfrac{1}{3}a_{22}\right)\begin{pmatrix}1 & 0\\ 1 & 1\end{pmatrix}$$

$$+\left(-\tfrac{2}{3}a_{11}+\tfrac{1}{3}a_{12}+\tfrac{1}{3}a_{21}+\tfrac{1}{3}a_{22}\right)\begin{pmatrix}0 & 1\\ 1 & 1\end{pmatrix}$$

$$=\begin{pmatrix}a_{11} & a_{12}\\ a_{21} & a_{22}\end{pmatrix}. \quad \blacksquare$$

EXERCISES

1. Label the following statements as being true or false.

(a) The zero vector is a linear combination of any nonempty set of vectors.

(b) The span of $\varnothing$ is $\varnothing$.

(c) If S is a subset of a vector space V, then span(S) equals the intersection of all subspaces of V that contain S.

(d) In solving a system of linear equations it is permissible to multiply an equation by any constant.

(e) In solving a system of linear equations it is permissible to add a multiple of one equation to another.

(f) Every system of linear equations has a solution.

2. Solve the following systems of linear equations by the method introduced in this section.

(a)
$$\begin{cases} 2x_1 - 2x_2 - 3x_3 \phantom{{}+5x_4} = -2\\ 3x_1 - 3x_2 - 2x_3 + 5x_4 = 7\\ x_1 - x_2 - 2x_3 - x_4 = -3 \end{cases}$$

(b)
$$\begin{cases} 3x_1 - 7x_2 + 4x_3 = 10\\ x_1 - 2x_2 + x_3 = 3\\ 2x_1 - x_2 - 2x_3 = 6 \end{cases}$$

(c)
$$\begin{cases} x_1 + 2x_2 - x_3 + x_4 = 5\\ x_1 + 4x_2 - 3x_3 - 3x_4 = 6\\ 2x_1 + 3x_2 - x_3 + 4x_4 = 8 \end{cases}$$

(d)
$$\begin{cases} x_1 + 2x_2 + 2x_3 = 2\\ x_1 + 8x_3 + 5x_4 = -6\\ x_1 + x_2 + 5x_3 + 5x_4 = 3 \end{cases}$$

(e)
$$\begin{cases} x_1 + 2x_2 - 4x_3 - x_4 + x_5 = 7\\ -x_1 + 10x_3 - 3x_4 - 4x_5 = -16\\ 2x_1 + 5x_2 - 5x_3 - 4x_4 - x_5 = 2\\ 4x_1 + 11x_2 - 7x_3 - 10x_4 - 2x_5 = 7 \end{cases}$$

(f) $\begin{cases} x_1 + 2x_2 + 6x_3 = -1 \\ 2x_1 + x_2 + x_3 = 8 \\ 3x_1 + x_2 - x_3 = 15 \\ x_1 + 3x_2 + 10x_3 = -5 \end{cases}$

3. For each of the following lists of vectors in R^3, determine whether or not the first vector can be expressed as a linear combination of the other two.
   (a) $(-2, 0, 3)$, $(1, 3, 0)$, $(2, 4, -1)$
   (b) $(1, 2, -3)$, $(-3, 2, 1)$, $(2, -1, -1)$
   (c) $(3, 4, 1)$, $(1, -2, 1)$, $(-2, -1, 1)$
   (d) $(2, -1, 0)$, $(1, 2, -3)$, $(1, -3, 2)$
   (e) $(5, 1, -5)$, $(1, -2, -3)$, $(-2, 3, -4)$
   (f) $(-2, 2, 2)$, $(1, 2, -1)$, $(-3, -3, 3)$
4. For each of the following lists of polynomials in $\mathsf{P}_3(R)$, determine whether or not the first polynomial can be expressed as a linear combination of the other two.
   (a) $x^3 - 3x + 5$, $x^3 + 2x^2 - x + 1$, $x^3 + 3x^2 - 1$
   (b) $4x^3 + 2x^2 - 6$, $x^3 - 2x^2 + 4x + 1$, $3x^3 - 6x^2 + x + 4$
   (c) $-2x^3 - 11x^2 + 3x + 2$, $x^3 - 2x^2 + 3x - 1$, $2x^3 + x^2 + 3x - 2$
   (d) $x^3 + x^2 + 2x + 13$, $2x^3 - 3x^2 + 4x + 1$, $x^3 - x^2 + 2x + 3$
   (e) $x^3 - 8x^2 + 4x$, $x^3 - 2x^2 + 3x - 1$, $x^3 - 2x + 3$
   (f) $6x^3 - 3x^2 + x + 2$, $x^3 - x^2 + 2x + 3$, $2x^3 + x^2 - 3x + 1$
5. In F^n let e_j denote the vector whose jth coordinate is 1 and whose other coordinates are 0. Prove that $\{e_1, e_2, \ldots, e_n\}$ generates F^n.
6. Show that $\mathsf{P}_n(F)$ is generated by $\{1, x, x^2, \ldots, x^n\}$.
7. Show that the matrices

$$\begin{pmatrix} 1 & 0 \\ 0 & 0 \end{pmatrix}, \begin{pmatrix} 0 & 1 \\ 0 & 0 \end{pmatrix}, \begin{pmatrix} 0 & 0 \\ 1 & 0 \end{pmatrix}, \quad \text{and} \quad \begin{pmatrix} 0 & 0 \\ 0 & 1 \end{pmatrix}$$

   generate $\mathsf{M}_{2\times 2}(F)$.
8. Show that if

$$M_1 = \begin{pmatrix} 1 & 0 \\ 0 & 0 \end{pmatrix}, \quad M_2 = \begin{pmatrix} 0 & 0 \\ 0 & 1 \end{pmatrix}, \quad \text{and} \quad M_3 = \begin{pmatrix} 0 & 1 \\ 1 & 0 \end{pmatrix},$$

   then the span of $\{M_1, M_2, M_3\}$ is the set of all symmetric 2×2 matrices.
9.† For any element x in a vector space, prove that $\text{span}(\{x\}) = \{ax\colon a \in F\}$. Interpret this result geometrically in R^3.
10. Show that a subset W of a vector space V is a subspace of V if and only if span(W) = W.
11.† Show that if S_1 and S_2 are subsets of a vector space V such that $S_1 \subseteq S_2$, then $\text{span}(S_1) \subseteq \text{span}(S_2)$. In particular, if $S_1 \subseteq S_2$ and $\text{span}(S_1) = \mathsf{V}$, deduce that $\text{span}(S_2) = \mathsf{V}$.

12. Show that if S_1 and S_2 are arbitrary subsets of a vector space V, then $\text{span}(S_1 \cup S_2) = \text{span}(S_1) + \text{span}(S_2)$.

13. Let S_1 and S_2 be subsets of a vector space V. Prove that $\text{span}(S_1 \cap S_2) \subseteq \text{span}(S_1) \cap \text{span}(S_2)$. Give an example in which $\text{span}(S_1 \cap S_2)$ and $\text{span}(S_1) \cap \text{span}(S_2)$ are equal and an example in which they are unequal.

1.5 LINEAR DEPENDENCE AND LINEAR INDEPENDENCE

At the beginning of Section 1.4 we remarked that the equation of a plane through three noncollinear points in space, one of which is the origin, is of the form $x = t_1u + t_2v$, where $u, v \in \mathsf{R}^3$ and t_1 and t_2 are scalars. Thus a vector x in R^3 is a linear combination of $u, v \in \mathsf{R}^3$ if and only if x lies in the plane containing u and v (see Figure 1.5). We see, therefore, that in R^3 the span of two nonparallel vectors has a simple geometric interpretation. A similar interpretation can be given for the span of a single nonzero vector in R^3 (see Exercise 9 of Section 1.4).

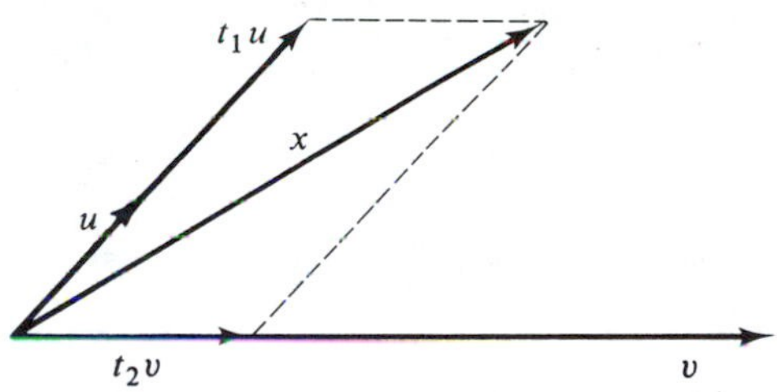

Figure 1.5

In the equation $x = t_1u + t_2v$, x depends on u and v in the sense that x is a linear combination of u and v. A set in which at least one vector is a linear combination of the others is called a linearly dependent set. Consider, for example, the set $S = \{x_1, x_2, x_3, x_4\} \subseteq \mathsf{R}^3$, where $x_1 = (2, -1, 4)$, $x_2 = (1, -1, 3)$, $x_3 = (1, 1, -1)$, and $x_4 = (1, -2, -1)$. To see if S is linearly dependent, we must check whether or not there is a vector in S that is a linear combination of the others. Now the vector x_4 is a linear combination of x_1, x_2, and x_3 if and only if there are scalars a, b, and c such that

$$x_4 = ax_1 + bx_2 + cx_3,$$

that is, if and only if

$$x_4 = (2a + b + c,\ -a - b + c,\ 4a + 3b - c).$$

Thus x_4 is a linear combination of x_1, x_2, and x_3 if and only if the system

$$\begin{cases} 2a + b + c = 1 \\ -a - b + c = -2 \\ 4a + 3b - c = -1 \end{cases}$$

has a solution. The reader should verify that no such solution exists. Notice, however, that this does not show that the set S is not linearly dependent, for we must now check whether or not x_1, x_2, or x_3 can be written as a linear combination of the other vectors in S. It can be shown, in fact, that x_3 is a linear combination of x_1, x_2, and x_4; specifically, $x_3 = 2x_1 - 3x_2 + 0x_4$. So S is indeed linearly dependent.

We see from this example that the condition for linear dependence given above is inconvenient to use because it may require checking several vectors to see if some vector in S is a linear combination of the others. By reformulating the definition in the following way, we obtain a definition of linear dependence that is easier to use.

Definition. *A subset* S *of a vector space* V *is said to be* linearly dependent *if there exist a finite number of distinct vectors* $\mathsf{x}_1, \mathsf{x}_2, \ldots, \mathsf{x}_n$ *in* S *and scalars* $\mathsf{a}_1, \mathsf{a}_2, \ldots, \mathsf{a}_n$, *not all zero, such that*

$$\mathsf{a}_1\mathsf{x}_1 + \mathsf{a}_2\mathsf{x}_2 + \cdots + \mathsf{a}_n\mathsf{x}_n = 0.$$

In this case we will also say that the elements of S *are linearly dependent.*

To show that the subset S of R^3 defined above is linearly dependent using this definition, we must find scalars a_1, a_2, a_3, and a_4, not all zero, such that

$$a_1x_1 + a_2x_2 + a_3x_3 + a_4x_4 = 0,$$

that is, such that

$$(2a_1 + a_2 + a_3 + a_4, -a_1 - a_2 + a_3 - 2a_4, 4a_1 + 3a_2 - a_3 - a_4) = (0, 0, 0).$$

Thus we must find a solution to the system

$$\begin{cases} 2a_1 + a_2 + a_3 + a_4 = 0 \\ -a_1 - a_2 + a_3 - 2a_4 = 0 \\ 4a_1 + 3a_2 - a_3 - a_4 = 0, \end{cases}$$

in which not all the unknowns are zero. Using the techniques discussed in Section 1.4, we find that $a_1 = 2$, $a_2 = -3$, $a_3 = -1$, $a_4 = 0$ is one such solution. Notice that by using the definition of linear dependence stated above, we are able to check that S is linearly dependent by solving only one system of equations. The reader should verify that the two conditions for linear dependence discussed above are, in fact, equivalent (see Exercise 11).

It is easily seen that in any vector space a subset S that contains the zero vector must be linearly dependent. For since $0 = 1 \cdot 0$, the zero vector is a linear combination of elements of S in which some coefficient is nonzero.

Example 1

In R^4 the set

$$S = \{(1, 3, -4, 2), (2, 2, -4, 0), (1, -3, 2, -4), (-1, 0, 1, 0)\}$$

is linearly dependent because

$$4(1, 3, -4, 2) - 3(2, 2, -4, 0) + 2(1, -3, 2, -4) + 0(-1, 0, 1, 0) = (0, 0, 0, 0).$$

Similarly, in $\mathsf{M}_{2\times 3}(R)$ the set

$$\left\{\begin{pmatrix} 1 & -3 & 2 \\ -4 & 0 & 5 \end{pmatrix}, \begin{pmatrix} -3 & 7 & 4 \\ 6 & -2 & -7 \end{pmatrix}, \begin{pmatrix} -2 & 3 & 11 \\ -1 & -3 & 2 \end{pmatrix}\right\}$$

is linearly dependent since

$$5\begin{pmatrix} 1 & -3 & 2 \\ -4 & 0 & 5 \end{pmatrix} + 3\begin{pmatrix} -3 & 7 & 4 \\ 6 & -2 & -7 \end{pmatrix} - 2\begin{pmatrix} -2 & 3 & 11 \\ -1 & -3 & 2 \end{pmatrix} = \begin{pmatrix} 0 & 0 & 0 \\ 0 & 0 & 0 \end{pmatrix}. \blacksquare$$

Definition. *A subset* S *of a vector space that is not linearly dependent is said to be* linearly independent. *As before, we will also say that the elements of* S *are linearly independent in this case.*

The following facts about linearly independent sets are true in any vector space.

1. The empty set is linearly independent, for linearly dependent sets must be nonempty.
2. A set consisting of a single nonzero vector is linearly independent. For if $\{x\}$ is linearly dependent, then $ax = 0$ for some nonzero scalar a. Thus
$$x = a^{-1}(ax) = a^{-1}0 = 0.$$
3. For any vectors $x_1, x_2, \ldots, x_n$, we have $a_1x_1 + a_2x_2 + \cdots + a_nx_n = 0$ if $a_1 = 0, a_2 = 0, \ldots, a_n = 0$. We call this the *trivial representation* of 0 as a linear combination of $x_1, x_2, \ldots, x_n$. *A set is linearly independent if and only if the only representations of* 0 *as linear combinations of its elements are the trivial representations.*

The condition in item 3 provides a very useful method for determining if a finite set is linearly independent. This technique is illustrated in the following example.

Example 2

Let x_k denote the vector in F^n whose first $k-1$ coordinates are zero and whose last $n-k+1$ coordinates are 1. Then $\{x_1, x_2, \ldots, x_n\}$ is linearly independent, for if $a_1x_1 + a_2x_2 + \cdots + a_nx_n = 0$, equating the corresponding coordinates of the left and right sides of this equality gives the following system of equations:

$$\begin{cases} a_1 & = 0 \\ a_1 + a_2 & = 0 \\ a_1 + a_2 + a_3 & = 0 \\ \quad\vdots & \\ a_1 + a_2 + a_3 + \cdots + a_n & = 0. \end{cases}$$

Clearly, the only solution of this system is $a_1 = \cdots = a_n = 0$. ∎

The following useful results are immediate consequences of the definitions of linear dependence and linear independence.

Theorem 1.6. *Let V be a vector space, and let $\mathsf{S}_1 \subseteq \mathsf{S}_2 \subseteq \mathsf{V}$. If S_1 is linearly dependent, then S_2 is linearly dependent.*

Proof. Exercise. ∎

Corollary. *Let V be a vector space, and let $\mathsf{S}_1 \subseteq \mathsf{S}_2 \subseteq \mathsf{V}$. If S_2 is linearly independent, then S_1 is linearly independent.*

Proof. Exercise. ∎

EXERCISES

1. Label the following statements as being true or false.
   (a) If S is a linearly dependent set, then each element of S is a linear combination of other elements of S.
   (b) Any set containing the zero vector is linearly dependent.
   (c) The empty set is linearly dependent.
   (d) Subsets of linearly dependent sets are linearly dependent.
   (e) Subsets of linearly independent sets are linearly independent.
   (f) If $a_1x_1 + a_2x_2 + \cdots + a_nx_n = 0$ and $x_1, x_2, \ldots, x_n$ are linearly independent, then all the scalars a_i equal zero.

2. In F^n let e_j denote the vector whose jth coordinate is 1 and whose other coordinates are 0. Prove that $\{e_1, e_2, \ldots, e_n\}$ is linearly independent.

3. Show that the set $\{1, x, x^2, \ldots, x^n\}$ is linearly independent in $\mathsf{P}_n(F)$.

4. Prove that the matrices

$$\begin{pmatrix} 1 & 0 \\ 0 & 0 \end{pmatrix}, \begin{pmatrix} 0 & 1 \\ 0 & 0 \end{pmatrix}, \begin{pmatrix} 0 & 0 \\ 1 & 0 \end{pmatrix}, \quad \text{and} \quad \begin{pmatrix} 0 & 0 \\ 0 & 1 \end{pmatrix}$$

are linearly independent in $\mathsf{M}_{2 \times 2}(F)$.

5. Find a set of linearly independent diagonal matrices that generate the vector space of 2×2 diagonal matrices.

6.† Show that $\{x, y\}$ is linearly dependent if and only if x or y is a multiple of the other.

7. Give an example of three linearly dependent vectors in R^3 such that none of the three is a multiple of another.

8. Let $S = \{x_1, x_2, \ldots, x_n\}$ be a linearly independent subset of a vector space V over the field Z_2. How many elements are there in span(S)? Justify your answer.

9. Prove Theorem 1.6 and its corollary.

10. Let V be a vector space over a field of characteristic not equal to 2.
    (a) Prove that $\{u, v\}$ is linearly independent if and only if $\{u + v, u - v\}$ is linearly independent.
    (b) Similarly, prove that $\{u, v, w\}$ is linearly independent if and only if $\{u + v, u + w, v + w\}$ is linearly independent.

11. Prove that a set S is linearly dependent if and only if $S = \{0\}$ or there exist distinct vectors $y, x_1, x_2, \ldots, x_n$ in S such that y is a linear combination of $x_1, x_2, \ldots, x_n$.

12. Let $S = \{x_1, x_2, \ldots, x_n\}$ be a finite set of vectors. Prove that S is linearly dependent if and only if $x_1 = 0$ or $x_{k+1} \in \text{span}(\{x_1, x_2, \ldots, x_k\})$ for some $k < n$.

13. Prove that a set S of vectors is linearly independent if and only if each finite subset of S is linearly independent.

14. Let M be a square upper triangular matrix (as defined in Exercise 12 of Section 1.3) having nonzero diagonal entries. Prove that the columns of M are linearly independent.

15. Let f and g be functions defined by $f(t) = e^{rt}$ and $g(t) = e^{st}$, where $r \neq s$. Prove that f and g are linearly independent in $\mathcal{F}(R, R)$. *Hint:* Suppose that $ae^{rt} + be^{st} = 0$. Let $t = 0$ to obtain an equation involving a and b. Then differentiate $ae^{rt} + be^{st} = 0$, and let $t = 0$ to obtain a second equation involving a and b. Solve these equations for a and b.

1.6 BASES AND DIMENSION

A subset S of a vector space V that is linearly independent and generates V possesses a very useful property—every element of V can be expressed in one and only one way as a linear combination of elements of S. (This property will be proved in Theorem 1.7.) It is this result that makes linearly independent generating sets the building blocks of vector spaces.

Definition. *A* basis β *for a vector space* V *is a linearly independent subset of* V *that generates* V. *(If β is a basis for* V, *we also say that the elements of β form a basis for* V.)

Example 1

Recalling that span($\varnothing$) $= \{0\}$ and that $\varnothing$ is linearly independent, we see that $\varnothing$ is a basis for the vector space $\{0\}$. ∎

Example 2

In F^n, let $e_1 = (1, 0, 0, \ldots, 0, 0)$, $e_2 = (0, 1, 0, \ldots, 0, 0)$, $\ldots$, $e_n = (0, 0, 0, \ldots, 0, 1)$; $\{e_1, e_2, \ldots, e_n\}$ is readily seen to be a basis for F^n and is called the *standard basis* for F^n. ∎

Example 3

In $\mathsf{M}_{m \times n}(F)$, let M^{ij} denote the matrix whose only nonzero entry is a 1 in the ith row and jth column. Then $\{M^{ij}: 1 \leqslant i \leqslant m,\ 1 \leqslant j \leqslant n\}$ is a basis for $\mathsf{M}_{m \times n}(F)$. ∎

Example 4

In $\mathsf{P}_n(F)$ the set $\{1, x, x^2, \ldots, x^n\}$ is a basis. We call this basis the *standard basis* for $\mathsf{P}_n(F)$. ∎

Example 5

In $\mathsf{P}(F)$ the set $\{1, x, x^2, \ldots\}$ is a basis. ∎

Observe that Example 5 shows that a basis need not be finite. In fact, we will see later in this section that no basis for $\mathsf{P}(F)$ can be finite. Hence not every vector space has a finite basis.

The following theorem, which will be used frequently in Chapter 2, shows the most significant property of a basis.

Theorem 1.7. *Let* V *be a vector space and* $\beta = \{x_1, \ldots, x_n\}$ *be a subset of* V. *Then* β *is a basis for* V *if and only if each vector* y *in* V *can be uniquely expressed as a linear combination of vectors in* β, *i.e., can be expressed in the form*

$$y = a_1 x_1 + \cdots + a_n x_n$$

for unique scalars $a_1, \ldots, a_n$.

Proof. Let β be a basis for V. If $y \in \mathsf{V}$, then $y \in \operatorname{span}(\beta)$ since $\operatorname{span}(\beta) = \mathsf{V}$. Thus y is a linear combination of the elements of β. Suppose that $y = a_1 x_1 + \cdots + a_n x_n$ and $y = b_1 x_1 + \cdots + b_n x_n$ are two such representations of y. Subtracting the second equality from the first gives

$$0 = (a_1 - b_1)x_1 + \cdots + (a_n - b_n)x_n.$$

Since β is linearly independent, it follows that $a_1 - b_1 = \cdots = a_n - b_n = 0$. Thus $a_1 = b_1, \ldots, a_n = b_n$, so that y is uniquely expressible as a linear combination of the elements of β.

The proof of the converse is an exercise. ∎

Theorem 1.7 shows that each vector v in a vector space V with basis $\beta = \{x_1, \ldots, x_n\}$ can be uniquely expressed in the form

$$v = a_1 x_1 + \cdots + a_n x_n$$

for appropriately chosen scalars $a_1, \ldots, a_n$. Thus v determines a unique n-tuple of scalars $(a_1, \ldots, a_n)$, and conversely, each n-tuple of scalars determines a unique vector v by using the entries of the n-tuple as the coefficients of a linear combination of the vectors in β. This fact suggests that V is like the vector space F^n, where n is the number of vectors in a basis for V. We will see in Section 2.4 that this is indeed the case.

Theorem 1.9 identifies a large class of vector spaces having finite bases. First, however, we must prove a preliminary result.

Theorem 1.8. *Let S be a linearly independent subset of a vector space V, and let x be an element of V that is not in S. Then $S \cup \{x\}$ is linearly dependent if and only if $x \in \text{span}(S)$.*

Proof. If $S \cup \{x\}$ is linearly dependent, then there are vectors $x_1, \ldots, x_n$ in $S \cup \{x\}$ and nonzero scalars $a_1, \ldots, a_n$ such that $a_1x_1 + \cdots + a_nx_n = 0$. Because S is linearly independent, one of the x_i, say x_1, equals x. Thus $a_1x + a_2x_2 + \cdots + a_nx_n = 0$, and so

$$x = a_1^{-1}(-a_2x_2 - \cdots - a_nx_n).$$

Since x is a linear combination of $x_2, \ldots, x_n$, which are elements of S, $x \in \text{span}(S)$.

Conversely, suppose that $x \in \text{span}(S)$. Then there exist vectors $x_1, \ldots, x_n$ in S and scalars $a_1, \ldots, a_n$ such that $x = a_1x_1 + \cdots + a_nx_n$. So $0 = a_1x_1 + \cdots + a_nx_n + (-1)x$, and since $x \neq x_i$ for $i = 1, \ldots, n$, $\{x_1, \ldots, x_n, x\}$ is linearly dependent. Thus $S \cup \{x\}$ is linearly dependent by Theorem 1.6. ∎

Theorem 1.9. *If a vector space V is generated by a finite set S_0, then a subset of S_0 is a basis for V. Hence V has a finite basis.*

Proof. If $S_0 = \varnothing$ or $S_0 = \{0\}$, then $V = \{0\}$ and $\varnothing$ is a subset of S_0 that is a basis for V. Otherwise S_0 contains a nonzero element x_1. Recall that $\{x_1\}$ is a linearly independent set. Continue, if possible, choosing elements $x_2, \ldots, x_r$ in S_0 so that $\{x_1, x_2, \ldots, x_r\}$ is linearly independent. Since S_0 is a finite set, we must eventually reach a stage at which $S = \{x_1, \ldots, x_r\}$ is a linearly independent subset of S_0, but adjoining to S any element of S_0 not in S produces a linearly dependent set. We will show that S is a basis for V. Because S is linearly independent, it suffices to prove that $\text{span}(S) = V$. Since $\text{span}(S_0) = V$, it suffices by Theorem 1.5 to show that $S_0 \subseteq \text{span}(S)$. Let $x \in S_0$. If $x \in S$, then clearly $x \in \text{span}(S)$. Otherwise, if $x \notin S$, then the construction above shows that $S \cup \{x\}$ is linearly dependent. So $x \in \text{span}(S)$ by Theorem 1.8. Thus $S_0 \subseteq \text{span}(S)$. ∎

The method by which the basis S was obtained in the proof of Theorem 1.9 is a useful way of obtaining bases. An example of this procedure is given below.

Example 6

The reader should check that the elements $(2, -3, 5)$, $(8, -12, 20)$, $(1, 0, -2)$, $(0, 2, -1)$, and $(7, 2, 0)$ generate R^3. We will select a basis for R^3 from among these elements. To start, select any nonzero element from the generating set, say $(2, -3, 5)$, as one of the elements of the basis. Since $4(2, -3, 5) = (8, -12, 20)$, the set $\{(2, -3, 5), (8, -12, 20)\}$ is linearly dependent (Exercise 6 of Section 1.5). Hence we do not include $(8, -12, 20)$ in our basis. Since $(1, 0, -2)$ is not a multiple of $(2, -3, 5)$, and vice versa, the set $\{(2, -3, 5), (1, 0, -2)\}$ is linearly independent. Thus we include $(1, 0, -2)$ in our basis. Proceeding to the next element in the generating set, we exclude from or include into our basis the element $(0, 2, -1)$ according to whether the set $\{(2, -3, 5), (1, 0, -2), (0, 2, -1)\}$ is linearly dependent or linearly independent. An easy calculation shows that the set is linearly independent; so we include $(0, 2, -1)$ in our basis. The final element of the generating set $(7, 2, 0)$ is excluded from or included into our basis according to whether $\{(2, -3, 5), (1, 0, -2), (0, 2, -1), (7, 2, 0)\}$ is linearly dependent or linearly independent. Since

$$2(2, -3, 5) + 3(1, 0, -2) + 4(0, 2, -1) - (7, 2, 0) = (0, 0, 0),$$

the set is linearly dependent and we exclude $(7, 2, 0)$ from the basis. So the set $\{(2, -3, 5), (1, 0, -2), (0, 2, -1)\}$ is a basis for R^3. ∎

The following theorem and its corollaries are perhaps the most significant results in Chapter 1.

Theorem 1.10 (Replacement Theorem). *Let* V *be a vector space having a basis* β *containing exactly* n *elements. Let* $S = \{y_1, \ldots, y_m\}$ *be a linearly independent subset of* V *containing exactly* m *elements, where* $m \leq n$. *Then there exists a subset* S_1 *of* β *containing exactly* $n - m$ *elements such that* $S \cup S_1$ *generates* V.

Proof. The proof will be by induction on m. The induction begins with $m = 0$; for in this case $S = \varnothing$, so $S_1 = \beta$ clearly satisfies the conclusion of the theorem.

Now assume that the theorem is true for some m, where $m < n$. We will prove that the theorem is true for $m + 1$. Let $S = \{y_1, \ldots, y_m, y_{m+1}\}$ be a linearly independent subset of V containing exactly $m + 1$ elements. Since $\{y_1, \ldots, y_m\}$ is linearly independent by the corollary to Theorem 1.6, we may apply the inductive hypothesis to conclude that there exists a subset $\{x_1, \ldots, x_{n-m}\}$ of β such that $\{y_1, \ldots, y_m\} \cup \{x_1, \ldots, x_{n-m}\}$ generates V. Thus there exist scalars $a_1, \ldots, a_m, b_1, b_2, \ldots, b_{n-m}$ such that

$$y_{m+1} = a_1 y_1 + \cdots + a_m y_m + b_1 x_1 + b_2 x_2 + \cdots + b_{n-m} x_{n-m}. \tag{9}$$

Observe that some b_i, say b_1, is nonzero, for otherwise (9) would imply that y_{m+1} is a linear combination of $y_1, \ldots, y_m$, in contradiction to the assumption that $\{y_1, \ldots, y_m, y_{m+1}\}$ is linearly independent. Solving (9) for x_1 gives

$$x_1 = (-b_1^{-1}a_1)y_1 + \cdots + (-b_1^{-1}a_m)y_m - (-b_1^{-1})y_{m+1} + (-b_1^{-1}b_2)x_2 + \cdots + (-b_1^{-1}b_{n-m})x_{n-m}.$$

Hence $x_1 \in \text{span}(\{y_1, \ldots, y_m, y_{m+1}, x_2, \ldots, x_{n-m}\})$. But since $y_1, \ldots, y_m, x_2, \ldots, x_{n-m}$ are clearly elements of $\text{span}(\{y_1, \ldots, y_m, y_{m+1}, x_2, \ldots, x_{n-m}\})$, it follows that

$$\{y_1, \ldots, y_m, x_1, x_2, \ldots, x_{n-m}\} \subseteq \text{span}(\{y_1, \ldots, y_m, y_{m+1}, x_2, \ldots, x_{n-m}\}).$$

Thus Theorem 1.5 implies that

$$\text{span}(\{y_1, \ldots, y_m, y_{m+1}, x_2, \ldots, x_{n-m}\}) = \mathsf{V}.$$

So the choice of $S_1 = \{x_2, \ldots, x_{n-m}\}$ proves that the theorem is true for $m + 1$.

This completes the proof. ∎

To illustrate the replacement theorem, note that $S = \{x^2 - 4, x + 6\}$ is a linearly independent subset of $\mathsf{P}_2(F)$. Since $\beta = \{1, x, x^2\}$ is a basis for $\mathsf{P}_2(F)$, there must be a subset S_1 of β containing $3 - 2 = 1$ element such that $S \cup S_1$ generates $\mathsf{P}_2(F)$. In this example any subset of β containing one element will suffice for S_1. Hence we see that the set S_1 in Theorem 1.10 need not be unique.

Corollary 1. *Let* V *be a vector space having a basis* β *containing exactly* n *elements. Then any linearly independent subset of* V *containing exactly* n *elements is a basis for* V.

Proof. Let $S = \{y_1, \ldots, y_n\}$ be a linearly independent subset of V containing exactly n elements. Applying the replacement theorem, we see that there exists a subset S_1 of β containing $n - n = 0$ elements such that $S \cup S_1$ generates V. Clearly, $S_1 = \varnothing$; so S generates V. Since S is also linearly independent, S is a basis for V. ∎

Example 7

The vectors $(1, -3, 2)$, $(4, 1, 0)$, and $(0, 2, -1)$ form a basis for R^3, for if

$$a_1(1, -3, 2) + a_2(4, 1, 0) + a_3(0, 2, -1) = (0, 0, 0),$$

then a_1, a_2, and a_3 must satisfy the system of equations

$$\begin{cases} a_1 + 4a_2 \qquad\quad = 0 \\ -3a_1 + a_2 + 2a_3 = 0 \\ 2a_1 \qquad\quad - a_3 = 0. \end{cases}$$

But it is easily seen that the only solution of this system is $a_1 = 0$, $a_2 = 0$, and $a_3 = 0$. Hence $(1, -3, 2)$, $(4, 1, 0)$, and $(0, 2, -1)$ are linearly independent and therefore form a basis for R^3 by Corollary 1. Note that we do *not* have to check that the given vectors span R^3. ∎

Corollary 2. *Let* V *be a vector space having a basis* β *containing exactly* n *elements. Then any subset of* V *containing more than* n *elements is linearly dependent. Consequently, any linearly independent subset of* V *contains at most* n *elements.*

Proof. Let S be a subset of V containing more than n elements. In order to reach a contradiction, assume that S is linearly independent. Let S_1 be any subset of S containing exactly n elements; then S_1 is a basis for V by Corollary 1. Because S_1 is a proper subset of S, we can select an element x of S that is not an element of S_1. Since S_1 is a basis for V, $x \in \text{span}(S_1) = \mathsf{V}$. Thus Theorem 1.8 implies that $S_1 \cup \{x\}$ is linearly dependent. But $S_1 \cup \{x\} \subseteq S$; so S is linearly dependent—a contradiction. We conclude therefore that S is linearly dependent. ∎

Example 8

Let $S = \{x^2 + 7, 8x^2 - 2x, 4x - 3, 7x + 2\}$. Although we can prove directly that S is a linearly dependent subset of $\mathsf{P}_2(F)$, this conclusion follows immediately from Corollary 2 since $\beta = \{1, x, x^2\}$ is a basis for $\mathsf{P}_2(F)$ containing fewer elements than S. ∎

Corollary 3. *Let* V *be a vector space having a basis* β *containing exactly* n *elements. Then every basis for* V *contains exactly* n *elements.*

Proof. Let S be a basis for V. Since S is linearly independent, S contains at most n elements by Corollary 2. Suppose that S contains exactly m elements; then $m \leqslant n$. But also S is a basis for V and β is a linearly independent subset of V. So Corollary 2 may be applied with the roles of β and S interchanged to yield $n \leqslant m$. Thus $m = n$. ∎

If a vector space has a basis containing a finite number of elements, then the corollary above asserts that the number of elements in each basis for the space is the same. This result makes the following definitions possible.

Definitions. *A vector space* V *is called* finite-dimensional *if it has a basis consisting of a finite number of elements; the unique number of elements in each basis for* V *is called the* dimension of V *and is denoted* dim(V). *If a vector space is not finite-dimensional, then it is called* infinite-dimensional.

The following results are consequences of Examples 1 through 5.

Example 9

The vector space $\{0\}$ has dimension zero. ∎

Example 10

The vector space F^n has dimension n. ∎

Example 11

The vector space $\mathsf{M}_{m \times n}(F)$ has dimension mn. ∎

Example 12

The vector space $\mathsf{P}_n(F)$ has dimension $n + 1$. ∎

Example 13

The vector space $\mathsf{P}(F)$ is infinite-dimensional. ∎

The following two examples show that the dimension of a vector space depends on its field of scalars.

Example 14

Over the field of complex numbers, the vector space of complex numbers has dimension 1. (A basis is $\{1\}$.) ∎

Example 15

Over the field of real numbers, the vector space of complex numbers has dimension 2. (A basis is $\{1, i\}$.) ∎

Corollary 4. *Let* V *be a vector space having dimension* n, *and let* S *be a subset of* V *that generates* V *and contains at most* n *elements. Then* S *is a basis for* V *and hence contains exactly* n *elements.*

Proof. There exists a subset S_1 of S such that S_1 is a basis for V (Theorem 1.9). By Corollary 3, S_1 contains exactly n elements. But $S_1 \subseteq S$ and S contains at most n elements. Hence $S = S_1$; so S is a basis for V. ∎

Example 16

It follows from Example 4 of Section 1.4 and Corollary 4 that

$$\{x^2 + 3x - 2,\ 2x^2 + 5x - 3,\ -x^2 - 4x + 4\}$$

is a basis for $\mathsf{P}_2(R)$. ∎

Example 17

It follows from Example 5 of Section 1.4 and Corollary 4 that

$$\left\{\begin{pmatrix}1 & 1\\ 1 & 0\end{pmatrix}, \begin{pmatrix}1 & 1\\ 0 & 1\end{pmatrix}, \begin{pmatrix}1 & 0\\ 1 & 1\end{pmatrix}, \begin{pmatrix}0 & 1\\ 1 & 1\end{pmatrix}\right\}$$

is a basis for $\mathsf{M}_{2\times 2}(R)$. ∎

Corollary 5. *Let β be a basis for a finite-dimensional vector space* V, *and let* S *be a linearly independent subset of* V. *There exists a subset* S_1 *of β such that* $S \cup S_1$ *is a basis for* V. *Thus every linearly independent subset of* V *can be extended to a basis for* V.

Proof. Let $\dim(V) = n$. By Corollary 2 we know that S must contain m elements, where $m \leqslant n$. Hence the replacement theorem guarantees that there is a subset S_1 of β containing exactly $n - m$ elements such that $S \cup S_1$ generates V. Clearly, $S \cup S_1$ contains at most n elements; so Corollary 4 implies that $S \cup S_1$ is a basis for V. ∎

Theorem 1.9, the replacement theorem, and the five corollaries of the replacement theorem contain a wealth of information about the relationships among linearly independent sets, bases, and generating sets. For this reason we will summarize here the main results of this section in order to put them into better perspective.

A basis for a vector space V is a linearly independent subset that generates V. If V has a finite basis, then every basis for V contains the same number of vectors. This number is called the dimension of V, and V is said to be finite-dimensional. Thus if the dimension of V is n, every basis for V contains exactly n vectors. Moreover, every linearly independent subset of V contains no more than n vectors and can be extended to a basis for V by including appropriately chosen vectors. Also, each generating set for V contains at least n vectors and can be reduced to a basis for V by excluding appropriately chosen vectors. The Venn diagram in Figure 1.6 depicts these relationships. We will see in Section 2.4 that every vector space over F of dimension n is essentially F^n.

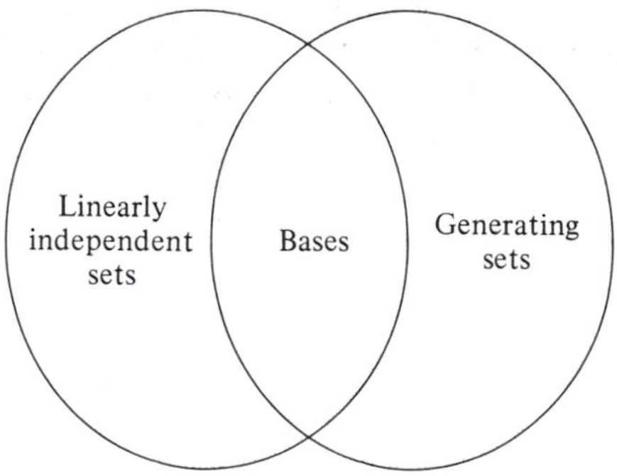

Figure 1.6

The Lagrange Interpolation Formula

The preceding results can be applied to obtain a useful formula. Let $c_0, c_1, \ldots, c_n$, be distinct elements in an infinite field F. The polynomials $f_0(x), f_1(x), \ldots, f_n(x)$, where

$$f_i(x) = \frac{(x - c_0)\cdots(x - c_{i-1})(x - c_{i+1})\cdots(x - c_n)}{(c_i - c_0)\cdots(c_i - c_{i-1})(c_i - c_{i+1})\cdots(c_i - c_n)}$$

$$= \prod_{\substack{j=0 \\ j \neq i}}^{n} \frac{x - c_j}{c_i - c_j}$$

are called the *Lagrange polynomials* (*associated with* $c_0, c_1, \ldots, c_n$). Note that each $f_i(x)$ is a polynomial of degree n and hence is an element of $\mathsf{P}_n(F)$. By regarding $f_i(x)$ as a polynomial function $f_i\colon F \to F$, we see that

$$f_i(c_j) = \begin{cases} 0 & \text{if } i \neq j \\ 1 & \text{if } i = j. \end{cases} \tag{10}$$

We will use this property of the Lagrange polynomials to show that $\beta = \{f_0, f_1, \ldots, f_n\}$ is a linearly independent subset of $\mathsf{P}_n(F)$. Since the dimension of $\mathsf{P}_n(F)$ is $n + 1$, it will follow from Corollary 1 of Theorem 1.10 that β is a basis for $\mathsf{P}_n(F)$. To show that β is linearly independent, suppose that

$$\sum_{i=0}^{n} a_i f_i = \mathit{0} \qquad \text{for some scalars } a_0, a_1, \ldots, a_n,$$

where $\mathit{0}$ denotes the zero function. Then

$$\sum_{i=0}^{n} a_i f_i(c_j) = 0 \qquad \text{for } j = 0, 1, \ldots, n.$$

But also

$$\sum_{i=0}^{n} a_i f_i(c_j) = a_j$$

by (10). Hence $a_j = 0$ for $j = 0, 1, \ldots, n$; so β is linearly independent.

Because β is a basis for $\mathsf{P}_n(F)$, every polynomial function g in $\mathsf{P}_n(F)$ is a linear combination of elements of β, say

$$g = \sum_{i=0}^{n} b_i f_i.$$

Then

$$g(c_j) = \sum_{i=0}^{n} b_i f_i(c_j) = b_j;$$

so

$$g = \sum_{i=0}^{n} g(c_i) f_i$$

is the unique representation of g as a linear combination of elements of β. This representation is called the *Lagrange interpolation formula*. Notice that the argument above shows that if $b_0, b_1, \ldots, b_n$ are any $n + 1$ elements of F (not necessarily distinct), then the polynomial function

$$g = \sum_{i=0}^{n} b_i f_i$$

is the unique element of $\mathsf{P}_n(F)$ such that $g(c_j) = b_j$. Thus we have found the unique polynomial of degree not exceeding n that has specified values b_j at given points c_j in its domain ($j = 0, 1, \ldots, n$). For example, let us construct the real polynomial g of degree at most 2 whose graph contains the points (1, 8), (2, 5), and (3, −4). (Thus in the notation above $c_0 = 1$, $c_1 = 2$, $c_2 = 3$, $b_0 = 8$, $b_1 = 5$, and $b_2 = -4$.) The Lagrange polynomials associated with c_0, c_1, and c_2 are

$$f_0(x) = \frac{(x-2)(x-3)}{(1-2)(1-3)} = \frac{1}{2}(x^2 - 5x + 6),$$

$$f_1(x) = \frac{(x-1)(x-3)}{(2-1)(2-3)} = -1(x^2 - 4x + 3),$$

and

$$f_2(x) = \frac{(x-1)(x-2)}{(3-1)(3-2)} = \frac{1}{2}(x^2 - 3x + 2).$$

Hence the desired polynomial is

$$\begin{aligned} g(x) &= \sum_{i=0}^{2} b_i f_i(x) = 8f_0(x) + 5f_1(x) - 4f_2(x) \\ &= 4(x^2 - 5x + 6) - 5(x^2 - 4x + 3) - 2(x^2 - 3x + 2) \\ &= -3x^2 + 6x + 5. \end{aligned}$$

An important consequence of the Lagrange interpolation formula is the following result: If $f \in \mathsf{P}_n(F)$ and $f(c_j) = 0$ for $n + 1$ distinct elements $c_0, c_1, \ldots, c_n$ in F, then f is the zero function.

The Dimension of Subspaces

Our next result relates the dimension of a subspace to the dimension of the vector space that contains it.

Theorem 1.11. *Let* W *be a subspace of a finite-dimensional vector space* V. *Then* W *is finite-dimensional and* $\dim(\mathsf{W}) \leqslant \dim(\mathsf{V})$. *Moreover, if* $\dim(\mathsf{W}) = \dim(\mathsf{V})$, *then* $\mathsf{W} = \mathsf{V}$.

Proof. Let $\dim(\mathsf{V}) = n$. If $\mathsf{W} = \{0\}$, then W is finite-dimensional and $\dim(\mathsf{W}) = 0 \leqslant n$. Otherwise, W contains a nonzero element x_1; so $\{x_1\}$ is a linearly independent set. Continuing in this way, choose elements $x_1, x_2, \ldots, x_k$ in W such that $\{x_1, x_2, \ldots, x_k\}$ is linearly independent. Since no linearly independent subset of V can contain more than n elements, this process must stop at a stage where $k \leqslant n$ and $\{x_1, x_2, \ldots, x_k\}$ is linearly independent but adjoining any other element of W produces a linearly dependent set. Theorem 1.8 now implies that $\{x_1, x_2, \ldots, x_k\}$ generates W, and hence it is a basis for W. Therefore, $\dim(\mathsf{W}) = k \leqslant n$.

If $\dim(\mathsf{W}) = n$, then a basis for W is a linearly independent subset of V containing n elements. But Corollary 1 of Theorem 1.10 implies that this basis for W is also a basis for V; so W = V. ∎

Example 18

Let

$$\mathsf{W} = \{(a_1, a_2, a_3, a_4, a_5) \in \mathsf{F}^5 \colon a_1 + a_3 + a_5 = 0, a_2 = a_4\}.$$

It is easily shown that W is a subspace of F^5 having

$$\{(1, 0, 0, 0, -1), (0, 0, 1, 0, -1), (0, 1, 0, 1, 0)\}$$

as a basis. Thus $\dim(\mathsf{W}) = 3$. ∎

Example 19

The set of diagonal $n \times n$ matrices is a subspace W of $\mathsf{M}_{n \times n}(F)$ (see Example 1 of Section 1.3). A basis for W is

$$\{M^{11}, M^{22}, \ldots, M^{nn}\},$$

where M^{ij} is the matrix in which the only nonzero entry is a 1 in the ith row and jth column. Thus $\dim(\mathsf{W}) = n$. ∎

Example 20

We saw in Section 1.3 that the set of symmetric $n \times n$ matrices is a subspace W of $\mathsf{M}_{n \times n}(F)$. A basis for W is

$$\{A^{ij} \colon 1 \leqslant i \leqslant j \leqslant n\},$$

where A^{ij} is the $n \times n$ matrix having 1 in the ith row and jth column, 1 in the jth row and ith column, and 0 elsewhere. It follows that

$$\dim(\mathsf{W}) = n + (n-1) + \cdots + 1 = \tfrac{1}{2}n(n-1). \quad ∎$$

Corollary. *If* W *is a subspace of a finite-dimensional vector space* V, *then* W *has a finite basis, and any basis for* W *is a subset of a basis for* V.

Proof. Theorem 1.11 shows that W has a finite basis S. If β is any basis for V, then the replacement theorem shows that there exists a subset S_1 of β such that $S \cup S_1$ is a basis for V. Hence S is a subset of the basis $S \cup S_1$ for V. ∎

Example 21

The set of all polynomials of the form

$$a_{18}x^{18} + a_{16}x^{16} + \cdots + a_2x^2 + a_0,$$

where $a_{18}, a_{16}, \ldots, a_2, a_0 \in \mathsf{F}$, is a subspace W of $\mathsf{P}_{18}(F)$. A basis for W is $\{1, x^2, \ldots, x^{16}, x^{18}\}$, which is a subset of the standard basis for $\mathsf{P}_{18}(F)$. ∎

We conclude this section by using Theorem 1.11 to determine the subspaces of R^2 and R^3. Since R^2 has dimension 2, subspaces of R^2 can be of dimensions 0, 1, or 2 only. The only subspaces of dimensions 0 or 2 are $\{0\}$ and R^2, respectively. Any subspace of R^2 having dimension 1 consists of all scalar multiples of some nonzero vector in R^2 (Exercise 9 of Section 1.4).

If a point of R^2 is identified in the natural way with a point on the Euclidean plane, then it is possible to describe the subspaces of R^2 geometrically: A subspace of R^2 having dimension 0 consists of the origin of the Euclidean plane, a subspace of R^2 with dimension 1 consists of a line through the origin, and a subspace of R^2 having dimension 2 is the entire Euclidean plane.

As above, the subspaces of R^3 must have dimensions 0, 1, 2, or 3. Interpreting these possibilities geometrically, we see that a subspace of dimension zero must be the origin of Euclidean 3-space, a subspace of dimension 1 is a line through the origin, a subspace of dimension 2 is a plane through the origin, and a subspace of dimension 3 is Euclidean 3-space itself.

EXERCISES

1. Label the following statements as being true or false.
   (a) The zero vector space has no basis.
   (b) Every vector space that is generated by a finite set has a basis.
   (c) Every vector space has a finite basis.
   (d) A vector space cannot have more than one basis.
   (e) If a vector space has a finite basis, then the number of vectors in every basis is the same.
   (f) The dimension of $\mathsf{P}_n(F)$ is n.
   (g) The dimension of $\mathsf{M}_{m\times n}(F)$ is $m + n$.
   (h) Suppose that V is a finite-dimensional vector space, that S_1 is a linearly independent subset of V, and that S_2 is a subset of V that generates V. Then S_1 cannot contain more elements than S_2.
   (i) If S generates the vector space V, then every vector in V can be written as a linear combination of elements of S in only one way.
   (j) Every subspace of a finite-dimensional space is finite-dimensional.
   (k) If V is a vector space having dimension n, then V has exactly one subspace with dimension 0 and exactly one subspace with dimension n.
2. Determine which of the following sets are bases for R^3.
   (a) $\{(1, 0, -1), (2, 5, 1), (0, -4, 3)\}$
   (b) $\{(2, -4, 1), (0, 3, -1), (6, 0, -1)\}$
   (c) $\{(1, 2, -1), (1, 0, 2), (2, 1, 1)\}$
   (d) $\{(-1, 3, 1), (2, -4, -3), (-3, 8, 2)\}$
   (e) $\{(1, -3, -2), (-3, 1, 3), (-2, -10, -2)\}$

3. Determine which of the following sets are bases for $\mathsf{P}_2(R)$.
   (a) $\{-1 - x + 2x^2, 2 + x - 2x^2, 1 - 2x + 4x^2\}$
   (b) $\{1 + 2x + x^2, 3 + x^2, x + x^2\}$
   (c) $\{1 + 4x - 2x^2, -2 + 3x - x^2, -3 - 12x + 6x^2\}$
   (d) $\{-1 + 2x + 4x^2, 3 - 4x - 10x^2, -2 - 5x - 6x^2\}$
   (e) $\{1 + 2x - x^2, 4 - 2x + x^2, -1 + 18x - 9x^2\}$
4. Do the polynomials $x^3 - 2x^2 + 1$, $4x^2 - x + 3$, and $3x - 2$ generate $\mathsf{P}_3(R)$? Justify your answer.
5. Is $\{(1, 4, -6), (1, 5, 8), (2, 1, 1), (0, 1, 0)\}$ a linearly independent subset of R^3? Justify your answer.
6. Give three different bases for F^2 and for $\mathsf{M}_{2\times 2}(F)$.
7. The vectors $x_1 = (2, -3, 1)$, $x_2 = (1, 4, -2)$, $x_3 = (-8, 12, -4)$, $x_4 = (1, 37, -17)$, and $x_5 = (-3, -5, 8)$ generate R^3. Find a subset of $\{x_1, x_2, x_3, x_4, x_5\}$ that is a basis for R^3.
8. Let V denote the vector space consisting of all vectors in R^5 for which the sum of the coordinates equals zero. The vectors

$$x_1 = (2, -3, 4, -5, 2), \qquad x_2 = (-6, 9, -12, 15, -6),$$
$$x_3 = (3, -2, 7, -9, 1), \qquad x_4 = (2, -8, 2, -2, 6),$$
$$x_5 = (-1, 1, 2, 1, -3), \qquad x_6 = (0, -3, -18, 9, 12),$$
$$x_7 = (1, 0, -2, 3, -2), \qquad x_8 = (2, -1, 1, -9, 7)$$

   generate V. Find a subset of $\{x_1, \ldots, x_8\}$ that is a basis for V.
9. The vectors $x_1 = (1, 1, 1, 1)$, $x_2 = (0, 1, 1, 1)$, $x_3 = (0, 0, 1, 1)$, and $x_4 = (0, 0, 0, 1)$ form a basis for F^4. Find the unique representation of an arbitrary vector (a_1, a_2, a_3, a_4) in F^4 as a linear combination of the vectors x_1, x_2, x_3, and x_4.
10. Let $\{x, y\}$ be a basis for a vector space V. Show that both $\{x + y, ax\}$ and $\{ax, by\}$ are bases for V, where a and b are arbitrary nonzero scalars.
11. Suppose that V is a vector space with a basis $\{x_1, x_2, x_3\}$. Show that $\{x_1 + x_2 + x_3, x_2 + x_3, x_3\}$ is also a basis for V.
12. The set of solutions to the system

$$\begin{cases} x_1 - 2x_2 + x_3 = 0 \\ 2x_1 - 3x_2 + x_3 = 0 \end{cases}$$

   is a subspace of R^3. Find a basis for this subspace.
13. Find bases for the following subspaces of F^5:

$$\mathsf{W}_1 = \{(a_1, a_2, a_3, a_4, a_5) \in \mathsf{F}^5 \colon a_1 - a_3 - a_4 = 0\}$$

   and

$$\mathsf{W}_2 = \{(a_1, a_2, a_3, a_4, a_5) \in \mathsf{F}^5 \colon a_2 = a_3 = a_4,\ a_1 + a_5 = 0\}.$$

   What are the dimensions of W_1 and W_2?

14. The set of all $n \times n$ matrices having trace equal to zero is a subspace W of $\mathsf{M}_{n\times n}(F)$ (see Example 4 of Section 1.3). Find a basis for W. What is the dimension of W?

15. The set of all upper triangular $n \times n$ matrices is a subspace W of $\mathsf{M}_{n\times n}(F)$ (see Exercise 12 of Section 1.3). Find a basis for W. What is the dimension of W?

16. The set of all skew-symmetric $n \times n$ matrices is a subspace W of $\mathsf{M}_{n\times n}(F)$ (see Exercise 26 of Section 1.3). Find a basis for W. What is the dimension of W?

17. Find a basis for the vector space in Example 5 of Section 1.2. Justify your answer.

18. Complete the proof of Theorem 1.7.

19.† Let V be a vector space having dimension n, and let S be a subset of V that generates V.

(a) Prove that a subset of S is a basis for V. (Be careful not to assume that S is finite.)

(b) Prove that S contains at least n elements.

Exercises 20 through 24 require knowledge of the sum and direct sum of subspaces, as defined in the exercises of Section 1.3.

20. **(a)** Prove that if W_1 and W_2 are finite-dimensional subspaces of a vector space V, then $W_1 + W_2$ is finite-dimensional, and $\dim(W_1 + W_2) = \dim(W_1) + \dim(W_2) - \dim(W_1 \cap W_2)$. *Hint:* Extend a basis for $W_1 \cap W_2$ to a basis for W_1 and a basis for W_2.

(b) Let W_1 and W_2 be finite-dimensional subspaces of a vector space V, and let $V = W_1 + W_2$. Deduce that V is the direct sum of W_1 and W_2 if and only if $\dim(V) = \dim(W_1) + \dim(W_2)$.

21. Let

$$V = \mathsf{M}_{2\times 2}(F), \qquad W_1 = \left\{\begin{pmatrix} a & b \\ c & a \end{pmatrix} \in V\colon a, b, c \in F\right\},$$

and

$$W_2 = \left\{\begin{pmatrix} 0 & a \\ -a & b \end{pmatrix} \in V\colon a, b \in F\right\}.$$

Prove that W_1 and W_2 are subspaces of V, and find the dimensions of W_1, W_2, $W_1 + W_2$, and $W_1 \cap W_2$.

22. Let W_1 and W_2 be subspaces of a vector space V having dimensions m and n, respectively, where $m \geqslant n$.

(a) Prove that $\dim(W_1 \cap W_2) \leqslant n$ and $\dim(W_1 + W_2) \leqslant m + n$.

(b) Give examples of subspaces W_1 and W_2 of R^3 for which $\dim(W_1 \cap W_2) = n$ and $\dim(W_1 + W_2) = m + n$.

(c) Give examples of subspaces W_1 and W_2 of R^3 for which $\dim(\mathsf{W}_1 \cap \mathsf{W}_2) < n$ and $\dim(\mathsf{W}_1 + \mathsf{W}_2) < m + n$.

23. **(a)** Let W_1 and W_2 be subspaces of a vector space V such that $\mathsf{V} = \mathsf{W}_1 \oplus \mathsf{W}_2$. If β_1 and β_2 are bases for W_1 and W_2, respectively, show that $\beta_1 \cap \beta_2 = \varnothing$ and $\beta_1 \cup \beta_2$ is a basis for V.

(b) Conversely, let β_1 and β_2 be disjoint bases for subspaces W_1 and W_2, respectively, of a vector space V. Prove that if $\beta_1 \cup \beta_2$ is a basis for V, then $\mathsf{V} = \mathsf{W}_1 \oplus \mathsf{W}_2$.

24. Prove that if W_1 is any subspace of a finite-dimensional vector space V, then there is a subspace W_2 of V such that $\mathsf{V} = \mathsf{W}_1 \oplus \mathsf{W}_2$.

25. Prove that a vector space is infinite-dimensional if and only if it contains an infinite linearly independent subset.

The following exercise requires familiarity with Exercise 29 of Section 1.3.

26. Let W be a subspace of a finite-dimensional vector space V. Let $\{x_1, x_2, \ldots, x_k\}$ be a basis for W, and extend this to a basis $\{x_1, x_2, \ldots, x_k, x_{k+1}, \ldots, x_n\}$ for V.

(a) Prove that $\{x_{k+1} + \mathsf{W}, x_{k+2} + \mathsf{W}, \ldots, x_n + \mathsf{W}\}$ is a basis for V/W.

(b) Derive a formula relating $\dim(\mathsf{V})$, $\dim(\mathsf{W})$, and $\dim(\mathsf{V}/\mathsf{W})$.

1.7* MAXIMAL LINEARLY INDEPENDENT SUBSETS

In this section several important results from Section 1.6 are extended to include infinite-dimensional vector spaces. Our principal goal is to prove that every vector space has a basis. This result is of importance in the study of infinite-dimensional vector spaces because it is often difficult to construct a basis for such a space explicitly.

The difficulty that arises in extending the theorems of the preceding section to infinite-dimensional spaces is that the *principle of mathematical induction*, which played a crucial role in many of the proofs of Section 1.6, is no longer adequate. We will use instead a more general result called the *maximal principle*, which requires the following terminology.

Definition. *Let $\mathscr{F}$ be a family of sets. A member* M *of $\mathscr{F}$ is called* maximal (with respect to set inclusion) *if no member of $\mathscr{F}$ properly contains* M.

Example 1

Let $\mathscr{F}$ be the family of all subsets of a nonempty set S. ($\mathscr{F}$ is called the *power set of S*.) The set S is easily seen to be a maximal element of $\mathscr{F}$. ∎

Definition. *A collection of sets $\mathscr{C}$ is called a* chain (*or* nest *or* tower) *if for each pair of sets* A *and* B *in $\mathscr{C}$, either* A ⊆ B *or* B ⊆ A.

Example 2

Let A_n denote the set consisting of the integers 1, 2, ..., n. Then $\mathscr{C} = \{A_n: n = 1, 2, 3, \ldots\}$ is a chain; in fact, $A_m \subseteq A_n$ if and only if $m \leqslant n$. ∎

With this terminology we can now state the maximal principle.

Maximal Principle. *Let $\mathscr{F}$ be a family of sets. If for each chain $\mathscr{C} \subseteq \mathscr{F}$ there exists a member of $\mathscr{F}$ that contains each member of $\mathscr{C}$, then $\mathscr{F}$ contains a maximal element.*

Because the maximal principle guarantees the existence of maximal elements in a family of sets, it will be useful to reformulate the definition of a basis in terms of a maximal property. We will subsequently show that this reformulation is equivalent to the original definition of a basis.

Definition. *Let* S *be a subset of a vector space* V. *A* maximal linearly independent subset *of* S *is a subset* B *of* S *satisfying both of the following conditions:*

(a) B *is linearly independent.*
(b) *Any subset of* S *that properly contains* B *is linearly dependent.*

Example 3

Example 2 of Section 1.4 shows that

$$\{x^3 - 2x^2 - 5x - 3, 3x^3 - 5x^2 - 4x - 9\}$$

is a maximal linearly independent subset of

$$S = \{2x^3 - 2x^2 + 12x - 6, x^3 - 2x^2 - 5x - 3, 3x^3 - 5x^2 - 4x - 9\}$$

in $\mathsf{P}_2(R)$. In this case, however, any two-element subset of S is easily shown to be a maximal linearly independent subset of S. Thus maximal linearly independent subsets of a set need not be unique. ∎

A basis β for a vector space V is a maximal linearly independent subset of V, for

(a) β is linearly independent by definition.
(b) If $x \in \mathsf{V}$ and $x \notin \beta$, then $\beta \cup \{x\}$ is linearly dependent by Theorem 1.8 because span(β) = V.

Our next result shows that the converse of this statement is also true.

Theorem 1.12. *Let* V *be a vector space and* S *a subset that generates* V. *If β is a maximal linearly independent subset of* S, *then β is a basis for* V.

Proof. Let β be a maximal linearly independent subset of S. Because β is linearly independent, it suffices to prove that β generates V. Suppose that $S \not\subseteq \text{span}(\beta)$; then there exists $x \in S$ such that $x \notin \text{span}(\beta)$. Since Theorem 1.8 implies that $\beta \cup \{x\}$ is linearly independent, we have contradicted the maximality of β. Hence $S \subseteq \text{span}(\beta)$. Because $\text{span}(S) = \mathsf{V}$, it follows from Exercise 11 of Section 1.4 that $\text{span}(\beta) = \mathsf{V}$. ∎

Corollary. *A subset β of a vector space* V *is a basis for* V *if and only if β is a maximal linearly independent subset of* V.

In view of the preceding corollary, we can accomplish our goal of proving that every vector space has a basis by proving that every vector space contains a maximal linearly independent subset. This result follows immediately from our next theorem.

Theorem 1.13. *Let* S *be a linearly independent subset of a vector space* V. *There exists a maximal linearly independent subset of* V *that contains* S.

Proof. Let $\mathscr{F}$ denote the family of all linearly independent subsets of V that contain S. We will use the maximal principle to show that $\mathscr{F}$ contains a maximal element. In order to apply the maximal principle, we must show that if $\mathscr{C}$ is a chain in $\mathscr{F}$, then there exists a member U of $\mathscr{F}$ that contains each member of $\mathscr{C}$. We will show that U, the union of the members of $\mathscr{C}$, is the desired set. Since U clearly contains each member of $\mathscr{C}$, it suffices to prove that $U \in \mathscr{F}$ (i.e., that U is a linearly independent subset of V that contains S). Now each element of $\mathscr{C}$ is a subset of V containing S; hence $S \subseteq U \subseteq \mathsf{V}$. To prove that U is linearly independent, let $u_1, \ldots, u_n$ be vectors in U and $c_1, \ldots, c_n$ be scalars such that $c_1u_1 + \cdots + c_nu_n = 0$. Because $u_i \in U$ for $i = 1, \ldots, n$, there exist sets A_i in $\mathscr{C}$ such that $u_i \in A_i$. But since $\mathscr{C}$ is a chain, one of the sets $A_1, \ldots, A_n$, say A_k, contains all the others. Thus $u_1, \ldots, u_n \in A_k$ for $i = 1, \ldots, n$. However, A_k is a linearly independent set; so $c_1u_1 + \cdots + c_nu_n = 0$ implies that $c_1 = \cdots = c_n = 0$. Therefore, U is linearly independent.

The maximal principle implies that $\mathscr{F}$ contains a maximal element. This maximal element is easily seen to be a maximal linearly independent subset of V that contains S. ∎

Corollary. *Every vector space has a basis.*

It can be shown, analogously to Corollary 3 to Theorem 1.10, that every basis for an infinite-dimensional vector space has the same cardinality (see, e.g., N. Jacobson, *Lectures in Linear Algebra*, Vol. 3, D. Van Nostrand Company, New York, 1964, p. 154).

Exercises 2 through 5 extend other results from Section 1.6 to include infinite-dimensional spaces.

EXERCISES

1. Label the following statements as being true or false.
   - **(a)** Every family of sets contains a maximal element.
   - **(b)** Every chain contains a maximal element.
   - **(c)** If a family of sets has a maximal element, then that maximal element is unique.
   - **(d)** If a chain of sets has a maximal element, then that maximal element is unique.
   - **(e)** A basis for a vector space is a maximal linearly independent subset of that vector space.
   - **(f)** A maximal linearly independent subset of a vector space is a basis for that vector space.
2. Let W be a subspace of a (not necessarily finite-dimensional) vector space V. Prove that any basis for W is a subset of a basis for V.
3. Prove the following infinite-dimensional version of Theorem 1.7: Let β be a subset of an infinite-dimensional vector space V. Then β is a basis for V if and only if, for each nonzero vector y in V, there exist unique vectors $x_1, \ldots, x_n$ in β and unique nonzero scalars $c_1, \ldots, c_n$ such that $y = c_1x_1 + \cdots + c_nx_n$.
4. Prove the following generalization of Theorem 1.9: Let S_1 and S_2 be subsets of a vector space V such that $S_1 \subseteq S_2$. If S_1 is linearly independent and S_2 generates V, then there exists a basis β for V such that $S_1 \subseteq \beta \subseteq S_2$. *Hint:* Apply the maximal principle to the family of all linearly independent subsets of S_2 that contain S_1, and proceed as in the proof of Theorem 1.13.
5. Prove the following generalization of the replacement theorem: Let β be a basis for a vector space V, and let S be a linearly independent subset of V. There exists a subset S_1 of β such that $S \cup S_1$ is a basis for V.

INDEX OF DEFINITIONS FOR CHAPTER 1

2

Linear Transformations and Matrices

In Chapter 1 we developed the theory of abstract vector spaces in considerable detail. It is now natural to consider those functions defined on vector spaces that in some sense "preserve" the structure. These special functions are called "linear transformations," and they abound in both pure and applied mathematics. In calculus the operations of differentiation and integration provide us with two of the most important examples of linear transformations (see Examples 1 and 2 of Section 2.1). These two examples allow us to reformulate many of the problems in differential and integral equations in terms of linear transformations on particular vector spaces (see Sections 2.7 and 5.2).

In geometry, rotations, reflections, and projections (see Examples 5, 6, and 7 of Section 2.1) provide us with another class of linear transformations. Later we use these transformations to study the rigid motions in R^n (Section 6.10).

In the remaining chapters we will see further examples of linear transformations occurring in both the physical and social sciences. Throughout this chapter we assume that all vector spaces are over a common field F.

2.1 LINEAR TRANSFORMATIONS, NULL SPACES, AND RANGES

In this section we consider a number of examples of linear transformations. Many of these transformations will be studied in more detail in later sections.

Definition. *Let* V *and* W *be vector spaces* (*over* F). *A function* $\mathsf{T}: \mathsf{V} \to \mathsf{W}$ *is called a* linear transformation from V into W *if for all* $x, y \in \mathsf{V}$ *and* $c \in F$ *we have*

(a) $T(x + y) = T(x) + T(y)$.
(b) $T(cx) = cT(x)$.

We often simply call T *linear*. The reader should verify the following facts about a function $T: V \to W$.

1. If T is linear, then $T(0) = 0$.
2. T is linear if and only if $T(ax + y) = aT(x) + T(y)$ for all $x, y \in V$ and $a \in F$.
3. T is linear if and only if for $x_1, \ldots, x_n \in V$ and $a_1, \ldots, a_n \in F$ we have

$$T\left(\sum_{i=1}^{n} a_i x_i\right) = \sum_{i=1}^{n} a_i T(x_i).$$

We generally use property 2 to prove that a given transformation is linear.

Example 1

Let $V = P_n(R)$ and $W = P_{n-1}(R)$. Define $T: V \to W$ by $T(f) = f'$, where f' denotes the derivative of f. To show that T is linear, let g and h be vectors in $P_n(R)$ and $a \in R$. Now

$$T(ag + h) = (ag + h)' = ag' + h' = aT(g) + T(h).$$

So by property 2 above, T is linear. ∎

Example 2

Let $V = C(R)$, the vector space of continuous real-valued functions on R. Let $a, b \in R$, $a < b$. Define $T: V \to R$ by

$$T(f) = \int_a^b f(t)\, dt$$

for all $f \in V$. Then T is a linear transformation by the elementary properties of the integral. ∎

Two very important examples of linear transformations that appear frequently in the remainder of the book, and therefore, deserve their own notation, are the identity and zero transformations.

For vector spaces V and W (over F) we define the *identity transformation* $I_V: V \to V$ by $I_V(x) = x$ for all $x \in V$ and the *zero transformation* $T_0: V \to W$ by $T_0(x) = 0$ for all $x \in V$. It is clear that both of these transformations are linear. We often write I instead of I_V.

We now look at some additional examples of linear transformations.

Example 3

Define

$$\mathsf{T}\colon \mathsf{R}^2 \to \mathsf{R}^2 \text{ by } \mathsf{T}(a_1, a_2) = (2a_1 + a_2, a_1).$$

To show that T is linear, let $c \in F$ and $x, y \in \mathsf{R}^2$, where $x = (b_1, b_2)$ and $y = (d_1, d_2)$. Since

$$cx + y = (cb_1 + d_1, cb_2 + d_2),$$

we have

$$\mathsf{T}(cx + y) = (2(cb_1 + d_1) + cb_2 + d_2, cb_1 + d_1).$$

Also,

$$\begin{aligned} c\mathsf{T}(x) + \mathsf{T}(y) &= c(2b_1 + b_2, b_1) + (2d_1 + d_2, d_1) \\ &= (2cb_1 + cb_2 + 2d_1 + d_2, cb_1 + d_1) \\ &= (2(cb_1 + d_1) + cb_2 + d_2, cb_1 + d_1). \end{aligned}$$

So T is linear. ∎

Example 4

Define $\mathsf{T}\colon \mathsf{M}_{m \times n}(F) \to \mathsf{M}_{n \times m}(F)$ by $\mathsf{T}(A) = A^t$, where A^t is as defined in Section 1.3. Then T is a linear transformation by Exercise 3 of Section 1.3. ∎

As we will see in Chapter 6, the applications of linear algebra to geometry are wide and varied. The main reason for this is that most of the important geometrical transformations are linear. Three particular transformations that we now consider are the rotation, reflection, and projection. We leave the proofs of linearity to the reader.

Example 5

For $0 \leqslant \theta < 2\pi$ we define $\mathsf{T}_\theta\colon \mathsf{R}^2 \to \mathsf{R}^2$ by

$$\mathsf{T}_\theta(a_1, a_2) = (a_1 \cos\theta - a_2 \sin\theta, a_1 \sin\theta + a_2 \cos\theta).$$

T_θ is called the *rotation by* θ [see Figure 2.1(a)]. ∎

Example 6

Define $\mathsf{T}\colon \mathsf{R}^2 \to \mathsf{R}^2$ by $\mathsf{T}(a_1, a_2) = (a_1, -a_2)$. T is called the *reflection about the* x-*axis* [see Figure 2.1(b)]. ∎

Example 7

Define $\mathsf{T}\colon \mathsf{R}^2 \to \mathsf{R}^2$ by $\mathsf{T}(a_1, a_2) = (a_1, 0)$. T is called the *projection on the* x-*axis* (see Figure 2.1(c)]. ∎

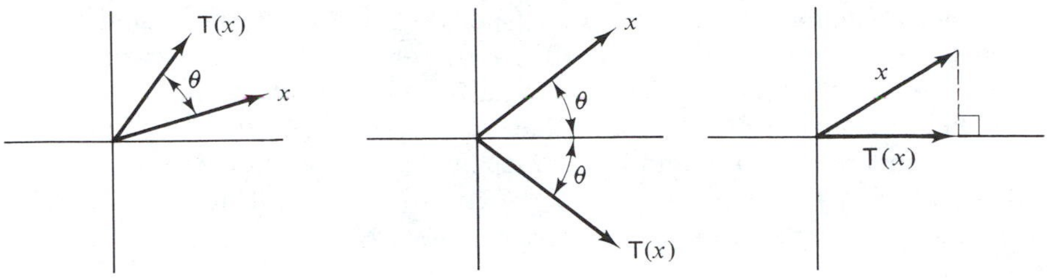

(a) Rotation (b) Reflection (c) Projection

Figure 2.1

We now turn our attention to two very important sets associated with linear transformations: the "range" and "null space." The determination of these sets allows us to examine more closely the intrinsic properties of a linear transformation.

Definitions. *Let* V *and* W *be vector spaces, and let* $\mathsf{T}: \mathsf{V} \to \mathsf{W}$ *be linear. We define the* null space (*or* kernel) $\mathsf{N}(\mathsf{T})$ *of* T *to be the set of all vectors* x *in* V *such that* $\mathsf{T}(x) = 0$; *that is,* $\mathsf{N}(\mathsf{T}) = \{x \in \mathsf{V}: \mathsf{T}(x) = 0\}$.

We define the range (*or* image) $\mathsf{R}(\mathsf{T})$ *of* T *to be the subset of* W *consisting of all images (under* T*) of elements of* V; *that is,* $\mathsf{R}(\mathsf{T}) = \{\mathsf{T}(x): x \in \mathsf{V}\}$.

Example 8

Let V and W be vector spaces, and let $\mathsf{I}: \mathsf{V} \to \mathsf{V}$ and $\mathsf{T}_0: \mathsf{V} \to \mathsf{W}$ be the identity and zero transformations, respectively, as defined above. Then $\mathsf{N}(\mathsf{I}) = \{0\}$, $\mathsf{R}(\mathsf{I}) = \mathsf{V}$, $\mathsf{N}(\mathsf{T}_0) = \mathsf{V}$, and $\mathsf{R}(\mathsf{T}_0) = \{0\}$. ■

Example 9

Define $\mathsf{T}: \mathsf{R}^3 \to \mathsf{R}^2$ by

$$\mathsf{T}(a_1, a_2, a_3) = (a_1 - a_2, 2a_3).$$

It is left as an exercise to verify that $\mathsf{N}(\mathsf{T}) = \{(a, a, 0): a \in R\}$ and $\mathsf{R}(\mathsf{T}) = \mathsf{R}^2$. ■

In Examples 8 and 9 we see that the range and null space of each of the linear transformations is a subspace. The next result shows that this is true in general.

Theorem 2.1. *Let* V *and* W *be vector spaces and* $\mathsf{T}: \mathsf{V} \to \mathsf{W}$ *be linear. Then* $\mathsf{N}(\mathsf{T})$ *and* $\mathsf{R}(\mathsf{T})$ *are subspaces of* V *and* W, *respectively.*

Proof. To clarify the notation, we use the symbols 0_{V} and 0_{W} to denote the zero vectors of V and W, respectively.

Since $\mathsf{T}(0_{\mathsf{V}}) = 0_{\mathsf{W}}$, we have that $0_{\mathsf{V}} \in \mathsf{N}(\mathsf{T})$. Let $x, y \in \mathsf{N}(\mathsf{T})$ and $c \in F$. Then $\mathsf{T}(x + y) = \mathsf{T}(x) + \mathsf{T}(y) = 0_{\mathsf{W}} + 0_{\mathsf{W}} = 0_{\mathsf{W}}$, and $\mathsf{T}(cx) = c\mathsf{T}(x) = c0_{\mathsf{W}} = 0_{\mathsf{W}}$. Hence $x + y \in \mathsf{N}(\mathsf{T})$ and $cx \in \mathsf{N}(\mathsf{T})$, so that $\mathsf{N}(\mathsf{T})$ is a subspace of V.

Because $\mathsf{T}(0_{\mathsf{V}}) = 0_{\mathsf{W}}$, we have that $0_{\mathsf{W}} \in \mathsf{R}(\mathsf{T})$. Now let $x, y \in \mathsf{R}(\mathsf{T})$ and $c \in F$. Then there exist v and w in V such that $\mathsf{T}(v) = x$ and $\mathsf{T}(w) = y$. So $\mathsf{T}(v + w) = \mathsf{T}(v) + \mathsf{T}(w) = x + y$, and $\mathsf{T}(cv) = c\mathsf{T}(v) = cx$. Thus $x + y \in \mathsf{R}(\mathsf{T})$ and $cx \in \mathsf{R}(\mathsf{T})$, so that $\mathsf{R}(\mathsf{T})$ is a subspace of W. ∎

The next theorem provides a method for finding a spanning set for the range of a linear transformation. With this accomplished, a basis for the range is easy to discover (see Example 6 of Section 1.6).

Theorem 2.2. *Let V and W be vector spaces, and let $\mathsf{T}: \mathsf{V} \to \mathsf{W}$ be linear. If V has a basis $\beta = \{x_1, \ldots, x_n\}$, then*

$$\mathsf{R}(\mathsf{T}) = \operatorname{span}\{\mathsf{T}(x_1), \ldots, \mathsf{T}(x_n)\}.$$

Proof. Clearly, $\mathsf{T}(x_i) \in \mathsf{R}(\mathsf{T})$ for each i. Because $\mathsf{R}(\mathsf{T})$ is a subspace, $\mathsf{R}(\mathsf{T})$ contains $\operatorname{span}\{\mathsf{T}(x_1), \ldots, \mathsf{T}(x_n)\} = \operatorname{span}(\mathsf{T}(\beta))$.

Now suppose that $y \in \mathsf{R}(\mathsf{T})$. Then $y = \mathsf{T}(x)$ for some $x \in \mathsf{V}$. Because β is a basis for V, we have

$$x = \sum_{i=1}^{n} a_i x_i \qquad \text{for some } a_1, \ldots, a_n \in F.$$

Since T is linear, it follows that

$$y = \mathsf{T}(x) = \sum_{i=1}^{n} a_i \mathsf{T}(x_i) \in \operatorname{span}(\mathsf{T}(\beta)). \quad ∎$$

The following example illustrates the usefulness of this result.

Example 10

Define the linear transformation $\mathsf{T}: \mathsf{P}_2(R) \to \mathsf{M}_{2 \times 2}(R)$ by

$$\mathsf{T}(f) = \begin{pmatrix} f(1) - f(2) & 0 \\ 0 & f(0) \end{pmatrix}.$$

Since $\beta = \{1, x, x^2\}$ is a basis for $\mathsf{P}_2(R)$, we have

$$\begin{aligned} \mathsf{R}(\mathsf{T}) &= \operatorname{span}(\mathsf{T}(\beta)) = \operatorname{span}(\{\mathsf{T}(1), \mathsf{T}(x), \mathsf{T}(x^2)\}) \\ &= \operatorname{span}\left(\left\{\begin{pmatrix} 0 & 0 \\ 0 & 1 \end{pmatrix}, \begin{pmatrix} -1 & 0 \\ 0 & 0 \end{pmatrix}, \begin{pmatrix} -3 & 0 \\ 0 & 0 \end{pmatrix}\right\}\right) \\ &= \operatorname{span}\left(\left\{\begin{pmatrix} 0 & 0 \\ 0 & 1 \end{pmatrix}, \begin{pmatrix} -1 & 0 \\ 0 & 0 \end{pmatrix}\right\}\right). \end{aligned}$$

Thus we have found a basis for $\mathsf{R}(\mathsf{T})$, and so $\dim(\mathsf{R}(\mathsf{T})) = 2$. ∎

As in Chapter 1, we measure the "size" of a subspace by its dimension. The null space and range are so important that we attach special names to their respective dimensions.

Definitions. *Let* V *and* W *be vector spaces, and let* $\mathsf{T}\colon \mathsf{V} \to \mathsf{W}$ *be linear. If* N(T) *and* R(T) *are finite-dimensional, then we define the* nullity of T, *denoted* nullity(T), *and the* rank of T, *denoted* rank(T), *to be the dimensions of* N(T) *and* R(T), *respectively.*

Reflecting on the action of a linear transformation, we see intuitively that the larger the nullity, the smaller the rank. In other words, the more vectors that are carried into 0, the smaller the range. The same heuristic reasoning will tell us that the larger the rank, the smaller the nullity. This balance between the rank and the nullity is made precise in the next theorem, appropriately called the *dimension theorem.*

Theorem 2.3 (Dimension Theorem). *Let* V *and* W *be vector spaces, and let* $\mathsf{T}\colon \mathsf{V} \to \mathsf{W}$ *be linear. If* V *is finite-dimensional, then*

$$\text{nullity}(\mathsf{T}) + \text{rank}(\mathsf{T}) = \dim(\mathsf{V}).$$

Proof. Suppose that $\dim(\mathsf{V}) = n$, and let $\{x_1, \ldots, x_k\}$ be a basis for N(T). By the corollary to Theorem 1.11 we may extend $\{x_1, \ldots, x_k\}$ to a basis $\beta = \{x_1, \ldots, x_n\}$ for V. We will show that the set $S = \{\mathsf{T}(x_{k+1}), \ldots, \mathsf{T}(x_n)\}$ is a basis for R(T).

First we prove that S generates R(T). Using Theorem 2.2 and the fact that $\mathsf{T}(x_i) = 0$ for $1 \leqslant i \leqslant k$, we have

$$\mathsf{R}(\mathsf{T}) = \text{span}\{\mathsf{T}(x_1), \ldots, \mathsf{T}(x_n)\} = \text{span}(S).$$

Now we prove that S is linearly independent. Suppose that

$$\sum_{i=k+1}^{n} b_i \mathsf{T}(x_i) = 0 \qquad \text{for } b_{k+1}, \ldots, b_n \in F.$$

Using the fact that T is linear, we have

$$\mathsf{T}\left(\sum_{i=k+1}^{n} b_i x_i\right) = 0.$$

So

$$\sum_{i=k+1}^{n} b_i x_i \in \mathsf{N}(\mathsf{T}).$$

Hence there exist $c_1, \ldots, c_k \in F$ such that

$$\sum_{i=k+1}^{n} b_i x_i = \sum_{i=1}^{k} c_i x_i \quad \text{or} \quad \sum_{i=1}^{k} (-c_i) x_i + \sum_{i=k+1}^{n} b_i x_i = 0.$$

Since β is a basis for V, we have $b_i = 0$ for all i. Hence S is linearly independent. ∎

If we apply the dimension theorem to the linear transformation T in Example 9, we have that nullity(T) + 2 = 3, so nullity(T) = 1.

The reader should review the concepts of "one-to-one" and "onto" presented in Appendix B. Interestingly, for a linear transformation both of these concepts are intimately connected with the rank and nullity of the transformation. This will be demonstrated in the next two theorems.

Theorem 2.4. *Let* V *and* W *be vector spaces, and let* $\mathsf{T}\colon \mathsf{V} \to \mathsf{W}$ *be linear. Then* T *is one-to-one if and only if* $\mathsf{N}(\mathsf{T}) = \{0\}$.

Proof. Suppose that T is one-to-one and $x \in \mathsf{N}(\mathsf{T})$. Then $\mathsf{T}(x) = 0 = \mathsf{T}(0)$. Since T is one-to-one, we have $x = 0$. Hence $\mathsf{N}(\mathsf{T}) = \{0\}$.

Now assume that $\mathsf{N}(\mathsf{T}) = \{0\}$, and suppose that $\mathsf{T}(x) = \mathsf{T}(y)$. Then $0 = \mathsf{T}(x) - \mathsf{T}(y) = \mathsf{T}(x - y)$. Hence $x - y \in \mathsf{N}(\mathsf{T}) = \{0\}$. So $x - y = 0$, or $x = y$. This means that T is one-to-one. ∎

The reader should observe that Theorem 2.4 allows us to conclude that the transformation defined in Example 9 is not one-to-one.

Surprisingly, the conditions of one-to-one and onto are equivalent in an important special case.

Theorem 2.5. *Let* V *and* W *be vector spaces of equal (finite) dimension, and let* $\mathsf{T}\colon \mathsf{V} \to \mathsf{W}$ *be linear. Then* T *is one-to-one if and only if* T *is onto.*

Proof. From the dimension theorem we have

$$\text{nullity}(\mathsf{T}) + \text{rank}(\mathsf{T}) = \dim(\mathsf{V}).$$

Now, with the use of Theorem 2.4, we have that T is one-to-one if and only if $\mathsf{N}(\mathsf{T}) = \{0\}$, if and only if nullity(T) = 0, if and only if rank(T) = dim(V), if and only if rank(T) = dim(W), and if and only if dim(R(T)) = dim(W). By Theorem 1.11 this equality is equivalent to R(T) = W, the definition of T being onto. ∎

The linearity of T in Theorems 2.4 and 2.5 is essential, for it is easy to construct examples of functions from R into R that are not one-to-one but are onto, and vice versa.

The following two examples make use of the theorems above in determining whether a given linear transformation is one-to-one or onto.

Example 11

Define

$$\mathsf{T}\colon \mathsf{P}_2(R) \to \mathsf{P}_3(R) \quad \text{by} \quad \mathsf{T}(f)(x) = 2f'(x) + \int_0^x 3f(t)\,dt.$$

Now

$$\mathsf{R}(\mathsf{T}) = \operatorname{span}(\{\mathsf{T}(1), \mathsf{T}(x), \mathsf{T}(x^2)\}) = \operatorname{span}(\{3x,\ 2 + \tfrac{3}{2}x^2,\ 4x + x^3\}).$$

Hence $\operatorname{rank}(\mathsf{T}) = 3$. Since $\dim(\mathsf{P}_3(R)) = 4$, T is not onto. From Theorem 2.3, $\operatorname{nullity}(\mathsf{T}) + 3 = 3$. So $\operatorname{nullity}(\mathsf{T}) = 0$, so $\mathsf{N}(\mathsf{T}) = \{0\}$. Thus by Theorem 2.4, T is one-to-one. ∎

Example 12

Define

$$\mathsf{T}\colon \mathsf{F}^2 \to \mathsf{F}^2 \quad \text{by} \quad \mathsf{T}(a_1, a_2) = (a_2 + a_1, a_1).$$

It is easy to see that $\mathsf{N}(\mathsf{T}) = \{0\}$; so T is one-to-one. Hence Theorem 2.5 tells us that T must be onto. ∎

In Exercise 14 it is stated that if T is linear and one-to-one, then a subset S is linearly independent if and only if $\mathsf{T}(S)$ is linearly independent. Example 13 illustrates the use of this result.

Example 13

Define

$$\mathsf{T}\colon \mathsf{P}_2(R) \to \mathsf{R}^3 \quad \text{by} \quad \mathsf{T}(a_0 + a_1x + a_2x^2) = (a_0, a_1, a_2).$$

Clearly, T is one-to-one. Let $S = \{2 - x + 3x^2,\ x + x^2,\ 1 - 2x^2\}$. Then S is linearly independent in $\mathsf{P}_2(R)$ if and only if

$$\mathsf{T}(S) = \{(2, -1, 3), (0, 1, 1), (1, 0, -2)\}$$

is linearly independent in R^3. ∎

In Example 13 we transferred a problem from the vector space of polynomials to a problem in the vector space of 3-tuples. This technique will be exploited more fully later.

One of the most important properties of linear transformations is that they are completely determined by their action on a basis. This result, which follows from the next theorem and corollary, will be used frequently throughout the book.

Theorem 2.6. *Let* V *and* W *be vector spaces over a common field* F, *and suppose that* V *is finite-dimensional with a basis* $\{x_1, \ldots, x_n\}$. *For any vectors* $y_1, \ldots, y_n$ *in* W *there exists exactly one linear transformation* $\mathsf{T}\colon \mathsf{V} \to \mathsf{W}$ *such that* $\mathsf{T}(x_i) = y_i$ *for* $i = 1, \ldots, n$.

Proof. Let $x \in \mathsf{V}$. Then

$$x = \sum_{i=1}^{n} a_i x_i,$$

where $a_1, \ldots, a_n$ are unique scalars. Define

$$\mathsf{T}: \mathsf{V} \to \mathsf{W} \quad \text{by} \quad \mathsf{T}(x) = \sum_{i=1}^{n} a_i y_i.$$

(a) T is linear: For suppose that $u, v \in \mathsf{V}$ and $d \in F$. Then we may write

$$u = \sum_{i=1}^{n} b_i x_i \quad \text{and} \quad v = \sum_{i=1}^{n} c_i x_i.$$

Now

$$du + v = \sum_{i=1}^{n} (db_i + c_i)x_i.$$

So

$$\mathsf{T}(du + v) = \sum_{i=1}^{n} (db_i + c_i)y_i = d \sum_{i=1}^{n} b_i y_i + \sum_{i=1}^{n} c_i y_i = d\mathsf{T}(u) + \mathsf{T}(v).$$

(b) Clearly,

$$\mathsf{T}(x_i) = y_i \quad \text{for } i = 1, \ldots, n.$$

(c) T is unique: For suppose that $\mathsf{U}: \mathsf{V} \to \mathsf{W}$ is linear and $\mathsf{U}(x_i) = y_i$ for $i = 1, \ldots, n$. Then for $x \in \mathsf{V}$ with

$$x = \sum_{i=1}^{n} a_i x_i$$

we have

$$\mathsf{U}(x) = \sum_{i=1}^{n} a_i \mathsf{U}(x_i) = \sum_{i=1}^{n} a_i y_i = \mathsf{T}(x).$$

Hence U = T. ∎

Corollary. *Let* V *and* W *be vector spaces, and suppose that* V *has a finite basis* $\{x_1, \ldots, x_n\}$. *If* U, T: V → W *are linear and* $\mathsf{U}(x_i) = \mathsf{T}(x_i)$ *for* $i = 1, \ldots, n$, *then* U = T.

Example 14

Define $\mathsf{T}: \mathsf{R}^2 \to \mathsf{R}^2$ by $\mathsf{T}(a_1, a_2) = (2a_2 - a_1, 3a_1)$, and suppose that $\mathsf{U}: \mathsf{R}^2 \to \mathsf{R}^2$ is linear. If we know that $\mathsf{U}(1, 2) = (3, 3)$ and $\mathsf{U}(1, 1) = (1, 3)$, then U = T. This follows from the corollary and from the fact that $\{(1, 2), (1, 1)\}$ is a basis for R^2. ∎

EXERCISES

√ **1.** Label the following statements as being true or false. For the following, V and W are finite-dimensional vector spaces (over F) and T is a function from V into W.

(a) If T is linear, then T preserves sums and scalar products.
(b) If $\mathsf{T}(x + y) = \mathsf{T}(x) + \mathsf{T}(y)$, then T is linear.
(c) T is one-to-one if and only if $\mathsf{N}(\mathsf{T}) = \{0\}$.
(d) If T is linear, then $\mathsf{T}(0_{\mathsf{V}}) = 0_{\mathsf{W}}$.
(e) If T is linear, then $\text{nullity}(\mathsf{T}) + \text{rank}(\mathsf{T}) = \dim(\mathsf{W})$.
(f) If T is linear, then T carries linearly independent subsets of V onto linearly independent subsets of W.
(g) If $\mathsf{T}, \mathsf{U}\colon \mathsf{V} \to \mathsf{W}$ are both linear and agree on a basis of V, then $\mathsf{T} = \mathsf{U}$.
(h) Given $x_1, x_2 \in \mathsf{V}$ and $y_1, y_2 \in \mathsf{W}$, there exists a linear transformation $\mathsf{T}\colon \mathsf{V} \to \mathsf{W}$ such that $\mathsf{T}(x_1) = y_1$ and $\mathsf{T}(x_2) = y_2$.

For Exercises 2 through 6, prove that T is a linear transformation and find bases for both N(T) and R(T). Then compute the nullity and rank of T and verify the dimension theorem. Finally, use the appropriate theorems in this section to determine whether T is one-to-one or onto.

2. $\mathsf{T}\colon \mathsf{R}^3 \to \mathsf{R}^2$; $\mathsf{T}(a_1, a_2, a_3) = (a_1 - a_2, 2a_3)$.
3. $\mathsf{T}\colon \mathsf{R}^2 \to \mathsf{R}^3$; $\mathsf{T}(a_1, a_2) = (a_1 + a_2, 0, 2a_1 - a_2)$.
4. $\mathsf{T}\colon \mathsf{M}_{2\times 3}(F) \to \mathsf{M}_{2\times 2}(F)$; $\mathsf{T}\begin{pmatrix} a_{11} & a_{12} & a_{13} \\ a_{21} & a_{22} & a_{23} \end{pmatrix} = \begin{pmatrix} 2a_{11} - a_{12} & a_{13} + 2a_{12} \\ 0 & 0 \end{pmatrix}$.
5. $\mathsf{T}\colon \mathsf{P}_2(R) \to \mathsf{P}_3(R)$; $\mathsf{T}(f(x)) = xf(x) + f'(x)$.
6. $\mathsf{T}\colon \mathsf{M}_{n\times n} \to F$; $\mathsf{T}(A) = \text{tr}(A)$. Recall that

$$\text{tr}(A) = \sum_{i=1}^{n} A_{ii}.$$

7. Verify statements 1, 2, and 3 at the beginning of this section.
8. Verify that the transformations defined in Examples 5 and 6 are linear.
9. For the following $\mathsf{T}\colon \mathsf{R}^2 \to \mathsf{R}^2$, state why T is *not* linear.
(a) $\mathsf{T}(a_1, a_2) = (1, a_2)$
(b) $\mathsf{T}(a_1, a_2) = (a_1, a_1^2)$
(c) $\mathsf{T}(a_1, a_2) = (\sin a_1, 0)$
(d) $\mathsf{T}(a_1, a_2) = (|a_1|, a_2)$
(e) $\mathsf{T}(a_1, a_2) = (a_1 + 1, a_2)$
10. Suppose that $\mathsf{T}\colon \mathsf{R}^2 \to \mathsf{R}^2$ is linear and that $\mathsf{T}(1, 0) = (1, 4)$ and $\mathsf{T}(1, 1) = (2, 5)$. What is $\mathsf{T}(2, 3)$? Is T one-to-one?
11. Prove that there exists a linear transformation $\mathsf{T}\colon \mathsf{R}^2 \to \mathsf{R}^3$ such that $\mathsf{T}(1, 1) = (1, 0, 2)$ and $\mathsf{T}(2, 3) = (1, -1, 4)$. What is $\mathsf{T}(8, 11)$?
12. Is there a linear transformation $\mathsf{T}\colon \mathsf{R}^3 \to \mathsf{R}^2$ such that $\mathsf{T}(1, 0, 3) = (1, 1)$ and $\mathsf{T}(-2, 0, -6) = (2, 1)$?
13. Let V and W be vector spaces and $\mathsf{T}\colon \mathsf{V} \to \mathsf{W}$ be linear. Let $\{y_1, \ldots, y_k\}$ be a linearly independent subset of R(T). If $S = \{x_1, \ldots, x_k\}$ is chosen so that $\mathsf{T}(x_i) = y_i$ for $i = 1, \ldots, k$, prove that S is linearly independent.

14. Let V and W be vector spaces and $\mathsf{T}: \mathsf{V} \to \mathsf{W}$ be linear.
   (a) Prove that T is one-to-one if and only if T carries linearly independent subsets of V onto linearly independent subsets of W.
   (b) Suppose that T is one-to-one and that S is a subset of V. Prove that S is linearly independent if and only if $\mathsf{T}(S)$ is linearly independent.

15. Recall the definition of $\mathsf{P}(R)$ in Section 1.2. Define

$$\mathsf{T}: \mathsf{P}(R) \to \mathsf{P}(R) \quad \text{by} \quad \mathsf{T}(f)(x) = \int_0^x f(t)\, dt.$$

Prove that T is one-to-one but not onto.

16. Let V and W be finite-dimensional vector spaces and $\mathsf{T}: \mathsf{V} \to \mathsf{W}$ be linear.
   (a) Prove that if $\dim(\mathsf{V}) < \dim(\mathsf{W})$, then T cannot be onto.
   (b) Prove that if $\dim(\mathsf{V}) > \dim(\mathsf{W})$, then T cannot be one-to-one.

17. Give an example of a linear transformation $\mathsf{T}: \mathsf{R}^2 \to \mathsf{R}^2$ such that $\mathsf{N}(\mathsf{T}) = \mathsf{R}(\mathsf{T})$.

18. Give an example of distinct linear transformations T and U such that $\mathsf{N}(\mathsf{T}) = \mathsf{N}(\mathsf{U})$ and $\mathsf{R}(\mathsf{T}) = \mathsf{R}(\mathsf{U})$.

19. Let V and W be vector spaces with subspaces V_1 and W_1, respectively. If $\mathsf{T}: \mathsf{V} \to \mathsf{W}$ is linear, prove that $\mathsf{T}(\mathsf{V}_1)$ is a subspace of W and that $\{x \in \mathsf{V}: \mathsf{T}(x) \in \mathsf{W}_1\}$ is a subspace of V.

20. Let $\mathsf{T}: \mathsf{R}^3 \to R$ be linear. Show that there exist scalars a, b, and c such that $\mathsf{T}(x, y, z) = ax + by + cz$ for all $(x, y, z) \in \mathsf{R}^3$. Can you generalize this result for $\mathsf{T}: \mathsf{F}^n \to F$? State and prove an analogous result for $\mathsf{T}: \mathsf{F}^n \to \mathsf{F}^m$.

21. Let $\mathsf{T}: \mathsf{R}^3 \to R$ be linear. Describe geometrically the possibilities for the null space of T. *Hint:* Use Exercise 20.

Definition. *Let* V *be a vector space and let* W_1 *be a subspace of* V. *A function* $\mathsf{T}: \mathsf{V} \to \mathsf{V}$ *is called a* projection *on* W_1 *if:*
   (a) *There exists a subspace* W_2 *of* V *such that* $\mathsf{V} = \mathsf{W}_1 \oplus \mathsf{W}_2$. *(Recall the definition of* direct sum *defined in the exercises of Section 1.3.)*
   (b) *For* $\mathrm{x} = \mathrm{x}_1 + \mathrm{x}_2$, *where* $\mathrm{x}_1 \in \mathsf{W}_1$ *and* $\mathrm{x}_2 \in \mathsf{W}_2$, *we have* $\mathsf{T}(\mathrm{x}) = \mathrm{x}_1$.

Exercises 22 through 24 assume the notation above.

22. Prove that T is linear and $\mathsf{W}_1 = \{x: \mathsf{T}(x) = x\}$.

23. Prove that $\mathsf{W}_1 = \mathsf{R}(\mathsf{T})$ and $\mathsf{W}_2 = \mathsf{N}(\mathsf{T})$.

24. Describe T if $\mathsf{W}_1 = \mathsf{V}$ or W_1 is the zero subspace.

25. Suppose that W is a subspace of a finite-dimensional vector space V. Prove that there exists a projection on W.

26. Let V be a vector space, and let $\mathsf{T}: \mathsf{V} \to \mathsf{V}$ be linear. A subspace W of V is said to be T-*invariant* if $\mathsf{T}(x) \in \mathsf{W}$ for every $x \in \mathsf{W}$, i.e., $\mathsf{T}(\mathsf{W}) \subseteq \mathsf{W}$.
   (a) Prove that the subspaces $\{0\}$, V, $\mathsf{R}(\mathsf{T})$, and $\mathsf{N}(\mathsf{T})$ are all T-invariant.

(b) If W is a T-invariant subspace of V, define $\mathsf{T_W}: \mathsf{W} \to \mathsf{W}$ by $\mathsf{T_W}(x) = \mathsf{T}(x)$ for all $x \in \mathsf{W}$. Prove that $\mathsf{T_W}$ is linear.

(c) If T is a projection on W, show that W is T-invariant and that $\mathsf{T_W} = \mathsf{I_W}$.

(d) If $\mathsf{V} = \mathsf{R}(\mathsf{T}) \oplus \mathsf{W}$ and W is T-invariant, prove that $\mathsf{W} \subseteq \mathsf{N}(\mathsf{T})$. Show that if V is also finite-dimensional, then $\mathsf{W} = \mathsf{N}(\mathsf{T})$.

(e) Prove that $\mathsf{N}(\mathsf{T_W}) = \mathsf{N}(\mathsf{T}) \cap \mathsf{W}$ and $\mathsf{R}(\mathsf{T_W}) = \mathsf{T}(\mathsf{W})$.

27. Prove the following generalization of Theorem 2.6 to infinite-dimensional spaces: Let V and W be vector spaces over a common field and β be a basis for V. Then for any function $f\colon \beta \to \mathsf{W}$ there exists exactly one linear transformation $\mathsf{T}: \mathsf{V} \to \mathsf{W}$ such that $\mathsf{T}(x) = f(x)$ for all $x \in \beta$.

28. A function $\mathsf{T}: \mathsf{V} \to \mathsf{W}$ between vector spaces V and W is called *additive* if $\mathsf{T}(x + y) = \mathsf{T}(x) + \mathsf{T}(y)$ for all $x, y \in \mathsf{V}$. Prove that if V and W are vector spaces over the field of rational numbers, then any additive function from V into W is a linear transformation.

29. Prove that there is an additive function $\mathsf{T}: \mathsf{R} \to \mathsf{R}$ (as defined in Exercise 28) that is not linear. *Hint:* Regard R as a vector space over the field of rational numbers Q. By the corollary to Theorem 1.13 this vector space has a basis β. Let x and y be distinct elements of β, and define $f\colon \beta \to \mathsf{R}$ by $f(x) = y$, $f(y) = x$, and $f(z) = z$ otherwise. By Exercise 28 there exists a linear transformation $\mathsf{T}: \mathsf{R} \to \mathsf{R}$, where R is regarded as a vector space over Q, such that $\mathsf{T}(z) = f(z)$ for all $z \in \beta$. Then T is additive, but for $c = y/x$, $\mathsf{T}(cx) \neq c\mathsf{T}(x)$.

The following exercise requires familiarity with the definition of *quotient space* in Exercise 29 of Section 1.3.

30. Let V be a vector and let W be a subspace of V. Define the mapping $\eta: \mathsf{V} \to \mathsf{V}/\mathsf{W}$ by $\eta(v) = v + \mathsf{W}$ for $v \in \mathsf{V}$.

(a) Prove that η is a linear transformation from V onto V/W and that $\mathsf{N}(\eta) = \mathsf{W}$.

(b) Suppose that V is finite-dimensional. Use part (a) and the dimension theorem to derive a formula relating $\dim(\mathsf{V})$, $\dim(\mathsf{W})$, and $\dim(\mathsf{V}/\mathsf{W})$.

(c) Read the proof of the dimension theorem. Compare the method of solving part (b) with the method of deriving the same result as outlined in Exercise 26 of Section 1.6.

2.2 THE MATRIX REPRESENTATION OF A LINEAR TRANSFORMATION

Until now we have studied linear transformations by examining their ranges and null spaces. We now embark upon one of the most useful approaches to the analysis of a linear transformation on a finite-dimensional vector space: the representation of a linear transformation by a matrix. In fact, we develop a one-to-one correspondence between matrices and transformations that allows us to utilize properties of one to study properties of the other.

We first need the concept of an "ordered basis" for a vector space.

Definition. *Let* V *be a finite-dimensional vector space. An* ordered basis *for* V *is a basis for* V *endowed with a specified order; that is, an ordered basis for* V *is a finite sequence of linearly independent elements of* V *that generate* V.

Example 1

Let V have $\beta = \{x_1, x_2, x_3\}$ as an ordered basis. Then $\gamma = \{x_2, x_1, x_3\}$ is also an ordered basis, but $\beta \neq \gamma$ as ordered bases. ∎

For the vector space F^n we call the set $\{e_1, \ldots, e_n\}$ the *standard ordered basis* for F^n. Similarly, for the vector space $\mathsf{P}_n(F)$ we call the set $\{1, x, \ldots, x^n\}$ the *standard ordered basis* for $\mathsf{P}_n(F)$.

Now that we have the concept of an ordered basis we will be able to identify abstract vectors in an n-dimensional vector space with n-tuples. This identification will be provided through the use of "coordinate vectors" as introduced below.

Definition. *Let* $\beta = \{x_1, \ldots, x_n\}$ *be an ordered basis for a finite-dimensional vector space* V. *For* $x \in$ V *we define the* coordinate vector of x relative to β, *denoted* $[x]_\beta$, by

$$[x]_\beta = \begin{pmatrix} a_1 \\ \vdots \\ a_n \end{pmatrix},$$

where

$$x = \sum_{i=1}^{n} a_i x_i.$$

Notice that $[x_i]_\beta = e_i$ in the definition above. It is left as an exercise to show that the correspondence $x \to [x]_\beta$ provides us with a linear transformation from V to F^n. We will study this transformation in Section 2.4 in more detail.

Example 2

Let $\mathsf{V} = \mathsf{P}_2(R)$, and let $\beta = \{1, x, x^2\}$ be the standard ordered basis for V. If $f(x) = 4 + 6x - 7x^2$, then

$$[f]_\beta = \begin{pmatrix} 4 \\ 6 \\ -7 \end{pmatrix}. \quad ∎$$

Let us now proceed with the promised matrix representation of a linear transformation. Suppose that V and W are finite-dimensional vector spaces with ordered bases $\beta = \{x_1, \ldots, x_n\}$ and $\gamma = \{y_1, \ldots, y_m\}$, respectively. Let $\mathsf{T}\colon \mathsf{V} \to \mathsf{W}$ be linear. Then there exist unique scalars $a_{ij} \in F$ ($i = 1, \ldots, m$ and $j = 1, \ldots, n$) such that

$$\mathsf{T}(x_j) = \sum_{i=1}^{m} a_{ij} y_i \qquad \text{for } 1 \leqslant j \leqslant n.$$

Definition. *Using the notation above, we call the* $m \times n$ *matrix* A *defined by* $A_{ij} = a_{ij}$ *the* matrix that represents T in the ordered bases β and γ *and will write* $A = [\mathsf{T}]_\beta^\gamma$. *If* $\mathsf{V} = \mathsf{W}$ *and* $\beta = \gamma$, *we write* $A = [\mathsf{T}]_\beta$.

Notice that the jth column of A is simply $[\mathsf{T}(x_j)]_\gamma$. Also observe that from the corollary to Theorem 2.6 it follows that if $\mathsf{U}\colon \mathsf{V} \to \mathsf{W}$ is a linear transformation such that $[\mathsf{U}]_\beta^\gamma = [\mathsf{T}]_\beta^\gamma$, then $\mathsf{U} = \mathsf{T}$.

We will illustrate the computation of $[\mathsf{T}]_\beta^\gamma$ in the next several examples.

Example 3

Define

$$\mathsf{T}\colon \mathsf{P}_3(R) \to \mathsf{P}_2(R) \quad \text{by} \quad \mathsf{T}(f) = f'.$$

Let β and γ be the standard ordered bases for $\mathsf{P}_3(R)$ and $\mathsf{P}_2(R)$, respectively. Then

$$\begin{aligned} \mathsf{T}(1) &= 0\cdot 1 + 0\cdot x + 0\cdot x^2 \\ \mathsf{T}(x) &= 1\cdot 1 + 0\cdot x + 0\cdot x^2 \\ \mathsf{T}(x^2) &= 0\cdot 1 + 2\cdot x + 0\cdot x^2 \\ \mathsf{T}(x^3) &= 0\cdot 1 + 0\cdot x + 3\cdot x^2 \end{aligned}$$

So

$$[\mathsf{T}]_\beta^\gamma = \begin{pmatrix} 0 & 1 & 0 & 0 \\ 0 & 0 & 2 & 0 \\ 0 & 0 & 0 & 3 \end{pmatrix}.$$

Note that the coefficients of $\mathsf{T}(x^j)$ when written as a linear combination of elements of γ give the entries of the jth column. ∎

Example 4

Define

$$\mathsf{T}\colon \mathsf{R}^2 \to \mathsf{R}^3 \text{ by } \mathsf{T}(a_1, a_2) = (a_1 + 3a_2, 0, 2a_1 - 4a_2).$$

Let β and γ be the standard ordered bases for R^2 and R^3, respectively. Now

$$\mathsf{T}(1, 0) = (1, 0, 2) = 1e_1 + 0e_2 + 2e_3$$

and

$$\mathsf{T}(0, 1) = (3, 0, -4) = 3e_1 + 0e_2 - 4e_3.$$

Hence

$$[\mathsf{T}]_\beta^\gamma = \begin{pmatrix} 1 & 3 \\ 0 & 0 \\ 2 & -4 \end{pmatrix}.$$

If we let $\gamma' = \{e_3, e_2, e_1\}$, then

$$[\mathsf{T}]_\beta^{\gamma'} = \begin{pmatrix} 2 & -4 \\ 0 & 0 \\ 1 & 3 \end{pmatrix}. \quad \blacksquare$$

Now that we have defined a procedure for associating matrices with linear transformations, we will see shortly that this association "preserves" addition. To make this more explicit, we need some preliminary discussion about the addition of linear transformations.

Definition. *Let* $\mathsf{T}, \mathsf{U}\colon \mathsf{V} \to \mathsf{W}$ *be arbitrary functions, where* V *and* W *are vector spaces, and let* $a \in F$. *We define* $\mathsf{T} + \mathsf{U}\colon \mathsf{V} \to \mathsf{W}$ *by* $(\mathsf{T} + \mathsf{U})(x) = \mathsf{T}(x) + \mathsf{U}(x)$ *for all* $x \in \mathsf{V}$, *and* $a\mathsf{T}\colon \mathsf{V} \to \mathsf{W}$ *by* $(a\mathsf{T})(x) = a\mathsf{T}(x)$ for all $x \in \mathsf{V}$.

Of course, this is just the usual definition of addition and scalar multiplication for functions. We are fortunate, however, to have the result that both sums and scalar multiples of linear transformations are also linear.

Theorem 2.7. *Let* V *and* W *be vector spaces, and let* $\mathsf{T}, \mathsf{U}\colon \mathsf{V} \to \mathsf{W}$ *be linear. Then for all* $a \in F$

(a) $a\mathsf{T} + \mathsf{U}$ *is linear.*

(b) *Using the operations of addition and scalar multiplication as defined above, the collection of all linear transformations from* V *into* W *is a vector space over* F. *This vector space is denoted by* $\mathcal{L}(\mathsf{V}, \mathsf{W})$.

Proof. (a) Let $x, y \in \mathsf{V}$ and $c \in F$. Then

$$\begin{aligned}(a\mathsf{T} + \mathsf{U})(cx + y) &= a\mathsf{T}(cx + y) + \mathsf{U}(cx + y) \\ &= a[c\mathsf{T}(x) + \mathsf{T}(y)] + c\mathsf{U}(x) + \mathsf{U}(y) \\ &= ac\mathsf{T}(x) + c\mathsf{U}(x) + a\mathsf{T}(y) + \mathsf{U}(y) \\ &= c[a\mathsf{T} + \mathsf{U}](x) + [a\mathsf{T} + \mathsf{U}](y).\end{aligned}$$

So $a\mathsf{T} + \mathsf{U}$ is linear.

(b) Noting that T_0, the zero transformation, plays the role of the zero element in $\mathcal{L}(\mathsf{V}, \mathsf{W})$, it is easy to show that $\mathcal{L}(\mathsf{V}, \mathsf{W})$ is a vector space over F. ∎

In the case where $\mathsf{V} = \mathsf{W}$ we will write $\mathcal{L}(\mathsf{V})$ instead of $\mathcal{L}(\mathsf{V}, \mathsf{V})$.

In Section 2.3 we see a complete identification of $\mathcal{L}(\mathsf{V}, \mathsf{W})$ with the vector space $\mathsf{M}_{m\times n}(F)$, where n and m are the dimensions of V and W, respectively. This identification is easily established by use of the next theorem.

Theorem 2.8. *Let* V *and* W *be finite-dimensional vector spaces with ordered bases* β *and* γ*, respectively, and let* $\mathsf{T}, \mathsf{U}\colon \mathsf{V} \to \mathsf{W}$ *be linear transformations. Then*

(a) $[\mathsf{T} + \mathsf{U}]_\beta^\gamma = [\mathsf{T}]_\beta^\gamma + [\mathsf{U}]_\beta^\gamma$.

(b) $[a\mathsf{T}]_\beta^\gamma = a[\mathsf{T}]_\beta^\gamma$ *for all* $a \in F$.

Proof. Let $\beta = \{x_1, \ldots, x_n\}$ and $\gamma = \{y_1, \ldots, y_m\}$. There exist unique scalars a_{ij} and b_{ij} in F $(1 \leqslant i \leqslant m,\ 1 \leqslant j \leqslant n)$ such that

$$\mathsf{T}(x_j) = \sum_{i=1}^{m} a_{ij} y_i \quad \text{and} \quad \mathsf{U}(x_j) = \sum_{i=1}^{m} b_{ij} y_i \qquad \text{for } 1 \leqslant j \leqslant n.$$

Hence

$$(\mathsf{T} + \mathsf{U})(x_j) = \sum_{i=1}^{m} (a_{ij} + b_{ij}) y_i.$$

Thus

$$([\mathsf{T} + \mathsf{U}]_\beta^\gamma)_{ij} = a_{ij} + b_{ij} = ([\mathsf{T}]_\beta^\gamma + [\mathsf{U}]_\beta^\gamma)_{ij}.$$

So (a) is proved, and the proof of (b) is similar. ∎

Example 5

Define

$$\mathsf{T}\colon \mathsf{R}^2 \to \mathsf{R}^3 \text{ by } \mathsf{T}(a_1, a_2) = (a_1 + 3a_2, 0, 2a_1 - 4a_2)$$

and

$$\mathsf{U}\colon \mathsf{R}^2 \to \mathsf{R}^3 \text{ by } \mathsf{U}(a_1, a_2) = (a_1 - a_2, 2a_1, 3a_1 + 2a_2).$$

Let β and γ be the standard ordered bases of R^2 and R^3, respectively. Then

$$[\mathsf{T}]_\beta^\gamma = \begin{pmatrix} 1 & 3 \\ 0 & 0 \\ 2 & -4 \end{pmatrix},$$

(as computed in Example 4) and

$$[\mathsf{U}]_\beta^\gamma = \begin{pmatrix} 1 & -1 \\ 2 & 0 \\ 3 & 2 \end{pmatrix}.$$

If we now compute T + U using the definitions above, we obtain

$$(\mathsf{T} + \mathsf{U})(a_1, a_2) = (2a_1 + 2a_2, 2a_1, 5a_1 - 2a_2).$$

So

$$[\mathsf{T} + \mathsf{U}]_\beta^\gamma = \begin{pmatrix} 2 & 2 \\ 2 & 0 \\ 5 & -2 \end{pmatrix},$$

which is simply $[\mathsf{T}]_\beta^\gamma + [\mathsf{U}]_\beta^\gamma$, illustrating Theorem 2.8. ∎

EXERCISES

1. Label the following statements as being true or false. For the following V and W denote finite-dimensional vector spaces with ordered bases β and γ, respectively. Assume that T, U: V → W are linear.
   - **(a)** For any scalar a, $a\mathsf{T} + \mathsf{U}$ is a linear transformation from V into W.
   - **(b)** $[\mathsf{T}]_\beta^\gamma = [\mathsf{U}]_\beta^\gamma$ implies that T = U.
   - **(c)** If $m = \dim(\mathsf{V})$ and $n = \dim(\mathsf{W})$, then $[\mathsf{T}]_\beta^\gamma$ is an $m \times n$ matrix.
   - **(d)** $[\mathsf{T} + \mathsf{U}]_\beta^\gamma = [\mathsf{T}]_\beta^\gamma + [\mathsf{U}]_\beta^\gamma$.
   - **(e)** $\mathscr{L}(\mathsf{V}, \mathsf{W})$ is a vector space.
   - **(f)** $\mathscr{L}(\mathsf{V}, \mathsf{W}) = \mathscr{L}(\mathsf{W}, \mathsf{V})$.

2. Let β and γ be the standard ordered bases for R^n and R^m, respectively. For the following transformations $\mathsf{T}: \mathsf{R}^n \to \mathsf{R}^m$, compute $[\mathsf{T}]_\beta^\gamma$.
   - **(a)** $\mathsf{T}: \mathsf{R}^2 \to \mathsf{R}^3$ defined by $\mathsf{T}(a_1, a_2) = (2a_1 - a_2, 3a_1 + 4a_2, a_1)$.
   - **(b)** $\mathsf{T}: \mathsf{R}^3 \to \mathsf{R}^2$ defined by $\mathsf{T}(a_1, a_2, a_3) = (2a_1 + 3a_2 - a_3, a_1 + a_3)$.
   - **(c)** $\mathsf{T}: \mathsf{R}^3 \to R$ defined by $\mathsf{T}(a_1, a_2, a_3) = 2a_1 + a_2 - 3a_3$.
   - **(d)** $\mathsf{T}: \mathsf{R}^3 \to \mathsf{R}^3$ defined by
     $\mathsf{T}(a_1, a_2, a_3) = (2a_2 + a_3, -a_1 + 4a_2 + 5a_3, a_1 + a_3)$.
   - **(e)** $\mathsf{T}: \mathsf{R}^n \to \mathsf{R}^n$ defined by $\mathsf{T}(a_1, a_2, \ldots, a_n) = (a_1, a_1, \ldots, a_1)$.
   - **(f)** $\mathsf{T}: \mathsf{R}^n \to \mathsf{R}^n$ defined by $\mathsf{T}(a_1, a_2, \ldots, a_n) = (a_n, a_{n-1}, \ldots, a_1)$.
   - **(g)** $\mathsf{T}: \mathsf{R}^n \to R$ defined by $\mathsf{T}(a_1, a_2, \ldots, a_n) = a_1 + a_n$.

3. Let $\mathsf{T}: \mathsf{R}^2 \to \mathsf{R}^3$ be defined as $\mathsf{T}(a_1, a_2) = (a_1 - a_2, a_1, 2a_1 + a_2)$. Let β be the standard ordered basis for R^2 and $\gamma = \{(1,1,0), (0,1,1), (2,2,3)\}$. Compute $[\mathsf{T}]_\beta^\gamma$. If $\alpha = \{(1,2), (2,3)\}$, compute $[\mathsf{T}]_\alpha^\gamma$.

4. Define

$$\mathsf{T}: \mathsf{M}_{2\times 2}(R) \to \mathsf{P}_2(R) \quad \text{by} \quad \mathsf{T}\begin{pmatrix} a & b \\ c & d \end{pmatrix} = (a + b) + (2d)x + bx^2.$$

Let

$$\beta = \left\{ \begin{pmatrix} 1 & 0 \\ 0 & 0 \end{pmatrix}, \begin{pmatrix} 0 & 1 \\ 0 & 0 \end{pmatrix}, \begin{pmatrix} 0 & 0 \\ 1 & 0 \end{pmatrix}, \begin{pmatrix} 0 & 0 \\ 0 & 1 \end{pmatrix} \right\} \quad \text{and} \quad \gamma = \{1, x, x^2\}.$$

Compute $[\mathsf{T}]_\beta^\gamma$.

5. For the following parts, let

$$\alpha = \left\{ \begin{pmatrix} 1 & 0 \\ 0 & 0 \end{pmatrix}, \begin{pmatrix} 0 & 1 \\ 0 & 0 \end{pmatrix}, \begin{pmatrix} 0 & 0 \\ 1 & 0 \end{pmatrix}, \begin{pmatrix} 0 & 0 \\ 0 & 1 \end{pmatrix} \right\},$$

$$\beta = \{1, x, x^2\},$$

and

$$\gamma = \{1\}.$$

(a) Define $\mathsf{T}\colon \mathsf{M}_{2\times 2}(F) \to \mathsf{M}_{2\times 2}(F)$ by $\mathsf{T}(A) = A^t$. Compute $[\mathsf{T}]_\alpha$.
(b) Define

$$\mathsf{T}\colon \mathsf{P}_2(R) \to \mathsf{M}_{2\times 2}(R) \quad \text{by} \quad \mathsf{T}(f) = \begin{pmatrix} f'(0) & 2f(1) \\ 0 & f''(3) \end{pmatrix},$$

where $'$ denotes differentiation. Compute $[\mathsf{T}]_\beta^\alpha$.
(c) Define $\mathsf{T}\colon \mathsf{M}_{2\times 2}(F) \to F$ by $\mathsf{T}(A) = \operatorname{tr}(A)$. Compute $[\mathsf{T}]_\alpha^\gamma$.
(d) Define $\mathsf{T}\colon \mathsf{P}_2(R) \to R$ by $\mathsf{T}(f) = f(2)$. Compute $[\mathsf{T}]_\beta^\gamma$.
(e) If

$$A = \begin{pmatrix} 1 & -2 \\ 0 & 4 \end{pmatrix},$$

compute $[A]_\alpha$.
(f) If $f(x) = 3 - 6x + x^2$, compute $[f]_\beta$.
(g) For $a \in F$, compute $[a]_\gamma$.

6. Prove (b) of Theorem 2.8.

7.† Let V be an n-dimensional vector space with an ordered basis β. Define $\mathsf{T}\colon \mathsf{V} \to \mathsf{F}^n$ by $\mathsf{T}(x) = [x]_\beta$. Prove that T is linear.

8. Let V be the vector space of complex numbers over the field R. If $\mathsf{T}\colon \mathsf{V} \to \mathsf{V}$ is defined by $\mathsf{T}(z) = \bar{z}$, where $\bar{z}$ is the complex conjugate of z, prove that T is linear, and compute $[\mathsf{T}]_\beta$, where $\beta = \{1, i\}$. Show that T is not linear if V is regarded as a vector space over the field C.

9. Let V be a vector space with the ordered basis $\beta = \{x_1, \ldots, x_n\}$. Define $x_0 = 0$. By Theorem 2.6 there exists a linear transformation $\mathsf{T}\colon \mathsf{V} \to \mathsf{V}$ defined by $\mathsf{T}(x_j) = x_j + x_{j-1}$ for $j = 1, \ldots, n$. Compute $[\mathsf{T}]_\beta$.

10. Let V be an n-dimensional vector space, and let $\mathsf{T}\colon \mathsf{V} \to \mathsf{V}$ be a linear transformation. Suppose that W is a T-invariant subspace of V (see Exercise 26 of Section 2.1) having dimension k. Show that there is a basis β

for V such that $[\mathsf{T}]_\beta$ has the form

$$\begin{pmatrix} A & B \\ O & C \end{pmatrix},$$

where A is a $k \times k$ matrix and O is an $(n-k) \times k$ zero matrix.

11. Let V be a finite-dimensional vector space, and let T be a projection on a subspace W of V. (See the definition of *projection* preceding Exercise 22 of Section 2.1.) Find an ordered basis β for V such that $[\mathsf{T}]_\beta$ is a diagonal matrix.

12. Let V and W be vector spaces, and let T and U be nonzero linear transformations from V into W. If $\mathsf{R}(\mathsf{T}) \cap \mathsf{R}(\mathsf{U}) = \{0\}$, prove that $\{\mathsf{T}, \mathsf{U}\}$ is a linearly independent subset of $\mathscr{L}(\mathsf{V}, \mathsf{W})$.

13. Let $\mathsf{V} = \mathsf{P}(R)$, and for $j \geqslant 0$ define $\mathsf{T}_j(f) = f^{(j)}$, where $f^{(j)}$ is the jth derivative of f. Prove that the set $\{\mathsf{T}_1, \mathsf{T}_2, \ldots, \mathsf{T}_n\}$ is a linearly independent subset of $\mathscr{L}(\mathsf{V})$ for any positive integer n.

14. Let V and W be vector spaces, and let S be a subset of V. Define $S^0 = \{\mathsf{T} \in \mathscr{L}(\mathsf{V}, \mathsf{W}) \colon \mathsf{T}(x) = 0 \text{ for all } x \in S\}$. Prove

(a) S^0 is a subspace of $\mathscr{L}(\mathsf{V}, \mathsf{W})$.

(b) If S_1 and S_2 are subsets of V and $S_1 \subseteq S_2$, then $S_2^0 \subseteq S_1^0$.

(c) If V_1 and V_2 are subspaces of V, then $(\mathsf{V}_1 + \mathsf{V}_2)^0 = \mathsf{V}_1^0 \cap \mathsf{V}_2^0$.

15. Let V and W be vector spaces such that $\dim(\mathsf{V}) = \dim(\mathsf{W})$, and let $\mathsf{T}\colon \mathsf{V} \to \mathsf{W}$ be linear. Find ordered bases β and γ for V and W, respectively, such that $[\mathsf{T}]_\beta^\gamma$ is a diagonal matrix.

2.3 COMPOSITION OF LINEAR TRANSFORMATIONS AND MATRIX MULTIPLICATION

In Section 2.2 we learned how to associate a matrix with a linear transformation in such a way that both sums and scalar multiples of matrices are associated with the corresponding sums and scalar multiples of transformations. The question now arises as to how the matrix representation of a composition of linear transformations is related to the matrix representations of each of the associated linear transformations. The attempt to answer this question will lead to a definition of matrix multiplication. We use the notation UT for composition of linear transformations U and T as contrasted with $g \circ f$ for arbitrary functions g and f. Specifically, we have the following definition.

Definition. *Let* V, W, *and* Z *be vector spaces, and let* $\mathsf{T}\colon \mathsf{V} \to \mathsf{W}$ *and* $\mathsf{U}\colon \mathsf{W} \to \mathsf{Z}$ *be linear. We define* $\mathsf{UT}\colon \mathsf{V} \to \mathsf{Z}$ *by* $(\mathsf{UT})(\mathsf{x}) = \mathsf{U}(\mathsf{T}(\mathsf{x}))$ *for all* $\mathsf{x} \in \mathsf{V}$.

Our first result shows that the composition of linear transformations is linear.

Theorem 2.9. *Let* V, W, *and* Z *be vector spaces and* $\mathsf{T}: \mathsf{V} \to \mathsf{W}$ *and* $\mathsf{U}: \mathsf{W} \to \mathsf{Z}$ *be linear. Then* $\mathsf{UT}: \mathsf{V} \to \mathsf{Z}$ *is linear.*

Proof. Let $x, y \in \mathsf{V}$ and $a \in F$. Then

$$\begin{aligned}\mathsf{UT}(ax + y) &= \mathsf{U}(\mathsf{T}(ax + y)) = \mathsf{U}(a\mathsf{T}(x) + \mathsf{T}(y))\\ &= a\mathsf{U}(\mathsf{T}(x)) + \mathsf{U}(\mathsf{T}(y)) = a(\mathsf{UT})(x) + \mathsf{UT}(y). \quad \blacksquare\end{aligned}$$

The following theorem lists some properties of the composition of linear transformations.

Theorem 2.10. *Let* V *be a vector space. Let* $\mathsf{T}, \mathsf{U}_1, \mathsf{U}_2 \in \mathcal{L}(\mathsf{V})$. *Then*
(a) $\mathsf{T}(\mathsf{U}_1 + \mathsf{U}_2) = \mathsf{TU}_1 + \mathsf{TU}_2$ *and* $(\mathsf{U}_1 + \mathsf{U}_2)\mathsf{T} = \mathsf{U}_1\mathsf{T} + \mathsf{U}_2\mathsf{T}$.
(b) $\mathsf{T}(\mathsf{U}_1\mathsf{U}_2) = (\mathsf{TU}_1)\mathsf{U}_2$.
(c) $\mathsf{TI} = \mathsf{IT} = \mathsf{T}$.
(d) $a(\mathsf{U}_1\mathsf{U}_2) = (a\mathsf{U}_1)(\mathsf{U}_2) = \mathsf{U}_1(a\mathsf{U}_2)$ *for all* $a \in F$.

Proof. Exercise. ∎

We are now in a position to define the product AB of two matrices A and B. Because of Theorem 2.8 it seems reasonable by analogy to require that if $A = [\mathsf{U}]_\beta^\gamma$ and $B = [\mathsf{T}]_\alpha^\beta$, where $\mathsf{T}: \mathsf{V} \to \mathsf{W}$ and $\mathsf{U}: \mathsf{W} \to \mathsf{Z}$, then $AB = [\mathsf{UT}]_\alpha^\gamma$.

Now let T, U, A, and B be as above, and let $\alpha = \{x_1, \ldots, x_n\}$, $\beta = \{y_1, \ldots, y_m\}$, and $\gamma = \{z_1, \ldots, z_p\}$ be ordered bases for V, W, and Z, respectively. For $1 \leqslant j \leqslant n$ we have

$$\begin{aligned}(\mathsf{UT})(x_j) = \mathsf{U}(\mathsf{T}(x_j)) &= \mathsf{U}\left(\sum_{k=1}^{m} B_{kj}y_k\right) = \sum_{k=1}^{m} B_{kj}\mathsf{U}(y_k)\\ &= \sum_{k=1}^{m} B_{kj}\left(\sum_{i=1}^{p} A_{ik}z_i\right) = \sum_{i=1}^{p}\left(\sum_{k=1}^{m} A_{ik}B_{kj}\right)z_i\\ &= \sum_{i=1}^{p} C_{ij}z_i,\end{aligned}$$

where

$$C_{ij} = \sum_{k=1}^{m} A_{ik}B_{kj}.$$

This computation suggests the following definition of matrix multiplication.

Definition. *Let* A *be an* $m \times n$ *matrix and* B *be an* $n \times p$ *matrix. We define the* product of A and B, *denoted* AB, *to be the* $m \times p$ *matrix such that*

$$(AB)_{ij} = \sum_{k=1}^{n} A_{ik}B_{kj} \qquad \text{for } 1 \leqslant i \leqslant m,\ 1 \leqslant j \leqslant p.$$

Note that $(AB)_{ij}$ is the sum of products of corresponding elements from the ith row of A and the jth column of B. At the end of this section the reader will see some interesting applications of this definition.

The reader should observe that in order for the product AB to be defined, there are restrictions regarding the relative sizes of A and B. The following mnemonic device is helpful: "$(m \times n)\cdot(n \times p) = (m \times p)$"; that is, in order for the product AB to be defined, the two "inner" dimensions must be equal, and the two "outer" dimensions yield the size of the product.

Example 1

$$\begin{pmatrix} 1 & 2 & 1 \\ 0 & 4 & -1 \end{pmatrix}\begin{pmatrix} 4 \\ 2 \\ 5 \end{pmatrix} = \begin{pmatrix} 4+4+5 \\ 0+8-5 \end{pmatrix} = \begin{pmatrix} 13 \\ 3 \end{pmatrix}$$

Notice again the symbolic relationship $(2 \times 3)\cdot(3 \times 1) = 2 \times 1$. ∎

As in the case with composition of functions, we have that matrix multiplication is not commutative. Consider the following two products.

$$\begin{pmatrix} 1 & 2 \\ 0 & 1 \end{pmatrix}\begin{pmatrix} 0 & 3 \\ 1 & 2 \end{pmatrix} = \begin{pmatrix} 2 & 7 \\ 1 & 2 \end{pmatrix} \quad \text{and} \quad \begin{pmatrix} 0 & 3 \\ 1 & 2 \end{pmatrix}\begin{pmatrix} 1 & 2 \\ 0 & 1 \end{pmatrix} = \begin{pmatrix} 0 & 3 \\ 1 & 4 \end{pmatrix}.$$

Hence we see that even if both of the matrix products AB and BA are defined, it need not be true that $AB = BA$.

Recalling the definition of the transpose of a matrix from Section 1.3, we will show that if A is an $m \times n$ matrix and B is an $n \times p$ matrix, then $(AB)^t = B^t A^t$. Since

$$(AB)^t_{ij} = (AB)_{ji} = \sum_{k=1}^{n} A_{jk} B_{ki}$$

and

$$(B^t A^t)_{ij} = \sum_{k=1}^{n} (B^t)_{ik}(A^t)_{kj} = \sum_{k=1}^{n} B_{ki} A_{jk},$$

we are done. Hence the transpose of a product is the product of the transposes *in the opposite order.*

The following theorem is an immediate consequence of our definition of matrix multiplication.

Theorem 2.11. *Let V, W, and Z be finite-dimensional vector spaces with ordered bases α, β, and γ, respectively. Let $\mathsf{T}: \mathsf{V} \to \mathsf{W}$ and $\mathsf{U}: \mathsf{W} \to \mathsf{Z}$ be linear transformations. Then*

$$[\mathsf{UT}]_\alpha^\gamma = [\mathsf{U}]_\beta^\gamma [\mathsf{T}]_\alpha^\beta.$$

Corollary. *Let* V *be a finite-dimensional vector space with an ordered basis* β. *Let* T, U $\in \mathscr{L}(\mathsf{V})$. *Then* $[\mathsf{UT}]_\beta = [\mathsf{U}]_\beta[\mathsf{T}]_\beta$.

We illustrate Theorem 2.11 in the following example.

Example 2

Define

$$\mathsf{U}\colon \mathsf{P}_3(R) \to \mathsf{P}_2(R) \quad \text{by} \quad \mathsf{U}(f) = f'$$

as in Example 3 of Section 2.2. Define

$$\mathsf{T}\colon \mathsf{P}_2(R) \to \mathsf{P}_3(R) \quad \text{by} \quad \mathsf{T}(f)(x) = \int_0^x f(t)\,dt.$$

Let $\alpha = \{1, x, x^2, x^3\}$ and $\beta = \{1, x, x^2\}$. Clearly, $\mathsf{UT} = \mathsf{I}$. To illustrate Theorem 2.11, observe that

$$[\mathsf{UT}]_\beta = [\mathsf{U}]_\alpha^\beta[\mathsf{T}]_\beta^\alpha = \begin{pmatrix} 0 & 1 & 0 & 0 \\ 0 & 0 & 2 & 0 \\ 0 & 0 & 0 & 3 \end{pmatrix} \begin{pmatrix} 0 & 0 & 0 \\ 1 & 0 & 0 \\ 0 & \frac{1}{2} & 0 \\ 0 & 0 & \frac{1}{3} \end{pmatrix} = \begin{pmatrix} 1 & 0 & 0 \\ 0 & 1 & 0 \\ 0 & 0 & 1 \end{pmatrix} = [\mathsf{I}]_\beta. \quad \blacksquare$$

The 3×3 diagonal matrix above is called an "identity matrix" and is defined below along with a very useful notation, the "Kronecker delta."

Definitions. *We define the* Kronecker delta δ_{ij} *by* $\delta_{ij} = 1$ *if* $i = j$ *and* $\delta_{ij} = 0$ *if* $i \neq j$ *and the* $n \times n$ identity matrix I_n *by* $(I_n)_{ij} = \delta_{ij}$.

Thus

$$I_1 = (1), \qquad I_2 = \begin{pmatrix} 1 & 0 \\ 0 & 1 \end{pmatrix}, \quad \text{and} \quad I_3 = \begin{pmatrix} 1 & 0 & 0 \\ 0 & 1 & 0 \\ 0 & 0 & 1 \end{pmatrix}.$$

We see in the next theorem that the identity matrix acts as a unity element in $\mathsf{M}_{n\times n}(F)$. When the context is sufficiently clear, we sometimes omit the subscript n from I_n.

Theorem 2.12. *For any* $n \times n$ *matrix* A *we have* $I_nA = AI_n = A$. *Furthermore, if* V *is a finite-dimensional vector space of dimension* n *with an ordered basis* β, *then* $[\mathsf{I_V}]_\beta = I_n$.

Proof.

$$(I_nA)_{ij} = \sum_{k=1}^{n} (I_n)_{ik}A_{kj} = \sum_{k=1}^{n} \delta_{ik}A_{kj} = A_{ij}.$$

Hence $I_n A = A$. Similarly, $AI_n = A$. Let $\beta = \{x_1, \ldots, x_n\}$. Then for each j we have

$$\mathsf{I_V}(x_j) = x_j = \sum_{i=1}^{n} \delta_{ij} x_i.$$

Hence $[\mathsf{I_V}]_\beta = I_n$. ∎

For an $n \times n$ matrix A we define $A^2 = AA$, $A^3 = A^2 A$, and, in general, $A^k = A^{k-1}A$ for $k = 2, 3, \ldots$. We define $A^0 = I_n$.

With this notation we see that if

$$A = \begin{pmatrix} 0 & 0 \\ 1 & 0 \end{pmatrix},$$

then $A^2 = O$ (the zero matrix) even though $A \neq O$. Thus the cancellation property for fields is not valid for matrices. The next theorem shows, however, that matrix multiplication does distribute over addition.

Theorem 2.13. *Let* A *be an* m × n *matrix, and let* B *and* C *be* n × p *matrices. Then*

$$\mathrm{A(B + C) = AB + AC},$$

and for any scalar a,

$$\mathrm{a(AB) = (aA)B = A(aB)}.$$

Proof.

$$[A(B + C)]_{ij} = \sum_{k=1}^{n} A_{ik}(B + C)_{kj} = \sum_{k=1}^{n} A_{ik}(B_{kj} + C_{kj})$$

$$= \sum_{k=1}^{n} (A_{ik}B_{kj} + A_{ik}C_{kj}) = \sum_{k=1}^{n} A_{ik}B_{kj} + \sum_{k=1}^{n} A_{ik}C_{kj}$$

$$= (AB)_{ij} + (AC)_{ij} = [AB + AC]_{ij}.$$

The remainder of the proof is left as an exercise. ∎

Corollary. *Let* A *be an* m × n *matrix,* $\mathrm{B_1, \ldots, B_k}$ *be* n × p *matrices, and* $\mathrm{a_1, \ldots, a_k \in F}$. *Then*

$$\mathrm{A\left(\sum_{i=1}^{k} a_i B_i\right) = \sum_{i=1}^{k} a_i AB_i}.$$

Proof. Exercise. ∎

If A is an $m \times n$ matrix, we sometimes write $A = (A^{(1)}, \ldots, A^{(n)})$, where $A^{(j)}$

is the jth column

$$\begin{pmatrix} A_{1j} \\ A_{2j} \\ \vdots \\ A_{mj} \end{pmatrix}$$

of the matrix A.

For the following theorem, e_j will denote the jth column of I_p.

Theorem 2.14. *Let* A *be an* m × n *matrix and* B *be an* n × p *matrix. Then*
(a) $(AB)^{(j)} = AB^{(j)}$.
(b) $B^{(j)} = Be_j$.

Proof.

$$(AB)^{(j)} = \begin{pmatrix} (AB)_{1j} \\ \vdots \\ (AB)_{mj} \end{pmatrix} = \begin{pmatrix} \sum_{k=1}^{n} A_{1k}B_{kj} \\ \vdots \\ \sum_{k=1}^{n} A_{mk}B_{kj} \end{pmatrix} = A \begin{pmatrix} B_{1j} \\ \vdots \\ B_{nj} \end{pmatrix} = AB^{(j)}.$$

Hence (a) is proved. The proof of (b) is left as an exercise. ∎

The next result justifies much of our past work. It utilizes both the matrix representation of a linear transformation and matrix multiplication in order to evaluate the transformation at any given vector.

Theorem 2.15. *Let* V *and* W *be finite-dimensional vector spaces having ordered bases* β *and* γ*, respectively, and let* T: V → W *be linear. Then for each* x ∈ V *we have*

$$[\mathsf{T}(x)]_\gamma = [\mathsf{T}]_\beta^\gamma [x]_\beta.$$

Proof. Let $\beta = \{x_1, \ldots, x_n\}$, and let $A = [\mathsf{T}]_\beta^\gamma$. Suppose that γ has exactly m elements. Define the two maps: $\phi, \psi: \mathsf{V} \to \mathsf{F}^m$ by

$$\phi(x) = [\mathsf{T}(x)]_\gamma \quad \text{and} \quad \psi(x) = A[x]_\gamma \qquad \text{for all } x \in \mathsf{V}.$$

To prove the theorem, we must show that $\phi = \psi$.

We first show that both maps are linear. By Exercise 7 of Section 2.2 ϕ is the composition of the two linear functions: $x \to \mathsf{T}(x)$ and $\mathsf{T}(x) \to [\mathsf{T}(x)]_\gamma$, so ϕ is linear by Theorem 2.9. Similarly, by Theorem 2.13 ψ is the composition of the two linear functions: $x \to [x]_\gamma$ and $[x]_\gamma \to A[x]_\gamma$, and hence ψ is linear. Now by the corollary to Theorem 2.6, we need only show that $\phi(x_j) = \psi(x_j)$ for all j. By the definition of A, we have $\phi(x_j) = [\mathsf{T}(x_j)]_\gamma = A^{(j)}$, and by Theorem 2.14 we have

$$\psi(x_j) = A[x_j]_\gamma = Ae_j = A^{(j)}. \quad \blacksquare$$

Example 3

Let $\mathsf{T}\colon \mathsf{P}_3(R) \to \mathsf{P}_2(R)$ be defined by $\mathsf{T}(f) = f'$, and let β and γ be the standard ordered bases for $\mathsf{P}_3(R)$ and $\mathsf{P}_2(R)$, respectively. If $A = [\mathsf{T}]_\beta^\gamma$, then we have from Example 3 of Section 2.2 that

$$A = \begin{pmatrix} 0 & 1 & 0 & 0 \\ 0 & 0 & 2 & 0 \\ 0 & 0 & 0 & 3 \end{pmatrix}.$$

We illustrate Theorem 2.15 by verifying that $[\mathsf{T}(p)]_\gamma = [\mathsf{T}]_\beta^\gamma [p]_\beta$, where $p \in \mathsf{P}_3(R)$ is the polynomial $p(x) = 2 - 4x + x^2 + 3x^3$. Let $q = \mathsf{T}(p)$; then $q(x) = p'(x) = -4 + 2x + 9x^2$. Hence

$$[\mathsf{T}(p)]_\gamma = [q]_\gamma = \begin{pmatrix} -4 \\ 2 \\ 9 \end{pmatrix}.$$

But also

$$[\mathsf{T}]_\beta^\gamma [p]_\beta = A[p]_\beta = \begin{pmatrix} 0 & 1 & 0 & 0 \\ 0 & 0 & 2 & 0 \\ 0 & 0 & 0 & 3 \end{pmatrix} \begin{pmatrix} 2 \\ -4 \\ 1 \\ 3 \end{pmatrix} = \begin{pmatrix} -4 \\ 2 \\ 9 \end{pmatrix}. \quad \blacksquare$$

We complete this section with the introduction of the "left-multiplication transformation" L_A, where A is an $m \times n$ matrix. This transformation is probably the most important tool for transferring properties about transformations to analogous properties about matrices and vice versa. For example, we use it to prove that matrix multiplication is associative.

Definition. *Let* A *be an* m × n *matrix with entries from a field* F. *We denote by* L_A *the mapping* $\mathsf{L}_A\colon \mathsf{F}^n \to \mathsf{F}^m$ *defined by* $\mathsf{L}_A(x) = Ax$ *(the matrix product of* A *and* x) *for each column vector* $x \in \mathsf{F}^n$. *We call* L_A *a* left-multiplication transformation.

Example 4

Let

$$A = \begin{pmatrix} 1 & 2 & 1 \\ 0 & 1 & 2 \end{pmatrix}.$$

If

$$x = \begin{pmatrix} 1 \\ 3 \\ -1 \end{pmatrix},$$

then

$$\mathsf{L}_A(x) = Ax = \begin{pmatrix} 1 & 2 & 1 \\ 0 & 1 & 2 \end{pmatrix} \begin{pmatrix} 1 \\ 3 \\ -1 \end{pmatrix} = \begin{pmatrix} 6 \\ 1 \end{pmatrix}. \quad \blacksquare$$

We see in the next theorem that not only is L_A linear, but, in fact, it has a great many other useful properties. These properties are all quite natural and so are easy to remember.

Theorem 2.16. *Let* A *be an* m × n *matrix with entries from* F. *Then the left-multiplication transformation* $\mathsf{L}_\mathrm{A}: \mathsf{F}^n \to \mathsf{F}^m$ *is linear. Furthermore, if* B *is any other* m × n *matrix (with entries from* F) *and* β *and* γ *are the standard ordered bases for* F^n *and* F^m, *respectively, we have the following properties,*

(a) $[\mathsf{L}_\mathrm{A}]_\beta^\gamma = \mathrm{A}$.

(b) $\mathsf{L}_\mathrm{A} = \mathsf{L}_\mathrm{B}$ *if and only if* A = B.

(c) $\mathsf{L}_{\mathrm{A}+\mathrm{B}} = \mathsf{L}_\mathrm{A} + \mathsf{L}_\mathrm{B}$ *and* $\mathsf{L}_{a\mathrm{A}} = a\mathsf{L}_\mathrm{A}$ *for all* $a \in \mathrm{F}$.

(d) *If* $\mathsf{T}: \mathsf{F}^n \to \mathsf{F}^m$ *is linear, then there exists a unique* m × n *matrix* C *such that* $\mathsf{T} = \mathsf{L}_\mathrm{C}$. *In fact,* $\mathrm{C} = [\mathsf{T}]_\beta^\gamma$.

(e) *If* E *is an* n × p *matrix, then* $\mathsf{L}_{\mathrm{AE}} = \mathsf{L}_\mathrm{A}\mathsf{L}_\mathrm{E}$.

(f) *If* m = n, *then* $\mathsf{L}_{\mathsf{I}_n} = \mathsf{I}_{\mathsf{F}^n}$.

Proof. The fact that L_A is linear follows immediately from Theorem 2.13 and its corollary.

(a) The jth column of $[\mathsf{L}_A]_\beta^\gamma$ is equal to $\mathsf{L}_A(e_j)$. But $\mathsf{L}_A(e_j) = Ae_j = A^{(j)}$, so $[\mathsf{L}_A]_\beta^\gamma = A$.

(b) If $\mathsf{L}_A = \mathsf{L}_B$, then we may use (a) to write $A = [\mathsf{L}_A]_\beta^\gamma = [\mathsf{L}_B]_\beta^\gamma = B$. Hence $A = B$. The proof of the converse is trivial.

The proof of (c) is left to the reader.

(d) Let $C = [\mathsf{T}]_\beta^\gamma$. By Theorem 2.15 we have $[\mathsf{T}(x)]_\gamma = [\mathsf{T}]_\beta^\gamma [x]_\beta$, or $\mathsf{T}(x) = Cx = \mathsf{L}_C(x)$ for all x. So $\mathsf{T} = \mathsf{L}_C$. The uniqueness of C follows from (b).

(e) For any j we have $\mathsf{L}_{AE}(e_j) = (AE)e_j = (AE)^{(j)} = AE^{(j)} = A(Ee_j) = \mathsf{L}_A(Ee_j) = \mathsf{L}_A(\mathsf{L}_E(e_j)) = (\mathsf{L}_A\mathsf{L}_E)(e_j)$. Hence $\mathsf{L}_{AE} = \mathsf{L}_A\mathsf{L}_E$ by the corollary to Theorem 2.6.

The proof of (f) is left to the reader. ∎

We now use left-multiplication transformations to establish an important property about matrices.

Theorem 2.17. *Let* A, B, *and* C *be matrices such that* A(BC) *is defined. Then* (AB)C *is defined and* A(BC) = (AB)C; *that is, matrix multiplication is associative.*

Proof. It is left to the reader to show that $(AB)C$ is defined. Using (e) of Theorem 2.16 and the associativity of functional composition, we have

$$\mathsf{L}_{A(BC)} = \mathsf{L}_A \mathsf{L}_{BC} = \mathsf{L}_A(\mathsf{L}_B \mathsf{L}_C) = (\mathsf{L}_A \mathsf{L}_B)\mathsf{L}_C = \mathsf{L}_{AB}\mathsf{L}_C = \mathsf{L}_{(AB)C}.$$

So from (b) of Theorem 2.16 we have $A(BC) = (AB)C$. ∎

Needless to say, this theorem could be proved directly from the definition of matrix multiplication. The proof above, however, provides a prototype of many other arguments that utilize the relationships between linear transformations and matrices.

An Application

A large and varied collection of interesting applications arises in connection with special matrices called "incidence matrices." An *incidence matrix* is a square matrix in which all the entries are either zero or one and, for convenience, all the diagonal entries are zero. If we have a relationship on a group of n objects that we denote by $1, 2, \ldots, n$, then we define the associated incidence matrix A by $A_{ij} = 1$ if i is related to j, and $A_{ij} = 0$ otherwise.

To make things concrete, suppose that we have four people each of whom owns a communication device. If the relationship on this group is "can transmit to," then $A_{ij} = 1$ if i can send (a message) to j and $A_{ij} = 0$ otherwise. Suppose that

$$A = \begin{pmatrix} 0 & 1 & 0 & 0 \\ 1 & 0 & 0 & 1 \\ 0 & 1 & 0 & 1 \\ 1 & 1 & 0 & 0 \end{pmatrix}.$$

Then since $A_{34} = 1$ and $A_{14} = 0$, we see that person 3 can send to 4 but 1 cannot send to 4.

We can obtain an interesting interpretation of the entries of A^2. Consider, for instance,

$$(A^2)_{31} = A_{31}A_{11} + A_{32}A_{21} + A_{33}A_{31} + A_{34}A_{41}.$$

Note that any term $A_{3k}A_{k1}$ equals 1 if and only if both A_{3k} and A_{k1} equal 1, that is, if and only if 3 can send to k and k can send to 1. Thus $(A^2)_{31}$ gives the number of ways in which 3 can send to 1 in two *stages* (or in one *relay*). Since

$$A^2 = \begin{pmatrix} 1 & 0 & 0 & 1 \\ 1 & 2 & 0 & 0 \\ 2 & 1 & 0 & 1 \\ 1 & 1 & 0 & 1 \end{pmatrix},$$

we see that 3 can send to 1 in two ways in two stages. In general, $(A + A^2 + \cdots + A^n)_{ij}$ is the number of ways in which i can send to j in at most n stages.

A maximal collection of three or more people with the property that any two can send to each other is called a *clique*. The problem of determining cliques seems at first to be quite difficult. However, if we define a new matrix B by $B_{ij} = 1$ if i and j can send to each other, and $B_{ij} = 0$ otherwise, then it can be shown (see Exercise 16) that person i belongs to a clique if and only if $(B^3)_{ii} > 0$. For example, suppose that the incidence matrix associated with some relationship is

$$A = \begin{pmatrix} 0 & 1 & 0 & 1 \\ 1 & 0 & 1 & 0 \\ 1 & 1 & 0 & 1 \\ 1 & 1 & 1 & 0 \end{pmatrix}.$$

To determine which people belong to cliques, we form the matrix B as above and compute B^3. In this case

$$B = \begin{pmatrix} 0 & 1 & 0 & 1 \\ 1 & 0 & 1 & 0 \\ 0 & 1 & 0 & 1 \\ 1 & 0 & 1 & 0 \end{pmatrix} \quad \text{and} \quad B^3 = \begin{pmatrix} 0 & 4 & 0 & 4 \\ 4 & 0 & 4 & 0 \\ 0 & 4 & 0 & 4 \\ 4 & 0 & 4 & 0 \end{pmatrix}.$$

Since all the diagonal entries of B^3 are zero, we conclude that there are no cliques in this relationship.

Our final example of the use of incidence matrices is concerned with the concept of dominance. A relation among a group of people is called a *dominance relation* if the associated incidence matrix A has the property that $A_{ij} = 1$ if and only if $A_{ji} = 0$ for all i and j, that is, given any two people, exactly one of them dominates (or, using the terminology of our first example, can send a message to) the other. For such a relation, it can be shown (see Exercise 18) that the matrix $A + A^2$ has a row [column] containing positive entries in every position except on the diagonal. In other words, there is at least one person who dominates [is dominated by] all the others in one or two stages. In fact, it can be shown that any person who dominates [is dominated by] the greatest number of people in the first stage has this property. Consider, for example, the matrix

$$A = \begin{pmatrix} 0 & 1 & 0 & 1 & 0 \\ 0 & 0 & 1 & 0 & 0 \\ 1 & 0 & 0 & 1 & 0 \\ 0 & 1 & 0 & 0 & 1 \\ 1 & 1 & 1 & 0 & 0 \end{pmatrix}.$$

The reader should verify that this matrix corresponds to a dominance relation. Now

$$A + A^2 = \begin{pmatrix} 0 & 2 & 1 & 1 & 1 \\ 1 & 0 & 1 & 1 & 0 \\ 1 & 2 & 0 & 2 & 1 \\ 1 & 2 & 2 & 0 & 1 \\ 2 & 2 & 2 & 2 & 0 \end{pmatrix}.$$

Thus persons 1, 3, 4, and 5 dominate (can send messages to) all the others in at most two stages, while persons 1, 2, 3, and 4 are dominated by (can receive messages from) all the others in at most two stages.

EXERCISES

√ **1.** Label the following statements as being true or false. For what occurs below, V, W, and Z denote vector spaces with ordered (finite) bases α, β, and γ, respectively; $\mathsf{T}: \mathsf{V} \to \mathsf{W}$ and $\mathsf{U}: \mathsf{W} \to \mathsf{Z}$ are linear; and A and B denote matrices.

(a) $[\mathsf{UT}]_\alpha^\gamma = [\mathsf{T}]_\alpha^\beta [\mathsf{U}]_\beta^\gamma$.
(b) $[\mathsf{T}(x)]_\beta = [\mathsf{T}]_\alpha^\beta [x]_\alpha$ for all $x \in \mathsf{V}$.
(c) $[\mathsf{U}(y)]_\beta = [\mathsf{U}]_\alpha^\beta [y]_\beta$ for all $y \in \mathsf{W}$.
(d) $[\mathsf{I}_\mathsf{V}]_\alpha = I$.
(e) $[\mathsf{T}^2]_\alpha^\beta = ([\mathsf{T}]_\alpha^\beta)^2$.
(f) $A^2 = I$ implies that $A = I$ or $A = -I$.
(g) $\mathsf{T} = \mathsf{L}_A$ for some matrix A.
(h) $A^2 = O$ implies that $A = O$, where O denotes the zero matrix.
(i) $\mathsf{L}_{A+B} = \mathsf{L}_A + \mathsf{L}_B$.
(j) If A is square and $A_{ij} = \delta_{ij}$ for all i and j, then $A = I$.

√ **2. (a)** Let

$$A = \begin{pmatrix} 1 & 3 \\ 2 & -1 \end{pmatrix}, \qquad B = \begin{pmatrix} 1 & 0 & -3 \\ 4 & 1 & 2 \end{pmatrix},$$

$$C = \begin{pmatrix} 1 & 1 & 4 \\ -1 & -2 & 0 \end{pmatrix}, \quad \text{and} \quad D = \begin{pmatrix} 2 \\ -2 \\ 3 \end{pmatrix}.$$

Compute $A(2B + 3C)$, $(AB)D$, and $A(BD)$.

(b) Let

$$A = \begin{pmatrix} 2 & 5 \\ -3 & 1 \\ 4 & 2 \end{pmatrix}, \qquad B = \begin{pmatrix} 3 & -2 & 0 \\ 1 & -1 & 4 \\ 5 & 5 & 3 \end{pmatrix}, \quad \text{and} \quad C = (4 \quad 0 \quad 3).$$

Compute A^t, A^tB, BC^t, CB, and CA.

3. Let $g(x) = 3 + x$. Define

$$\mathsf{T}\colon \mathsf{P}_2(R) \to \mathsf{P}_2(R) \quad \text{by} \quad \mathsf{T}(f) = f'g + 2f.$$

Define

$$\mathsf{U}\colon \mathsf{P}_2(R) \to \mathsf{R}^3 \quad \text{by} \quad \mathsf{U}(a + bx + cx^2) = (a + b, c, a - b).$$

Let $\beta = \{1, x, x^2\}$ and $\gamma = \{e_1, e_2, e_3\}$.

(a) Compute $[\mathsf{U}]_\beta^\gamma$, $[\mathsf{T}]_\beta$, and $[\mathsf{UT}]_\beta^\gamma$ directly. Then use Theorem 2.11 to verify your result.

(b) Let $h(x) = 3 - 2x + x^2$. Compute $[h]_\beta$ and $[\mathsf{U}(h)]_\gamma$. Then use $[\mathsf{U}]_\beta^\gamma$ from part (a) and Theorem 2.15 to verify your result.

4. For each of the following parts, let T be the linear transformation defined in the corresponding part of Exercise 5 of Section 2.2. Use Theorem 2.15 to compute the following:

(a) $[\mathsf{T}(A)]_\alpha$, where $A = \begin{pmatrix} 1 & 4 \\ -1 & 6 \end{pmatrix}$.

(b) $[\mathsf{T}(f)]_\alpha$, where $f(x) = 4 - 6x + 3x^2$.

(c) $[\mathsf{T}(A)]_\gamma$, where $A = \begin{pmatrix} 1 & 3 \\ 2 & 4 \end{pmatrix}$.

(d) $[\mathsf{T}(f)]_\gamma$, where $f(x) = 6 - x + 2x^2$.

5. Complete the proof of Theorem 2.13 and its corollary.

6. Prove (b) of Theorem 2.14.

7. Prove Theorem 2.10. Also, state and prove a more general result.

8. Find linear transformations $\mathsf{U}, \mathsf{T}\colon \mathsf{F}^2 \to \mathsf{F}^2$ such that $\mathsf{UT} = \mathsf{T}_0$ (the zero transformation) but $\mathsf{TU} \neq \mathsf{T}_0$. Use your answer to find matrices A and B such that $AB = O$ but $BA \neq O$.

9. Let A be an $n \times n$ matrix. Prove that A is a diagonal matrix if and only if $A_{ij} = \delta_{ij}A_{ii}$ for all i and j.

10. Let V be a vector space, and let $\mathsf{T}\colon \mathsf{V} \to \mathsf{V}$ be linear. Prove that $\mathsf{T}^2 = \mathsf{T}_0$ if and only if $\mathsf{R}(\mathsf{T}) \subseteq \mathsf{N}(\mathsf{T})$.

11. Let V, W, and Z be vector spaces, and let $\mathsf{T}\colon \mathsf{V} \to \mathsf{W}$ and $\mathsf{U}\colon \mathsf{W} \to \mathsf{Z}$ be linear.

(a) If UT is one-to-one, prove that T is one-to-one. Must U also be one-to-one?

(b) If UT is onto, prove that U is onto. Must T also be onto?

(c) If U and T are one-to-one and onto, prove that UT is also.

12. Let A and B be $n \times n$ matrices. Recall that the trace of A, written $\operatorname{tr}(A)$, equals

$$\sum_{i=1}^{n} A_{ii}.$$

Prove that $\operatorname{tr}(AB) = \operatorname{tr}(BA)$ and $\operatorname{tr}(A) = \operatorname{tr}(A^t)$.

13. Let V be a finite-dimensional vector space, and let T: V → V be linear.

(a) If rank(T) = rank(T^2), prove that R(T) ∩ N(T) = {0}. Deduce that V = R(T) ⊕ N(T) (see the exercises of Section 1.3).

(b) Prove that there exists a positive integer k such that V = R(T^k) ⊕ N(T^k).

14. Let V be a vector space. Determine all linear transformations T: V → V such that T = T^2. *Hint:* Note that x = T(x) + (x − T(x)) for every x in V, and show that V = {y: T(y) = y} ⊕ N(T) (see the exercises of Section 1.3).

15. Using only the definition of matrix multiplication, prove that multiplication of matrices is associative.

16. For an incidence matrix A with the related matrix B defined by $B_{ij} = 1$ if i is related to j and j is related to i, and $B_{ij} = 0$ otherwise, prove that i belongs to a clique if and only if $(B^3)_{ii} > 0$.

17. Use Exercise 16 to determine the cliques in the relations corresponding to the following incidence matrices.

(a) $\begin{pmatrix} 0 & 1 & 0 & 1 \\ 1 & 0 & 0 & 0 \\ 0 & 1 & 0 & 1 \\ 1 & 0 & 1 & 0 \end{pmatrix}$ **(b)** $\begin{pmatrix} 0 & 0 & 1 & 1 \\ 1 & 0 & 0 & 1 \\ 1 & 0 & 0 & 1 \\ 1 & 0 & 1 & 0 \end{pmatrix}$

18. Let A be an incidence matrix that is associated with a dominance relation. Prove that the matrix $A + A^2$ has a row [column] that contains positive entries in all positions except on the diagonal.

19. Prove that the matrix A given below corresponds to a dominance relation, and use Exercise 18 to determine which person(s) dominate(s) [is dominated by] each of the others within two stages.

$$A = \begin{pmatrix} 0 & 1 & 0 \\ 0 & 0 & 1 \\ 1 & 0 & 0 \end{pmatrix}.$$

20. Let A be an $n \times n$ incidence matrix that corresponds to a dominance relation. Determine the number of nonzero entries of A.

2.4 INVERTIBILITY AND ISOMORPHISMS

The concept of invertibility is introduced quite early in the study of functions. Fortunately, many of the intrinsic properties of functions are shared by their inverses. For example, in calculus we learned that the properties of being continuous or differentiable are generally retained by the inverse functions. We will see in this section (Theorem 2.18) that the inverse of a linear transformation is also linear. This result will greatly aid us in the study of "inverses" of matrices. As one might expect from Section 2.3, the inverse of the left-multiplication transformation L_A (when it exists) can be used to determine properties of the inverse of the matrix A.

In the remainder of this section we apply many of the results about invertibility to the concept of "isomorphism." We will see that finite-dimensional vector spaces (over F) of equal dimension may be identified. These ideas will be made more precise shortly.

The facts about inverse functions presented in Appendix B are, of course, true for linear transformations. Nevertheless, we repeat some of these definitions for use in this section.

Definition. *Let* V *and* W *be vector spaces, and let* $\mathsf{T}: \mathsf{V} \to \mathsf{W}$ *be linear.* T *has an* inverse $\mathsf{U}: \mathsf{W} \to \mathsf{V}$ *if* $\mathsf{TU} = \mathsf{I}_\mathsf{W}$ *and* $\mathsf{UT} = \mathsf{I}_\mathsf{V}$. *As noted in Appendix B, inverses are unique, and we write* $\mathsf{U} = \mathsf{T}^{-1}$. *We say* T *is* invertible *if* T *has an inverse.*

The following facts hold for invertible functions T and U.

1. $(\mathsf{TU})^{-1} = \mathsf{U}^{-1}\mathsf{T}^{-1}$.
2. $(\mathsf{T}^{-1})^{-1} = \mathsf{T}$; in particular, T^{-1} is invertible.

We also use the fact that a function is invertible if and only if it is one-to-one and onto.

Example 1

Define $\mathsf{T}: \mathsf{P}_1(R) \to R^2$ by $\mathsf{T}(a + bx) = (a, a + b)$. The reader can verify directly that $\mathsf{T}^{-1}: \mathsf{R}^2 \to \mathsf{P}_1(R)$ is defined by $\mathsf{T}^{-1}(c, d) = c + (d - c)x$. Observe that T^{-1} is also linear. As Theorem 2.18 demonstrates, this is true in general. ∎

Theorem 2.18. *Let* V *and* W *be vector spaces, and let* $\mathsf{T}: \mathsf{V} \to \mathsf{W}$ *be linear and invertible. Then* $\mathsf{T}^{-1}: \mathsf{W} \to \mathsf{V}$ *is linear.*

Proof. Let $y_1, y_2 \in \mathsf{W}$ and $c \in F$. Since T is onto and one-to-one, there exist unique vectors x_1 and x_2 such that $\mathsf{T}(x_1) = y_1$ and $\mathsf{T}(x_2) = y_2$. Thus $x_1 = \mathsf{T}^{-1}(y_1)$ and $x_2 = \mathsf{T}^{-1}(y_2)$, so

$$\mathsf{T}^{-1}(cy_1 + y_2) = \mathsf{T}^{-1}[c\mathsf{T}(x_1) + \mathsf{T}(x_2)] = \mathsf{T}^{-1}[\mathsf{T}(cx_1 + x_2)]$$
$$= cx_1 + x_2 = c\mathsf{T}^{-1}(y_1) + \mathsf{T}^{-1}(y_2). \quad ∎$$

It now follows immediately from Theorem 2.5 that if T is a linear transformation between vector spaces of equal (finite) dimension, then the conditions of being invertible, one-to-one, and onto are all equivalent.

We are now ready to define the inverse of a matrix. The reader should note the analogy with the inverse of a linear transformation.

Definition. *Let A be an* n × n *matrix. Then A is* invertible *if there exists an* n × n *matrix B such that* AB = BA = I.

The matrix B is unique and is called the *inverse* of A and written $B = A^{-1}$. (If C were another such matrix, then $C = CI = C(AB) = (CA)B = IB = B$.)

Example 2

The reader should verify that the inverse of

$$\begin{pmatrix} 5 & 7 \\ 2 & 3 \end{pmatrix} \quad \text{is} \quad \begin{pmatrix} 3 & -7 \\ -2 & 5 \end{pmatrix}. \quad \blacksquare$$

In Section 3.2 we will learn a technique for actually computing the inverse of a matrix. At this point we would like to develop a number of results that relate the inverses of matrices with the inverses of linear transformations.

Lemma. *Let* V *and* W *be finite-dimensional vector spaces and let* $\mathsf{T}: \mathsf{V} \to \mathsf{W}$ *be linear. If* T *is invertible, then* $\dim(\mathsf{V}) = \dim(\mathsf{W})$.

Proof. Because T is one-to-one and onto, we have

$$\text{nullity}(\mathsf{T}) = 0 \quad \text{and} \quad \text{rank}(\mathsf{T}) = \dim(\mathsf{R}(\mathsf{T})) = \dim(\mathsf{W}).$$

So by the dimension theorem it follows that $\dim(\mathsf{V}) = \dim(\mathsf{W})$. ∎

Theorem 2.19. *Let* V *and* W *be finite-dimensional vector spaces with ordered bases* β *and* γ*, respectively. Let* $\mathsf{T}: \mathsf{V} \to \mathsf{W}$ *be linear. Then* T *is invertible if and only if* $[\mathsf{T}]_\beta^\gamma$ *is invertible. Furthermore,* $[\mathsf{T}^{-1}]_\gamma^\beta = ([\mathsf{T}]_\beta^\gamma)^{-1}$.

Proof. Suppose that T is invertible. By the lemma we have $\dim(\mathsf{V}) = \dim(\mathsf{W})$. Let $n = \dim(\mathsf{V})$. Then $[\mathsf{T}]_\beta^\gamma$ is an $n \times n$ matrix. Now $\mathsf{T}^{-1}: \mathsf{W} \to \mathsf{V}$ satisfies $\mathsf{T}\mathsf{T}^{-1} = \mathsf{I}_\mathsf{W}$ and $\mathsf{T}^{-1}\mathsf{T} = \mathsf{I}_\mathsf{V}$. Thus

$$I_n = [\mathsf{I}_\mathsf{V}]_\beta = [\mathsf{T}^{-1}\mathsf{T}]_\beta = [\mathsf{T}^{-1}]_\gamma^\beta[\mathsf{T}]_\beta^\gamma.$$

Similarly, $[\mathsf{T}]_\beta^\gamma[\mathsf{T}^{-1}]_\gamma^\beta = I_n$, and hence $([\mathsf{T}]_\beta^\gamma)^{-1} = [\mathsf{T}^{-1}]_\gamma^\beta$.

Now let $A = [\mathsf{T}]_\beta^\gamma$ be invertible. There exists an $n \times n$ matrix B such that $AB = BA = I_n$. By Theorem 2.6 there exists $\mathsf{U} \in \mathscr{L}(\mathsf{W}, \mathsf{V})$ such that

$$\mathsf{U}(x_j) = \sum_{i=1}^{n} B_{ij} y_i \qquad \text{for } j = 1, 2, \ldots, n,$$

where $\gamma = \{x_1, \ldots, x_n\}$ and $\beta = \{y_1, \ldots, y_n\}$. Then $[\mathsf{U}]_\gamma^\beta = B$. To show that $\mathsf{U} = \mathsf{T}^{-1}$, observe that

$$[\mathsf{U}\mathsf{T}]_\beta = [\mathsf{U}]_\gamma^\beta[\mathsf{T}]_\beta^\gamma = BA = I_n = [\mathsf{I}_\mathsf{V}]_\beta$$

by Theorem 2.11. So $\mathsf{U}\mathsf{T} = \mathsf{I}_\mathsf{V}$, and similarly, $\mathsf{T}\mathsf{U} = \mathsf{I}_\mathsf{W}$. ∎

Example 3

For the vector spaces $P_1(R)$ and R^2, choose the bases $\beta = \{1, x\}$ and $\gamma = \{e_1, e_2\}$, respectively. In the notation of Example 1, we have that

$$[T]_\beta^\gamma = \begin{pmatrix} 1 & 0 \\ 1 & 1 \end{pmatrix} \quad \text{and} \quad [T^{-1}]_\gamma^\beta = \begin{pmatrix} 1 & 0 \\ -1 & 1 \end{pmatrix}.$$

It can be verified by matrix multiplication that each matrix is the inverse of the other. ∎

Corollary 1. *Let V be a finite-dimensional vector space with an ordered basis β, and let $T: V \to V$ be linear. Then T is invertible if and only if $[T]_\beta$ is invertible. Furthermore, $[T^{-1}]_\beta = [T]_\beta^{-1}$.*

Proof. Exercise. ∎

Corollary 2. *Let A be an $n \times n$ matrix. Then A is invertible if and only if L_A is invertible. Furthermore, $(L_A)^{-1} = L_{A^{-1}}$.*

Proof. Exercise. ∎

The notion of invertibility may be used to formalize what may already have been observed by the reader, that is, that certain vector spaces strongly resemble one another except for the form of their elements. For example, in the case of $M_{2\times 2}(F)$ and F^4, if we associate to each matrix

$$\begin{pmatrix} a & b \\ c & d \end{pmatrix}$$

the 4-tuple (a, b, c, d), we see that sums and scalar products associate in a similar manner; that is, in terms of the vector space structure, these two vector spaces may be considered identical or "isomorphic."

Definition. *Let V and W be vector spaces. We say that V is* isomorphic *to W if there exists a linear transformation $T: V \to W$ that is invertible. Such a linear transformation is called an* isomorphism *from V onto W.*

We leave the proof of the fact that "is isomorphic to" is an equivalence relation as an exercise.

Example 4

Define $T: F^2 \to P_1(F)$ by $T(a_1, a_2) = a_1 + a_2 x$. Clearly T is invertible; so F^2 is isomorphic to $P_1(F)$. ∎

Example 5

Define

$$\mathsf{T}\colon \mathsf{P}_3(R) \to \mathsf{M}_{2\times 2}(R) \qquad \text{by } \mathsf{T}(f) = \begin{pmatrix} f(1) & f(2) \\ f(3) & f(4) \end{pmatrix}.$$

It is easily verified that T is linear. By use of the Lagrange interpolation formula in Section 1.6, it can be shown (compare with Exercise 20) that $\mathsf{T}(f) = O$ only when f is the zero polynomial. Thus T is one-to-one and so T is invertible. We may now conclude that $\mathsf{P}_3(R)$ is isomorphic to $\mathsf{M}_{2\times 2}(R)$. ∎

In each of Examples 4 and 5 the reader may have observed that the isomorphic vector spaces have equal dimensions. As the next theorem shows, this is no coincidence.

Theorem 2.20. *Let* V *and* W *be finite-dimensional vector spaces (over the same field* F*). Then* V *is isomorphic to* W *if and only if* $\dim(\mathsf{V}) = \dim(\mathsf{W})$.

Proof. Suppose that V is isomorphic to W and that $\mathsf{T}\colon \mathsf{V} \to \mathsf{W}$ is an isomorphism from V onto W. By the lemma preceding Theorem 2.19, we have that $\dim(\mathsf{V}) = \dim(\mathsf{W})$.

Now suppose that $\dim(\mathsf{V}) = \dim(\mathsf{W})$, and let $\beta = \{x_1, \ldots, x_n\}$ and $\gamma = \{y_1, \ldots, y_n\}$ be bases for V and W, respectively. By Theorem 2.6 there exists $\mathsf{T}\colon \mathsf{V} \to \mathsf{W}$ such that T is linear and $\mathsf{T}(x_i) = y_i$ for $i = 1, \ldots, n$. Using Theorem 2.2, we have

$$\mathsf{R}(\mathsf{T}) = \operatorname{span}(\mathsf{T}(\beta)) = \operatorname{span}(\gamma) = \mathsf{W}.$$

So T is onto. From Theorem 2.5 we have that T is also one-to-one. ∎

Corollary. *If* V *is a vector space over* F *of dimension* n, *then* V *is isomorphic to* F^n.

Until now we have associated transformations with their matrix representations. We are now in a position to prove that as a vector space the collection of all transformations between two given vector spaces may be identified with the appropriate vector space of $m \times n$ matrices.

Theorem 2.21. *Let* V *and* W *be finite-dimensional vector spaces over* F *of dimensions* n *and* m, *respectively, and let* β *and* γ *be ordered bases for* V *and* W, *respectively. Then the function* $\Phi\colon \mathscr{L}(\mathsf{V}, \mathsf{W}) \to \mathsf{M}_{m\times n}(\mathsf{F})$, *defined by* $\Phi(\mathsf{T}) = [\mathsf{T}]_\beta^\gamma$ *for* $\mathsf{T} \in \mathscr{L}(\mathsf{V}, \mathsf{W})$, *is an isomorphism.*

Proof. Theorem 2.8 allows us to conclude that Φ is linear. Hence we must show that Φ is one-to-one and onto. This will be accomplished if we can show that for every $m \times n$ matrix A, there exists a unique linear transformation

$T\colon V \to W$ such that $\Phi(T) = A$. Let $\beta = \{x_1, \ldots, x_n\}$ and $\gamma = \{y_1, \ldots, y_m\}$. Let A be a given $m \times n$ matrix. By Theorem 2.6 there exists a unique linear transformation $T\colon V \to W$ such that

$$T(x_j) = \sum_{i=1}^{m} A_{ij} y_i \qquad \text{for } 1 \leqslant j \leqslant n.$$

But this means that $[T]_\beta^\gamma = A$, or $\Phi(T) = A$. Thus Φ is an isomorphism. ∎

Corollary. *Let* V *and* W *be finite-dimensional vector spaces of dimensions* n *and* m, *respectively. Then* $\mathscr{L}(V, W)$ *is finite-dimensional of dimension* mn.

Proof. The proof follows from Theorems 2.21 and 2.20 and the fact that $\dim(M_{m\times n}(F)) = mn$. ∎

We conclude this section with a result that allows us to see more clearly the relationship between linear transformations defined on abstract finite-dimensional vector spaces and linear transformations defined on F^n.

We begin by naming the transformation $x \to [x]_\beta$ introduced in Section 2.2.

Definition. *Let β be an ordered basis for an* n*-dimensional vector space* V *over the field* F. *The* standard representation of V with respect to β *is the function* $\phi_\beta\colon V \to F^n$ *defined by* $\phi_\beta(x) = [x]_\beta$ *for each* $x \in V$.

Example 6

Let $V = R^2$, $\beta = \{(1, 0), (0, 1)\}$, and $\gamma = \{(1, 2), (3, 4)\}$. For $x = (1, -2)$ we have

$$\phi_\beta(x) = [x]_\beta = \begin{pmatrix} 1 \\ -2 \end{pmatrix} \quad \text{and} \quad \phi_\gamma(x) = [x]_\gamma = \begin{pmatrix} -5 \\ 2 \end{pmatrix}. \quad ∎$$

We have observed earlier that ϕ_β is a linear transformation. The following theorem tells us much more.

Theorem 2.22. *For any finite-dimensional vector space* V *with ordered basis β, ϕ_β is an isomorphism.*

Proof. Exercise. ∎

This theorem provides us with an alternate proof that an n-dimensional vector space is isomorphic to F^n (see the corollary to Theorem 2.20).

We are now ready to use the standard representation of a vector space along with the matrix representation of a linear transformation to study the relationship between the linear transformation $T\colon V \to W$, where V and W are abstract finite-dimensional vector spaces, and $L_A\colon F^n \to F^m$, where $A = [T]_\beta^\gamma$ and β and γ are arbitrary ordered bases of V and W, respectively.

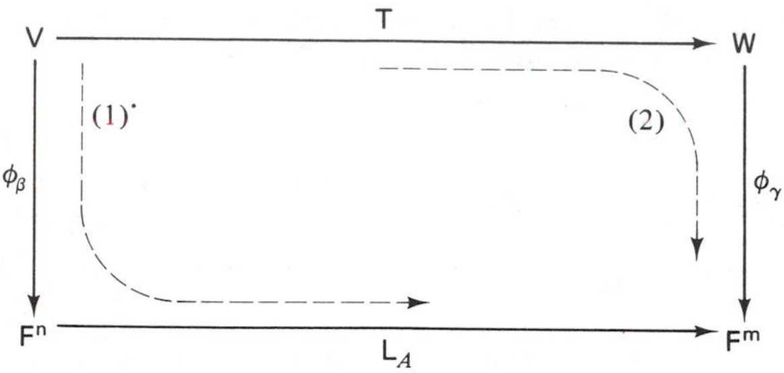

Figure 2.2

Let us first consider Figure 2.2. Notice that there are two compositions of linear transformations that map V into F^m:

1. Map V into F^n with ϕ_β and follow this transformation with L_A; this yields the composition $L_A\phi_\beta$.
2. Map V into W with T and follow it by ϕ_γ to obtain the composition $\phi_\gamma T$.

These two compositions are depicted by the dashed arrows in the diagram. By a simple reformulation of Theorem 2.15, we may conclude that

$$L_A\phi_\beta = \phi_\gamma T$$

Heuristically, this relationship indicates that after V and W are identified with F^n and F^m via ϕ_β and ϕ_γ, respectively, we may "identify" T with L_A.

Example 7

Recall the transformation $T: P_3(R) \to P_2(R)$ defined in Example 3 of Section 2.2. ($T(f) = f'$.) Let β and γ be the standard ordered bases for $P_3(R)$ and $P_2(R)$, respectively, and let $\phi_\beta: P_3(R) \to R^4$ and $\phi_\gamma: P_2(R) \to R^3$ be the corresponding standard representations of $P_3(R)$ and $P_2(R)$. Let $A = [T]_\beta^\gamma$; then

$$A = \begin{pmatrix} 0 & 1 & 0 & 0 \\ 0 & 0 & 2 & 0 \\ 0 & 0 & 0 & 3 \end{pmatrix}.$$

Consider the polynomial $p(x) = 2 + x - 3x^2 + 5x^3$. We will show that $L_A\phi_\beta(p) = \phi_\gamma T(p)$.

Now

$$L_A\phi_\beta(p) = \begin{pmatrix} 0 & 1 & 0 & 0 \\ 0 & 0 & 2 & 0 \\ 0 & 0 & 0 & 3 \end{pmatrix}\begin{pmatrix} 2 \\ 1 \\ -3 \\ 5 \end{pmatrix} = \begin{pmatrix} 1 \\ -6 \\ 15 \end{pmatrix}.$$

But since

$$\mathsf{T}(p) = p' = 1 - 6x + 15x^2,$$

we have

$$\phi_\gamma \mathsf{T}(p) = \begin{pmatrix} 1 \\ -6 \\ 15 \end{pmatrix}.$$

So $\mathsf{L}_A \phi_\beta(p) = \phi_\gamma \mathsf{T}(p)$.

Try repeating this example with different polynomials $p(x)$. ∎

EXERCISES

1. Label the following statements as being true or false. For the following, V and W are vector spaces with ordered (finite) bases α and β, respectively, and $\mathsf{T}: \mathsf{V} \to \mathsf{W}$ is linear. A and B are matrices.
   (a) $([\mathsf{T}]_\alpha^\beta)^{-1} = [\mathsf{T}^{-1}]_\alpha^\beta$.
   (b) T is invertible if and only if T is one-to-one and onto.
   (c) $\mathsf{T} = \mathsf{L}_A$, where $A = [\mathsf{T}]_\alpha^\beta$.
   (d) $\mathsf{M}_{2\times 3}(F)$ is isomorphic to F^5.
   (e) $\mathsf{P}_n(F)$ is isomorphic to $\mathsf{P}_m(F)$ if and only if $n = m$.
   (f) $AB = I$ implies that A and B are invertible.
   (g) $(A^{-1})^{-1} = A$.
   (h) A is invertible if and only if L_A is invertible.
   (i) A must be square in order to possess an inverse.
2.† Let A and B be $n \times n$ invertible matrices. Prove that AB is invertible and $(AB)^{-1} = B^{-1}A^{-1}$.
3.† Let A be invertible. Prove that A^t is invertible and $(A^t)^{-1} = (A^{-1})^t$.
4. Prove that if A is invertible and $AB = O$, then $B = O$.
5. If $A^2 = O$, prove that A cannot be invertible.
6. Prove Corollaries 1 and 2 of Theorem 2.19.
7. Let A and B be $n \times n$ matrices such that AB is invertible. Prove that A and B are invertible. Show that, in general, this result is false if at least one of the matrices is not square.
8.† Let A and B be $n \times n$ matrices such that $AB = I_n$. Prove $A = B^{-1}$ (and hence $B = A^{-1}$). (We are in effect saying that for square matrices, a one-sided inverse is a two-sided inverse.)
9. Prove that the transformation defined in Example 5 is one-to-one.
10. Prove Theorem 2.22.
11. Let $\sim$ mean "is isomorphic to." Prove that $\sim$ is an equivalence relation on the class of vector spaces over F as defined in Appendix A.

12. Let

$$\mathsf{V} = \left\{ \begin{pmatrix} a & a+b \\ 0 & c \end{pmatrix} : a, b, c \in F \right\}.$$

Construct an isomorphism from V onto F^3.

13. Let V and W be finite-dimensional vector spaces, and let $\mathsf{T}: \mathsf{V} \to \mathsf{W}$ be an isomorphism. If β is a basis for V, prove that $\mathsf{T}(\beta)$ is a basis for W.

14. Let B be an $n \times n$ invertible matrix. Define $\Phi: \mathsf{M}_{n \times n}(F) \to \mathsf{M}_{n \times n}(F)$ by $\Phi(A) = B^{-1}AB$. Prove that Φ is an isomorphism.

15.† Let V and W be finite-dimensional vector spaces and $\mathsf{T}: \mathsf{V} \to \mathsf{W}$ be an isomorphism. Let V_0 be a subspace of V.

- **(a)** Prove that $\mathsf{T}(\mathsf{V}_0)$ is a subspace of W.
- **(b)** Prove that $\dim(\mathsf{V}_0) = \dim(\mathsf{T}(\mathsf{V}_0))$.

16. Repeat Example 7 with the polynomial $p(x) = 1 + x + 2x^2 + x^3$.

17. Let $\mathsf{V} = \mathsf{M}_{2 \times 2}(R)$, the four-dimensional vector space of 2×2 matrices having real entries. Recall from Example 4 of Section 2.1 that the mapping $\mathsf{T}: \mathsf{V} \to \mathsf{V}$ defined by $\mathsf{T}(A) = A^t$ for each $A \in \mathsf{V}$ is a linear transformation.

- **(a)** Let $\beta = \{E^{11}, E^{12}, E^{21}, E^{22}\}$, where E^{ij} is the 2×2 matrix having the i, jth entry equal to one and all other entries zero. Prove that β is an ordered basis for V.
- **(b)** Let $A = [\mathsf{T}]_\beta$. Compute A.
- **(c)** Let ϕ denote the standard representation of V with respect to β. Verify $\mathsf{L}_A\phi = \phi\mathsf{T}$ for the matrix

$$M = \begin{pmatrix} 1 & 2 \\ 3 & 4 \end{pmatrix};$$

that is, prove that $\mathsf{L}_A\phi(M) = \phi\mathsf{T}(M)$.

18.† Let $\mathsf{T}: \mathsf{V} \to \mathsf{W}$ be a linear transformation from an n-dimensional vector space V to an m-dimensional vector space W. Let β and γ be ordered bases for V and W, respectively. Prove that $\text{rank}(\mathsf{T}) = \text{rank}(\mathsf{L}_A)$ and that $\text{nullity}(\mathsf{T}) = \text{nullity}(\mathsf{L}_A)$, where $A = [\mathsf{T}]_\beta^\gamma$. *Hint:* Apply Exercise 15 to Figure 2.2.

19. Let V and W be finite-dimensional vector spaces with ordered bases $\beta = \{x_1, \ldots, x_n\}$ and $\gamma = \{y_1, \ldots, y_m\}$, respectively. By Theorem 2.6 there exists a linear transformation $\mathsf{T}_{ij}: \mathsf{V} \to \mathsf{W}$ such that

$$\mathsf{T}_{ij}(x_k) = \begin{cases} y_i & \text{if } k = j \\ 0 & \text{if } k \neq j. \end{cases}$$

First prove that $\{\mathsf{T}_{ij}: 1 \leqslant i \leqslant m,\ 1 \leqslant j \leqslant n\}$ is a basis for $\mathcal{L}(\mathsf{V}, \mathsf{W})$. Then let E^{ij} be the $m \times n$ matrix with 1 in the ith row and jth column and 0 elsewhere, and prove that $[\mathsf{T}_{ij}]_\beta^\gamma = E^{ij}$. Again by Theorem 2.6 there exists a linear transformation $\Phi\colon \mathcal{L}(\mathsf{V},\mathsf{W}) \to \mathsf{M}_{m \times n}(F)$ such that $\Phi(\mathsf{T}_{ij}) = E^{ij}$. Prove that Φ is an isomorphism.

20. Let $c_0, c_1, \ldots, c_n$ be distinct elements of an infinite field F. Define $\mathsf{T}\colon \mathsf{P}_n(F) \to \mathsf{F}^{n+1}$ by $\mathsf{T}(f) = (f(c_0), \ldots, f(c_n))$. Prove that T is an isomorphism. *Hint:* Use the Lagrange polynomials associated with $c_0, \ldots, c_n$.

21. Let V denote the vector space defined in Example 5 of Section 1.2, and let $\mathsf{W} = \mathsf{P}(F)$. Define

$$\mathsf{T}\colon \mathsf{V} \to \mathsf{W} \quad \text{by} \quad \mathsf{T}(\sigma) = \sum_{i=0}^{n} \sigma(i)x^i,$$

where n is the largest integer such that $\sigma(n) \neq 0$. Prove that T is an isomorphism.

The following exercise requires familiarity with the definition of *quotient space* defined in Exercise 29 of Section 1.3 and Exercise 30 of Section 2.1.

22. Let $\mathsf{T}\colon \mathsf{V} \to \mathsf{Z}$ be a linear transformation of a vector space V onto a vector space Z. Define the mapping

$$\bar{\mathsf{T}}\colon \mathsf{V}/\mathsf{N}(\mathsf{T}) \to \mathsf{Z} \quad \text{by} \quad \bar{\mathsf{T}}(v + \mathsf{N}(\mathsf{T})) = \mathsf{T}(v)$$

for any coset $v + \mathsf{N}(\mathsf{T})$ in $\mathsf{V}/\mathsf{N}(\mathsf{T})$.

(a) Prove that $\bar{\mathsf{T}}$ is well-defined; that is, prove that if $v + \mathsf{N}(\mathsf{T}) = v' + \mathsf{N}(\mathsf{T})$, then $\mathsf{T}(v) = \mathsf{T}(v')$.

(b) Prove that $\bar{\mathsf{T}}$ is linear.

(c) Prove that $\bar{\mathsf{T}}$ is an isomorphism.

(d) Prove that the diagram shown in Figure 2.3 commutes; that is, prove that $\mathsf{T} = \bar{\mathsf{T}}\eta$.

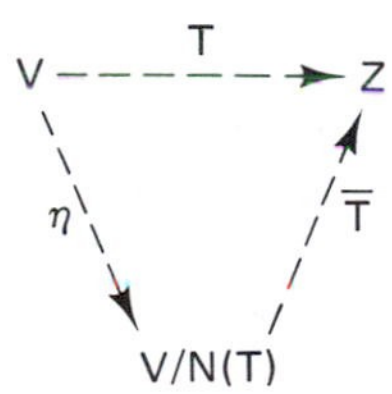

Figure 2.3

2.5 THE CHANGE OF COORDINATE MATRIX

In many areas of mathematics a change of variable is used to simplify the appearance of an expression. For example, in calculus an antiderivative of $2xe^{x^2}$ would be found by making the change of variable $u = x^2$. The resulting expression is of such a simple form that an antiderivative is easily recognized:

$$\int 2xe^{x^2}\,dx = \int e^u\,du = e^u = e^{x^2}.$$

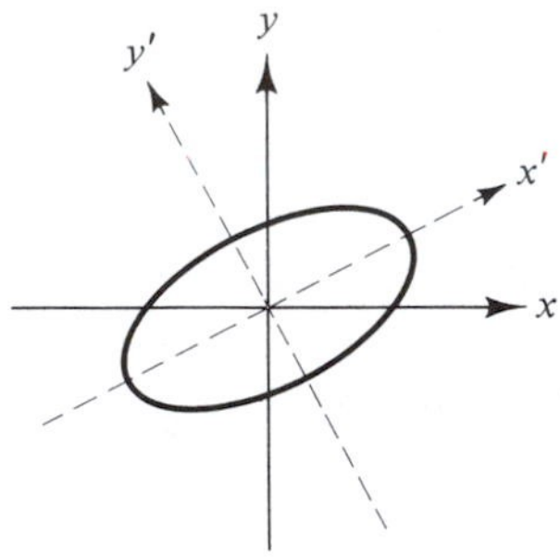

Figure 2.4

Similarly, in plane geometry the change of variable

$$\begin{cases} x = \dfrac{2}{\sqrt{5}}x' - \dfrac{1}{\sqrt{5}}y' \\ y = \dfrac{1}{\sqrt{5}}x' + \dfrac{2}{\sqrt{5}}y' \end{cases}$$

can be used to transform the equation $2x^2 - 4xy + 5y^2 = 1$ into the simpler equation $(x')^2 + 6(y')^2 = 1$, in which form it is easily seen to be the equation of an ellipse (see Figure 2.4). We will see how this change of variable is determined in Section 6.5. Geometrically, the change of variable

$$\begin{pmatrix} x \\ y \end{pmatrix} \to \begin{pmatrix} x' \\ y' \end{pmatrix}$$

is a change in the way that the position of a point P in the plane is described. This is done by introducing a new frame of reference, an $x'y'$-coordinate system with coordinate axes rotated from the original xy-coordinate axes. In this case the new coordinate axes are chosen to lie in the directions of the axes of the ellipse. The unit vectors along the x'-axis and the y'-axis form an ordered basis

$$\beta' = \left\{ \begin{pmatrix} \dfrac{2}{\sqrt{5}} \\ \dfrac{1}{\sqrt{5}} \end{pmatrix}, \begin{pmatrix} \dfrac{-1}{\sqrt{5}} \\ \dfrac{2}{\sqrt{5}} \end{pmatrix} \right\}$$

for R^2, and the change of variable is actually a change from the coordinate vector of P relative to the standard ordered basis $\beta = \{e_1, e_2\}$, $[\mathsf{P}]_\beta = \begin{pmatrix} x \\ y \end{pmatrix}$, to the coordinate vector relative to the new rotated basis β', $[\mathsf{P}]_{\beta'} = \begin{pmatrix} x' \\ y' \end{pmatrix}$.

A natural question arises: How can a coordinate vector relative to one basis be changed into a coordinate vector relative to the other? Notice that the system of equations relating the new and old coordinates can be represented by the matrix equation

$$\begin{pmatrix} x \\ y \end{pmatrix} = \begin{pmatrix} \frac{2}{\sqrt{5}} & \frac{-1}{\sqrt{5}} \\ \frac{1}{\sqrt{5}} & \frac{2}{\sqrt{5}} \end{pmatrix} \begin{pmatrix} x' \\ y' \end{pmatrix}.$$

Notice also that the matrix

$$Q = \begin{pmatrix} \frac{2}{\sqrt{5}} & \frac{-1}{\sqrt{5}} \\ \frac{1}{\sqrt{5}} & \frac{2}{\sqrt{5}} \end{pmatrix}$$

equals $[\mathsf{I}]_{\beta'}^{\beta}$, where I denotes the identity transformation on R^2. Thus $[v]_\beta = Q[v]_{\beta'}$ for all $v \in \mathsf{R}^2$. A similar result is true in general.

Theorem 2.23. *Let β and β' be two ordered bases for a finite-dimensional vector space V, and let $\mathsf{Q} = [\mathsf{I}_\mathsf{V}]_{\beta'}^{\beta}$. Then*

(a) Q *is invertible.*

(b) *For any $v \in \mathsf{V}$, $[v]_\beta = \mathsf{Q}[v]_{\beta'}$.*

Proof. (a) Since I_V is invertible, Q is invertible by Theorem 2.19.

(b) For any $v \in \mathsf{V}$,

$$[v]_\beta = [\mathsf{I}_\mathsf{V}(v)]_\beta = [\mathsf{I}_\mathsf{V}]_{\beta'}^{\beta}[v]_{\beta'} = Q[v]_{\beta'}$$

by Theorem 2.15. ∎

The matrix Q defined in Theorem 2.23 is called a *change of coordinate matrix*. Because of part (b) of the theorem we shall say that Q *changes β'-coordinates into β-coordinates*. Observe that if $\beta = \{x_1, x_2, \ldots, x_n\}$ and $\beta' = \{x_1', x_2', \ldots, x_n'\}$, then

$$x_j' = \sum_{i=1}^{n} Q_{ij} x_i$$

for $j = 1, 2, \ldots, n$; that is, the jth column of Q is $[x_j']_\beta$.

Notice that if Q changes β'-coordinates into β-coordinates, then Q^{-1} changes β-coordinates into β'-coordinates (see Exercise 10).

Example 1

Let $\mathsf{V} = \mathsf{R}^2$, $\beta = \{(1, 1), (1, -1)\}$, and $\beta' = \{(2, 4), (3, 1)\}$. Since $(2, 4) = 3(1, 1) - 1(1, -1)$ and $(3, 1) = 2(1, 1) + 1(1, -1)$, the matrix that changes β'-coordinates into β-coordinates is

$$Q = \begin{pmatrix} 3 & 2 \\ -1 & 1 \end{pmatrix}.$$

Thus, for instance,

$$[(2, 4)]_\beta = Q[(2, 4)]_{\beta'} = Q\begin{pmatrix}1\\0\end{pmatrix} = \begin{pmatrix}3\\-1\end{pmatrix}. \quad \blacksquare$$

Suppose now that $\mathsf{T}\colon \mathsf{V} \to \mathsf{V}$ is a linear transformation on a vector space V and that β and β' are ordered bases for V. Then T can be represented by the matrices $[\mathsf{T}]_\beta$ and $[\mathsf{T}]_{\beta'}$. What is the relationship between these matrices? The next theorem provides a simple answer using a change of coordinate matrix.

Theorem 2.24. *Let $\mathsf{T}\colon \mathsf{V} \to \mathsf{V}$ be a linear transformation on the finite-dimensional vector space V, and let β and β' be ordered bases for V. Let Q be the change of coordinate matrix which changes β'-coordinates into β-coordinates. Then*

$$[\mathsf{T}]_{\beta'} = Q^{-1}[\mathsf{T}]_\beta Q.$$

Proof. Let I be the identity transformation on V. Then $\mathsf{T} = \mathsf{IT} = \mathsf{TI}$ and hence, by Theorem 2.11,

$$\begin{aligned} Q[\mathsf{T}]_{\beta'} &= [\mathsf{I}]_{\beta'}^{\beta}[\mathsf{T}]_{\beta'}^{\beta'} \\ &= [\mathsf{IT}]_{\beta'}^{\beta} \\ &= [\mathsf{TI}]_{\beta'}^{\beta} \\ &= [\mathsf{T}]_{\beta}^{\beta}[\mathsf{I}]_{\beta'}^{\beta} \\ &= [\mathsf{T}]_\beta Q. \end{aligned}$$

Therefore, $[\mathsf{T}]_{\beta'} = Q^{-1}[\mathsf{T}]_\beta Q$. $\blacksquare$

Example 2

Let $\mathsf{V} = \mathsf{R}^3$, and let $\mathsf{T}\colon \mathsf{V} \to \mathsf{V}$ be defined by

$$\mathsf{T}\begin{pmatrix}a_1\\a_2\\a_3\end{pmatrix} = \begin{pmatrix}2a_1 + a_2\\a_1 + a_2 + 3a_3\\-a_2\end{pmatrix}.$$

Let β be the standard ordered basis for R^3, and let

$$\beta' = \left\{\begin{pmatrix}-1\\0\\0\end{pmatrix}, \begin{pmatrix}2\\1\\0\end{pmatrix}, \begin{pmatrix}1\\1\\1\end{pmatrix}\right\},$$

which is an ordered basis for R^3. To illustrate Theorem 2.24, we note that

$$[\mathsf{T}]_\beta = \begin{pmatrix}2 & 1 & 0\\1 & 1 & 3\\0 & -1 & 0\end{pmatrix}.$$

Let Q be the matrix that changes β'-coordinates into β-coordinates. Since β is the standard ordered basis for R^3, the columns of Q are simply the elements of β' written in the same order (see Exercise 11). Thus

$$Q = \begin{pmatrix} -1 & 2 & 1 \\ 0 & 1 & 1 \\ 0 & 0 & 1 \end{pmatrix}.$$

In Section 3.2 a method for computing Q^{-1} will be described. It can easily be verified that

$$Q^{-1} = \begin{pmatrix} -1 & 2 & -1 \\ 0 & 1 & -1 \\ 0 & 0 & 1 \end{pmatrix}.$$

By Theorem 2.24 $[\mathsf{T}]_{\beta'} = Q^{-1}[\mathsf{T}]_\beta Q$. A straightforward multiplication shows that

$$[\mathsf{T}]_{\beta'} = \begin{pmatrix} 0 & 2 & 8 \\ -1 & 4 & 6 \\ 0 & -1 & -1 \end{pmatrix}.$$

To show that this is the correct matrix, we can verify that the image under T of the jth element of β' is the linear combination of the elements of β' with the entries of the jth column as its coefficients. For example, for $j = 2$ we have

$$\mathsf{T}\begin{pmatrix} 2 \\ 1 \\ 0 \end{pmatrix} = 2\begin{pmatrix} -1 \\ 0 \\ 0 \end{pmatrix} + 4\begin{pmatrix} 2 \\ 1 \\ 0 \end{pmatrix} + (-1)\begin{pmatrix} 1 \\ 1 \\ 1 \end{pmatrix} = \begin{pmatrix} 5 \\ 3 \\ -1 \end{pmatrix}.$$

Notice that the coefficients are the entries of the second column of $[\mathsf{T}]_{\beta'}$. ∎

It is often useful to apply Theorem 2.24 in the reverse direction, as the next example shows.

Example 3

Recall the reflection about the x-axis in Example 6 of Section 2.1. The rule $(x, y) \to (x, -y)$ is easy to obtain. We now derive the less obvious rule for the reflection T about the line $y = 2x$ (see Figure 2.5). We wish to find an expression for $\mathsf{T}(a, b)$ for any (a, b) in R^2. Since T is linear, it is completely determined by its values on a basis for R^2. Clearly, $\mathsf{T}(1, 2) = (1, 2)$ and $\mathsf{T}(-2, 1) = -(-2, 1) = (2, -1)$. Therefore, if we let

$$\beta' = \left\{\begin{pmatrix} 1 \\ 2 \end{pmatrix}, \begin{pmatrix} -2 \\ 1 \end{pmatrix}\right\},$$

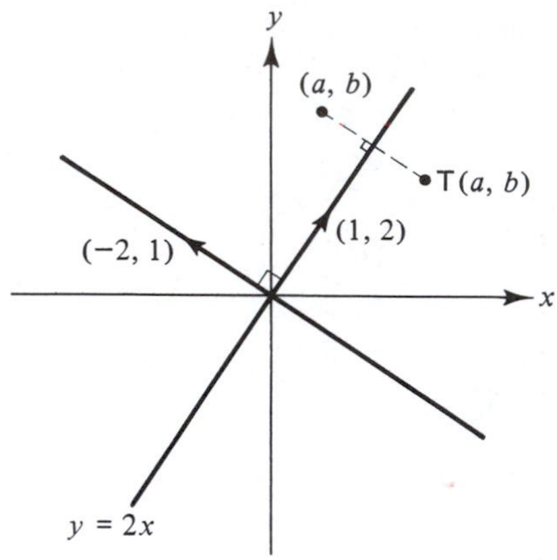

Figure 2.5

then β' is an ordered basis for R^2 and

$$[\mathsf{T}]_{\beta'} = \begin{pmatrix} 1 & 0 \\ 0 & -1 \end{pmatrix}.$$

Let β be the standard ordered basis for R^2, and let Q be the matrix that changes β'-coordinates into β-coordinates. Then

$$Q = \begin{pmatrix} 1 & -2 \\ 2 & 1 \end{pmatrix}$$

and $Q^{-1}[\mathsf{T}]_\beta Q = [\mathsf{T}]_{\beta'}$. We can solve this equation for $[\mathsf{T}]_\beta$ to obtain that $[\mathsf{T}]_\beta = Q[\mathsf{T}]_{\beta'}Q^{-1}$. Because

$$Q^{-1} = \frac{1}{5}\begin{pmatrix} 1 & 2 \\ -2 & 1 \end{pmatrix},$$

the reader can verify that

$$[\mathsf{T}]_\beta = \frac{1}{5}\begin{pmatrix} -3 & 4 \\ 4 & 3 \end{pmatrix}.$$

Since β is the standard ordered basis, it follows that T is left-multiplication by $[\mathsf{T}]_\beta$. Thus for any (a, b) in R^2, we have

$$\mathsf{T}\begin{pmatrix} a \\ b \end{pmatrix} = \frac{1}{5}\begin{pmatrix} -3a + 4b \\ 4a + 3b \end{pmatrix}. \quad \blacksquare$$

The relationship between the matrices $[\mathsf{T}]_{\beta'}$ and $[\mathsf{T}]_\beta$ in Theorem 2.24 will be the subject of further study in Chapters 5, 6, and 7. At this time, however, we shall introduce the name for this relationship.

Definition. *Let* A *and* B *be elements of* $\mathsf{M}_{n\times n}(F)$. *We say that* B *is* similar *to* A *if there exists an invertible matrix* $Q \in \mathsf{M}_{n\times n}(F)$ *such that* $B = Q^{-1}AQ$.

Observe that the relation of similarity is an equivalence relation (see Exercise 8).

Notice also that in this terminology Theorem 2.24 can be stated as follows: If $\mathsf{T}\colon \mathsf{V} \to \mathsf{V}$ is a linear transformation on a finite-dimensional vector space V, and if β and β' are any ordered bases for V, then $[\mathsf{T}]_{\beta'}$ is similar to $[\mathsf{T}]_\beta$. Theorem 2.24 can be generalized to allow $\mathsf{T}\colon \mathsf{V} \to \mathsf{W}$, where V is distinct from W. In this case we can change bases in V as well as in W (see Exercise 7).

EXERCISES

1. Label the following statements as being true or false.

(a) If Q is the change of coordinate matrix that changes β'-coordinates into β-coordinates, where $\beta' = \{x'_1, \ldots, x'_n\}$ and $\beta = \{x_1, \ldots, x_n\}$ are ordered bases for a vector space, then the jth column of Q is $[x_j]_{\beta'}$.

(b) Every change of coordinate matrix is invertible.

(c) Let $\mathsf{T}\colon \mathsf{V} \to \mathsf{V}$ be a linear transformation on a finite-dimensional vector space V, and let β and β' be ordered bases for V. Then $[\mathsf{T}]_\beta = Q[\mathsf{T}]_{\beta'}Q^{-1}$, where Q is the change of coordinate matrix that changes β'-coordinates into β-coordinates.

(d) The matrices $A, B \in \mathsf{M}_{n\times n}(F)$ are called similar if $B = Q^tAQ$ for some $Q \in \mathsf{M}_{n\times n}(F)$.

(e) Let $\mathsf{T}\colon \mathsf{V} \to \mathsf{V}$ be a linear transformation on a finite-dimensional vector space V. Then for any ordered bases β and γ for V, $[\mathsf{T}]_\beta$ is similar to $[\mathsf{T}]_\gamma$.

2. For each of the following pairs of ordered bases β and β' for R^2, find the change of coordinate matrix that changes β'-coordinates into β-coordinates.

(a) $\beta = \{e_1, e_2\}$ and $\beta' = \{(a_1, a_2), (b_1, b_2)\}$

(b) $\beta = \{(-1, 3), (2, -1)\}$ and $\beta' = \{(0, 10), (5, 0)\}$

(c) $\beta = \{(2, 5), (-1, -3)\}$ and $\beta' = \{e_1, e_2\}$

(d) $\beta = \{(-4, 3), (2, -1)\}$ and $\beta' = \{(2, 1), (-4, 1)\}$

3. For each of the following pairs of ordered bases β and β' for $\mathsf{P}_2(R)$, find the change of coordinate matrix that changes β'-coordinates into β-coordinates.

(a) $\beta = \{x^2, x, 1\}$ and
$\beta' = \{a_2x^2 + a_1x + a_0, b_2x^2 + b_1x + b_0, c_2x^2 + c_1x + c_0\}$

(b) $\beta = \{1, x, x^2\}$ and
$\beta' = \{a_2x^2 + a_1x + a_0, b_2x^2 + b_1x + b_0, c_2x^2 + c_1x + c_0\}$

(c) $\beta = \{2x^2 - x, 3x^2 + 1, x^2\}$ and $\beta' = \{1, x, x^2\}$

(d) $\beta = \{x^2 - x + 1, x + 1, x^2 + 1\}$ and
$\beta' = \{x^2 + x + 4, 4x^2 - 3x + 2, 2x^2 + 3\}$

(e) $\beta = \{x^2 - x, x^2 + 1, x - 1\}$ and
$\beta' = \{5x^2 - 2x - 3, -2x^2 + 5x + 5, 2x^2 - x - 3\}$

(f) $\beta = \{2x^2 - x + 1, x^2 + 3x - 2, -x^2 + 2x + 1\}$ and
$\beta' = \{9x - 9, x^2 + 21x - 2, 3x^2 + 5x + 2\}$

4. Let $\mathsf{T}: \mathsf{R}^2 \to \mathsf{R}^2$ be defined by

$$\mathsf{T}\begin{pmatrix} a \\ b \end{pmatrix} = \begin{pmatrix} 2a + b \\ a - 3b \end{pmatrix},$$

let β be the standard ordered basis for R^2, and let

$$\beta' = \left\{\begin{pmatrix} 1 \\ 1 \end{pmatrix}, \begin{pmatrix} 1 \\ 2 \end{pmatrix}\right\}.$$

Use the fact that

$$\begin{pmatrix} 1 & 1 \\ 1 & 2 \end{pmatrix}^{-1} = \begin{pmatrix} 2 & -1 \\ -1 & 1 \end{pmatrix}$$

and Theorem 2.24 to find $[\mathsf{T}]_{\beta'}$.

5. Let $\mathsf{T}: \mathsf{P}_1(R) \to \mathsf{P}_1(R)$ be defined by $\mathsf{T}(p) = p'$, the derivative of $p \in \mathsf{P}_1(R)$. Let $\beta = \{1, x\}$ and $\beta' = \{1 + x, 1 - x\}$. Use the fact that

$$\begin{pmatrix} 1 & 1 \\ 1 & -1 \end{pmatrix}^{-1} = \begin{pmatrix} \frac{1}{2} & \frac{1}{2} \\ \frac{1}{2} & -\frac{1}{2} \end{pmatrix}$$

and Theorem 2.24 to find $[\mathsf{T}]_{\beta'}$.

6. Let L be the line $y = mx$, where $m \neq 0$, and let T be the reflection of R^2 about L. Find an expression for $\mathsf{T}(x, y)$.

7. Prove the following generalization of Theorem 2.24. Let $\mathsf{T}: \mathsf{V} \to \mathsf{W}$ be a linear transformation from a finite-dimensional vector space V to a finite-dimensional vector space W, let β and β' be ordered bases for V, and let γ and γ' be ordered bases for W. Then $[\mathsf{T}]_{\beta'}^{\gamma'} = P^{-1}[\mathsf{T}]_{\beta}^{\gamma}Q$, where Q is the matrix that changes β'-coordinates into β-coordinates and P is the matrix that changes γ'-coordinates into γ coordinates.

8. For A and B in $\mathsf{M}_{n \times n}(F)$, define $A \sim B$ to mean A is similar to B. Prove that $\sim$ is an equivalence relation on $\mathsf{M}_{n \times n}(F)$.

9. Prove that if A and B are similar $n \times n$ matrices, then $\operatorname{tr}(A) = \operatorname{tr}(B)$. *Hint:* Use Exercise 12 of Section 2.3.

10. Let V be a finite-dimensional vector space with ordered bases α, β, and γ.

(a) Prove that if Q and R are the change of coordinate matrices which change α-coordinates into β-coordinates and β-coordinates into γ-coordinates, respectively, then RQ is the change of coordinate matrix which changes α-coordinates into γ-coordinates.

(b) Prove that if Q changes α-coordinates into β-coordinates, then Q^{-1} changes β-coordinates into α-coordinates.

11. Let A be an $n \times n$ matrix with entries in a field F, let β be an ordered basis for F^n, and let $B = [\mathsf{L}_A]_\beta$. Prove that $B = Q^{-1}AQ$, where Q is the $n \times n$ matrix whose jth column equals the jth vector of β.

12.† Let V be a finite-dimensional vector space over a field F, and let $\beta = \{x_1, \ldots, x_n\}$ be an ordered basis for V. Let Q be an $n \times n$ invertible matrix with entries from F. Define

$$x'_j = \sum_{i=1}^{n} Q_{ij} x_i \qquad \text{for } 1 \leqslant j \leqslant n,$$

and set $\beta' = \{x'_1, \ldots, x'_n\}$. Prove that β' is a basis for V and hence that Q is the change of coordinate matrix changing β'-coordinates into β-coordinates.

13. Prove the converse of Exercise 7: If A and B are each $m \times n$ matrices over a field F, and if there exist invertible $m \times m$ and $n \times n$ matrices P and Q, respectively, such that $B = PAQ$, then there exist an n-dimensional vector space V and an m-dimensional vector space W (both over F), ordered bases β and β' for V and γ and γ' for W, and a linear transformation $\mathsf{T}: \mathsf{V} \to \mathsf{W}$ such that

$$A = [\mathsf{T}]_\beta^\gamma \quad \text{and} \quad B = [\mathsf{T}]_{\beta'}^{\gamma'}.$$

Hints: Let $\mathsf{V} = \mathsf{F}^n$, $\mathsf{W} = \mathsf{F}^m$, $\mathsf{T} = \mathsf{L}_A$, and β and γ be the standard ordered bases for F^n and F^m, respectively. Let β' be the ordered basis for V obtained from β via Q (according to the definition on page 95 and justified by Exercise 12), and let γ' be the basis for W obtained from γ via P^{-1}.

2.6* DUAL SPACES

In this section we are concerned exclusively with linear transformations from a vector space V into its field of scalars F, which is itself a vector space of dimension 1 over F. Such a linear transformation is called a *linear functional on* V. We generally use the letters f, g, h, ... to denote linear functionals. As we will see in Example 1, the definite integral provides us with one of the most important examples of a linear functional in mathematics.

Example 1

Let V be the vector space of continuous complex- (or real-) valued functions on the interval $[a, b]$. The function $\mathsf{f}: \mathsf{V} \to C$ (or R) defined by

$$\mathsf{f}(x) = \int_a^b x(t)\,dt$$

is a linear functional on V. If the interval is $[0, 2\pi]$ and n is an integer, the function defined by

$$h_n(x) = \frac{1}{2\pi}\int_0^{2\pi} x(t)e^{-int}\,dt$$

is also a linear functional. In analysis texts the scalar $h_n(x)$ is called the nth *Fourier coefficient of* x. ▮

Example 2

Let $V = M_{n\times n}(F)$, and define $f\colon V \to F$ by $f(A) = \operatorname{tr}(A)$, the trace of A. By Exercise 6 of Section 1.3, we have that f is a linear functional. ▮

Example 3

Let V be a finite-dimensional vector space with the ordered basis $\beta = \{x_1, x_2, \dots, x_n\}$. For each $i = 1, \dots, n$, define $f_i(x) = a_i$, where

$$[x]_\beta = \begin{pmatrix} a_1 \\ a_2 \\ \vdots \\ a_n \end{pmatrix}$$

is the coordinate vector of x relative to β. Then f_i is a linear functional on V called the *i*th *coordinate function with respect to the basis* β. Note that $f_i(x_j) = \delta_{ij}$. These linear functionals play a very important role in the theory of dual spaces (see Theorem 2.25). ▮

Definition. *For a vector space* V *over* F, *we define the* dual space of V *to be the vector space* $\mathscr{L}(V, F)$, *denoted by* V^*.

Thus V^* is the vector space consisting of all linear functionals on V with the operations of addition and scalar multiplication as defined in Section 2.2. Note that if V is finite-dimensional, then

$$\dim(V^*) = \dim(\mathscr{L}(V, F)) = \dim(V)\cdot\dim(F) = \dim(V).$$

Hence, by Theorem 2.20, V and V^* are isomorphic. We also define the *double dual* V^{**} of V to be the dual of V^*. In Theorem 2.27 we show, in fact, that there is a natural identification of V and V^{**}.

Theorem 2.25. *Suppose that* V *is a finite-dimensional vector space with the ordered basis* $\beta = \{x_1, \dots, x_n\}$. *Let* f_i $(1 \leqslant i \leqslant n)$ *be the coordinate functions with respect to* β *as defined above, and let* $\beta^* = \{f_1, \dots, f_n\}$. *Then* β^* *is an ordered basis for* V^*, *and for any* $f \in V^*$ *we have*

$$f = \sum_{i=1}^{n} f(x_i)f_i.$$

We call β^* *the* dual basis of β.

Proof. Let $\mathsf{f} \in \mathsf{V}^*$. Since $\dim(\mathsf{V}^*) = n$, we need only show that

$$\mathsf{f} = \sum_{i=1}^{n} \mathsf{f}(x_i)\mathsf{f}_i,$$

for then β^* will generate V^*. Let

$$\mathsf{g} = \sum_{i=1}^{n} \mathsf{f}(x_i)\mathsf{f}_i.$$

For $1 \leqslant j \leqslant n$, we have

$$\mathsf{g}(x_j) = \left(\sum_{i=1}^{n} \mathsf{f}(x_i)\mathsf{f}_i\right)(x_j) = \sum_{i=1}^{n} \mathsf{f}(x_i)\mathsf{f}_i(x_j)$$

$$= \sum_{i=1}^{n} \mathsf{f}(x_i)\delta_{ij} = \mathsf{f}(x_j).$$

Hence $\mathsf{g} = \mathsf{f}$ by the corollary to Theorem 2.6, and we are done. ∎

Example 4

Let $\beta = \{(2, 1), (3, 1)\}$ be an ordered basis for R^2. To explicitly determine the dual basis $\beta^* = \{\mathsf{f}_1, \mathsf{f}_2\}$ of β, we need to consider the equations

$$1 = \mathsf{f}_1(2, 1) = \mathsf{f}_1(2e_1 + e_2) = 2\mathsf{f}_1(e_1) + \mathsf{f}_1(e_2)$$

$$0 = \mathsf{f}_1(3, 1) = \mathsf{f}_1(3e_1 + e_2) = 3\mathsf{f}_1(e_1) + \mathsf{f}_1(e_2).$$

Solving these equations, we obtain that $\mathsf{f}_1(e_1) = -1$ and $\mathsf{f}_1(e_2) = 3$, i.e., $\mathsf{f}_1(x, y) = -x + 3y$. Similarly, it can be shown that $\mathsf{f}_2(x, y) = x - 2y$. ∎

We now assume that V and W are finite-dimensional vector spaces over F with ordered bases β and γ, respectively. In Section 2.4 we proved that there exists a one-to-one correspondence between linear transformations $\mathsf{T}: \mathsf{V} \to \mathsf{W}$ and $m \times n$ matrices (over F) via the correspondence $\mathsf{T} \leftrightarrow [\mathsf{T}]_\beta^\gamma$. For a matrix of the form $A = [\mathsf{T}]_\beta^\gamma$, the question arises as to whether or not there exists a linear transformation U associated with T in some natural way such that U may be represented in some basis as A^t. Of course, if $m \neq n$, it would be impossible for U to be a linear transformation from V into W. We now answer this question by applying what we have already learned about dual spaces.

Theorem 2.26. *Let* V *and* W *be finite-dimensional vector spaces over* F *with ordered bases* β *and* γ, *respectively. For any linear transformation* $\mathsf{T}: \mathsf{V} \to \mathsf{W}$, *the mapping* $\mathsf{T}^t: \mathsf{W}^* \to \mathsf{V}^*$ *defined by* $\mathsf{T}^t(\mathsf{g}) = \mathsf{g}\mathsf{T}$ *for all* $\mathsf{g} \in \mathsf{W}^*$ *is a linear transformation with the property that* $[\mathsf{T}^t]_{\gamma^*}^{\beta^*} = ([\mathsf{T}]_\beta^\gamma)^t$.

Proof. For $\mathsf{g} \in \mathsf{W}^*$, it is clear that $\mathsf{T}^t(\mathsf{g}) = \mathsf{g}\mathsf{T}$ is a linear functional on V and hence is an element of V^*. Thus T^t maps W^* into V^*. We leave the proof that T^t is linear to the reader.

To complete the proof, let $\beta = \{x_1, \ldots, x_n\}$ and $\gamma = \{y_1, \ldots, y_m\}$ with dual bases $\beta^* = \{\mathsf{f}_1, \ldots, \mathsf{f}_n\}$ and $\gamma^* = \{\mathsf{g}_1, \ldots, \mathsf{g}_m\}$, respectively. For convenience, let $A = [\mathsf{T}]_\beta^\gamma$ and $B = [\mathsf{T}^t]_{\gamma^*}^{\beta^*}$. Then

$$\mathsf{T}(x_i) = \sum_{k=1}^{m} A_{ki} y_k \qquad \text{for } 1 \leqslant i \leqslant n,$$

and

$$\mathsf{T}^t(\mathsf{g}_j) = \sum_{i=1}^{n} B_{ij} \mathsf{f}_i \qquad \text{for } 1 \leqslant j \leqslant m.$$

We must show that $B = A^t$. Theorem 2.25 shows that

$$\mathsf{T}^t(\mathsf{g}_j) = \mathsf{g}_j \mathsf{T} = \sum_{i=1}^{n} (\mathsf{g}_j \mathsf{T})(x_i) \mathsf{f}_i,$$

so

$$B_{ij} = (\mathsf{g}_j \mathsf{T})(x_i) = \mathsf{g}_j(\mathsf{T}(x_i)) = \mathsf{g}_j\left(\sum_{k=1}^{m} A_{ki} y_k\right)$$

$$= \sum_{k=1}^{m} A_{ki} \mathsf{g}_j(y_k) = \sum_{k=1}^{m} A_{ki} \delta_{jk} = A_{ji} = (A^t)_{ij}.$$

Hence $B = A^t$. ∎

The linear transformation T^t defined in Theorem 2.26 is called the *transpose of* T. It is clear that T^t is the unique linear transformation U such that $[\mathsf{U}]_{\gamma^*}^{\beta^*} = ([\mathsf{T}]_\beta^\gamma)^t$.

We now concern ourselves with demonstrating that any finite-dimensional vector space V can be identified in a very natural way with its double dual V^{**}. There is, in fact, an isomorphism between V and V^{**} that does not depend on any choice of bases for the two vector spaces.

For a vector $x \in \mathsf{V}$ we define $\hat{x}: \mathsf{V}^* \to F$ by $\hat{x}(\mathsf{f}) = \mathsf{f}(x)$ for every $\mathsf{f} \in \mathsf{V}^*$. It is easy to verify that $\hat{x}$ is a linear functional on V^*, so $\hat{x} \in \mathsf{V}^{**}$. The correspondence $x \leftrightarrow \hat{x}$ allows us to define the desired isomorphism between V and V^{**}.

Lemma. *Let* V *be a finite-dimensional vector space, and let* $\mathrm{x} \in \mathsf{V}$. *If* $\hat{\mathrm{x}}(\mathsf{f}) = 0$ *for all* $\mathsf{f} \in \mathsf{V}^*$, *then* $\mathrm{x} = 0$.

Proof. Let $x \neq 0$. We will show that there exists $\mathsf{f} \in \mathsf{V}^*$ such that $\hat{x}(\mathsf{f}) \neq 0$. Choose an ordered basis $\beta = \{x_1, \ldots, x_n\}$ for V such that $x_1 = x$. Let $\{\mathsf{f}_1, \ldots, \mathsf{f}_n\}$ be the dual basis of β. Then $\mathsf{f}_1(x_1) = 1 \neq 0$. Let $\mathsf{f} = \mathsf{f}_1$. ∎

Theorem 2.27. *Let* V *be a finite-dimensional vector space, and let* $\psi: \mathsf{V} \to \mathsf{V}^{**}$ *be defined by* $\psi(\mathrm{x}) = \hat{\mathrm{x}}$. *Then* ψ *is an isomorphism.*

Proof. (a) ψ is linear: Let $x, y \in \mathsf{V}$ and $a \in F$. For $\mathsf{f} \in \mathsf{V}^*$ we have

$$\begin{aligned}\psi(x + ay)(\mathsf{f}) &= \mathsf{f}(x + ay) = \mathsf{f}(x) + a\mathsf{f}(y) = \hat{x}(\mathsf{f}) + a\hat{y}(\mathsf{f}) \\ &= (\hat{x} + a\hat{y})(\mathsf{f}).\end{aligned}$$

Hence

$$\psi(x + ay) = \hat{x} + a\hat{y} = \psi(x) + a\psi(y).$$

(b) ψ is one-to-one: Suppose that $\psi(x)$ is the zero functional on V^* for some $x \in \mathsf{V}$. Then $\hat{x}(\mathsf{f}) = 0$ for every $\mathsf{f} \in \mathsf{V}^*$. By the previous lemma we conclude that $x = 0$.

(c) ψ is an isomorphism: This follows from (b) and the fact that $\dim(\mathsf{V}) = \dim(\mathsf{V}^{**})$. ∎

Corollary. *Let* V *be a finite-dimensional vector space with dual space* V^*. *Then every ordered basis of* V^* *is the dual basis of some basis of* V.

Proof. Let $\{\mathsf{f}_1, \ldots, \mathsf{f}_n\}$ be an ordered basis of V^*. We may combine Theorems 2.25 and 2.27 to conclude that for this basis of V^* there exists a dual basis $\{\hat{x}_1, \ldots, \hat{x}_n\}$ in V^{**}, that is, $\delta_{ij} = \hat{x}_i(\mathsf{f}_j) = \mathsf{f}_j(x_i)$. Thus $\{\mathsf{f}_1, \ldots, \mathsf{f}_n\}$ is the dual basis of $\{x_1, \ldots, x_n\}$. ∎

Although many of the ideas of this section can be extended to the case where V is not finite-dimensional, for example the existence of a dual space, only a finite-dimensional vector space is isomorphic to its double dual via the map $x \to \hat{x}$. In fact, for infinite-dimensional vector spaces, V and V^* are never isomorphic.

EXERCISES

1. Label the following statements as being true or false. Assume that all vector spaces are finite-dimensional.

(a) Every linear transformation is a linear functional.

(b) A linear functional defined on a field may be represented as a 1×1 matrix.

(c) Every vector space is isomorphic to its dual space.

(d) Every vector space is the dual of some other vector space.

(e) If T is an isomorphism from V onto V^* and β is a finite ordered basis of V, then $\mathsf{T}(\beta) = \beta^*$.

(f) If T is a linear transformation from V into W, then the domain of $(\mathsf{T}^t)^t$ is V^{**}.

(g) If V is isomorphic to W, then V^* is isomorphic to W^*.

(h) The derivative of a function may be considered as a linear functional on the vector space of differentiable functions.

2. For the following functions on a vector space V, determine which are linear functionals.
   (a) $V = P(R)$; $f(p) = 2p'(0) + p''(1)$, where $'$ denotes differentiation
   (b) $V = R^2$; $f(x, y) = (2x, 4y)$
   (c) $V = M_{2\times 2}(F)$; $f(A) = \text{tr}(A)$
   (d) $V = R^3$; $f(x, y, z) = x^2 + y^2 + z^2$
   (e) $V = P(R)$; $f(p) = \int_0^1 p(t)\,dt$
   (f) $V = M_{2\times 2}(R)$; $f(A) = A_{11}$
3. For each vector space V and basis β below, find the dual basis β^* for V^* as in Example 4.
   (a) $V = R^3$; $\beta = \{(1, 0, 1), (1, 2, 1), (0, 0, 1)\}$
   (b) $V = P_2(R)$; $\beta = \{1, x, x^2\}$
4. Let $V = R^3$ and define $f_1, f_2, f_3 \in V^*$ as follows:

   $$f_1(x, y, z) = x - 2y, \qquad f_2(x, y, z) = x + y + z, \quad \text{and} \quad f_3(x, y, z) = y - 3z.$$

   Prove that $\{f_1, f_2, f_3\}$ is a basis for V^*, and then find a basis for V for which it is the dual.
5. Let $V = P_1(R)$, and for $p \in V$ define $f_1, f_2 \in V^*$ by

   $$f_1(p) = \int_0^1 p(t)\,dt$$

   and

   $$f_2(p) = \int_0^2 p(t)\,dt.$$

   Prove that $\{f_1, f_2\}$ is a basis for V^*, and find a basis for V for which it is the dual.
6. Define $f \in (R^2)^*$ by $f(x, y) = 2x + y$ and $T: R^2 \to R^2$ by $T(x, y) = (3x + 2y, x)$.
   (a) Compute $T^t(f)$.
   (b) Compute $[T^t]_{\beta^*}$, where β is the standard ordered basis for R^2 and $\beta^* = \{f_1, f_2\}$, by finding scalars a, b, c, and d such that $T^t(f_1) = af_1 + bf_2$ and $T^t(f_2) = cf_1 + df_2$.
   (c) Compute $[T]_\beta$ and $[T]_\beta^t$, and compare your results with part (b).
7. Let $V = P_1(R)$ and $W = R^2$ with respective ordered bases $\beta = \{1, x\}$ and $\gamma = \{e_1, e_2\}$. Define $T: V \to W$ by $T(p) = (p(0) - 2p(1), p(0) + p'(0))$, where p' is the derivative of p.
   (a) If $f \in W^*$ is defined by

   $$f(a, b) = a - 2b,$$

   compute $T^t(f)$.
   (b) Compute $[T^t]_{\gamma^*}^{\beta^*}$ without appealing to Theorem 2.26.
   (c) Compute $[T]_\beta^\gamma$ and its transpose, and compare your result with part (b).

8. Show that every plane through the origin in R^3 may be identified with the null space of an element in $(\mathsf{R}^3)^*$. State an analogous result in R^2.

9. Let T be a function from F^n into F^m. Prove that T is linear if and only if there exist $\mathsf{f}_1, \ldots, \mathsf{f}_m \in (\mathsf{F}^n)^*$ such that $\mathsf{T}(x) = (\mathsf{f}_1(x), \ldots, \mathsf{f}_m(x))$ for all $x \in \mathsf{F}^n$. *Hint:* If T is linear, define $\mathsf{f}_i(x) = (\mathsf{g}_i\mathsf{T})(x)$ for $x \in \mathsf{F}^n$, i.e., $\mathsf{f}_i = \mathsf{T}^t(\mathsf{g}_i)$ for $1 \leqslant i \leqslant m$, where $\{\mathsf{g}_1, \ldots, \mathsf{g}_m\}$ is the dual basis of the standard ordered basis of F^m.

10. Let $\mathsf{V} = \mathsf{P}_n(F)$, and let $c_0, \ldots, c_n$ be distinct scalars in F.

(a) For $0 \leqslant i \leqslant n$, define $\mathsf{f}_i \in \mathsf{V}^*$ by $\mathsf{f}_i(p) = p(c_i)$. Prove that $\{\mathsf{f}_0, \ldots, \mathsf{f}_n\}$ is a basis of V^*. *Hint:* Apply any linear combination of this set that equals the zero transformation to $p(t) = (t - c_1)(t - c_2)\cdots(t - c_n)$ and deduce that the first coefficient is zero.

(b) Use the corollary to Theorem 2.27 and part (a) to show that there exist unique polynomials $p_0, \ldots, p_n$ such that $p_i(c_j) = \delta_{ij}$ for $0 \leqslant i \leqslant n$. These polynomials are the Lagrange polynomials defined in Section 1.6.

(c) For any scalars $a_0, \ldots, a_n$ (not necessarily distinct), deduce that there exists a unique polynomial q of degree at most n such that $q(c_i) = a_i$ for $0 \leqslant i \leqslant n$. In fact,

$$q = \sum_{i=0}^{n} a_i p_i.$$

(d) Deduce the Lagrange interpolation formula:

$$p = \sum_{i=0}^{n} p(c_i) p_i$$

for any $p \in \mathsf{V}$.

(e) Prove that

$$\int_a^b p(t)\,dt = \sum_{i=0}^{n} p(c_i) d_i,$$

where

$$d_i = \int_a^b p_i(t)\,dt.$$

Suppose now that

$$c_i = a + \frac{i(b-a)}{n} \qquad \text{for } i = 0, \ldots, n.$$

For $n = 1$, the above yields the trapezoidal rule for polynomials. For $n = 2$, this result is Simpson's rule for polynomials.

11. Let V and W be finite-dimensional vector spaces over F, and let ψ_1 and ψ_2 be the isomorphisms between V and V^{**} and W and W^{**}, respectively, as defined in Theorem 2.27. Let $\mathsf{T}\colon \mathsf{V} \to \mathsf{W}$ be linear, and define $\mathsf{T}^{tt} = (\mathsf{T}^t)^t$.

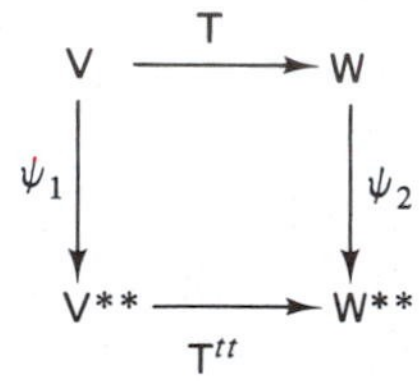

Figure 2.6

Prove that the diagram depicted in Figure 2.6 commutes, i.e., that $\psi_2\mathsf{T} = \mathsf{T}^{tt}\psi_1$.

12. Let V be a finite-dimensional vector space with the ordered basis β. Prove that $\psi(\beta) = \beta^{**}$, where ψ is as defined in Theorem 2.27.

In Exercises 13 through 17 V denotes a finite-dimensional vector space over F. For every subset S of V define the *annihilator* S^0 of S as

$$S^0 = \{\mathsf{f} \in \mathsf{V}^*: \mathsf{f}(x) = 0 \text{ for all } x \in S\}.$$

13. **(a)** Prove that S^0 is a subspace of V^*.
(b) If W is a subspace of V and $x \notin \mathsf{W}$, prove that there exists $\mathsf{f} \in \mathsf{W}^0$ such that $\mathsf{f}(x) \neq 0$.
(c) Prove that $S^{00} = \text{span}(\psi(S))$, where ψ is as defined in Theorem 2.27.
(d) For subspaces W_1 and W_2, prove that $\mathsf{W}_1 = \mathsf{W}_2$ if and only if $\mathsf{W}_1^0 = \mathsf{W}_2^0$.
(e) For subspaces W_1 and W_2, show that $(\mathsf{W}_1 + \mathsf{W}_2)^0 = \mathsf{W}_1^0 \cap \mathsf{W}_2^0$.

14. If W is a subspace of V, prove that $\dim(\mathsf{W}) + \dim(\mathsf{W}^0) = \dim(\mathsf{V})$. *Hint:* Extend an ordered basis $\{x_1, \ldots, x_k\}$ of W to an ordered basis $\beta = \{x_1, \ldots, x_n\}$ of V. Let $\beta^* = \{\mathsf{f}_1, \ldots, \mathsf{f}_n\}$. Prove that $\{\mathsf{f}_{k+1}, \ldots, \mathsf{f}_n\}$ is a basis of W^0.

15. Suppose that W is a finite-dimensional vector space over F and that $\mathsf{T}: \mathsf{V} \to \mathsf{W}$ is linear. Prove that $\mathsf{N}(\mathsf{T}^t) = (\mathsf{R}(\mathsf{T}))^0$.

16. Use Exercises 14 and 15 to deduce that $\text{rank}(\mathsf{L}_{A^t}) = \text{rank}(\mathsf{L}_A)$ for any $A \in \mathsf{M}_{m \times n}(F)$.

17. Let $\mathsf{T}: \mathsf{V} \to \mathsf{V}$ be a linear transformation and W be a subspace of V. Prove that W is T-invariant (as defined in Exercise 26 of Section 2.1) if and only if W^0 is T^t-invariant.

2.7* HOMOGENEOUS LINEAR DIFFERENTIAL EQUATIONS WITH CONSTANT COEFFICIENTS

As an introduction to this section, let us consider the following physical problem. A weight of mass m is attached to a vertically suspended spring that is allowed to stretch until the forces acting on the weight are in equilibrium. Let us suppose that the weight is now motionless and impose an XY-coordinate system

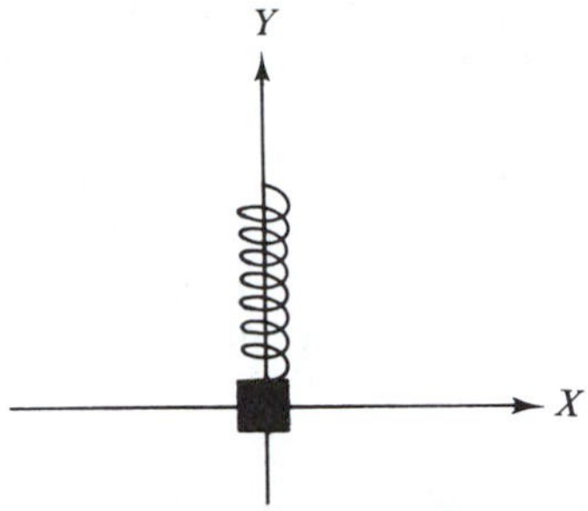

Figure 2.7

with the weight at the origin and the spring lying on the upper part of the Y-axis (see Figure 2.7).

Suppose that at a certain time, say $t = 0$, the weight is lowered a distance s along the Y-axis and released. The spring will then begin to oscillate.

Let us describe the motion of the spring. At any time $t \geqslant 0$, let $F(t)$ denote the force acting on the weight and $y(t)$ denote the coordinate of the weight along the Y-axis. For example, $y(0) = -s$. The second derivative of y with respect to time, $y''(t)$, is the acceleration of the weight at time t, and hence by Newton's second law

$$F(t) = my''(t). \tag{1}$$

It is reasonable to assume that the force acting on the weight is totally due to the tension of the spring and that this force satisfies Hooke's law: *The force acting on the weight is proportional to its displacement from the equilibrium position but acts in the opposite direction.* If $k > 0$ is the proportionality constant, then Hooke's law states that

$$F(t) = -ky(t). \tag{2}$$

Combining (1) and (2), we obtain

$$my'' = -ky$$

or

$$y'' + \frac{k}{m}y = 0. \tag{3}$$

The expression in (3) is an example of a "differential equation." A *differential equation* in an unknown function $y = y(t)$ is an equation involving y, t, and derivatives of y. If the differential equation is of the form

$$a_n y^{(n)} + a_{n-1} y^{(n-1)} + \cdots + a_1 y^{(1)} + a_0 y = f, \tag{4}$$

where $a_0, a_1, \ldots, a_n$ and f are functions of t and $y^{(k)}$ denotes the kth derivative of y, then the equation is said to be *linear*. The functions a_i are called the *coefficients* of the differential equation (4). Thus (3) is an example of a linear differential equation in which the coefficients are constants and the function f is identically zero. When the function f in (4) is identically zero, the linear differential equation is called *homogeneous.*

In this section we apply the linear algebra we have studied to solve homogeneous linear differential equations with constant coefficients. If $a_n \neq 0$, we say that the differential equation in (4) is of *order n.* In this case we divide both sides by a_n to obtain a new, but equivalent, equation

$$y^{(n)} + b_{n-1}y^{(n-1)} + \cdots + b_1 y^{(1)} + b_0 y = 0,$$

where $b_i = a_i/a_n$ for $i = 0, 1, \ldots, n-1$. Because of this observation, we shall always assume that the leading coefficient a_n in (4) is 1.

A *solution* to (4) is a function that when substituted for y reduces (4) to an identity.

Example 1

The function $y(t) = \sin\sqrt{k/m}\,t$ is a solution to (3) since

$$y''(t) + \frac{k}{m}\,y(t) = -\frac{k}{m}\sin\sqrt{\frac{k}{m}}\,t + \frac{k}{m}\sin\sqrt{\frac{k}{m}}\,t = 0$$

for all t. Notice, however, that substituting $y(t) = t$ into (3) yields

$$y''(t) + \frac{k}{m}\,y(t) = \frac{k}{m}\,t,$$

which is not identically zero. Thus $y(t) = t$ is not a solution to (3). ∎

While attempting to solve differential equations, we will discover that it is useful to view solutions as complex-valued functions of a real variable even though the solutions that are meaningful to us in a physical sense are real-valued functions of a real variable. The convenience of this viewpoint will become clear later. Thus we will be concerned with the vector space $\mathcal{F}(R, C)$ (as defined in Example 3 of Section 1.2). In order to consider complex-valued functions of a real variable as solutions to differential equations, we must define what it means to differentiate such functions. Given a complex-valued function $x \in \mathcal{F}(R, C)$ of a real variable t, there exist unique real-valued functions x_1 and x_2 of t, such that

$$x(t) = x_1(t) + ix_2(t) \qquad \text{for } t \in R,$$

where i is the purely imaginary number such that $i^2 = -1$. We say that x_1 is the *real part* and x_2 is the *imaginary part* of x.

Definition. *Given a function* $x \in \mathcal{F}(R, C)$ *with real part* x_1 *and imaginary part* x_2, *we say that* x *is* differentiable *if* x_1 *and* x_2 *are differentiable. If* x *is differentiable, we define the derivative of* x, x', *to be*

$$x' = x_1' + ix_2'.$$

In Example 2 we illustrate some computations with complex-valued functions.

Example 2

If $x(t) = \cos 2t + i \sin 2t$, then

$$x'(t) = -2 \sin 2t + i(2 \cos 2t).$$

We next find the real and imaginary parts of x^2. Since

$$\begin{aligned} x^2(t) &= (\cos 2t + i \sin 2t)^2 = (\cos^2 2t - \sin^2 2t) + i(2 \sin 2t \cos 2t) \\ &= \cos 4t + i \sin 4t, \end{aligned}$$

the real part of $x^2(t)$ is $\cos 4t$, and the imaginary part is $\sin 4t$. ∎

The following theorem indicates that we may limit our investigations to a vector space considerably smaller than $\mathscr{F}(R, C)$. Its proof, which is illustrated by Example 3, involves a simple induction argument, which we omit.

Theorem 2.28. *Any solution to a homogeneous linear differential equation with constant coefficients has derivatives of all orders; that is, if* x *is a solution to such an equation, then* $x_{(k)}$ *exists for every positive integer* k.

Example 3

As an illustration of Theorem 2.28, consider the equation

$$y^{(2)} + 4y = 0.$$

Clearly, to qualify as a solution, a function y must have two derivatives. If y is a solution, however, then

$$y^{(2)} = -4y.$$

Thus since $y^{(2)}$ is a constant multiple of a function that has two derivatives, namely y, $y^{(2)}$ must have two derivatives, so $y^{(4)}$ exists. In fact,

$$y^{(4)} = -4y^{(2)}.$$

Since $y^{(4)}$ is a constant multiple of a function that we have shown has at least two derivatives, it also has at least two derivatives, and hence $y^{(6)}$ exists. Continuing in this manner, we can show that any solution has derivatives of all orders. ∎

Definition. *We use* C^∞ *to denote the set of all functions in* $\mathscr{F}$(R, C) *that have derivatives of all orders.*

It is a simple exercise to show that C^∞ is a subspace of $\mathscr{F}(R, C)$ and hence a vector space over C. In view of Theorem 2.28 it is this vector space that is of interest to us. For $x \in \mathsf{C}^\infty$ the derivative x' of x also lies in C^∞. We can use the derivative operation to define a mapping $\mathsf{D}: \mathsf{C}^\infty \to \mathsf{C}^\infty$ by

$$\mathsf{D}(x) = x' \qquad \text{for } x \in \mathsf{C}^\infty.$$

It is easy to show that D is a linear transformation. More generally, consider any polynomial over C of the form

$$p(t) = a_n t^n + a_{n-1} t^{n-1} + \cdots + a_1 t + a_0.$$

Then if we define

$$p(\mathsf{D}) = a_n \mathsf{D}^n + a_{n-1} \mathsf{D}^{n-1} + \cdots + a_1 \mathsf{D} + a_0 \mathsf{I},$$

$p(\mathsf{D})$ is a linear transformation (see Appendix E).

Definitions. *For any polynomial* $p(t)$ *over* C, $p(\mathsf{D})$ *is called a* differential operator. *The* order *of the differential operator* $p(\mathsf{D})$ *is the degree of the polynomial* $p(t)$.

Differential operators are useful since they provide us with a means of reformulating a differential equation in the context of linear algebra. Any homogeneous linear differential equation with constant coefficients

$$y^{(n)} + a_{n-1} y^{(n-1)} + \cdots + a_1 y^{(1)} + a_0 y = 0$$

can be rewritten by means of differential operators as

$$(\mathsf{D}^n + a_{n-1} \mathsf{D}^{n-1} + \cdots + a_1 \mathsf{D} + a_0 \mathsf{I})(y) = 0.$$

Definition. *Given the differential equation above, the complex polynomial*

$$p(t) = t^n + a_{n-1} t^{n-1} + \cdots + a_1 t + a_0$$

is called the auxiliary polynomial *associated with the equation.*

For example, (3) has auxiliary polynomial

$$p(t) = t^2 + \frac{k}{m}.$$

Any homogeneous linear differential equation with constant coefficients can be rewritten as

$$p(\mathsf{D})(y) = 0,$$

where $p(t)$ is the auxiliary polynomial associated with the equation. Clearly, this equation implies the following.

Theorem 2.29. *The set of all solutions to a homogeneous linear differential equation with constant coefficients coincides with the null space of* $p(\mathsf{D})$, *where* $p(t)$ *is the auxiliary polynomial associated with the equation.*

Corollary. *The set of all solutions to a homogeneous linear differential equation with constant coefficients is a subspace of* C^∞.

In view of the corollary above we call the set of solutions to a homogeneous linear differential equation with constant coefficients the *solution space* of the equation. A practical way of describing such a space is to find a basis for it. We will examine a certain class of functions that is of use in finding bases for these solution spaces.

For a real number s we are familiar with the real number e^s, where e is the unique number whose natural logarithm is 1 (that is, $\ln(e) = 1$). We know, for instance, certain properties of exponentiation:

$$e^{s+t} = e^s e^t \quad \text{and} \quad e^{-t} = \frac{1}{e^t}$$

for any real numbers s and t. We now extend the definition of powers of e to include complex numbers in such a way that these properties remain true.

Definition. *Let* $c = a + ib$ *be any complex number with real part* a *and imaginary part* b. *Define*

$$e^c = e^a(\cos b + i \sin b).$$

For example, for $c = 2 + i(\pi/3)$,

$$e^c = e^2\left(\cos\frac{\pi}{3} + i\sin\frac{\pi}{3}\right) = e^2\left(\frac{1}{2} + i\frac{\sqrt{3}}{2}\right).$$

Clearly, if c is real ($b = 0$), we obtain the usual result: $e^c = e^a$. It can be shown with the use of trigonometric identities that

$$e^{c+d} = e^c e^d \quad \text{and} \quad e^{-c} = \frac{1}{e^c}$$

for any complex numbers c and d.

Definition. *Let* c *be any complex number. The function* $f\colon R \to C$ *defined by* $f(t) = e^{ct}$ *for all* t *in* R *is called an* exponential function.

The derivative of an exponential function, as described in the following theorem, is as one would expect. The proof involves a straightforward computation, which we leave as an exercise.

Theorem 2.30. *For any exponential function* $f(t) = e^{ct}$, $f'(t) = ce^{ct}$.

We will use exponential functions to describe all solutions of a homogeneous linear differential equation of order 1. Recall that the *order* of such an equation is the degree of its auxiliary polynomial. Thus an equation of order 1 is of the form

$$y' + a_0 y = 0. \tag{5}$$

Theorem 2.31. *The solution space for* (5) *is of dimension 1 and has* $\{e^{-a_0 t}\}$ *as a basis.*

Proof. Clearly (5) has $e^{-a_0 t}$ as a solution. Suppose that $x(t)$ is any solution to (5). Then

$$x'(t) = -a_0 x(t) \qquad \text{for all } t \in R.$$

Define

$$z(t) = e^{a_0 t} x(t).$$

Differentiating z yields

$$z'(t) = (e^{a_0 t})' x(t) + e^{a_0 t} x'(t) = a_0 e^{a_0 t} x(t) - a_0 e^{a_0 t} x(t) = 0.$$

(Notice that the familiar product rule for differentiation holds for complex-valued functions of a real variable. A justification involves a lengthy, although direct, computation.)

Since z' is identically zero, z is a constant function. (Again, this fact, well-known for real-valued functions of a real variable, is also true for complex-valued functions. The proof, which relies on the real case, involves looking separately at the real and imaginary parts of z.) Thus there exists a complex number c such that

$$z(t) = e^{a_0 t} x(t) = c \qquad \text{for all } t \in R.$$

So

$$x(t) = ce^{-a_0 t}.$$

We conclude that any member of the solution space of (5) is a linear combination of $e^{-a_0 t}$. ∎

Another way of formulating Theorem 2.31 is as follows.

Corollary. *For any complex number* c *the null space of the differential operator* $\mathsf{D} - c\mathsf{I}$ *has* $\{e^{ct}\}$ *as a basis.*

We next concern ourselves with differential equations of order greater than one. Given an nth order homogeneous linear differential equation with constant coefficients

$$y^{(n)} + a_{n-1} y^{(n-1)} + \cdots + a_1 y^{(1)} + a_0 y = 0,$$

its auxiliary polynomial

$$p(t) = t^n + a_{n-1}t^{n-1} + \cdots + a_1 t + a_0$$

factors into a product of polynomials of degree 1:

$$p(t) = (t - c_1)(t - c_2)\cdots(t - c_n),$$

where $c_1, c_2, \ldots, c_n$ are (not necessarily distinct) complex numbers. (This follows from the fundamental theorem of algebra in Appendix D.) Thus

$$p(\mathsf{D}) = (\mathsf{D} - c_1\mathsf{I})(\mathsf{D} - c_2\mathsf{I})\cdots(\mathsf{D} - c_n\mathsf{I}).$$

Now the operators $\mathsf{D} - c_i\mathsf{I}$ commute, and so by Exercise 9 we have that

$$\mathsf{N}(\mathsf{D} - c_i\mathsf{I}) \subset \mathsf{N}(p(\mathsf{D})) \qquad \text{for all } i.$$

Since $\mathsf{N}(p(\mathsf{D}))$ coincides with the solution space of the given differential equation, we can conclude the following result by the corollary to Theorem 2.31.

Theorem 2.32. *Let* p(t) *be the auxiliary polynomial for a homogeneous linear differential equation with constant coefficients. For any complex number* c, *if* c *is a zero of* p(t), *then* e^{ct} *is a solution to the differential equation.*

Example 4

Given the differential equation

$$y'' - 3y' + 2y = 0,$$

its auxiliary polynomial $p(t) = t^2 - 3t + 2$ factors as

$$p(t) = (t - 1)(t - 2).$$

Hence by Theorem 2.32 e^t and e^{2t} are solutions to the equation above because $c = 1$ and $c = 2$ are zeros of $p(t)$. Since the solution space of the equation above is a subspace of C^∞, span($\{e^t, e^{2t}\}$) lies in the solution space. It is a simple matter to show that $\{e^t, e^{2t}\}$ is linearly independent. Thus if we show that the solution space is two-dimensional, we can conclude that $\{e^t, e^{2t}\}$ is a basis for the solution space. This result follows from the following theorem. ∎

Theorem 2.33. *For any differential operator* $p(\mathsf{D})$ *of order* n, *the null space of* $p(\mathsf{D})$ *is an* n*-dimensional subspace of* C^∞.

As a preliminary to the proof of Theorem 2.33 we must establish two lemmas.

Lemma 1. *The differential operator* $\mathsf{D} - c\mathsf{I}\colon \mathsf{C}^\infty \to \mathsf{C}^\infty$ *is onto for any complex number* c.

Proof. Let $x \in \mathsf{C}^\infty$. We wish to find a $y \in \mathsf{C}^\infty$ such that $(\mathsf{D} - c\mathsf{I})y = x$. Define a function w by $w(t) = x(t)e^{-ct}$ for $t \in R$.

Clearly, $w \in \mathsf{C}^\infty$ because x and e^{-ct} lie in C^∞. Let w_1 and w_2 be the real and imaginary parts of w. Since $w \in \mathsf{C}^\infty$, w_1 and w_2 are differentiable and hence continuous. Thus they have antiderivatives, say W_1 and W_2, such that $W'_1 = w_1$ and $W'_2 = w_2$. Define $W\colon R \to C$ by

$$W(t) = W_1(t) + iW_2(t) \qquad \text{for } t \in R.$$

Then $W \in \mathsf{C}^\infty$, and the real and imaginary parts of W are W_1 and W_2, respectively. Also $W' = w$. Finally, define $y\colon R \to C$ by $y(t) = W(t)e^{ct}$ for $t \in R$.

Clearly, $y \in \mathsf{C}^\infty$, and since

$$\begin{aligned}(\mathsf{D} - c\mathsf{I})y(t) &= y'(t) - cy(t)\\ &= W'(t)e^{ct} + W(t)ce^{ct} - cW(t)e^{ct}\\ &= w(t)e^{ct}\\ &= x(t)e^{-ct}e^{ct}\\ &= x(t),\end{aligned}$$

$(\mathsf{D} - c\mathsf{I})y = x$. ∎

Lemma 2. *Let* V *be a vector space, and suppose that* T *and* U *are linear operators on* V *such that*

(a) U *is onto.*

(b) *The null spaces of* T *and* U *are finite-dimensional.*

Then the null space of TU *is finite-dimensional, and*

$$\dim(\mathsf{N}(\mathsf{TU})) = \dim(\mathsf{N}(\mathsf{T})) + \dim(\mathsf{N}(\mathsf{U})).$$

Proof. Let $p = \dim(\mathsf{N}(\mathsf{T}))$, $q = \dim(\mathsf{N}(\mathsf{U}))$, and $\{u_1, u_2, \ldots, u_p\}$ and $\{v_1, v_2, \ldots, v_q\}$ be bases for $\mathsf{N}(\mathsf{T})$ and $\mathsf{N}(\mathsf{U})$, respectively. Since U is onto, we can choose for each i ($i = 1, \ldots, p$) an element $w_i \in \mathsf{V}$ such that $\mathsf{U}(w_i) = u_i$. Thus we obtain a set of p elements $\{w_1, w_2, \ldots, w_p\}$. Note that for any i and j, $w_i \neq v_j$, for otherwise $u_i = \mathsf{U}(w_i) = \mathsf{U}(v_j) = 0$—a contradiction. Hence the set

$$\beta = \{w_1, w_2, \ldots, w_p, v_1, \ldots, v_q\}$$

contains $p + q$ distinct elements. To prove the lemma, it suffices to show that β is a basis for $\mathsf{N}(\mathsf{TU})$.

We first show that β generates $\mathsf{N}(\mathsf{TU})$. Since for any w_i and v_j in β

$$\mathsf{TU}(w_i) = \mathsf{T}(u_i) = 0 \quad \text{and} \quad \mathsf{TU}(v_j) = \mathsf{T}(0) = 0,$$

$$\beta \subseteq \mathsf{N}(\mathsf{TU}).$$

Now suppose that $v \in \mathsf{N}(\mathsf{TU})$. Then

$$0 = \mathsf{TU}(v) = \mathsf{T}(\mathsf{U}(v)).$$

Thus $\mathsf{U}(v) \in \mathsf{N}(\mathsf{T})$. So there exist scalars $a_1, a_2, \ldots, a_p$ such that

$$\begin{aligned}\mathsf{U}(v) &= a_1 u_1 + a_2 u_2 + \cdots + a_p u_p \\ &= \mathsf{U}(a_1 w_1 + a_2 w_2 + \cdots + a_p w_p).\end{aligned}$$

Hence

$$\mathsf{U}(v - (a_1 w_1 + a_2 w_2 + \cdots + a_p w_p)) = 0.$$

We conclude that $v - (a_1 w_1 + \cdots + a_p w_p)$ lies in $\mathsf{N}(\mathsf{U})$. It follows that there exist scalars $b_1, b_2, \ldots, b_q$ such that

$$v - (a_1 w_1 + a_2 w_2 + \cdots + a_p w_p) = b_1 v_1 + b_2 v_2 + \cdots + b_q v_q$$

or

$$v = a_1 w_1 + a_2 w_2 + \cdots + a_p w_p + b_1 v_1 + b_2 v_2 + \cdots + b_q v_q.$$

Therefore, β spans $\mathsf{N}(\mathsf{TU})$.

We next show that β is linearly independent. Let $a_1, a_2, \ldots, a_p$, $b_1, b_2, \ldots, b_q$ be any scalars such that

$$a_1 w_1 + a_2 w_2 + \cdots + a_p w_p + b_1 v_1 + b_2 v_2 + \cdots + b_q v_q = 0. \tag{6}$$

Applying U to both sides of (6), we obtain

$$a_1 u_1 + a_2 u_2 + \cdots + a_p u_p = 0.$$

Since $\{u_1, u_2, \ldots, u_p\}$ is linearly independent, the a_i's are all zero. Thus (6) reduces to

$$b_1 v_1 + b_2 v_2 + \cdots + b_q v_q = 0.$$

Again, the linear independence of $\{v_1, v_2, \ldots, v_q\}$ implies that the b_i's are all zero. We conclude that β is a basis for $\mathsf{N}(\mathsf{TU})$. Hence $\mathsf{N}(\mathsf{TU})$ is finite-dimensional and $\dim(\mathsf{N}(\mathsf{TU})) = p + q = \dim(\mathsf{N}(\mathsf{T})) + \dim(\mathsf{N}(\mathsf{U}))$. ∎

Proof of Theorem 2.33. The proof is by mathematical induction on the order of the differential operator $p(\mathsf{D})$. The first-order case coincides with Theorem 2.31. For some integer $n > 1$ suppose that Theorem 2.33 holds for any differential operator of order less than n, and suppose we are given a differential operator $p(\mathsf{D})$ of order n. The polynomial $p(t)$ can be factored into a product of two polynomials

$$p(t) = q(t)(t - c)$$

for some polynomial $q(t)$ of degree $n - 1$ and for some complex number c. Thus the given differential operator may be rewritten as

$$p(\mathsf{D}) = q(\mathsf{D})(\mathsf{D} - c\mathsf{I}).$$

By Lemma 1, $\mathsf{D} - c\mathsf{I}$ is onto; by the corollary to Theorem 2.31, $\dim(\mathsf{N}(\mathsf{D} - c\mathsf{I})) = 1$; and by the induction hypothesis, $\dim(\mathsf{N}(q(\mathsf{D})) =$

$n-1$. Thus by applying Lemma 2 we conclude that

$$\begin{aligned}\dim(\mathsf{N}(p(\mathsf{D}))) &= \dim(\mathsf{N}(q(\mathsf{D}))) + \dim(\mathsf{N}(\mathsf{D} - c\mathsf{I}))\\ &= (n-1) + 1 = n. \quad \blacksquare\end{aligned}$$

Corollary. *For any* n*th-order homogeneous linear differential equation with constant coefficients, the solution space is an* n*-dimensional subspace of* C^∞.

The corollary to Theorem 2.33 reduces the problem of finding all solutions to an nth-order homogeneous linear differential equation with constant coefficients to finding a set of n linearly independent solutions to the equation. By the results of Chapter 1 any such set must be a basis for the solution space. The following theorem enables us to find a basis quickly for many such equations. Hints for its proof are provided in the exercises.

Theorem 2.34. *Given* n *distinct complex numbers* $c_1, c_2, \ldots, c_n$, *the set of exponential functions* $\{e^{c_1 t}, e^{c_2 t}, \ldots, e^{c_n t}\}$ *is linearly independent.*

Corollary. *For any* n*th-order homogeneous linear differential equation with constant coefficients, if its auxiliary polynomial* $p(t)$ *has* n *distinct zeros* $c_1, c_2, \ldots, c_n$, *then the set* $\{e^{c_1 t}, e^{c_2 t}, \ldots, e^{c_n t}\}$ *is a basis for the solution space of the equation.*

Proof. Exercise. $\blacksquare$

Example 5

We will find all solutions to the differential equation

$$y'' + 5y' + 4y = 0.$$

Since the auxiliary polynomial $p(t)$ factors as $(t+4)(t+1)$, $p(t)$ has two distinct zeros: -1 and -4. Thus $\{e^{-t}, e^{-4t}\}$ is a basis for the solution space. So any solution to the given equation is of the form

$$y(t) = b_1 e^{-t} + b_2 e^{-4t} \qquad \text{for some constants } b_1 \text{ and } b_2. \quad \blacksquare$$

Example 6

We find all solutions to the differential equation

$$y'' + 9y = 0.$$

The auxiliary polynomial $p(t) = t^2 + 9$ factors as $p(t) = (t-3i)(t+3i)$ and hence has distinct zeros $c_1 = 3i$, $c_2 = -3i$. Thus $\{e^{3it}, e^{-3it}\}$ is a basis for the solution space. A more useful basis is obtained by applying Exercise 7. Since

$$\cos 3t = \frac{1}{2}(e^{3it} + e^{-3it}) \quad \text{and} \quad \sin 3t = \frac{1}{2i}(e^{3it} - e^{-3it}),$$

it follows that $\{\cos 3t, \sin 3t\}$ is also a basis. This basis has an advantage over the original one in that it consists of the familiar sine and cosine functions and makes no reference to the imaginary number i. ∎

Next consider the differential equation

$$y'' + 2y' + y = 0,$$

for which the auxiliary polynomial is $p(t) = (t + 1)^2$. By Theorem 2.32, e^{-t} is a solution to the equation above. By the corollary to Theorem 2.33 its solution space is two-dimensional. In order to find a basis for the solution space we need to find a solution that is linearly independent of e^{-t}. The reader can verify that te^{-t} will do. Thus $\{e^{-t}, te^{-t}\}$ is a basis for the solution space. This result can be generalized as follows.

Theorem 2.35. *Let* $\mathrm{p(t) = (t - c)^n}$, *where* c *is a complex number and* n *is a positive integer, be the auxiliary polynomial of a homogeneous linear differential equation with constant coefficients. The set*

$$\beta = \{\mathrm{e^{ct}, te^{ct}, \ldots, t^{n-1}e^{ct}}\}$$

is a basis for the solution space.

Proof. Since the solution space is n-dimensional, we need only show that β is linearly independent and lies in the solution space. First, observe that for any positive integer k

$$(\mathsf{D} - c\mathsf{I})(t^k e^{ct}) = kt^{k-1}e^{ct} + ct^k e^{ct} - ct^k e^{ct}$$
$$= kt^{k-1}e^{ct}.$$

Hence for $k < n$,

$$(\mathsf{D} - c\mathsf{I})^n(t^k e^{ct}) = 0.$$

It follows that β is a subset of the solution space.

We next show that β is linearly independent. Consider any linear combination of β such that

$$b_1 t^{n-1}e^{ct} + b_2 t^{n-2}e^{ct} + \cdots + b_{n-1}te^{ct} + b_n e^{ct} = 0 \tag{7}$$

for some scalars $b_1, \ldots, b_n$. Dividing by e^{ct} in (7), we obtain

$$b_1 t^{n-1} + b_2 t^{n-2} + \cdots + b_{n-1}t + b_n = 0. \tag{8}$$

Thus the left-hand side of (8) must be the zero polynomial function. We conclude that the coefficients $b_1, b_2, \ldots, b_n$ are all zero. Thus β is linearly independent and hence is a basis for the solution space. ∎

Example 7

Given the differential equation

$$y^{(4)} - 4y^{(3)} + 6y^{(2)} - 4y^{(1)} + y = 0,$$

we wish to find a basis for the solution space. Since its auxiliary polynomial is

$$p(t) = t^4 - 4t^3 + 6t^2 - 4t + 1 = (t - 1)^4,$$

we can immediately conclude by Theorem 2.35 that $\{e^t, te^t, t^2e^t, t^3e^t\}$ is a basis for the solution space. So any solution to the given equation is of the form

$$y(t) = b_1e^t + b_2te^t + b_3t^2e^t + b_4t^3e^t$$

for some scalars b_1, b_2, b_3, and b_4. ∎

The most general situation (whose proof we leave as an exercise) is stated in the following theorem.

Theorem 2.36. *For a homogeneous linear differential equation with constant coefficients whose auxiliary polynomial is*

$$p(t) = (t - c_1)^{n_1}(t - c_2)^{n_2} \cdots (t - c_k)^{n_k},$$

where $n_1, n_2, \ldots, n_k$ *are positive integers and* $c_1, c_2, \ldots, c_k$ *are distinct complex numbers, the following set is a basis for the solution space of the equation:*

$$\{e^{c_1t}, te^{c_1t}, \ldots, t^{n_1-1}e^{c_1t}, \ldots, e^{c_kt}, te^{c_kt}, \ldots, t^{n_k-1}e^{c_kt}\}.$$

Example 8

Consider the differential equation

$$y^{(3)} - 4y^{(2)} + 5y^{(1)} - 2y = 0.$$

We will find a basis for its solution space. Since the auxiliary polynomial $p(t)$ factors as

$$p(t) = t^3 - 4t^2 + 5t - 2 = (t - 1)^2(t - 2),$$

we conclude that a basis for the solution space to the differential equation is

$$\{e^t, te^t, e^{2t}\}.$$

Thus any solution of the given equation is of the form

$$y(t) = b_1e^t + b_2te^t + b_3e^{2t}$$

for some scalars b_1, b_2, and b_3. ∎

EXERCISES

1. Label the following statements as being true or false.
   (a) The set of solutions to an nth-order homogeneous linear differential equation with constant coefficients is an n-dimensional subspace of C^∞.
   (b) The solution space of a homogeneous linear differential equation is the null space of a differential operator.

(c) The auxiliary polynomial of a homogeneous linear differential equation with constant coefficients is a solution to the differential equation.

(d) Any solution to a homogeneous linear differential equation with constant coefficients is of the form ae^{ct} or $at^k e^{ct}$, where a and c are complex numbers and k is a positive integer.

(e) Any linear combination of solutions to a given homogeneous linear differential equation with constant coefficients is also a solution to the given equation.

(f) For any homogeneous linear differential equation with constant coefficients having auxiliary polynomial $p(t)$, if $c_1, c_2, \ldots, c_k$ are the distinct zeros of $p(t)$, then $\{e^{c_1 t}, e^{c_2 t}, \ldots, e^{c_k t}\}$ is a basis for the solution space of the given differential equation.

(g) Given any polynomial $p(t) \in \mathsf{P}(C)$, there exists a homogeneous linear differential equation with constant coefficients whose auxiliary polynomial is $p(t)$.

2. For each of the following, determine whether the statement is true or false. Justify your claim with either a proof or counterexample, whichever is appropriate.

(a) Any finite-dimensional subspace of C^∞ is the solution space of a homogeneous linear differential equation with constant coefficients.

(b) There exists a homogeneous linear differential equation with constant coefficients whose solution space has $\{t, t^2\}$ as a basis.

(c) For any homogeneous linear differential equation with constant coefficients, if x is a solution to the equation, so is its derivative x'.

Given two polynomials $p(t)$ and $g(t)$ in $\mathsf{P}(C)$, if $x \in \mathsf{N}(p(\mathsf{D}))$ and $y \in \mathsf{N}(q(\mathsf{D}))$, then

(d) $x + y \in \mathsf{N}(p(\mathsf{D})q(\mathsf{D}))$.

(e) $xy \in \mathsf{N}(p(\mathsf{D})q(\mathsf{D}))$.

3. Find bases for the solution spaces of the following differential equations.

(a) $y'' + 2y' + y = 0$

(b) $y''' = y'$

(c) $y^{(4)} - 2y^{(2)} + y = 0$

(d) $y'' + 2y' + y = 0$

(e) $y^{(3)} - y^{(2)} + 3y^{(1)} + 5y = 0$

4. Find bases for the following subspaces of C^∞.

(a) $\mathsf{N}(\mathsf{D}^2 - \mathsf{D} - \mathsf{I})$

(b) $\mathsf{N}(\mathsf{D}^3 - 3\mathsf{D}^2 + 3\mathsf{D} - \mathsf{I})$

(c) $\mathsf{N}(\mathsf{D}^3 + 6\mathsf{D}^2 + 8\mathsf{D})$

5. Show that C^∞ is a subspace of $\mathscr{F}(R, C)$.

6. (a) Show that $\mathsf{D}\colon \mathsf{C}^\infty \to \mathsf{C}^\infty$ is a linear transformation.

(b) Show that any differential operator is a linear transformation on C^∞.

7. Prove that if $\{x, y\}$ is a basis for a vector space over C, then so is

$$\left\{\frac{1}{2}(x+y), \frac{1}{2i}(x-y)\right\}.$$

8. Given a second-order homogeneous linear differential equation with constant coefficients, suppose that the auxiliary polynomial has distinct conjugate complex roots $a+ib$ and $a-ib$, where $a, b \in R$. Show that $\{e^{at}\cos bt, e^{at}\sin bt\}$ is a basis for the solution space.

9. Given a collection of pairwise commutative linear transformations $\{\mathsf{U}_1, \mathsf{U}_2, \ldots, \mathsf{U}_n\}$ of a vector space V (i.e., transformations such that $\mathsf{U}_i\mathsf{U}_j = \mathsf{U}_j\mathsf{U}_i$ for all i, j), prove that for any $i = 1, 2, \ldots, n$

$$\mathsf{N}(\mathsf{U}_i) \subseteq \mathsf{N}(\mathsf{U}_1\mathsf{U}_2\cdots\mathsf{U}_n).$$

10. Prove Theorem 2.34 and its corollary. *Hint:* Suppose that

$$b_1e^{c_1t} + b_2e^{c_2t} + \cdots + b_ne^{c_nt} = 0 \quad \text{(where the } c_i\text{'s are distinct).}$$

To show the b_i's are zero, apply mathematical induction on n. Verify the theorem for $n = 1$. Assuming the theorem is true for any $n-1$ such functions, apply the operator $\mathsf{D} - c_n\mathsf{I}$ to both sides of the equation above to establish the theorem for n distinct exponential functions.

11. Prove Theorem 2.36. *Hint:* First verify that the alleged basis lies in the solution space. Then verify that this set is linearly independent by mathematical induction on k. The case $k = 1$ is Theorem 2.35. Assuming that the theorem holds for $k-1$ distinct c_i's, apply the operator $(\mathsf{D} - c_k\mathsf{I})^{n_k}$ to any linear combination of the alleged basis that equals 0.

12. Let V be the solution space of an nth-order homogeneous linear differential equation with constant coefficients having auxiliary polynomial $p(t)$. Prove that if $p(t) = g(t)h(t)$, where $g(t)$ and $h(t)$ are polynomials of positive degree, then

$$\mathsf{N}(h(\mathsf{D})) = \mathsf{R}(g(\mathsf{D}_\mathsf{V})) = g(\mathsf{D})(\mathsf{V}),$$

where $\mathsf{D}_\mathsf{V}\colon \mathsf{V} \to \mathsf{V}$ is defined by $\mathsf{D}_\mathsf{V}(x) = x'$ for $x \in \mathsf{V}$. *Hint:* First prove $g(\mathsf{D})(\mathsf{V}) \subseteq \mathsf{N}(h(\mathsf{D}))$. Then prove that the two spaces have the same finite dimension.

13. A differential equation

$$y^{(n)} + a_{n-1}y^{(n-1)} + \cdots + a_1y^{(1)} + a_0y = x$$

is called a *nonhomogeneous* linear differential equation with constant coefficients if the coefficients a_i are constant and the right side of the equation, x, is a function that is not identically zero.

(a) Prove that for any $x \in \mathsf{C}^\infty$ there exists a $y \in \mathsf{C}^\infty$ such that y is a solution to the equation above. *Hint:* Use Lemma 1 to Theorem 2.33 to show that if

$$p(t) = t^n + a_{n-1}t^{n-1} + \cdots + a_1 t + a_0,$$

then $p(\mathsf{D})\colon \mathsf{C}^\infty \to \mathsf{C}^\infty$ is onto.

(b) Let V be the solution space for the homogeneous linear equation

$$y^{(n)} + a_{n-1}y^{(n-1)} + \cdots + a_1 y^{(1)} + a_0 y = 0.$$

Prove that if z is any solution to the nonhomogeneous linear differential equation above, then the set of all solutions to the nonhomogeneous linear differential equation is

$$\{z + y \colon y \in \mathsf{V}\}.$$

14. Given any nth-order homogeneous linear differential equation with constant coefficients, prove that for any solution x and any $t_0 \in R$ if $x(t_0) = x'(t_0) = \cdots = x^{(n-1)}(t_0) = 0$, then $x = 0$ (the zero function). *Hint:* Use mathematical induction on n. First prove the conclusion for the case $n = 1$. Next suppose it to be true for equations of order $n - 1$, and consider an nth-order equation with auxiliary polynomial $p(t)$. Factor $p(t)$ as $p(t) = q(t)(t - c)$ for some complex number c and polynomial $q(t)$ of degree $n - 1$. Let $z = q(\mathsf{D})x$. Show that $z(t_0) = 0$ and z is a solution to the equation $y' - cy = 0$. Conclude that $z = 0$. Now apply the induction hypothesis.

15. Let V be the solution space of an nth-order homogeneous linear differential equation with constant coefficients. Fix $t_0 \in R$, and define a mapping $\Phi\colon \mathsf{V} \to \mathsf{C}^n$ by

$$\Phi(x) = \begin{pmatrix} x(t_0) \\ x'(t_0) \\ \vdots \\ x^{(n-1)}(t_0) \end{pmatrix} \qquad \text{for each } x \in \mathsf{V}.$$

(a) Prove that Φ is linear and its null space is trivial. Deduce that Φ is an isomorphism. *Hint:* Use Exercise 14.

(b) Prove the following: For any nth-order homogeneous linear differential equation with constant coefficients, any $t_0 \in R$, and any complex numbers $c_0, c_1, \ldots, c_{n-1}$ (not necessarily distinct), there exists exactly one solution, x, to the given differential equation such that $x^{(k)}(t_0) = c_k$ for $k = 0, 1, \ldots, n - 1$.

16. *Pendular Motion.* It is well-known that the motion of a pendulum is approximated by the differential equation

$$\theta'' + \frac{g}{l}\theta = 0,$$

where $\theta(t)$ is the angle in radians that the pendulum makes with a vertical

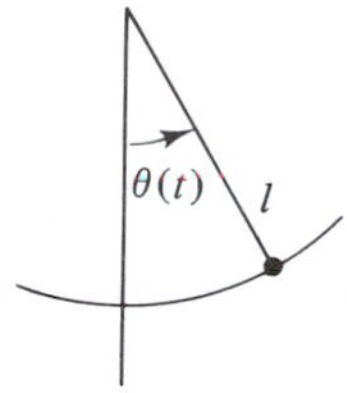

Figure 2.8

line at time t (see Figure 2.8) interpreted so that θ is positive if the pendulum is to the right and negative if the pendulum is to the left of the vertical line as viewed by the reader. Here l is the length of the pendulum and g is the magnitude of acceleration due to gravity. The variable t and constants l and g must be in compatible units (e.g., t in seconds, l in meters, and g in meters per second per second).

(a) Express an arbitrary solution to this equation as a linear combination of two fixed real-valued functions.

(b) Find the unique solution to the equation that satisfies the conditions

$$\theta(0) = \theta_0 > 0 \quad \text{and} \quad \theta'(0) = 0.$$

(The significance of the above is that at time $t = 0$ the pendulum is displaced from the vertical by θ_0 radians and has zero velocity.)

(c) Prove that it takes $2\pi\sqrt{l/g}$ units of time for the pendulum to make one circuit back and forth. (This time is called the *period* of the pendulum.)

17. *Periodic Motion of a Spring with Damping.* At the beginning of this section we discussed the motion of an oscillating spring under the assumption that the only force acting on the spring was the force due to the tension of the spring. We found in this case that (3) describes the motion of the spring.

(a) Find the general form of all solutions to (3).

If we analyze the behavior of the general solution in part (a), we see that the solution is a periodic function. Hence (3) indicates that the spring will never stop oscillating. We know from experience, however, that the amplitude of the oscillation will decrease until motion finally ceases. The reason that the solutions in part (a) do not exhibit this behavior is that we ignored the effect of friction on the moving weight. At low speeds such as those under consideration the resistance of the air provides an example of viscous damping—the resistance is proportional to the velocity of the moving weight but opposite in direction. To correct for air resistance, we must add the term $-ry'$ to (2). The constant $r > 0$ depends on the medium in which the motion takes place (in this case, air), and the term $-ry'$ has a negative sign because the resistance is always opposite to the direction of the motion. Thus the differential equation of motion is $my'' = -ry' - ky$; that is,

$$my'' + ry' + ky = 0.$$

(b) Find the general solution to this equation.

(c) Find the unique solution in part (b) that satisfies the initial conditions $y(0) = 0$ and $y'(0) = v_0$.

(d) For $y(t)$ as in part (c), show that the amplitude of the oscillation decreases to zero; that is, prove that $\lim_{t \to \infty} y(t) = 0$.

18. At the beginning of this section it was stated that it is useful to view solutions to differential equations as complex-valued functions of a real variable even though solutions that are meaningful to us in a physical sense are real-valued. Justify this point of view.

19. The following set of exercises does not involve linear algebra. We list them for the sake of completeness.

(a) Prove Theorem 2.28. *Hint:* Use mathematical induction on the number of derivatives possessed by a solution.

(b) For any $c, d \in C$, prove

(i) $e^{c+d} = e^c e^d$.

(ii) $e^{-c} = \dfrac{1}{e^c}$.

(c) Prove Theorem 2.30.

(d) Verify the product rule of differentiation for complex-valued functions of a real variable: For any differentiable functions x and y in $\mathscr{F}(R, C)$ the product xy is differentiable and

$$(xy)' = x'y + xy'.$$

Hint: Find the real and imaginary parts of xy in terms of those of x and y; then differentiate.

(e) Prove that if $x \in \mathscr{F}(R, C)$ and $x' = 0$, then x is a constant function.

INDEX OF DEFINITIONS FOR CHAPTER 2

3

Elementary Matrix Operations and Systems of Linear Equations

This chapter is devoted to two related objectives:

1. The study of certain "rank-preserving" operations on matrices
2. The application of these operations and the theory of linear transformations to the solution of systems of linear equations

As a consequence of objective 1 we will obtain a simple method for computing the rank of a linear transformation between finite-dimensional vector spaces by applying these rank-preserving matrix operations to a matrix that represents that transformation.

The solution of systems of linear equations is probably the most important application of linear algebra. The familiar method of elimination for solving systems of linear equations, which was discussed in Section 1.4, involves the elimination of variables so that a simpler system can be obtained. The technique by which the variables are eliminated utilizes three types of operations:

1. Interchanging any two equations in the system.
2. Multiplying any equation in the system by a nonzero constant.
3. Adding a multiple of one equation to another equation.

We will see in Section 3.3 that a system of linear equations can be expressed as a single matrix equation. In this representation of the system the three operations above are the "elementary row operations" for matrices. These operations will provide a convenient computational method for determining all solutions of a system of linear equations.

3.1 ELEMENTARY MATRIX OPERATIONS AND ELEMENTARY MATRICES

In this section we define the elementary matrix operations that will be used throughout the chapter. In subsequent sections we will use these operations to obtain simple computational methods for determining the rank of a linear transformation and the solution to a system of linear equations. There are two types of elementary matrix operations—row operations and column operations. As we will see, the row operations are more useful. They arise from the three operations that can be used to eliminate variables in a system of linear equations.

Let A be an $m \times n$ matrix over a field F. Recall that A can be considered as an array of m rows,

$$A = \begin{pmatrix} A_{(1)} \\ A_{(2)} \\ \vdots \\ A_{(m)} \end{pmatrix}$$

or as an array of n columns, $A = (A^{(1)}, A^{(2)}, \ldots, A^{(n)})$.

Definitions. *Let* A *be an* m × n *matrix, as above. Any one of the following three operations on the rows [columns] of* A *is called an* elementary row [column] operation:

(a) *Interchanging any two rows [columns] of* A.

(b) *Multiplying any row [column] of* A *by a nonzero constant.*

(c) *Adding any constant multiple of a row [column] of* A *to another row [column].*

Any of the three operations above will be called elementary operations. *Elementary operations are either of* type 1, type 2, *or* type 3 *depending on whether they are obtained by (a), (b), or (c).*

Example 1

Let

$$A = \begin{pmatrix} 1 & 2 & 3 & 4 \\ 2 & 1 & -1 & 3 \\ 4 & 0 & 1 & 2 \end{pmatrix}.$$

Interchanging $A_{(2)}$ with $A_{(1)}$ is an example of an elementary row operation of type 1. The resulting matrix is

$$B = \begin{pmatrix} 2 & 1 & -1 & 3 \\ 1 & 2 & 3 & 4 \\ 4 & 0 & 1 & 2 \end{pmatrix}.$$

Again, multiplying $A^{(2)}$ by 3 is an example of an elementary column operation of type 2. The resulting matrix is

$$C = \begin{pmatrix} 1 & 6 & 3 & 4 \\ 2 & 3 & -1 & 3 \\ 4 & 0 & 1 & 2 \end{pmatrix}.$$

Finally, adding to $A_{(1)}$ four times $A_{(3)}$ is an example of an elementary row operation of type 3. The resulting matrix is

$$D = \begin{pmatrix} 17 & 2 & 7 & 12 \\ 2 & 1 & -1 & 3 \\ 4 & 0 & 1 & 2 \end{pmatrix}. \quad \blacksquare$$

Definition. *An* n × n elementary matrix *is a matrix obtained by performing an elementary operation on* I_n. *The elementary matrix is said to be of* type 1, 2, *or* 3 *according to whether the elementary operation performed on* I_n *is a type 1, 2, or 3 operation, respectively.*

For example, interchanging the first two rows of I_3 produces the elementary matrix

$$E = \begin{pmatrix} 0 & 1 & 0 \\ 1 & 0 & 0 \\ 0 & 0 & 1 \end{pmatrix}.$$

Note that E can also be obtained by interchanging the first two columns of I_3. In fact *any elementary matrix can be obtained in at least two ways*—either by performing an elementary row operation on I_n or by performing an elementary column operation on I_n. Similarly,

$$\begin{pmatrix} 1 & 0 & -2 \\ 0 & 1 & 0 \\ 0 & 0 & 1 \end{pmatrix}$$

is an elementary matrix since it can be obtained from I_3 by an elementary column operation of type 3 (adding -2 times the first column of I_3 to the third column) or by an elementary row operation of type 3 (adding -2 times the third row to the first row).

Our first theorem shows that performing an elementary operation on a matrix is equivalent to multiplying the matrix by an elementary matrix.

Theorem 3.1. *Let* $A \in M_{m \times n}(F)$, *and suppose that* B *is obtained from* A *by performing an elementary row [column] operation. Then there exists an* $m \times m$ $[n \times n]$ *elementary matrix* E *such that* $B = EA$ $[B = AE]$. *In fact,* E *is obtained by performing the corresponding row [column] operation on* I_m $[I_n]$. *Conversely, if* E *is an elementary* $m \times m$ $[n \times n]$ *matrix, then* EA [AE] *is a matrix that can be obtained by performing an elementary row* [column] *operation on* A.

The proof, which we omit, requires verifying Theorem 3.1 for each type of elementary row operation. The proof for column operations can then be obtained by using the matrix transpose to transform a column operation into a row operation. The details are left as an exercise.

The following example illustrates the use of the theorem.

Example 2

Consider the matrix B in Example 1. This matrix was obtained from A (in Example 1) by interchanging the first two rows of A. Performing this same operation on I_3, we obtain the elementary matrix

$$E = \begin{pmatrix} 0 & 1 & 0 \\ 1 & 0 & 0 \\ 0 & 0 & 1 \end{pmatrix}.$$

Note that $EA = B$.

In the second part of Example 1, C is obtained from A by multiplying the second column of A by 3. Performing this same operation on I_4, we obtain the elementary matrix

$$E = \begin{pmatrix} 1 & 0 & 0 & 0 \\ 0 & 3 & 0 & 0 \\ 0 & 0 & 1 & 0 \\ 0 & 0 & 0 & 1 \end{pmatrix}.$$

Observe that $AE = C$. ∎

It is a useful fact that the inverse of an elementary matrix is also an elementary matrix.

Theorem 3.2. *Elementary matrices are invertible, and the inverse of an elementary matrix is an elementary matrix of the same type.*

Proof. In view of the fact that any elementary $n \times n$ matrix can be obtained by an elementary row operation on I_n, we need consider only three cases—one for each type of operation.

Let E be an elementary $n \times n$ matrix.

CASE 1. Suppose that E is obtained by interchanging the pth and the qth rows of I_n $(p \neq q)$, an elementary row operation of type 1. It is easy to verify that $E^2 = I_n$. Hence E is invertible, and in fact $E = E^{-1}$.

CASE 2. Suppose that E is obtained by multiplying the pth row of I_n by a nonzero constant c, an elementary row operation of type 2. Since $c \neq 0$, c has a multiplicative inverse. Let $\bar{E}$ be the elementary matrix obtained from I_n by the elementary row operation of multiplying the pth row of I_n by c^{-1}. It is easily shown that $E\bar{E} = \bar{E}E = I_n$, and hence $\bar{E} = E^{-1}$.

CASE 3. Suppose that E is obtained by adding to the pth row of I_n c times the qth row of I_n, where $p \neq q$ and c is any scalar. Thus E can be obtained from I_n by an elementary row operation of type 3.

Observe that I_n can be obtained from E via an elementary row operation of type 3, namely, by adding to the pth row $-c$ times the qth row of E. By Theorem 3.1 there is an elementary matrix $\bar{E}$ (of type 3) such that $\bar{E}E = I_n$. Thus by Exercise 8 of Section 2.4 E is invertible and $E^{-1} = \bar{E}$. ∎

EXERCISES

1. Label the following statements as being true or false.
   (a) An elementary matrix is always square.
   (b) The only entries of an elementary matrix are zeros and ones.
   (c) The $n \times n$ identity matrix is an elementary matrix.
   (d) The product of two $n \times n$ elementary matrices is an elementary matrix.
   (e) The inverse of an elementary matrix is an elementary matrix.
   (f) The sum of two $n \times n$ elementary matrices is an elementary matrix.
   (g) The transpose of an elementary matrix is an elementary matrix.
   (h) If B is a matrix that can be obtained by performing an elementary row operation on a matrix A, then B can also be obtained by performing an elementary column operation on A.
   (i) If B is a matrix that can be obtained by performing an elementary row operation on a matrix A, then A can be obtained by performing an elementary row operation on B.
2. Let

$$A = \begin{pmatrix} 1 & 2 & 3 \\ 1 & 0 & 1 \\ 1 & -1 & 1 \end{pmatrix}, \quad B = \begin{pmatrix} 1 & 0 & 3 \\ 1 & -2 & 1 \\ 1 & -3 & 1 \end{pmatrix}, \quad \text{and} \quad C = \begin{pmatrix} 1 & 0 & 3 \\ 0 & -2 & -2 \\ 1 & -3 & 1 \end{pmatrix}.$$

Find an elementary operation which will transform A into B and an elementary operation which will transform B into C. By means of several additional elementary operations, transform C into I_3.

3. Prove the assertion made on page 129: Any elementary $n \times n$ matrix can be obtained in at least two ways—either by performing an elementary row operation on I_n or by performing an elementary column operation on I_n.
4. Prove that E is an elementary matrix if and only if E^t is.
5. Let A be an $m \times n$ matrix. Prove that if B can be obtained from A by an elementary row [column] operation, then B^t can be obtained from A^t by the corresponding elementary column [row] operation.
6. Prove Theorem 3.1.
7. Verify the assertion made in case 1 of the proof of Theorem 3.2: If E is an elementary $n \times n$ matrix of type 1, then $E^2 = I_n$.
8. Verify that for the matrix $\bar{E}$ defined in the proof of case 2 of Theorem 3.2 $E\bar{E} = \bar{E}E = I_n$.
9. Prove that any elementary row [column] operation of type 1 can be obtained by a succession of three elementary row [column] operations of type 3 followed by one elementary row [column] operation of type 2.
10. Prove that any elementary row [column] operation of type 2 can be obtained by *dividing* some row [column] by a nonzero scalar.
11. Prove that any elementary row [column] operation of type 3 can be obtained by *subtracting* a multiple of some row [column] from another row [column].

3.2 THE RANK OF A MATRIX AND MATRIX INVERSES

In this section we define the rank of a matrix. We will then use elementary operations to compute the rank of a matrix or a linear transformation. The section concludes with a procedure for computing the inverse of an invertible matrix.

Definition. *If* $A \in M_{m \times n}(F)$, *we define the* rank of A, *denoted* rank(A), *to be the rank of the linear transformation* $L_A: F^n \to F^m$.

Many results about the rank of matrices follow immediately from the corresponding facts about linear transformations. An important result of this type, which follows from Theorem 2.5 and Corollary 2 of Theorem 2.19, is that an $n \times n$ matrix is invertible if and only if its rank is n.

We would like the definition above to satisfy the condition that the rank of a linear transformation is equal to the rank of any matrix representing that transformation. Our first theorem shows that this condition is, in fact, fulfilled.

Theorem 3.3. *Let* $\mathsf{T}\colon \mathsf{V} \to \mathsf{W}$ *be a linear transformation between finite-dimensional vector spaces, and let* β *and* γ *be ordered bases for* V *and* W, *respectively. Then* $\operatorname{rank}(\mathsf{T}) = \operatorname{rank}([\mathsf{T}]_\beta^\gamma)$.

Proof. This is a restatement of Exercise 18 of Section 2.4. ∎

Now that the problem of finding the rank of a linear transformation has been reduced to the problem of finding the rank of a matrix, we need a result that will allow us to perform rank-preserving operations on matrices. The next theorem and its corollary tell us how to do this.

Theorem 3.4. *Let* A *be an* $m \times n$ *matrix. If* P *and* Q *are invertible* $m \times m$ *and* $n \times n$ *matrices, respectively, then*

(a) $\operatorname{rank}(AQ) = \operatorname{rank}(A)$,

(b) $\operatorname{rank}(PA) = \operatorname{rank}(A)$,

and therefore,

(c) $\operatorname{rank}(PAQ) = \operatorname{rank}(A)$.

Proof. First observe that $\mathsf{R}(\mathsf{L}_{AQ}) = \mathsf{R}(\mathsf{L}_A\mathsf{L}_Q) = \mathsf{L}_A\mathsf{L}_Q(\mathsf{F}^n) = \mathsf{L}_A(\mathsf{L}_Q(\mathsf{F}^n)) = \mathsf{L}_A(\mathsf{F}^n) = \mathsf{R}(\mathsf{L}_A)$ since L_Q is onto. Therefore,

$$\begin{aligned}\operatorname{rank}(AQ) &= \dim(\mathsf{R}(\mathsf{L}_{AQ}))\\ &= \dim(\mathsf{R}(\mathsf{L}_A))\\ &= \operatorname{rank}(A).\end{aligned}$$

This establishes (a). To establish (b), apply Exercise 15 of Section 2.4 to $\mathsf{T} = \mathsf{L}_P$. We omit the details. Finally, applying (a) and (b), we have that

$$\operatorname{rank}(PAQ) = \operatorname{rank}(PA) = \operatorname{rank}(A). \quad ∎$$

Corollary. *Elementary row and column operations on a matrix are rank-preserving.*

Proof. If B is obtained from the matrix A by an elementary row operation, then there exists an elementary matrix E such that $B = EA$. By Theorem 3.2, E is invertible, and hence $\operatorname{rank}(B) = \operatorname{rank}(A)$ by Theorem 3.4. The proof that elementary column operations are rank-preserving is left as an exercise. ∎

Now that we have a class of matrix operations that preserve rank we need a way of examining a transformed matrix to ascertain its rank. The next theorem is the first of several that enable us to determine the rank of a matrix.

Theorem 3.5. *The rank of any matrix equals the maximum number of its linearly independent columns; that is, the rank of a matrix is the dimension of the subspace generated by its columns.*

Proof. For any $A \in \mathsf{M}_{m \times n}(F)$,

$$\text{rank}(A) = \text{rank}(\mathsf{L}_A) = \dim(\mathsf{R}(\mathsf{L}_A)).$$

Let $\beta = \{e_1, e_2, \ldots, e_n\}$ be the standard ordered basis for F^n. Then β spans F^n and hence by Theorem 2.2

$$\mathsf{R}(\mathsf{L}_A) = \text{span}\{\mathsf{L}_A(e_1), \mathsf{L}_A(e_2), \ldots, \mathsf{L}_A(e_n)\}.$$

But we have seen that $\mathsf{L}_A(e_j) = A^{(j)}$. Hence

$$\mathsf{R}(\mathsf{L}_A) = \text{span}\{A^{(1)}, A^{(2)}, \ldots, A^{(n)}\}.$$

Thus

$$\text{rank}(A) = \dim(\mathsf{R}(\mathsf{L}_A)) = \dim(\text{span}\{A^{(1)}, A^{(2)}, \ldots, A^{(n)}\}). \quad \blacksquare$$

Example 1

Let

$$A = \begin{pmatrix} 1 & 0 & 1 \\ 0 & 1 & 1 \\ 1 & 0 & 1 \end{pmatrix}.$$

Observe that the first and second columns of A are linearly independent and that the third column is a linear combination of the first two. Thus

$$\text{rank}(A) = \dim\left(\text{span}\left\{\begin{pmatrix} 1 \\ 0 \\ 1 \end{pmatrix}, \begin{pmatrix} 0 \\ 1 \\ 0 \end{pmatrix}, \begin{pmatrix} 1 \\ 1 \\ 1 \end{pmatrix}\right\}\right) = 2. \quad \blacksquare$$

To compute the rank of a matrix A, it is frequently useful to postpone the use of Theorem 3.5 until A has been suitably modified by means of appropriate elementary row and column operations so that the number of linearly independent columns is obvious. The corollary to Theorem 3.4 guarantees that the rank of the modified matrix is the same as the rank of A. One such modification of A can be obtained by using elementary row and column operations to introduce zero entries. The following example illustrates this procedure.

Example 2

Let

$$A = \begin{pmatrix} 1 & 2 & 1 \\ 1 & 0 & 3 \\ 1 & 1 & 2 \end{pmatrix}.$$

If we subtract the first row of A from rows 2 and 3 (type 3 elementary row operations), the result is

$$\begin{pmatrix} 1 & 2 & 1 \\ 0 & -2 & 2 \\ 0 & -1 & 1 \end{pmatrix}.$$

If we now subtract twice the first column from the second and subtract the first column from the third (type 3 elementary column operations), we obtain

$$\begin{pmatrix} 1 & 0 & 0 \\ 0 & -2 & 2 \\ 0 & -1 & 1 \end{pmatrix}.$$

It is now obvious that the maximum number of linearly independent columns of this matrix is 2. Hence the rank of A is 2. ∎

The next theorem uses this process of modifying a matrix by means of elementary row and column operations to transform it into a particularly simple form. The power of this theorem can be seen in its corollaries.

Theorem 3.6. *Let* A *be an* $m \times n$ *matrix of rank* r. *Then* $r \leqslant m$, $r \leqslant n$, *and by means of a finite number of elementary row and column operations* A *can be transformed into a matrix* D *such that*

(a) $D_{ij} = 0$ *for* $i \neq j$,
(b) $D_{ii} = 1$ *for* $i \leqslant r$,
(c) $D_{ii} = 0$ *for* $i > r$.

Theorem 3.6 and its corollaries are quite important. Its proof, though easy to understand, is tedious to read. As an aid in following the proof we first consider an example.

Example 3

Consider the matrix

$$A = \begin{pmatrix} 0 & 2 & 4 & 2 & 2 \\ 4 & 4 & 4 & 8 & 0 \\ 8 & 2 & 0 & 10 & 2 \\ 6 & 3 & 2 & 9 & 1 \end{pmatrix}.$$

By means of a succession of elementary row and column operations we will transform A into a matrix D as in Theorem 3.6. We list many of the intermediate matrices, but on several occasions a matrix is transformed from the preceding one by means of several elementary operations. The number above each arrow indicates how many operations are involved. Try to identify the nature of each operation (row or column and type).

$$\begin{pmatrix} 0 & 2 & 4 & 2 & 2 \\ 4 & 4 & 4 & 8 & 0 \\ 8 & 2 & 0 & 10 & 2 \\ 6 & 3 & 2 & 9 & 1 \end{pmatrix} \xrightarrow{1} \begin{pmatrix} 4 & 4 & 4 & 8 & 0 \\ 0 & 2 & 4 & 2 & 2 \\ 8 & 2 & 0 & 10 & 2 \\ 6 & 3 & 2 & 9 & 1 \end{pmatrix} \xrightarrow{1} \begin{pmatrix} 1 & 1 & 1 & 2 & 0 \\ 0 & 2 & 4 & 2 & 2 \\ 8 & 2 & 0 & 10 & 2 \\ 6 & 3 & 2 & 9 & 1 \end{pmatrix} \xrightarrow{2}$$

$$\begin{pmatrix} 1 & 1 & 1 & 2 & 0 \\ 0 & 2 & 4 & 2 & 2 \\ 0 & -6 & -8 & -6 & 2 \\ 0 & -3 & -4 & -3 & 1 \end{pmatrix} \xrightarrow{3} \begin{pmatrix} 1 & 0 & 0 & 0 & 0 \\ 0 & 2 & 4 & 2 & 2 \\ 0 & -6 & -8 & -6 & 2 \\ 0 & -3 & -4 & -3 & 1 \end{pmatrix} \xrightarrow{1}$$

$$\begin{pmatrix} 1 & 0 & 0 & 0 & 0 \\ 0 & 1 & 2 & 1 & 1 \\ 0 & -6 & -8 & -6 & 2 \\ 0 & -3 & -4 & -3 & 1 \end{pmatrix} \xrightarrow{2} \begin{pmatrix} 1 & 0 & 0 & 0 & 0 \\ 0 & 1 & 2 & 1 & 1 \\ 0 & 0 & 4 & 0 & 8 \\ 0 & 0 & 2 & 0 & 4 \end{pmatrix} \xrightarrow{3} \begin{pmatrix} 1 & 0 & 0 & 0 & 0 \\ 0 & 1 & 0 & 0 & 0 \\ 0 & 0 & 4 & 0 & 8 \\ 0 & 0 & 2 & 0 & 4 \end{pmatrix} \xrightarrow{1}$$

$$\begin{pmatrix} 1 & 0 & 0 & 0 & 0 \\ 0 & 1 & 0 & 0 & 0 \\ 0 & 0 & 1 & 0 & 2 \\ 0 & 0 & 2 & 0 & 4 \end{pmatrix} \xrightarrow{1} \begin{pmatrix} 1 & 0 & 0 & 0 & 0 \\ 0 & 1 & 0 & 0 & 0 \\ 0 & 0 & 1 & 0 & 2 \\ 0 & 0 & 0 & 0 & 0 \end{pmatrix} \xrightarrow{1} \begin{pmatrix} 1 & 0 & 0 & 0 & 0 \\ 0 & 1 & 0 & 0 & 0 \\ 0 & 0 & 1 & 0 & 0 \\ 0 & 0 & 0 & 0 & 0 \end{pmatrix} = D.$$

By the corollary to Theorem 3.4, $\text{rank}(A) = \text{rank}(D)$. But clearly $\text{rank}(D) = 3$; so $\text{rank}(A) = 3$. ∎

Note that the first two elementary operations in Example 3 result in a 1 in the 1, 1 position, and the next several operations (type 3) result in 0's everywhere in the first row and first column except for the 1, 1 position. Subsequent elementary operations do not change the first row and first column. With this example in mind we proceed with the proof of Theorem 3.6.

Proof of Theorem 3.6. If A is the zero matrix, $r = 0$ by Exercise 3. In this case the conclusion follows with $D = A$.

Now suppose $A \neq O$ and $r = \text{rank}(A)$; then $r > 0$. The proof will be by mathematical induction on m, the number of rows of A.

Suppose that $m = 1$. By means of at most one type 1 column operation and at most one type 2 column operation, A can be transformed into a matrix with a 1 in the 1, 1 position. By means of at most $n - 1$ type 3 column operations this matrix, in turn, can be transformed into the matrix

$$D = (1 \quad 0 \quad 0 \quad 0).$$

Note that there is a maximum of one linearly independent column in D. So $\text{rank}(D) = \text{rank}(A) = 1$ by the corollary to Theorem 3.4 and by Theorem 3.5. Thus the theorem is established for $m = 1$.

Next assume that the theorem holds for any matrix with at most $m - 1$ rows (for some $m > 1$). We will prove that the theorem holds for any matrix with m rows.

Suppose that A is any $m \times n$ matrix. If $n = 1$, Theorem 3.6 can be established in a manner analogous to that for $m = 1$ (see Exercise 10).

We now suppose that $n > 1$. Since $A \neq O$, $A_{ij} \neq 0$ for some i, j. By means of at most one row and one column operation (each of type 1) we can move the nonzero entry into the 1, 1 position (just as was done in Example 3). By means of at most one additional type 2 operation we can assure a 1 in the 1, 1 position. (Look at the second operation in Example 3.) By means of at most $m - 1$ type 3 row operations and $n - 1$ type 3 column operations, we can eliminate all nonzero entries in the first row and the first column with the exception of the 1 in the 1, 1 position. (In Example 3 we used two row and three column operations to do this.)

Thus with a finite number of elementary operations A can be transformed into a matrix

$$B = \left(\begin{array}{c|ccc} 1 & 0 & \cdots & 0 \\ \hline 0 & & & \\ \vdots & & B' & \\ 0 & & & \end{array}\right),$$

where B' is an $(m - 1) \times (n - 1)$ matrix. In Example 3

$$B' = \begin{pmatrix} 2 & 4 & 2 & 2 \\ -6 & -8 & -6 & 2 \\ -3 & -4 & -3 & 1 \end{pmatrix}.$$

By Exercise 11, B' has rank one less than B. Since $\text{rank}(A) = \text{rank}(B) = r$, $\text{rank}(B') = r - 1$. By the induction hypothesis $r - 1 \leqslant n - 1$ and $r - 1 \leqslant m - 1$. Hence $r \leqslant m$ and $r \leqslant n$.

Also by the induction hypothesis B' can be transformed by means of a finite number of elementary row and column operations into an $(m - 1) \times (n - 1)$ matrix D' such that

$$\begin{aligned} (D')_{i,j} &= 0 \qquad \text{if } i \neq j, \\ (D')_{i,i} &= 1 \qquad \text{if } i \leqslant r - 1, \\ (D')_{i,i} &= 0 \qquad \text{if } i \geqslant r. \end{aligned}$$

That is, D' consists of all zeros except for ones in the first $r - 1$ positions of the main diagonal. Let

$$D = \left(\begin{array}{c|ccc} 1 & 0 & \cdots & 0 \\ \hline 0 & & & \\ \vdots & & D' & \\ 0 & & & \end{array}\right).$$

We see that the theorem now follows once we show that D can be obtained from B by means of a finite number of elementary row and column operations. But this follows by repeated applications of Exercise 12.

Thus since A can be transformed into B and B can be transformed into D, each by a finite number of elementary operations, A can be transformed into D by a finite number of elementary operations.

Finally, since D' contains ones in its first $r-1$ positions along the main diagonal, D contains ones in the first r positions along its main diagonal and zeros elsewhere. Thus $D_{ii} = 1$ if $i \leqslant r$, $D_{ii} = 0$ if $i > r$, and $D_{ij} = 0$ if $i \neq j$. This establishes the theorem. ∎

Corollary 1. *Let* A *be an* $m \times n$ *matrix of rank* r. *Then there exist invertible matrices* B *and* C *of dimensions* $m \times m$ *and* $n \times n$, *respectively, such that* $D = BAC$, *where* D *is an* $m \times n$ *matrix satisfying*

(a) $D_{ij} = 0$ *if* $i \neq j$,
(b) $D_{ii} = 1$ *if* $i \leqslant r$,
(c) $D_{ii} = 0$ *if* $i > r$.

Proof. By Theorem 3.6, A can be transformed by means of a finite number of elementary row and column operations into the matrix D. We can appeal to Theorem 3.1 each time we perform an elementary operation. Thus there exist elementary $m \times m$ matrices $E_1, E_2, \ldots, E_p$ and elementary $n \times n$ matrices $G_1, G_2, \ldots, G_q$ such that

$$D = E_p E_{p-1} \cdots E_2 E_1 A G_1 G_2 \cdots G_q.$$

By Theorem 3.2 each E_j and G_j is invertible. Let $B = E_p E_{p-1} \cdots E_1$ and $C = G_1 \cdots G_q$. Then B and C are invertible by Exercise 2 of Section 2.4, and $D = BAC$. ∎

Corollary 2. *Let* A *be any* $m \times n$ *matrix.*

(a) $\text{rank}(A^t) = \text{rank}(A)$.
(b) *The rank of any matrix equals the maximum number of its linearly independent rows; that is, the rank of a matrix is the dimension of the subspace generated by its rows.*
(c) *The rows and columns of any matrix generate subspaces of the same dimension, numerically equal to the rank of the matrix.*

Proof. (a) By Corollary 1 there exist invertible matrices B and C such that

$D = BAC$, where D satisfies the stated conditions of the corollary. Taking transposes, we have

$$D^t = C^t A^t B^t.$$

Since B and C are invertible, so are B^t and C^t by Exercise 3 of Section 2.4. Hence by Theorem 3.4

$$\text{rank}(A^t) = \text{rank}(C^t A^t B^t) = \text{rank}(D^t).$$

Suppose that $r = \text{rank}(A)$. Then D^t is an $n \times m$ matrix satisfying the conditions of Corollary 1, and hence $\text{rank}(D^t) = r$ by Theorem 3.5. Thus

$$\text{rank}(A^t) = \text{rank}(D^t) = r = \text{rank}(A).$$

This establishes (a).

The proofs of (b) and (c) are left as exercises. ∎

Corollary 3. *Any invertible matrix is a product of elementary matrices.*

Proof. If A is an invertible $n \times n$ matrix, then $\text{rank}(A) = n$. Hence by Corollary 1 there exist invertible matrices B and C such that $D = BAC$, where $D_{ij} = 0$ for $i \neq j$ and $D_{ii} = 1$ for $1 \leqslant i \leqslant n$. Thus $D = I_n$; that is, $I_n = BAC$.

As in the proof of Corollary 1, note also that $B = E_p E_{p-1} \cdots E_1$ and $C = G_1 G_2 \cdots G_q$, where the E_i's and the G_i's are elementary matrices. Thus $A = B^{-1} I_n C^{-1} = B^{-1} C^{-1}$, so that $A = E_1^{-1} E_2^{-1} \cdots E_p^{-1} G_q^{-1} G_{q-1}^{-1} \cdots G_1^{-1}$. But the inverse of an elementary matrix is elementary, and hence A is the product of elementary matrices. ∎

We now use Corollary 2 to relate the rank of a matrix product to the rank of each factor. Notice how the proof exploits the relationship between the rank of a matrix and the rank of a linear transformation.

Proposition 3.7. *Let* $\mathsf{T}: \mathsf{V} \to \mathsf{W}$ *and* $\mathsf{U}: \mathsf{W} \to \mathsf{Z}$ *be linear transformations on finite-dimensional vector spaces* V, W, *and* Z, *and let* A *and* B *be matrices such that the product* AB *is defined. Then*

(a) $\text{rank}(\mathsf{UT}) \leqslant \text{rank}(\mathsf{U})$.
(b) $\text{rank}(\mathsf{UT}) \leqslant \text{rank}(\mathsf{T})$.
(c) $\text{rank}(AB) \leqslant \text{rank}(A)$.
(d) $\text{rank}(AB) \leqslant \text{rank}(B)$.

Proof. Clearly, $\mathsf{R}(\mathsf{T}) \subseteq \mathsf{W}$. Hence

$$\mathsf{R}(\mathsf{UT}) = \mathsf{UT}(\mathsf{V}) = \mathsf{U}(\mathsf{R}(\mathsf{T})) \subseteq \mathsf{U}(\mathsf{W}) = \mathsf{R}(\mathsf{U}).$$

Thus

$$\text{rank}(\mathsf{UT}) = \dim(\mathsf{R}(\mathsf{UT})) \leqslant \dim(\mathsf{R}(\mathsf{U})) = \text{rank}(\mathsf{U}).$$

This establishes part (a).

By part (a)

$$\text{rank}(AB) = \text{rank}(\mathsf{L}_{AB}) = \text{rank}(\mathsf{L}_A \mathsf{L}_B) \leqslant \text{rank}(\mathsf{L}_A) = \text{rank}(A).$$

This establishes part (c).

By part (c) and Corollary 2 of Theorem 3.6

$$\text{rank}(AB) = \text{rank}((AB)^t) = \text{rank}(B^t A^t) \leqslant \text{rank}(B^t) = \text{rank}(B).$$

This establishes part (d).

Let α, β, and γ be ordered bases for V, W, and Z, respectively, and let $A' = [\mathsf{U}]_\beta^\gamma$ and $B' = [\mathsf{T}]_\alpha^\beta$. Then $A'B' = [\mathsf{UT}]_\alpha^\gamma$ by Theorem 2.11. Hence by Theorem 3.3 and part (d)

$$\text{rank}(\mathsf{UT}) = \text{rank}(A'B') \leqslant \text{rank}(B') = \text{rank}(\mathsf{T}).$$

This establishes part (b). ∎

We will see later that it is important to be able to compute the rank of any matrix. We can use the corollary to Theorem 3.4, Theorems 3.5 and 3.6, and Corollary 2 of Theorem 3.6 to accomplish this goal.

The object is to use elementary row and column operations on a matrix to "simplify" it (so that the transformed matrix has lots of zero entries) to the point where a simple observation enables us to determine how many linearly independent rows or columns the matrix has and thus to determine its rank.

Example 4

(a) Let

$$A = \begin{pmatrix} 1 & 2 & 1 & 3 \\ 1 & 1 & -1 & 1 \end{pmatrix}.$$

Note that the first and second rows of A are linearly independent since one is not a multiple of the other. Thus $\text{rank}(A) = 2$.

(b) Let

$$A = \begin{pmatrix} 1 & 3 & 1 & 1 \\ 1 & 0 & 1 & 1 \\ 0 & 3 & 0 & 0 \end{pmatrix}.$$

In this case there are several ways to proceed. Suppose we begin with an elementary row operation to obtain a zero in the 2, 1 position. Subtracting the first row from the second row, we obtain

$$\begin{pmatrix} 1 & 3 & 1 & 1 \\ 0 & -3 & 0 & 0 \\ 0 & 3 & 0 & 0 \end{pmatrix}.$$

Now note that the third row is a multiple of the second row, and the first and second rows are linearly independent. Thus rank$(A) = 2$.

As an alternative method, note that the first, third, and fourth columns of A are identical and that the first and second columns of A are linearly independent. Hence rank$(A) = 2$.

(c) Let

$$A = \begin{pmatrix} 1 & 2 & 3 & 1 \\ 2 & 1 & 1 & 1 \\ 1 & -1 & 1 & 0 \end{pmatrix}.$$

Using various elementary row and column operations, we obtain the following sequence of matrices:

$$\begin{pmatrix} 1 & 2 & 3 & 1 \\ 0 & -3 & -5 & -1 \\ 0 & -3 & -2 & -1 \end{pmatrix}, \begin{pmatrix} 1 & 2 & 3 & 1 \\ 0 & -3 & -5 & -1 \\ 0 & 0 & 3 & 0 \end{pmatrix},$$

$$\begin{pmatrix} 1 & 0 & 0 & 0 \\ 0 & -3 & -5 & -1 \\ 0 & 0 & 3 & 0 \end{pmatrix}, \text{ and } \begin{pmatrix} 1 & 0 & 0 & 0 \\ 0 & -3 & 0 & 0 \\ 0 & 0 & 3 & 0 \end{pmatrix}.$$

It is clear that the last matrix has three linearly independent rows and hence has rank 3. ∎

In summary, perform row and column operations until the matrix is simplified enough so that the maximum number of linearly independent rows or columns is obvious.

The Inverse of a Matrix

We have remarked that an $n \times n$ matrix is invertible if and only if its rank is n. Since we know how to compute the rank of any matrix, we can always test a matrix to determine if it is invertible. We now provide a simple technique for computing the inverse of a matrix that utilizes elementary row operations.

Definition. *Let* A *and* B *be* $m \times n$ *and* $m \times p$ *matrices, respectively. By the* augmented matrix $(A\,|\,B)$ *we mean the* $m \times (n + p)$ *matrix*

$$(A^{(1)}, \ldots, A^{(n)}, B^{(1)}, \ldots, B^{(p)}),$$

where $A^{(i)}$ *and* $B^{(j)}$ *denote the* i*th column of* A *and the* j*th column of* B, *respectively.*

Let A be an invertible $n \times n$ matrix, and consider the $n \times 2n$ augmented matrix $C = (A\,|\,I_n)$. By Exercise 15 we have

$$A^{-1}C = (A^{-1}A\,|\,A^{-1}I_n) = (I_n\,|\,A^{-1}). \tag{1}$$

By Corollary 3 to Theorem 3.6, A^{-1} is the product of elementary matrices, say $A^{-1} = E_pE_{p-1} \cdots E_1$. Thus (1) becomes

$$E_pE_{p-1} \cdots E_1(A \mid I_n) = A^{-1}C = (I_n \mid A^{-1}).$$

Because multiplying a matrix on the left by an elementary matrix transforms the matrix by an elementary row operation (Theorem 3.1), we have the following result: *If* A *is an invertible* n × n *matrix, then it is possible to transform the matrix* $(A \mid I_n)$ *into the matrix* $(I_n \mid A^{-1})$ *by means of a finite number of elementary row operations.*

Conversely, suppose that A is invertible and that the matrix $(A \mid I_n)$ can be transformed into the matrix $(I_n \mid B)$ by a finite number of elementary row operations. Let $E_1, E_2, \ldots, E_p$ be the elementary matrices associated with these elementary row operations as in Theorem 3.1; then

$$E_p \cdots E_2E_1(A \mid I_n) = (I_n \mid B). \tag{2}$$

Letting $M = E_p \cdots E_2E_1$, we have from (2) that

$$M(A \mid I_n) = (MA \mid M) = (I_n \mid B).$$

Hence $MA = I_n$ and $M = B$. It follows that $M = A^{-1}$. So $B = M = A^{-1}$. Thus we have the following result: *If* A *is an invertible* n × n *matrix and if the matrix* $(A \mid I_n)$ *is transformed into a matrix of the form* $(I_n \mid B)$ *by means of a finite number of elementary row operations, then* $B = A^{-1}$.

The following example demonstrates this procedure.

Example 5

Let us compute the inverse of the matrix

$$A = \begin{pmatrix} 0 & 2 & 4 \\ 2 & 4 & 2 \\ 3 & 3 & 1 \end{pmatrix}.$$

[The reader may wish to verify that rank$(A) = 3$ to be assured that A is invertible.] To compute A^{-1}, we must use elementary row operations to transform

$$(A \mid I) = \left(\begin{array}{ccc|ccc} 0 & 2 & 4 & 1 & 0 & 0 \\ 2 & 4 & 2 & 0 & 1 & 0 \\ 3 & 3 & 1 & 0 & 0 & 1 \end{array}\right)$$

into $(I \mid A^{-1})$. An efficient method for accomplishing this transformation is to change each column of A successively, beginning with the first column, into the corresponding column of I. Since we need a nonzero entry in the 1, 1 position, we begin by interchanging rows 1 and 2. The result is

$$\left(\begin{array}{ccc|ccc} 2 & 4 & 2 & 0 & 1 & 0 \\ 0 & 2 & 4 & 1 & 0 & 0 \\ 3 & 3 & 1 & 0 & 0 & 1 \end{array}\right).$$

In order to place a 1 in the 1, 1 position, we must multiply the first row by $\frac{1}{2}$; this operation yields

$$\left(\begin{array}{ccc|ccc} 1 & 2 & 1 & 0 & \frac{1}{2} & 0 \\ 0 & 2 & 4 & 1 & 0 & 0 \\ 3 & 3 & 1 & 0 & 0 & 1 \end{array}\right).$$

We now complete work in the first column by adding -3 times row 1 to row 3 to obtain

$$\left(\begin{array}{ccc|ccc} 1 & 2 & 1 & 0 & \frac{1}{2} & 0 \\ 0 & 2 & 4 & 1 & 0 & 0 \\ 0 & -3 & -2 & 0 & -\frac{3}{2} & 1 \end{array}\right).$$

In order to change the second column of the matrix above into the second column of I, we multiply row 2 by $\frac{1}{2}$ to obtain a 1 in the 2, 2 position. This operation produces

$$\left(\begin{array}{ccc|ccc} 1 & 2 & 1 & 0 & \frac{1}{2} & 0 \\ 0 & 1 & 2 & \frac{1}{2} & 0 & 0 \\ 0 & -3 & -2 & 0 & -\frac{3}{2} & 1 \end{array}\right).$$

We can now complete work on the second column by adding -2 times row 2 to row 1 and 3 times row 2 to row 3. The result is

$$\left(\begin{array}{ccc|ccc} 1 & 0 & -3 & -1 & \frac{1}{2} & 0 \\ 0 & 1 & 2 & \frac{1}{2} & 0 & 0 \\ 0 & 0 & 4 & \frac{3}{2} & -\frac{3}{2} & 1 \end{array}\right).$$

Only the third column remains to be changed. In order to place a 1 in the 3, 3 position, we multiply row 3 by $\frac{1}{4}$; this operation yields

$$\left(\begin{array}{ccc|ccc} 1 & 0 & -3 & -1 & \frac{1}{2} & 0 \\ 0 & 1 & 2 & \frac{1}{2} & 0 & 0 \\ 0 & 0 & 1 & \frac{3}{8} & -\frac{3}{8} & \frac{1}{4} \end{array}\right).$$

Adding appropriate multiples of row 3 to rows 1 and 2 completes the process and gives

$$\left(\begin{array}{ccc|ccc} 1 & 0 & 0 & \frac{1}{8} & -\frac{5}{8} & \frac{3}{4} \\ 0 & 1 & 0 & -\frac{1}{4} & \frac{3}{4} & -\frac{1}{2} \\ 0 & 0 & 1 & \frac{3}{8} & -\frac{3}{8} & \frac{1}{4} \end{array}\right).$$

Thus

$$A^{-1} = \begin{pmatrix} \frac{1}{8} & -\frac{5}{8} & \frac{3}{4} \\ -\frac{1}{4} & \frac{3}{4} & -\frac{1}{2} \\ \frac{3}{8} & -\frac{3}{8} & \frac{1}{4} \end{pmatrix}. \quad \blacksquare$$

Being able to compute the inverse of a matrix allows us to compute the inverse of a linear transformation. The following example demonstrates this technique.

Example 6

Let $\mathsf{T}: \mathsf{P}_2(R) \to \mathsf{P}_2(R)$ be defined by $\mathsf{T}(f) = f + f' + f''$, where f' and f'' denote the first and second derivatives of f. It is easily shown that $\mathsf{N}(\mathsf{T}) = \{0\}$, so that T is invertible. Taking $\beta = \{1, x, x^2\}$, we have

$$[\mathsf{T}]_\beta = \begin{pmatrix} 1 & 1 & 2 \\ 0 & 1 & 2 \\ 0 & 0 & 1 \end{pmatrix}.$$

Now the inverse of this matrix is

$$\begin{pmatrix} 1 & -1 & 0 \\ 0 & 1 & -2 \\ 0 & 0 & 1 \end{pmatrix}.$$

But $([\mathsf{T}]_\beta)^{-1} = [\mathsf{T}^{-1}]_\beta$ by Corollary 1 to Theorem 2.19. Hence by Theorem 2.15 we have

$$[\mathsf{T}^{-1}(a_0 + a_1x + a_2x^2)]_\beta = \begin{pmatrix} 1 & -1 & 0 \\ 0 & 1 & -2 \\ 0 & 0 & 1 \end{pmatrix} \begin{pmatrix} a_0 \\ a_1 \\ a_2 \end{pmatrix}$$

$$= \begin{pmatrix} a_0 - a_1 \\ a_1 - 2a_2 \\ a_2 \end{pmatrix}.$$

Therefore,

$$\mathsf{T}^{-1}(a_0 + a_1x + a_2x^2) = (a_0 - a_1) + (a_1 - 2a_2)x + a_2x^2. \quad \blacksquare$$

EXERCISES

1. Label the following statements as being true or false.
   (a) The rank of a matrix is equal to the number of its nonzero columns.
   (b) The product of two matrices always has rank equal to the lesser of the ranks of the two matrices.
   (c) The $m \times n$ zero matrix is the only $m \times n$ matrix having rank 0.
   (d) Elementary row operations preserve rank.
   (e) Elementary column operations do not necessarily preserve rank.
   (f) The rank of a matrix is equal to the maximum number of linearly independent rows in the matrix.

(g) The inverse of a matrix can be computed exclusively by means of elementary row operations.
(h) An $n \times n$ matrix is of rank at most n.
(i) An $n \times n$ matrix having rank n is invertible.

2. Find the rank of the following matrices.

(a) $\begin{pmatrix} 1 & 1 & 0 \\ 0 & 1 & 1 \\ 1 & 1 & 0 \end{pmatrix}$ **(b)** $\begin{pmatrix} 1 & 1 & 0 \\ 2 & 1 & 1 \\ 1 & 1 & 1 \end{pmatrix}$ **(c)** $\begin{pmatrix} 1 & 0 & 2 \\ 1 & 1 & 4 \end{pmatrix}$

(d) $\begin{pmatrix} 1 & 2 & 1 \\ 2 & 4 & 2 \end{pmatrix}$ **(e)** $\begin{pmatrix} 1 & 2 & 3 & 1 & 1 \\ 1 & 4 & 0 & 1 & 2 \\ 0 & 2 & -3 & 0 & 1 \\ 1 & 0 & 0 & 0 & 0 \end{pmatrix}$

(f) $\begin{pmatrix} 1 & 2 & 0 & 1 & 1 \\ 2 & 4 & 1 & 3 & 0 \\ 3 & 6 & 2 & 5 & 1 \\ -4 & -8 & 1 & -3 & 1 \end{pmatrix}$ **(g)** $\begin{pmatrix} 1 & 1 & 0 & 1 \\ 2 & 2 & 0 & 2 \\ 1 & 1 & 0 & 1 \\ 1 & 1 & 0 & 1 \end{pmatrix}$

3. Prove that for any $m \times n$ matrix A, $\operatorname{rank}(A) = 0$ if and only if A is the zero matrix.

4. Use elementary row and column operations to transform each of the following matrices into a matrix D satisfying the conditions of Theorem 3.6, and then determine the rank of each matrix.

(a) $\begin{pmatrix} 1 & 1 & 1 & 2 \\ 2 & 0 & -1 & 2 \\ 1 & 1 & 1 & 2 \end{pmatrix}$ **(b)** $\begin{pmatrix} 2 & 1 \\ -1 & 2 \\ 2 & 1 \end{pmatrix}$

5. For each of the following matrices, compute the rank and compute the inverse if it exists.

(a) $\begin{pmatrix} 1 & 2 \\ 1 & 1 \end{pmatrix}$ **(b)** $\begin{pmatrix} 1 & 2 \\ 2 & 4 \end{pmatrix}$ **(c)** $\begin{pmatrix} 1 & 2 & 1 \\ 1 & 3 & 4 \\ 2 & 3 & -1 \end{pmatrix}$

(d) $\begin{pmatrix} 0 & -2 & 4 \\ 1 & 1 & -1 \\ 2 & 4 & -5 \end{pmatrix}$ **(e)** $\begin{pmatrix} 1 & 2 & 1 \\ -1 & 1 & 2 \\ 1 & 0 & 1 \end{pmatrix}$ **(f)** $\begin{pmatrix} 1 & 2 & 1 \\ 1 & 0 & 1 \\ 1 & 1 & 1 \end{pmatrix}$

(g) $\begin{pmatrix} 1 & 2 & 1 & 0 \\ 2 & 5 & 5 & 1 \\ -2 & -3 & 0 & 3 \\ 3 & 4 & -2 & -3 \end{pmatrix}$ **(h)** $\begin{pmatrix} 1 & 0 & 1 & 1 \\ 1 & 1 & -1 & 2 \\ 2 & 0 & 1 & 0 \\ 0 & -1 & 1 & -3 \end{pmatrix}$

6. For each of the following linear transformations T, determine if T is invertible, and compute T^{-1} if it exists.
(a) $\mathsf{T}: \mathsf{P}_2(R) \to \mathsf{P}_2(R)$ defined by $\mathsf{T}(f) = f'' + 2f' - f$.
(b) $\mathsf{T}: \mathsf{P}_2(R) \to \mathsf{P}_2(R)$ defined by $\mathsf{T}(f)(x) = (x+1)f'(x)$.
(c) $\mathsf{T}: \mathsf{R}^3 \to \mathsf{R}^3$ defined by

$$\mathsf{T}(a_1, a_2, a_3) = (a_1 + 2a_2 + a_3,\ -a_1 + a_2 + 2a_3,\ a_1 + a_3).$$

(d) $\mathsf{T}: \mathsf{R}^3 \to \mathsf{P}_2(R)$ defined by

$$\mathsf{T}(a_1, a_2, a_3) = (a_1 + a_2 + a_3) + (a_1 - a_2 + a_3)x + a_1x^2.$$

(e) $\mathsf{T}: \mathsf{P}_2(R) \to \mathsf{R}^3$ defined by $\mathsf{T}(f) = (f(-1), f(0), f(1))$.
(f) $\mathsf{T}: \mathsf{M}_{2\times 2}(R) \to \mathsf{R}^4$ defined by

$$\mathsf{T}(A) = (\operatorname{tr}(A), \operatorname{tr}(A^t), \operatorname{tr}(EA), \operatorname{tr}(AE))$$

where

$$E = \begin{pmatrix} 0 & 1 \\ 1 & 0 \end{pmatrix},$$

7. Express the invertible matrix

$$\begin{pmatrix} 1 & 2 & 1 \\ 1 & 0 & 1 \\ 1 & 1 & 2 \end{pmatrix}$$

as a product of elementary matrices.

8. Let A be an $m \times n$ matrix. Prove that if c is any nonzero scalar, then $\operatorname{rank}(cA) = \operatorname{rank}(A)$.

9. Complete the proof of the corollary to Theorem 3.4 by showing that elementary column operations preserve rank.

10. Prove Theorem 3.6 for the case that A is an $m \times 1$ matrix.

11. Let

$$B = \left(\begin{array}{c|ccc} 1 & 0 & \cdots & 0 \\ \hline 0 & & & \\ \vdots & & B' & \\ 0 & & & \end{array}\right),$$

where B' is an $m \times n$ submatrix of B. Prove that if $\operatorname{rank}(B) = r$, then $\operatorname{rank}(B') = r - 1$.

12. Let B' and D' be $m \times n$ matrices, and let B and D be the $(m+1) \times (n+1)$ matrices defined by

$$B = \left(\begin{array}{c|ccc} 1 & 0 & \cdots & 0 \\ \hline 0 & & & \\ \vdots & & B' & \\ 0 & & & \end{array}\right) \quad \text{and} \quad D = \left(\begin{array}{c|ccc} 1 & 0 & \cdots & 0 \\ \hline 0 & & & \\ \vdots & & D' & \\ 0 & & & \end{array}\right).$$

Prove that if B' can be transformed into D' by an elementary row [column] operation, then B can be transformed into D by an elementary row [column] operation.

13. Prove parts (b) and (c) of Corollary 2 of Theorem 3.6.

14. Let T, U: V → W be linear transformations. Prove that
   (a) $\mathsf{R}(\mathsf{T}+\mathsf{U}) \subseteq \mathsf{R}(\mathsf{T}) + \mathsf{R}(\mathsf{U})$.
   (b) If W is finite-dimensional, then $\text{rank}(\mathsf{T}+\mathsf{U}) \leqslant \text{rank}(\mathsf{T}) + \text{rank}(\mathsf{U})$.
   (c) Deduce from part (b) that, for any $m \times n$ matrices A and B, $\text{rank}(A+B) \leqslant \text{rank}(A) + \text{rank}(B)$.

15. If A and B are matrices containing n rows, prove that $M(A\,|\,B) = (MA\,|\,MB)$ for any $m \times n$ matrix M.

16. Prove that if B is a 3×1 matrix and C is a 1×3 matrix, then the 3×3 matrix BC has rank at most 1. Conversely, show that if A is any 3×3 matrix having rank 1, then there exist a 3×1 matrix B and a 1×3 matrix C such that $A = BC$.

17. Supply the details to the proof of part (b) of Theorem 3.4.

3.3 SYSTEMS OF LINEAR EQUATIONS— THEORETICAL ASPECTS

This section and the next are devoted to the study of systems of linear equations, which arise naturally in both the physical and social sciences. In this section we apply results from Chapter 2 to describe the solution sets of systems of linear equations as subsets of a vector space. In Section 3.4, elementary row operations are used to provide a computational method for finding all solutions to such systems.

The system of equations

$$(S) \qquad \begin{cases} a_{11}x_1 + a_{12}x_2 + \cdots + a_{1n}x_n = b_1 \\ a_{21}x_1 + a_{22}x_2 + \cdots + a_{2n}x_n = b_2 \\ \vdots \\ a_{m1}x_1 + a_{m2}x_2 + \cdots + a_{mn}x_n = b_m, \end{cases}$$

where a_{ij} and b_i ($1 \leqslant i \leqslant m$ and $1 \leqslant j \leqslant n$) are elements of a field F and $x_1, x_2, \ldots, x_n$ are n variables taking values in F, is called *a system of* m *linear equations in* n *unknowns over the field* F.

The $m \times n$ matrix

$$A = \begin{pmatrix} a_{11} & \cdots & a_{1n} \\ \vdots & & \vdots \\ a_{m1} & \cdots & a_{mn} \end{pmatrix}$$

is called the *coefficient matrix* of the system (S).

If we let

$$X = \begin{pmatrix} x_1 \\ \vdots \\ x_n \end{pmatrix} \quad \text{and} \quad B = \begin{pmatrix} b_1 \\ \vdots \\ b_m \end{pmatrix},$$

then the system (*S*) may be rewritten as a single matrix equation

$$AX = B.$$

To exploit the results that we have developed, we will frequently consider a system of equations as a single matrix equation.

A *solution* to system (*S*) is an *n*-tuple

$$s = \begin{pmatrix} s_1 \\ \vdots \\ s_n \end{pmatrix} \in \mathsf{F}^n$$

such that $As = B$. The set of all solutions to system (*S*) is called the *solution set* of the system.

Example 1

(a) Consider the system

$$\begin{cases} x_1 + x_2 = 3 \\ x_1 - x_2 = 1. \end{cases}$$

By use of familiar techniques we can solve the system above and conclude that there is only one solution: $x_1 = 2$, $x_2 = 1$; that is,

$$s = \begin{pmatrix} 2 \\ 1 \end{pmatrix}.$$

In matrix form the system can be written

$$\begin{pmatrix} 1 & 1 \\ 1 & -1 \end{pmatrix}\begin{pmatrix} x_1 \\ x_2 \end{pmatrix} = \begin{pmatrix} 3 \\ 1 \end{pmatrix};$$

so

$$A = \begin{pmatrix} 1 & 1 \\ 1 & -1 \end{pmatrix} \quad \text{and} \quad B = \begin{pmatrix} 3 \\ 1 \end{pmatrix}.$$

(b) Consider

$$\begin{cases} 2x_1 + 3x_2 + x_3 = 1 \\ x_1 - x_2 + 2x_3 = 6; \end{cases}$$

i.e.,

$$\begin{pmatrix} 2 & 3 & 1 \\ 1 & -1 & 2 \end{pmatrix}\begin{pmatrix} x_1 \\ x_2 \\ x_3 \end{pmatrix} = \begin{pmatrix} 1 \\ 6 \end{pmatrix}.$$

This system has many solutions such as

$$s = \begin{pmatrix} -6 \\ 2 \\ 7 \end{pmatrix} \quad \text{and} \quad s = \begin{pmatrix} 8 \\ -4 \\ -3 \end{pmatrix}.$$

(c) Consider

$$\begin{cases} x_1 + x_2 = 0 \\ x_1 + x_2 = 1; \end{cases}$$

i.e.,

$$\begin{pmatrix} 1 & 1 \\ 1 & 1 \end{pmatrix} \begin{pmatrix} x_1 \\ x_2 \end{pmatrix} = \begin{pmatrix} 0 \\ 1 \end{pmatrix}.$$

no solution.

It is evident that this system has no solutions. Thus we see that a system of linear equations can have one, many, or no solutions. ∎

We must be able to recognize when a system has solutions and then be able to describe all its solutions. This section and the next are devoted to this end.

We begin our study of systems of equations by examining the class of "homogeneous" systems of linear equations. As we will see (Theorem 3.8), the set of solutions to a homogeneous system of m linear equations in n unknowns forms a subspace of F^n. We can then apply the theory of vector spaces to this set of solutions. For example, a basis for the solution space can be found, and any solution can be expressed as a linear combination of the basis vectors.

Definitions. *A system* $AX = B$ *of* m *equations in* n *unknowns is said to be* homogeneous *if* $B = 0$. *Otherwise the system is said to be* nonhomogeneous.

Any homogeneous system has at least one solution, namely,

$$s = \begin{pmatrix} 0 \\ \vdots \\ 0 \end{pmatrix}.$$

This solution is called the *trivial solution.* The next result gives further information about the set of solutions to a homogeneous system.

Theorem 3.8. *Let* $AX = 0$ *be a homogeneous system of* m *linear equations in* n *unknowns over a field* F. *Let* K *denote the set of all solutions to* $AX = 0$. *Then* $\mathsf{K} = \mathsf{N}(\mathsf{L}_A)$; *hence* K *is a subspace of* F^n *of dimension* $n - \text{rank}(\mathsf{L}_A) = n - \text{rank}(A)$.

Proof. $\mathsf{K} = \{s \in \mathsf{F}^n : As = 0\} = \mathsf{N}(\mathsf{L}_A)$. The second part now follows from the dimension theorem. ∎

Corollary. *If* m < n, *the system* AX = 0 *has a nontrivial solution.*

Proof. Suppose that $m < n$. Then $\text{rank}(A) = \text{rank}(\mathsf{L}_A) \leqslant m$. Hence

$$\dim(\mathsf{K}) = n - \text{rank}(\mathsf{L}_A) \geqslant n - m > 0,$$

where $\mathsf{K} = \mathsf{N}(\mathsf{L}_A)$. Since $\dim(\mathsf{K}) > 0$, $\mathsf{K} \neq \{0\}$. Thus there exists $s \in \mathsf{K}$, $s \neq 0$; so s is a nontrivial solution to $AX = 0$. ∎

Example 2

(a) Consider the system

$$\begin{cases} x_1 + 2x_2 + x_3 = 0 \\ x_1 - x_2 - x_3 = 0. \end{cases}$$

Let

$$A = \begin{pmatrix} 1 & 2 & 1 \\ 1 & -1 & -1 \end{pmatrix}$$

be the coefficient matrix. It is clear that $\text{rank}(A) = 2$. If K is the solution set of the system, then $\dim(\mathsf{K}) = 3 - 2 = 1$. Thus any nonzero solution will constitute a basis for K. For example, since

$$\begin{pmatrix} 1 \\ -2 \\ 3 \end{pmatrix}$$

is a solution,

$$\left\{\begin{pmatrix} 1 \\ -2 \\ 3 \end{pmatrix}\right\}$$

is a basis. Thus any element of K is of the form

$$t\begin{pmatrix} 1 \\ -2 \\ 3 \end{pmatrix} = \begin{pmatrix} t \\ -2t \\ 3t \end{pmatrix},$$

where $t \in R$.

(b) Consider the system $x_1 - 2x_2 + x_3 = 0$ of one equation in three unknowns. If $A = (1, -2, 1)$ is the coefficient matrix, $\text{rank}(A) = 1$. Hence if K is the solution set, $\dim(\mathsf{K}) = 3 - 1 = 2$. Note that

$$\begin{pmatrix} 2 \\ 1 \\ 0 \end{pmatrix} \quad \text{and} \quad \begin{pmatrix} 1 \\ 0 \\ -1 \end{pmatrix}$$

are linearly independent elements of K. Thus they constitute a basis for K, so that

$$\mathsf{K} = \left\{ t_1 \begin{pmatrix} 2 \\ 1 \\ 0 \end{pmatrix} + t_2 \begin{pmatrix} 1 \\ 0 \\ -1 \end{pmatrix} : \ t_1, t_2 \in R \right\}. \quad \blacksquare$$

In Section 3.4 we will discuss explicit computational methods for finding a basis for the solution set of a homogeneous system.

We now turn to the study of nonhomogeneous systems. Our next result shows that the solution set of a nonhomogeneous system $AX = B$ can be described in terms of the solution set of the homogeneous system $AX = 0$. We refer to the equation $AX = 0$ as the *homogeneous system corresponding to* AX = B.

Theorem 3.9. *Let* K *be the solution set of a system of linear equations* AX = B, *and let* K_H *be the solution set of the corresponding homogeneous system* AX = 0. *Then for any solution* s *to* AX = B

$$\mathsf{K} = \{\mathsf{s}\} + \mathsf{K}_\mathsf{H} = \{\mathsf{s} + \mathsf{k} \colon \ \mathsf{k} \in \mathsf{K}_\mathsf{H}\}.$$

Proof. Let s be any solution to $AX = B$. We must show that $K = \{s\} + \mathsf{K}_\mathsf{H}$. If $w \in K$, then $Aw = B$. Hence

$$A(w - s) = Aw - As = B - B = 0.$$

So $w - s \in \mathsf{K}_\mathsf{H}$. Thus there exists $k \in \mathsf{K}_\mathsf{H}$ such that $w - s = k$. So $w = s + k \in \{s\} + \mathsf{K}_\mathsf{H}$, and therefore

$$K \subseteq \{s\} + \mathsf{K}_\mathsf{H}.$$

Conversely, suppose that $w \in \{s\} + \mathsf{K}_\mathsf{H}$; then $w = s + k$ for some $k \in \mathsf{K}_\mathsf{H}$. But then $Aw = A(s + k) = As + Ak = B + 0 = B$; so $w \in K$. Therefore $\{s\} + \mathsf{K}_\mathsf{H} \subseteq K$, and thus $K = \{s\} + \mathsf{K}_\mathsf{H}$. $\blacksquare$

Example 3

(a) Consider the system

$$\begin{cases} x_1 + 2x_2 + x_3 = \ \ 7 \\ x_1 - \ \ x_2 - x_3 = -4. \end{cases}$$

The homogeneous system corresponding to the one above is the system given in Example 2(a). It is easily verified that

$$s = \begin{pmatrix} 1 \\ 1 \\ 4 \end{pmatrix}$$

is a solution to the nonhomogeneous system above. So the solution set to the system is

$$K = \left\{ \begin{pmatrix} 1 \\ 1 \\ 4 \end{pmatrix} + t \begin{pmatrix} 1 \\ -2 \\ 3 \end{pmatrix} : \quad t \in R \right\}$$

by Theorem 3.9.

(b) Consider the system $x_1 - 2x_2 + x_3 = 4$. The homogeneous system corresponding to this system is given in Example 2(b). Since

$$s = \begin{pmatrix} 4 \\ 0 \\ 0 \end{pmatrix}$$

is a solution to this system, the solution set K can be written as

$$K = \left\{ \begin{pmatrix} 4 \\ 0 \\ 0 \end{pmatrix} + t_1 \begin{pmatrix} 2 \\ 1 \\ 0 \end{pmatrix} + t_2 \begin{pmatrix} 1 \\ 0 \\ -1 \end{pmatrix} : \quad t_1, t_2 \in R \right\}. \quad \blacksquare$$

The following theorem provides us with a means of computing solutions to certain systems of equations.

Theorem 3.10. *Let* $AX = B$ *be a system of* n *equations in* n *unknowns. If* A *is invertible, then the system has* exactly one *solution, namely* $A^{-1}B$. *Conversely, if the system has exactly one solution, then* A *is invertible.*

Proof. Suppose that A is invertible. Substituting $A^{-1}B$ into the system, we have $A(A^{-1}B) = (AA^{-1})B = B$. Thus $A^{-1}B$ is a solution. If s is an arbitrary solution, then $As = B$. Multiplying both sides by A^{-1} gives $s = A^{-1}B$. Thus the system has one and only one solution, namely $A^{-1}B$.

Conversely, suppose the system has exactly one solution s. Let K_H denote the solution set for the corresponding homogeneous system $AX = 0$. By Theorem 3.9, $\{s\} = \{s\} + K_H$. But this can only occur if $K_H = \{0\}$. Thus $N(L_A) = \{0\}$, and hence A is invertible. $\blacksquare$

Example 4

Consider the system of three equations in three unknowns:

$$\begin{cases} \qquad\quad 2x_2 + 4x_3 = 2 \\ 2x_1 + 4x_2 + 2x_3 = 3 \\ 3x_1 + 3x_2 + x_3 = 1. \end{cases}$$

In Example 5 of Section 3.2 we computed the inverse of the coefficient matrix A

of this system. Thus it has exactly one solution:

$$\begin{pmatrix} x_1 \\ x_2 \\ x_3 \end{pmatrix} = A^{-1}B = \begin{pmatrix} \frac{1}{8} & -\frac{5}{8} & \frac{3}{4} \\ -\frac{1}{4} & \frac{3}{4} & -\frac{1}{2} \\ \frac{3}{8} & -\frac{3}{8} & \frac{1}{4} \end{pmatrix} \begin{pmatrix} 2 \\ 3 \\ 1 \end{pmatrix} = \begin{pmatrix} -\frac{7}{8} \\ \frac{5}{4} \\ -\frac{1}{8} \end{pmatrix}. \quad \blacksquare$$

We use this technique for solving systems of linear equations having invertible coefficient matrices in the application that concludes this section.

In Example 1(c) we saw a system of linear equations that has no solutions. We now establish a criterion for determining when a system has solutions. This criterion involves the rank of the coefficient matrix of the system $AX = B$ and the rank of the matrix $(A \mid B)$. The matrix $(A \mid B)$ is called the *augmented matrix of the system* $\mathrm{AX} = \mathrm{B}$.

Theorem 3.11. *Let* $\mathrm{AX} = \mathrm{B}$ *be a system of linear equations. Then the system has at least one solution if and only if* $\operatorname{rank}(\mathrm{A}) = \operatorname{rank}(\mathrm{A} \mid \mathrm{B})$.

Proof. To say that $AX = B$ has a solution is equivalent to saying that $B \in \mathsf{R}(\mathsf{L}_A)$. In the proof of Theorem 3.5 we saw that

$$\mathsf{R}(\mathsf{L}_A) = \operatorname{span}\{A^{(1)}, A^{(2)}, \ldots, A^{(n)}\},$$

the span of the columns of A. Thus $AX = B$ has a solution if and only if $B \in \operatorname{span}\{A^{(1)}, A^{(2)}, \ldots, A^{(n)}\}$. But $B \in \operatorname{span}\{A^{(1)}, A^{(2)}, \ldots, A^{(n)}\}$ if and only if $\operatorname{span}\{A^{(1)}, A^{(2)}, \ldots, A^{(n)}\} = \operatorname{span}\{A^{(1)}, A^{(2)}, \ldots, A^{(n)}, B\}$. This last statement is equivalent to

$$\dim(\operatorname{span}\{A^{(1)}, A^{(2)}, \ldots, A^{(n)}\}) = \dim(\operatorname{span}\{A^{(1)}, A^{(2)}, \ldots, A^{(n)}, B\}).$$

So by Theorem 3.5 the equation above reduces to

$$\operatorname{rank}(A) = \operatorname{rank}(A \mid B). \quad \blacksquare$$

Example 5

Recall the system of equations

$$\begin{cases} x_1 + x_2 = 0 \\ x_1 + x_2 = 1 \end{cases}$$

in Example 1(c).

Since

$$A = \begin{pmatrix} 1 & 1 \\ 1 & 1 \end{pmatrix} \quad \text{and} \quad (A \mid B) = \begin{pmatrix} 1 & 1 & 0 \\ 1 & 1 & 1 \end{pmatrix},$$

$\operatorname{rank}(A) = 1$ and $\operatorname{rank}(A \mid B) = 2$. Because the two ranks are unequal, the system has no solutions. $\blacksquare$

Example 6

We will use Theorem 3.11 to determine if (3, 3, 2) is in the range of the linear transformation $\mathsf{T}: \mathsf{R}^3 \to \mathsf{R}^3$ defined by

$$\mathsf{T}(a_1, a_2, a_3) = (a_1 + a_2 + a_3,\ a_1 - a_2 + a_3,\ a_1 + a_3).$$

Now $(3, 3, 2) \in \mathsf{R}(\mathsf{T})$ if and only if there exists a vector $s = (x_1, x_2, x_3)$ in R^3 such that $\mathsf{T}(s) = (3, 3, 2)$. Such a vector s must be a solution to the system

$$\begin{cases} x_1 + x_2 + x_3 = 3 \\ x_1 - x_2 + x_3 = 3 \\ x_1 \qquad\ + x_3 = 2. \end{cases}$$

Since the ranks of the coefficient matrix and the augmented matrix of this system are 2 and 3, respectively, it follows that this system has no solutions. Hence $(3, 3, 2) \notin \mathsf{R}(\mathsf{T})$. ∎

An Application

In 1973, Wassily Leontief won the Nobel Prize in economics for his work in developing a mathematical model that may be used to describe various economic phenomena. We close this section by applying some of the ideas we have studied to illustrate two special cases of his work.

We begin by considering a simple society composed of three people (industries)—a farmer who grows all the food, a tailor who makes all the clothing, and a carpenter who builds all the housing. We assume that each person sells to and buys from a central pool and that everything produced is consumed. Since no commodities either enter or leave the system, this case is referred to as the *closed model*.

Each of the three individuals consumes all three of the commodities produced in the society. Suppose that the proportion of each of the commodities consumed by each person is given in the following table. Notice that each of the columns of the table must sum to 1.

	Food	Clothing	Housing
Farmer	0.40	0.20	0.20
Tailor	0.10	0.70	0.20
Carpenter	0.50	0.10	0.60

Let p_1, p_2, and p_3 denote the incomes of the farmer, tailor, and carpenter, respectively. To assure that this society survives, we require that the consumption of each individual equals his or her income. In the case of the farmer, this requirement translates into the equation $0.40p_1 + 0.20p_2 + 0.20p_3 =$

p_1. Thus we need to consider the system of linear equations

$$\begin{cases} 0.40p_1 + 0.20p_2 + 0.20p_3 = p_1 \\ 0.10p_1 + 0.70p_2 + 0.20p_3 = p_2 \\ 0.50p_1 + 0.10p_2 + 0.60p_3 = p_3 \end{cases}$$

or, equivalently, $AP = P$, where

$$P = \begin{pmatrix} p_1 \\ p_2 \\ p_3 \end{pmatrix}$$

and A is the coefficient matrix of the system. In this context A is called the *input–output* (or *consumption*) *matrix*, and $AP = P$ is called the *equilibrium condition.*

For matrices B and C of the same size we use the notation $B \geqslant C$ $[B > C]$ to mean $B_{ij} \geqslant C_{ij}$ $[B_{ij} > C_{ij}]$ for all i and j. B is called *nonnegative* [*positive*] if $B \geqslant O$ $[B > O]$, where O is the zero matrix.

At first it may seem reasonable to replace the equilibrium condition by the inequality $AP \leqslant P$, that is, the requirement that consumption not exceed production. But in fact $AP \leqslant P$ implies that $AP = P$ in the closed model. For otherwise there exists a k for which

$$p_k > \sum_j A_{kj} p_j.$$

Hence, since the columns of A sum to 1,

$$\sum_i p_i > \sum_i \sum_j A_{ij} p_j = \sum_j \left(\sum_i A_{ij} \right) p_j = \sum_j p_j,$$

which is a contradiction.

One solution to the homogeneous system $(I - A)X = 0$, which is equivalent to the equilibrium condition, is

$$P = \begin{pmatrix} 0.25 \\ 0.35 \\ 0.40 \end{pmatrix}.$$

We may interpret this to mean that the society will survive if the farmer, tailor, and carpenter have incomes in the proportions 25:35:40 (or 5:7:8).

Notice that we are not simply interested in any nontrivial solution to the system but in one that is nonnegative. Thus we must consider the question of whether or not the system $(I - A)X = 0$ has a nonnegative solution, where A is a nonnegative matrix whose columns sum to 1. A useful theorem in this direction (whose proof may be found in "Applications of Matrices to Economic Models and Social Science Relationships," by Ben Noble, *Proceedings of the Summer Conference for College Teachers on Applied Mathematics*, 1971, CUPM, Berkeley, California) is stated below.

Theorem 3.12. *Let* A *be an* n × n *input–output matrix having the form*

$$\mathrm{A} = \begin{pmatrix} \mathrm{B} & \mathrm{C} \\ \mathrm{D} & \mathrm{E} \end{pmatrix},$$

where D *is a* 1 × (n − 1) *positive matrix and* C *is an* (n − 1) × 1 *positive matrix. Then* (I − A)X = 0 *has a one-dimensional solution set that is generated by a nonnegative vector.*

Observe that any positive input–output matrix satisfies the hypothesis of this theorem. The matrix below does also.

$$\begin{pmatrix} 0.75 & 0.50 & 0.65 \\ 0 & 0.25 & 0.35 \\ 0.25 & 0.25 & 0 \end{pmatrix}$$

In the *open model* we assume that there is an outside demand for each of the commodities produced. Returning to our simple society, let x_1, x_2, and x_3 be the monetary values of food, clothing, and housing produced with respective outside demands d_1, d_2, and d_3. Let A be the 3×3 matrix such that A_{ij} represents the amount (in a fixed monetary unit such as the dollar) of commodity i required in producing one monetary unit of commodity j. Then the value of the surplus of food in the society is

$$x_1 - (A_{11}x_1 + A_{12}x_2 + A_{13}x_3),$$

that is, the value of food produced minus the value of food consumed in producing the three commodities. The assumption that everything produced is consumed gives us a similar equilibrium condition for the open model, namely, that the surplus of each of the three commodities must equal the corresponding outside demands. Hence

$$x_i - \sum_{j=1}^{3} A_{ij}x_j = d_i \qquad \text{for } i = 1, 2, \text{ and } 3.$$

In general, we must find a nonnegative solution to $(I - A)X = D$, where A and D are nonnegative matrices and the sum of the entries of each column of A does not exceed one. It is easy to see that if $(I - A)^{-1}$ exists and is nonnegative, then the desired solution will be $(I - A)^{-1}D$.

Recall that for a real number a the series $1 + a + a^2 + \cdots$ converges to $(1 - a)^{-1}$ if $|a| < 1$. Similarly, it can be shown (using the concept of convergence of matrices developed in Section 5.3) that the series $I + A + A^2 + \cdots$ converges to $(I - A)^{-1}$ if $\{A^n\}$ converges to the zero matrix. In this case $(I - A)^{-1}$ will be nonnegative since the matrices $I, A, A^2, \ldots$ are nonnegative.

To illustrate the open model, suppose that 30 cents worth of food, 10 cents worth of clothing, and 30 cents worth of housing are required for the production of $1 worth of food. Similarly, suppose that 20 cents worth of food, 40 cents worth of clothing, and 20 cents worth of housing are required for the production of \$1 worth of clothing. Finally, suppose that 30 cents worth of food, 10 cents worth of clothing, and 30 cents worth of housing are required for the production of \$1 worth of housing. Then the input–output matrix is

$$A = \begin{pmatrix} 0.30 & 0.20 & 0.30 \\ 0.10 & 0.40 & 0.10 \\ 0.30 & 0.20 & 0.30 \end{pmatrix},$$

so

$$I - A = \begin{pmatrix} 0.70 & -0.20 & -0.30 \\ -0.10 & 0.60 & -0.10 \\ -0.30 & -0.20 & 0.70 \end{pmatrix} \quad \text{and} \quad (I - A)^{-1} = \begin{pmatrix} 2.0 & 1.0 & 1.0 \\ 0.5 & 2.0 & 0.5 \\ 1.0 & 1.0 & 2.0 \end{pmatrix}.$$

Since $(I - A)^{-1}$ is nonnegative, we can find a (unique) nonnegative solution to $(I - A)X = D$ for any demand D. For example, suppose that there are outside demands of \$30 billion in food, \$20 billion in clothing, and \$10 billion in housing. If we set

$$D = \begin{pmatrix} 30 \\ 20 \\ 10 \end{pmatrix},$$

then

$$X = (I - A)^{-1}D = \begin{pmatrix} 90 \\ 60 \\ 70 \end{pmatrix}.$$

So a gross production of \$90 billion of food, \$60 billion of clothing, and \$70 billion of housing is necessary to meet the required demands.

EXERCISES

1. Label the following statements as being true or false.
   - **(a)** Any system of linear equations has at least one solution.
   - **(b)** Any system of linear equations has at most one solution.
   - **(c)** Any homogeneous system of linear equations has at least one solution.
   - **(d)** Any system of n linear equations in n unknowns has at most one solution.
   - **(e)** Any system of n linear equations in n unknowns has at least one solution.
   - **(f)** If the homogeneous system corresponding to a given system of linear equations has a solution, then the given system has a solution.

(g) If the coefficient matrix of a homogeneous system of n linear equations in n unknowns is invertible, then the system has no nontrivial solutions.

(h) The solution set of any system of m linear equations in n unknowns is a subspace of F^n.

2. For each of the following systems of linear equations, find the dimension of and a basis for the solution set.

(a) $\begin{cases} x_1 + 3x_2 = 0 \\ 2x_1 + 6x_2 = 0 \end{cases}$

(b) $\begin{cases} x_1 + x_2 - x_3 = 0 \\ 4x_1 + x_2 - 2x_3 = 0 \end{cases}$

(c) $\begin{cases} x_1 + 2x_2 - x_3 = 0 \\ 2x_1 + x_2 + x_3 = 0 \end{cases}$

(d) $\begin{cases} 2x_1 + x_2 - x_3 = 0 \\ x_1 - x_2 + x_3 = 0 \\ x_1 + 2x_2 - 2x_3 = 0 \end{cases}$

(e) $x_1 + 2x_2 - 3x_3 + x_4 = 0$

(f) $\begin{cases} x_1 + 2x_2 = 0 \\ x_1 - x_2 = 0 \end{cases}$

(g) $\begin{cases} x_1 + 2x_2 + x_3 + x_4 = 0 \\ x_2 - x_3 + x_4 = 0 \end{cases}$

3. Using the results of Exercise 2, find all solutions to the following systems.

(a) $\begin{cases} x_1 + 3x_2 = 5 \\ 2x_1 + 6x_2 = 10 \end{cases}$

(b) $\begin{cases} x_1 + x_2 - x_3 = 1 \\ 4x_1 + x_2 - 2x_3 = 3 \end{cases}$

(c) $\begin{cases} x_1 + 2x_2 - x_3 = 3 \\ 2x_1 + x_2 + x_3 = 6 \end{cases}$

(d) $\begin{cases} 2x_1 + x_2 - x_3 = 5 \\ x_1 - x_2 + x_3 = 1 \\ x_1 + 2x_2 - 2x_3 = 4 \end{cases}$

(e) $x_1 + 2x_2 - 3x_3 + x_4 = 1$

(f) $\begin{cases} x_1 + 2x_2 = 5 \\ x_1 - x_2 = -1 \end{cases}$

(g) $\begin{cases} x_1 + 2x_2 + x_3 + x_4 = 1 \\ x_2 - x_3 + x_4 = 1 \end{cases}$

4. For each system of linear equations with coefficient matrix A:

(1) Show that A is invertible.
(2) Compute A^{-1}.
(3) Use A^{-1} to solve the system.

(a) $\begin{cases} x_1 + 3x_2 = 4 \\ 2x_1 + 5x_2 = 3 \end{cases}$

(b) $\begin{cases} x_1 + 2x_2 - x_3 = 5 \\ x_1 + x_2 + x_3 = 1 \\ 2x_1 - 2x_2 + x_3 = 4 \end{cases}$

5. Give an example of a system of n linear equations in n unknowns with infinitely many solutions.

6. Let $\mathsf{T}\colon \mathsf{R}^3 \to \mathsf{R}^2$ be defined by $\mathsf{T}(a, b, c) = (a + b, 2a - c)$. Describe $\mathsf{T}^{-1}(1, 11)$.

7. Determine which of the following systems of linear equations has a solution.

(a) $\begin{cases} x_1 + x_2 - x_3 + 2x_4 = 2 \\ x_1 + x_2 + 2x_3 = 1 \\ 2x_1 + 2x_2 + x_3 + 2x_4 = 4 \end{cases}$ **(b)** $\begin{cases} x_1 + x_2 - x_3 = 1 \\ 2x_1 + x_2 + 3x_3 = 2 \end{cases}$

(c) $\begin{cases} x_1 + 2x_2 + 3x_3 = 1 \\ x_1 + x_2 - x_3 = 0 \\ x_1 + 2x_2 + x_3 = 3 \end{cases}$ **(d)** $\begin{cases} x_1 + x_2 + 3x_3 - x_4 = 0 \\ x_1 + x_2 + x_3 + x_4 = 1 \\ x_1 - 2x_2 + x_3 - x_4 = 1 \\ 4x_1 + x_2 + 8x_3 - x_4 = 0 \end{cases}$

(e) $\begin{cases} x_1 + 2x_2 - x_3 = 1 \\ 2x_1 + x_2 + 2x_3 = 3 \\ x_1 - 4x_2 + 7x_3 = 4 \end{cases}$

8. Let $\mathsf{T}\colon \mathsf{R}^3 \to \mathsf{R}^3$ be defined by $\mathsf{T}(a, b, c) = (a + b, b - 2c, a + 2c)$. For each vector B in R^3 determine whether or not $B \in \mathsf{R}(\mathsf{T})$.

(a) $B = (1, 3, -2)$ **(b)** $B = (2, 1, 1)$

9. Prove that a system $AX = B$ of m linear equations in n unknowns has a solution if and only if $B \in \mathsf{R}(\mathsf{L}_A)$.

10. Prove or give a counterexample to the following statement: If the coefficient matrix of a system of m linear equations in n unknowns has rank m, then the system has a solution.

11. In the closed model of Leontief with food, clothing, and housing as the basic industries, suppose that the input–output matrix is

$$A = \begin{pmatrix} \frac{7}{16} & \frac{1}{2} & \frac{3}{16} \\ \frac{5}{16} & \frac{1}{6} & \frac{5}{16} \\ \frac{1}{4} & \frac{1}{3} & \frac{1}{2} \end{pmatrix}.$$

At what ratio must the farmer, tailor, and carpenter produce in order for equilibrium to be attained?

12. A certain closed economy consists of two sectors: goods and services. Suppose that 60% of all goods and 30% of all services are used in the production of goods. What proportion of the total economic output is used in the production of goods?

13. In the notation of the open model of Leontief, suppose that

$$A = \begin{pmatrix} \frac{1}{2} & \frac{1}{5} \\ \frac{1}{3} & \frac{1}{5} \end{pmatrix}$$

and that the demand vector is $D = \begin{pmatrix} 2 \\ 5 \end{pmatrix}$. How much of each commodity must be produced to satisfy this demand?

14. A certain economy consisting of the two sectors of goods and services supports a defense system that consumes \$90 billion worth of goods and \$20 billion worth of services from the economy but does not contribute to economic production. Suppose that 0.5 unit of goods and 0.2 unit of services are required to produce 1 unit of goods and that 0.3 unit of goods and 0.6 unit of services are required to produce 1 unit of services. What must the total output of the economic system be to support this defense system?

3.4 SYSTEMS OF LINEAR EQUATIONS—COMPUTATIONAL ASPECTS

In Section 3.3 we obtained a necessary and sufficient condition for a system of linear equations to have solutions (Theorem 3.11) and learned how to express the solutions to a nonhomogeneous system in terms of solutions to the corresponding homogeneous system (Theorem 3.9). The latter result enables us to determine all the solutions to a given system if we can find one solution to the given system and a basis for the solution set of the corresponding homogeneous system. In this section we use elementary row operations to accomplish these two objectives. The essence of this technique is to transform a given system of linear equations into a system having the same solutions but which is easier to solve (as in Section 1.4).

Definition. *Two systems of* m *linear equations in* n *unknowns are called* equivalent *if they have the same solution set.*

The following theorem and corollary give a useful method for obtaining equivalent systems.

Theorem 3.13. *Let* (S): $AX = B$ *be a system of* m *linear equations in* n *unknowns, and let* C *be any invertible* $m \times m$ *matrix. Then the system* (S′): $(CA)X = CB$ *is equivalent to* (S).

Proof. Let K be the solution set for (S) and K' the solution set for (S'). If $w \in K$, then $Aw = B$. So $CAw = CB$, and hence $w \in K'$. Thus $K \subseteq K'$.

Conversely, if $w \in K'$, then $CAw = CB$. Hence

$$Aw = C^{-1}(CAw) = C^{-1}(CB) = B;$$

so $w \in K$. Thus $K' \subseteq K$, and therefore $K = K'$. ∎

Corollary. *Let* $AX = B$ *be a system of* m *linear equations in* n *unknowns. If* $(A' \mid B')$ *is obtained from* $(A \mid B)$ *by a finite number of elementary row operations, then the system* $A'X = B'$ *is equivalent to the original system.*

Proof. Suppose that $(A' \mid B')$ is obtained from $(A \mid B)$ by elementary row operations. These may be executed by multiplying by elementary $m \times m$ matrices $E_1, \ldots, E_p$. Let $C = E_p \cdots E_1$; then

$$(A' \mid B') = C(A \mid B) = (CA \mid CB).$$

Since each E_i is invertible, so is C. Now $A' = CA$ and $B' = CB$. Thus by Theorem 3.13 the system $A'X = B'$ is equivalent to the system $AX = B$. ∎

We now describe a method for solving any system of linear equations. We illustrate this method with the system of equations

$$\begin{cases} 3x_1 + 2x_2 + 3x_3 - 2x_4 = 1 \\ x_1 + x_2 + x_3 = 3 \\ x_1 + 2x_2 + x_3 - x_4 = 2. \end{cases}$$

First we form the augmented matrix

$$\left(\begin{array}{cccc|c} 3 & 2 & 3 & -2 & 1 \\ 1 & 1 & 1 & 0 & 3 \\ 1 & 2 & 1 & -1 & 2 \end{array}\right).$$

We will use elementary row operations to reduce the augmented matrix into an upper triangular matrix in which the first nonzero entry of each row is 1 and it occurs in a column to the right of the first nonzero entry of each preceding row. (Recall that matrix A is upper triangular if $A_{ij} = 0$ whenever $i > j$.)

1. *Put a 1 in the first row, first column.* In our example we can accomplish this step by interchanging the first and third rows. The resulting matrix is

$$\left(\begin{array}{cccc|c} 1 & 2 & 1 & -1 & 2 \\ 1 & 1 & 1 & 0 & 3 \\ 3 & 2 & 3 & -2 & 1 \end{array}\right).$$

2. *By means of type 3 operations, use the first row to obtain zeros in the remaining positions of the first column.* In our example we must add -1 times the first row to the second row and then add -3 times the first row to the third row to obtain

$$\left(\begin{array}{cccc|c} 1 & 2 & 1 & -1 & 2 \\ 0 & -1 & 0 & 1 & 1 \\ 0 & -4 & 0 & 1 & -5 \end{array}\right).$$

3. *Put a 1 in the next row in the leftmost possible column, without using previous row(s).* In our example the second column is the leftmost possible column, and we can make the second row, second column entry a 1 by multiplying the second row by -1. This operation produces

$$\left(\begin{array}{rrrr|r} 1 & 2 & 1 & -1 & 2 \\ 0 & 1 & 0 & -1 & -1 \\ 0 & -4 & 0 & 1 & -5 \end{array}\right).$$

4. *Now use type 3 operations to obtain zeros below the 1 created in the preceding step.* In our example we must add four times the second row to the third row. The resulting matrix is

$$\left(\begin{array}{rrrr|r} 1 & 2 & 1 & -1 & 2 \\ 0 & 1 & 0 & -1 & -1 \\ 0 & 0 & 0 & -3 & -9 \end{array}\right).$$

5. *Repeat steps 3 and 4 on each succeeding row until no nonzero rows remain.* In our example this can be accomplished by multiplying the third row by $-\frac{1}{3}$. This operation produces

$$\left(\begin{array}{rrrr|r} 1 & 2 & 1 & -1 & 2 \\ 0 & 1 & 0 & -1 & -1 \\ 0 & 0 & 0 & 1 & 3 \end{array}\right).$$

We have now obtained the desired matrix. To complete the simplification of the augmented matrix, we must make the first nonzero entry in each row the only nonzero entry in its column. (This corresponds to eliminating certain unknowns from all but one of the equations.)

6. *To accomplish this objective, we work upward, beginning with the last nonzero row, and add multiples of each row to the rows above.* In our example the third row is the last nonzero row, and the first nonzero entry of this row lies in column 4. Hence we add the third row to the first and second rows to obtain zeros in row one, column four and row two, column four. The resulting matrix is

$$\left(\begin{array}{rrrr|r} 1 & 2 & 1 & 0 & 5 \\ 0 & 1 & 0 & 0 & 2 \\ 0 & 0 & 0 & 1 & 3 \end{array}\right).$$

7. *Make the first nonzero entry in the next-to-last row the only nonzero entry in its column.* In our example we must add -2 times the second row to the first row in order to make the first row, second column entry become zero. This operation produces

$$\left(\begin{array}{rrrr|r} 1 & 0 & 1 & 0 & 1 \\ 0 & 1 & 0 & 0 & 2 \\ 0 & 0 & 0 & 1 & 3 \end{array}\right).$$

8. *Repeat the process described in step 7 for each preceding row until it is performed with the second row, at which time the reduction process is complete.*

Since we have already reached the point described in step 8, the preceding matrix is the desired reduction of the augmented matrix. This matrix corresponds to the system of linear equations

$$\begin{cases} x_1 + \quad x_3 \qquad = 1 \\ \qquad x_2 \qquad\quad = 2 \\ \qquad\qquad\quad x_4 = 3. \end{cases}$$

Recall that by the corollary to Theorem 3.13 this system is equivalent to the original system. But this system is easily solved. Obviously, $x_2 = 2$ and $x_4 = 3$. Moreover, x_1 and x_3 can have any values provided that their sum is 1. Letting $x_3 = t$, we then have $x_1 = 1 - t$. Thus an arbitrary solution to the original system has the form

$$\begin{pmatrix} 1-t \\ 2 \\ t \\ 3 \end{pmatrix} = \begin{pmatrix} 1 \\ 2 \\ 0 \\ 3 \end{pmatrix} + t\begin{pmatrix} -1 \\ 0 \\ 1 \\ 0 \end{pmatrix}.$$

Observe that

$$\left\{\begin{pmatrix} -1 \\ 0 \\ 1 \\ 0 \end{pmatrix}\right\}$$

is a basis for the homogeneous system of equations corresponding to the given system.

In the example above we performed elementary row operations on the augmented matrix of the system until we obtained the augmented matrix of a system having properties 1, 2, and 3 on pages 24–25. Such a matrix has a special name.

Definition. *A matrix is said to be in* row echelon form *if the following three conditions are satisfied:*

(a) *Any row containing a nonzero entry precedes any row in which all the entries are zero (if any).*

(b) *The first nonzero entry in each row is the only nonzero entry in its column.*

(c) *The first nonzero entry in each row is 1 and it occurs in a column to the right of the leading 1 in any preceding row.*

Example 1

(a) The last matrix on page 162 is in row echelon form. Note that the first nonzero entry of each row is 1 and that the column containing this first entry has all zeros otherwise. Also note that each time we move downward to a new row, we must move to the right one or more columns to find the first nonzero entry of the new row.

(b) The following are *not* in row echelon form:

$$\begin{pmatrix} 1 & 1 & 0 \\ 0 & 1 & 0 \\ 1 & 0 & 1 \end{pmatrix},$$

because the first column contains more than one nonzero entry;

$$\begin{pmatrix} 0 & 1 & 0 & 2 \\ 1 & 0 & 0 & 1 \\ 0 & 0 & 1 & 1 \end{pmatrix},$$

because the first nonzero entry of the second row is not to the right of the first nonzero entry of the first row; and

$$\begin{pmatrix} 2 & 0 & 0 \\ 0 & 1 & 0 \end{pmatrix},$$

because the first nonzero entry of the first row is not 1. ∎

It can be shown (see Exercise 7) that the row echelon form of a matrix is unique; that is, if different sequences of elementary row operations are used to transform a matrix into matrices Q and Q' in row echelon form, then $Q = Q'$. Thus although there are many different sequences of elementary row operations that can be used to transform a given matrix into row echelon form, they all produce the same result.

The procedure used above to reduce an augmented matrix to row echelon form is called *Gaussian elimination.* It consists of two separate parts.

1. In the *forward pass*, the augmented matrix is transformed into an upper triangular matrix in which the first nonzero entry of each row is 1, and it occurs in a column to the right of the first nonzero entry of each preceding row.
2. In the *backward pass* (or *back-substitution*), the upper triangular matrix is transformed into row echelon form.

Of all the methods for transforming a matrix into its row echelon form, Gaussian elimination requires the fewest arithmetic operations. (For large matrices it requires approximately 50% fewer operations than the Gauss–

Jordan method, in which the matrix is transformed into row echelon form by using the first nonzero entry in each row to make zero all other entries in its column.) Because of this efficiency, Gaussian elimination is the preferred method when solving systems of linear equations on a computer. In this context the Gaussian elimination procedure is usually modified in order to minimize round-off errors. Since discussion of these techniques is inappropriate here, readers who are interested in such matters are referred to books on numerical analysis.

When a matrix is in row echelon form, the corresponding system of linear equations is easy to solve. We present below a procedure for solving any system of linear equations for which the augmented matrix is in row echelon form. First, however, we note that every matrix can be transformed into row echelon form by Gaussian elimination. In the forward pass we satisfy conditions (a) and (c) in the definition of row echelon form and also make zero all entries below the first nonzero entry in each row. Then in the backward pass we make zero all entries above the first nonzero entry in each row, thereby satisfying condition (b) in the definition of row echelon form.

Theorem 3.14. *Gaussian elimination transforms any matrix into its row echelon form.*

We now describe a method for solving a system in which the augmented matrix is in row echelon form. To illustrate this procedure, we consider the system

$$\begin{cases} 2x_1 + 3x_2 + x_3 + 4x_4 - 9x_5 = 17 \\ x_1 + x_2 + x_3 + x_4 - 3x_5 = 6 \\ x_1 + x_2 + x_3 + 2x_4 - 5x_5 = 8 \\ 2x_1 + 2x_2 + 2x_3 + 3x_4 - 8x_5 = 14 \end{cases}$$

for which the augmented matrix is

$$\left(\begin{array}{ccccc|c} 2 & 3 & 1 & 4 & -9 & 17 \\ 1 & 1 & 1 & 1 & -3 & 6 \\ 1 & 1 & 1 & 2 & -5 & 8 \\ 2 & 2 & 2 & 3 & -8 & 14 \end{array}\right).$$

Applying Gaussian elimination to the augmented matrix of the system produces the following sequence of matrices.

$$\left(\begin{array}{ccccc|c} 2 & 3 & 1 & 4 & -9 & 17 \\ 1 & 1 & 1 & 1 & -3 & 6 \\ 1 & 1 & 1 & 2 & -5 & 8 \\ 2 & 2 & 2 & 3 & -8 & 14 \end{array}\right), \qquad \left(\begin{array}{ccccc|c} 1 & 1 & 1 & 1 & -3 & 6 \\ 2 & 3 & 1 & 4 & -9 & 17 \\ 1 & 1 & 1 & 2 & -5 & 8 \\ 2 & 2 & 2 & 3 & -8 & 14 \end{array}\right),$$

$$\begin{pmatrix} 1 & 1 & 1 & 1 & -3 & \vert & 6 \\ 0 & 1 & -1 & 2 & -3 & \vert & 5 \\ 0 & 0 & 0 & 1 & -2 & \vert & 2 \\ 0 & 0 & 0 & 1 & -2 & \vert & 2 \end{pmatrix}, \quad \begin{pmatrix} 1 & 1 & 1 & 1 & -3 & \vert & 6 \\ 0 & 1 & -1 & 2 & -3 & \vert & 5 \\ 0 & 0 & 0 & 1 & -2 & \vert & 2 \\ 0 & 0 & 0 & 0 & 0 & \vert & 0 \end{pmatrix},$$

$$\begin{pmatrix} 1 & 1 & 1 & 0 & -1 & \vert & 4 \\ 0 & 1 & -1 & 0 & 1 & \vert & 1 \\ 0 & 0 & 0 & 1 & -2 & \vert & 2 \\ 0 & 0 & 0 & 0 & 0 & \vert & 0 \end{pmatrix}, \quad \begin{pmatrix} 1 & 0 & 2 & 0 & -2 & \vert & 3 \\ 0 & 1 & -1 & 0 & 1 & \vert & 1 \\ 0 & 0 & 0 & 1 & -2 & \vert & 2 \\ 0 & 0 & 0 & 0 & 0 & \vert & 0 \end{pmatrix}.$$

The system of linear equations corresponding to this last matrix is

$$\begin{cases} x_1 \quad + 2x_3 \quad - 2x_5 = 3 \\ \quad x_2 - \ x_3 \quad + \ x_5 = 1 \\ \qquad\qquad x_4 - 2x_5 = 2. \end{cases}$$

Notice that we have ignored the last row since it consists entirely of zeros.

To solve a system for which the augmented matrix is in row echelon form, divide the variables $x_1, x_2, \ldots, x_5$ into two sets. The first set consists of those variables that appear as leftmost variables in one of the equations of the system (in this case the set is $\{x_1, x_2, x_4\}$). The second set consists of all the remaining variables (in this case, $\{x_3, x_5\}$). To each variable in the second set, assign a parametric value $t_1, t_2, \ldots (x_3 = t_1, x_5 = t_2)$, and then solve for the variables of the first set in terms of those in the second set:

$$\begin{aligned} x_1 &= -2x_3 + 2x_5 + 3 = -2t_1 + 2t_2 + 3 \\ x_2 &= \quad x_3 - \ x_5 + 1 = \quad t_1 - \ t_2 + 1 \\ x_4 &= \qquad\quad 2x_5 + 2 = \qquad\quad 2t_2 + 2. \end{aligned}$$

Thus an arbitrary solution, s, is of the form

$$s = \begin{pmatrix} x_1 \\ x_2 \\ x_3 \\ x_4 \\ x_5 \end{pmatrix} = \begin{pmatrix} -2t_1 + 2t_2 + 3 \\ t_1 - t_2 + 1 \\ t_1 \\ 2t_2 + 2 \\ t_2 \end{pmatrix} = \begin{pmatrix} 3 \\ 1 \\ 0 \\ 2 \\ 0 \end{pmatrix} + t_1 \begin{pmatrix} -2 \\ 1 \\ 1 \\ 0 \\ 0 \end{pmatrix} + t_2 \begin{pmatrix} 2 \\ -1 \\ 0 \\ 2 \\ 1 \end{pmatrix},$$

where $t_1, t_2 \in R$. Notice that

$$\left\{ \begin{pmatrix} -2 \\ 1 \\ 1 \\ 0 \\ 0 \end{pmatrix}, \begin{pmatrix} 2 \\ -1 \\ 0 \\ 2 \\ 1 \end{pmatrix} \right\}$$

is a basis for the solution set of the corresponding homogeneous system of equations.

Therefore, in simplifying the augmented matrix of the system to row echelon form, we are in effect simultaneously finding a particular solution to the original system and a basis for the solution set of the associated homogeneous system. Moreover, this procedure detects when a system has no solutions, for by Exercise 3, solutions exist if and only if, in the reduction of the augmented matrix to row echelon form, we do not obtain a row in which the only nonzero entry lies in the last column.

Thus to use this procedure for solving a system $AX = B$ of m linear equations in n unknowns, we need only begin to transform the augmented matrix $(A \mid B)$ into its row echelon form $(A' \mid B')$ by means of Gaussian elimination. If a row is obtained in which the only nonzero entry lies in the last column, then the original system has no solutions. Otherwise, discard any zero rows from $(A' \mid B')$, and write the corresponding system of equations. Solve this system as described above to obtain an arbitrary solution of the form

$$s = s_0 + t_1 u_1 + t_2 u_2 + \cdots + t_{n-m'} u_{n-m'},$$

where m' is the number of nonzero rows in A' ($m' \leqslant m$). The equation above suggests that an arbitrary solution, s, can be expressed in terms of $n - m'$ parameters. The following theorem states that s cannot be expressed in fewer than $n - m'$ parameters.

Theorem 3.15. *Let* $\mathrm{AX} = \mathrm{B}$ *be a system of* m *nonzero equations in* n *unknowns. Suppose that* $\mathrm{rank(A)} = \mathrm{rank(A \mid B)}$ *and that* $(\mathrm{A} \mid \mathrm{B})$ *is in row echelon form. Then*

(a) $\mathrm{rank(A)} = \mathrm{m}$.

(b) *If the general solution as obtained by the procedure above is of the form*

$$\mathrm{s} = \mathrm{s}_0 + \mathrm{t}_1 \mathrm{u}_1 + \mathrm{t}_2 \mathrm{u}_2 + \cdots + \mathrm{t}_{\mathrm{n-m}} \mathrm{u}_{\mathrm{n-m}},$$

then $\{\mathrm{u}_1, \mathrm{u}_2, \ldots, \mathrm{u}_{\mathrm{n-m}}\}$ *is a basis for the solution set of the corresponding homogeneous system and* s_0 *is a solution to the original system.*

Proof. Since $(A \mid B)$ is in row echelon form, $\mathrm{rank}(A \mid B) = \mathrm{rank}(A) = m$ by Exercises 5 and 6.

Let K be the solution set for $AX = B$, and let K_H be the solution set for $AX = 0$. Setting $t_1 = t_2 = \cdots = t_{n-m} = 0$, $s = s_0 \in K$. But by Theorem 3.9, $K = \{s_0\} + \mathsf{K}_\mathsf{H}$. Hence

$$\mathsf{K}_\mathsf{H} = K - \{s_0\} = \mathrm{span}(\{u_1, u_2, \ldots, u_{n-m}\}).$$

Since $\mathrm{rank}(A) = m$, $\dim(\mathsf{K}_\mathsf{H}) = n - m$. Thus since $\dim(\mathsf{K}_\mathsf{H}) = n - m$ and K_H is generated by a set $\{u_1, u_2, \ldots, u_{n-m}\}$, containing at most $n - m$ elements, we conclude that the set above is a basis for K_H. ∎

EXERCISES

1. Label the following statements as being true or false.
   - **(a)** If $(A' \mid B')$ is obtained from $(A \mid B)$ by a finite sequence of elementary column operations, then the systems $AX = B$ and $A'X = B'$ are equivalent.
   - **(b)** If $(A' \mid B')$ is obtained from $(A \mid B)$ by a finite sequence of elementary row operations, then the systems $AX = B$ and $A'X = B'$ are equivalent.
   - **(c)** If A is an $n \times n$ matrix with rank n, then the row echelon form of A is I_n.
   - **(d)** Any matrix can be put in row echelon form by means of a finite sequence of elementary row operations.
   - **(e)** If $(A \mid B)$ is in row echelon form, then the system $AX = B$ must have a solution.
   - **(f)** Let $AX = B$ be a system of m linear equations in n unknowns for which the augmented matrix is in row echelon form. If this system has solutions, then the dimension of the solution set of $AX = 0$ is $n - m'$, where m' equals the number of nonzero rows in A.
   - **(g)** If a matrix A is transformed by elementary row operations into a matrix A' in row echelon form, then the number of nonzero rows in A' is equal to the rank of A.
2. Solve the following systems of linear equations using Gaussian elimination.

(a)
$$\begin{cases} x_1 + 2x_2 - x_3 = -1 \\ 2x_1 + 2x_2 + x_3 = 1 \\ 3x_1 + 5x_2 - 2x_3 = -1 \end{cases}$$

(b)
$$\begin{cases} x_1 - 2x_2 - x_3 = 1 \\ 2x_1 - 3x_2 + x_3 = 6 \\ 3x_1 - 5x_2 = 7 \\ x_1 + 5x_3 = 9 \end{cases}$$

(c)
$$\begin{cases} x_1 + 2x_2 + 2x_4 = 6 \\ 3x_1 + 5x_2 - x_3 + 6x_4 = 17 \\ 2x_1 + 4x_2 + x_3 + 2x_4 = 12 \\ 2x_1 - 7x_3 + 11x_4 = 7 \end{cases}$$

(d)
$$\begin{cases} x_1 - x_2 + 2x_3 + 3x_4 = -7 \\ 2x_1 - x_2 + 6x_3 + 6x_4 = -2 \\ -2x_1 + x_2 - 4x_3 - 3x_4 = 0 \\ 3x_1 - 2x_2 + 9x_3 + 10x_4 = -5 \end{cases}$$

(e)
$$\begin{cases} x_1 - 4x_2 - x_3 + x_4 = 3 \\ 2x_1 - 8x_2 + x_3 - 4x_4 = 9 \\ -x_1 + 4x_2 - 2x_3 + 5x_4 = -6 \end{cases}$$

(f)
$$\begin{cases} x_1 + 2x_2 - x_3 + 3x_4 = 2 \\ 2x_1 + 4x_2 - x_3 + 6x_4 = 5 \\ x_2 + 2x_4 = 3 \end{cases}$$

(g) $$\begin{cases} 2x_1 - 2x_2 - x_3 + 6x_4 - 2x_5 = 1 \\ x_1 - x_2 + x_3 + 2x_4 - x_5 = 2 \\ 4x_1 - 4x_2 + 5x_3 + 7x_4 - x_5 = 6 \end{cases}$$

(h) $$\begin{cases} 3x_1 - x_2 + x_3 - x_4 + 2x_5 = 5 \\ x_1 - x_2 - x_3 - 2x_4 - x_5 = 2 \\ 5x_1 - 2x_2 + x_3 - 3x_4 + 3x_5 = 10 \\ 2x_1 - x_2 \qquad - 2x_4 + x_5 = 5 \end{cases}$$

(i) $$\begin{cases} 3x_1 - x_2 + 2x_3 + 4x_4 + x_5 = 2 \\ x_1 - x_2 + 2x_3 + 3x_4 + x_5 = -1 \\ 2x_1 - 3x_2 + 6x_3 + 9x_4 + 4x_5 = -5 \\ 7x_1 - 2x_2 + 4x_3 + 8x_4 + x_5 = 6 \end{cases}$$

(j) $$\begin{cases} 2x_1 \qquad + 3x_3 \qquad - 4x_5 = 5 \\ 3x_1 - 4x_2 + 8x_3 + 3x_4 \qquad = 8 \\ x_1 - x_2 + 2x_3 + x_4 - x_5 = 2 \\ -2x_1 + 5x_2 - 9x_3 - 3x_4 - 5x_5 = -8 \end{cases}$$

3. Suppose that the augmented matrix of the system $AX = B$ is transformed into a matrix $(A' \mid B')$ in row echelon form by a finite sequence of elementary row operations.

(a) Prove that $\text{rank}(A') \neq \text{rank}((A' \mid B'))$ if and only if $(A' \mid B')$ contains a row in which the only nonzero entry lies in the last column.

(b) Deduce that $AX = B$ has solutions if and only if $(A' \mid B')$ contains no row in which the only nonzero entry lies in the last column.

4. For each of the following systems, apply Exercise 3 to determine if the system has solutions. If there are solutions, find all of them. Finally, find a basis for the corresponding homogeneous system.

(a) $$\begin{cases} x_1 + 2x_2 - x_3 + x_4 = 2 \\ 2x_1 + x_2 + x_3 - x_4 = 3 \\ x_1 + 2x_2 - 3x_3 + 2x_4 = 2 \end{cases}$$

(b) $$\begin{cases} x_1 + x_2 - 3x_3 + x_4 = -2 \\ x_1 + x_2 + x_3 - x_4 = 2 \\ x_1 + x_2 - x_3 \qquad = 0 \end{cases}$$

(c) $$\begin{cases} x_1 + x_2 - 3x_3 + x_4 = 1 \\ x_1 + x_2 + x_3 - x_4 = 2 \\ x_1 + x_2 - x_3 \qquad = 0 \end{cases}$$

5. Prove that if A is a matrix in row echelon form, then $\text{rank}(A)$ equals the number of nonzero rows in A.

6. If $(A \mid B)$ is in row echelon form, prove that A is also in row echelon form.

7. **(a)** Prove that if Q and Q' are $m \times n$ matrices each in row echelon form such that Q can be transformed into Q' by means of a finite number of elementary row operations, then $Q = Q'$. *Hint:* Use induction on n.

   (b) Deduce that if A is any matrix, then there is a unique matrix in row echelon form that can be obtained from A by a finite number of elementary row operations.

INDEX OF DEFINITIONS FOR CHAPTER 3

4

Determinants

At one time determinants played a major role in the study of linear algebra. Now, however, they are of much less importance. In fact, virtually our only use of the determinant is in the computation of "eigenvalues." For this reason the important facts about the determinant needed for later chapters are summarized in Section 4.4. The reader who is not interested in pursuing a development of the theory of determinants may proceed immediately to that section.

The determinant of a square matrix with entries from a field F is a scalar (element of F). Thus we may regard the determinant as a function having domain $\mathsf{M}_{n \times n}(F)$ and taking values in F. Although the determinant of a square matrix can be defined in terms of the entries of the matrix, the resulting definition is cumbersome to use for computations. Instead of defining the determinant in this manner, in Section 4.2 we define a determinant as a function $\delta\colon \mathsf{M}_{n \times n}(F) \to F$ possessing three important properties. In that section we also verify that the familiar method of evaluating a determinant by expansion along a column is, in fact, a determinant in the sense of our definition. Section 4.3 contains further properties of a determinant and contains a proof that there is a unique determinant on $\mathsf{M}_{n \times n}(F)$, i.e., that the three defining properties of a determinant are satisfied by one and only one function from $\mathsf{M}_{n \times n}(F)$ into F.

The chapter begins with a discussion of the general theory in a simple setting. In Section 4.1 we also investigate the geometric significance of the determinant in terms of area and orientation. Readers who have studied advanced calculus will recall that a change of coordinates in multiple integrals necessitated the use of a determinant called the *Jacobian*.

4.1 DETERMINANTS OF ORDER 2

Eventually we will assign to each $n \times n$ matrix with entries from a field F a scalar called the "determinant" of the matrix, but first we consider an easy special case.

Definition. *The* determinant *of a* 2×2 *matrix* A *with entries from a field* F *is the scalar* $A_{11}A_{22} - A_{12}A_{21}$, *which we denote by* det(A).

Example 1

Consider the following element of $\mathsf{M}_{2\times 2}(R)$:

$$A = \begin{pmatrix} 1 & 2 \\ 3 & 4 \end{pmatrix}.$$

Then

$$\det(A) = 1 \cdot 4 - 2 \cdot 3 = -2. \quad \blacksquare$$

In the discussion that follows it will be convenient to represent a 2×2 matrix A in terms of its rows; as before, we will write

$$A = \begin{pmatrix} A_{(1)} \\ A_{(2)} \end{pmatrix}$$

and denote its determinant by

$$\det \begin{pmatrix} A_{(1)} \\ A_{(2)} \end{pmatrix}.$$

The determinant has the following important properties.

Theorem 4.1. *The determinant of a 2×2 matrix satisfies the following three conditions:*

(a) *The determinant is a linear function of each row when the other row is held fixed; that is,*

$$\det \begin{pmatrix} cA_{(1)} + A'_{(1)} \\ A_{(2)} \end{pmatrix} = c \det \begin{pmatrix} A_{(1)} \\ A_{(2)} \end{pmatrix} + \det \begin{pmatrix} A'_{(1)} \\ A_{(2)} \end{pmatrix}$$

and

$$\det \begin{pmatrix} A_{(1)} \\ cA_{(2)} + A'_{(2)} \end{pmatrix} = c \det \begin{pmatrix} A_{(1)} \\ A_{(2)} \end{pmatrix} + \det \begin{pmatrix} A_{(1)} \\ A'_{(2)} \end{pmatrix}$$

for all scalars c *in* F.

(b) *If* $A \in \mathsf{M}_{2\times 2}(F)$ *has identical rows, then* $\det(A) = 0$.

(c) *If* I *is the* 2×2 *identity matrix, then* $\det(I) = 1$.

Proof. (a) Let $A_{(1)} = (A_{11} \;\; A_{12})$, $A'_{(1)} = (A'_{11} \;\; A'_{12})$, and $A_{(2)} = (A_{21} \;\; A_{22})$; then

$$\det \begin{pmatrix} cA_{(1)} + A'_{(1)} \\ A_{(2)} \end{pmatrix} = \det \begin{pmatrix} cA_{11} + A'_{11} & cA_{12} + A'_{12} \\ A_{21} & A_{22} \end{pmatrix}$$

$$= (cA_{11} + A'_{11})A_{22} - (cA_{12} + A'_{12})A_{21}$$

$$= c(A_{11}A_{22} - A_{12}A_{21}) + (A'_{11}A_{22} - A'_{12}A_{21})$$

$$= c \det\begin{pmatrix} A_{11} & A_{12} \\ A_{21} & A_{22} \end{pmatrix} + \det\begin{pmatrix} A'_{11} & A'_{12} \\ A_{21} & A_{22} \end{pmatrix}$$

$$= c \det\begin{pmatrix} A_{(1)} \\ A_{(2)} \end{pmatrix} + \det\begin{pmatrix} A'_{(1)} \\ A_{(2)} \end{pmatrix}.$$

A similar argument proves that the determinant is also a linear function of the second row.

(b) If the rows of A are identical, then A has the form

$$\begin{pmatrix} A_{11} & A_{12} \\ A_{11} & A_{12} \end{pmatrix}.$$

So $\det(A) = A_{11}A_{12} - A_{12}A_{11} = 0$.

(c) Since

$$I = \begin{pmatrix} 1 & 0 \\ 0 & 1 \end{pmatrix},$$

$$\det(I) = 1 \cdot 1 - 0 \cdot 0 = 1. \quad \blacksquare$$

The next result shows that the three properties mentioned in Theorem 4.1 completely characterize the determinant as defined above.

Theorem 4.2. *Let $\delta\colon \mathsf{M}_{2\times 2}(F) \to F$ be any function having the following three properties:*

(a) *δ is a linear function of each row when the other row is held fixed.*

(b) *If $A \in \mathsf{M}_{2\times 2}(F)$ has identical rows, then $\delta(A) = 0$.*

(c) *If I is the 2×2 identity matrix, then $\delta(I) = 1$.*

Then $\delta = \det$; that is, $\delta(A) = A_{11}A_{22} - A_{12}A_{21}$ for each $A \in \mathsf{M}_{2\times 2}(F)$.

Proof. Let I denote the 2×2 identity matrix, and let

$$M_1 = \begin{pmatrix} 1 & 0 \\ 1 & 0 \end{pmatrix}, \qquad M_2 = \begin{pmatrix} 0 & 1 \\ 0 & 1 \end{pmatrix}, \quad \text{and} \quad M_3 = \begin{pmatrix} 0 & 1 \\ 1 & 0 \end{pmatrix}.$$

Observe that $\delta(M_1) = \delta(M_2) = 0$ by property (b). We will first prove that $\delta(M_3) = -1$. Using properties (b) and (a), we have

$$0 = \delta\begin{pmatrix} 1 & 1 \\ 1 & 1 \end{pmatrix} = \delta\begin{pmatrix} 1+0 & 0+1 \\ 1 & 1 \end{pmatrix}$$

$$= \delta\begin{pmatrix} 1 & 0 \\ 1 & 1 \end{pmatrix} + \delta\begin{pmatrix} 0 & 1 \\ 1 & 1 \end{pmatrix}$$

$$= \delta\begin{pmatrix} 1 & 0 \\ 0+1 & 1+0 \end{pmatrix} + \delta\begin{pmatrix} 0 & 1 \\ 0+1 & 1+0 \end{pmatrix}$$

$$= \delta\begin{pmatrix} 1 & 0 \\ 0 & 1 \end{pmatrix} + \delta\begin{pmatrix} 1 & 0 \\ 1 & 0 \end{pmatrix} + \delta\begin{pmatrix} 0 & 1 \\ 0 & 1 \end{pmatrix} + \delta\begin{pmatrix} 0 & 1 \\ 1 & 0 \end{pmatrix}$$
$$= \delta(I) + \delta(M_1) + \delta(M_2) + \delta(M_3)$$
$$= 1 + 0 + 0 + \delta(M_3).$$

Thus $\delta(M_3) = -1$.

Now let A be an arbitrary element of $\mathsf{M}_{2\times 2}(F)$; then

$$\delta(A) = \delta\begin{pmatrix} A_{11} & A_{12} \\ A_{21} & A_{22} \end{pmatrix} = \delta\begin{pmatrix} A_{11}+0 & 0+A_{12} \\ A_{21} & A_{22} \end{pmatrix}$$
$$= \delta\begin{pmatrix} A_{11} & 0 \\ A_{21} & A_{22} \end{pmatrix} + \delta\begin{pmatrix} 0 & A_{12} \\ A_{21} & A_{22} \end{pmatrix}$$
$$= \delta\begin{pmatrix} A_{11} & 0 \\ 0+A_{21} & A_{22}+0 \end{pmatrix} + \delta\begin{pmatrix} 0 & A_{12} \\ 0+A_{21} & A_{22}+0 \end{pmatrix}$$
$$= \delta\begin{pmatrix} A_{11} & 0 \\ 0 & A_{22} \end{pmatrix} + \delta\begin{pmatrix} A_{11} & 0 \\ A_{21} & 0 \end{pmatrix} + \delta\begin{pmatrix} 0 & A_{12} \\ 0 & A_{22} \end{pmatrix} + \delta\begin{pmatrix} 0 & A_{12} \\ A_{21} & 0 \end{pmatrix}$$
$$= A_{11}A_{22}\cdot\delta\begin{pmatrix} 1 & 0 \\ 0 & 1 \end{pmatrix} + A_{11}A_{21}\cdot\delta\begin{pmatrix} 1 & 0 \\ 1 & 0 \end{pmatrix} + A_{12}A_{22}\cdot\delta\begin{pmatrix} 0 & 1 \\ 0 & 1 \end{pmatrix}$$
$$+ A_{12}A_{21}\cdot\delta\begin{pmatrix} 0 & 1 \\ 1 & 0 \end{pmatrix}$$
$$= A_{11}A_{22}\cdot\delta(I) + A_{11}A_{21}\cdot\delta(M_1) + A_{12}A_{22}\cdot\delta(M_2) + A_{12}A_{21}\cdot\delta(M_3)$$
$$= A_{11}A_{22}(1) + A_{11}A_{21}(0) + A_{12}A_{22}(0) + A_{12}A_{21}(-1)$$
$$= A_{11}A_{22} - A_{12}A_{21} = \det(A).$$

So $\delta = \det$. ∎

The Area of a Parallelogram

Motivated by this characterization of the determinant of a 2×2 matrix, in Section 4.2 we define a determinant on $\mathsf{M}_{n\times n}(F)$ as a function possessing the three properties of Theorem 4.1. But first we will use this uniqueness property to study the geometric significance of the determinant of a 2×2 matrix. In particular, we will find that the sign of the determinant is of geometric importance in the study of orientation.

By the *angle between two vectors* in R^2 we mean the angle θ with measure such that $0 \leqslant \theta < \pi$ that is formed by the vectors having the same magnitude and direction as the given vectors but emanating from the origin (see Figure 4.1). Given three vectors u, v, and w emanating from the same point, we say that v lies *between* u and w if the angle between u and w equals the sum of the angles between u and v and between v and w (see Figure 4.2).

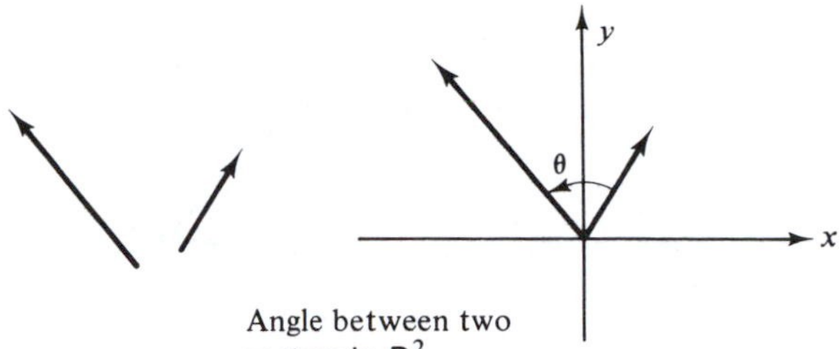

Angle between two vectors in R^2

Figure 4.1

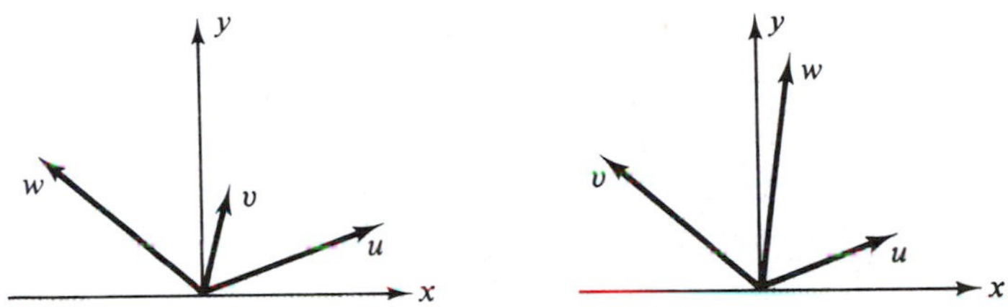

v lies between u and w.

v does not lie between u and w. (Here w lies between u and v.)

Figure 4.2

Given an ordered basis $\beta = \{u, v\}$ for R^2, where $u = (a_1, a_2)$ and $v = (b_1, b_2)$, we denote by

$$\det\begin{pmatrix} u \\ v \end{pmatrix}$$

the scalar

$$\det\begin{pmatrix} a_1 & a_2 \\ b_1 & b_2 \end{pmatrix},$$

and define the *orientation* of β to be the real number

$$\mathrm{O}\begin{pmatrix} u \\ v \end{pmatrix} = \frac{\det\begin{pmatrix} u \\ v \end{pmatrix}}{\left|\det\begin{pmatrix} u \\ v \end{pmatrix}\right|}.$$

(It follows from Exercise 10 that the denominator is not zero.) Clearly,

$$\mathrm{O}\begin{pmatrix} u \\ v \end{pmatrix} = \pm 1.$$

Notice that

$$\mathrm{O}\begin{pmatrix} e_1 \\ e_2 \end{pmatrix} = 1 \quad \text{and} \quad \mathrm{O}\begin{pmatrix} e_1 \\ -e_2 \end{pmatrix} = -1.$$

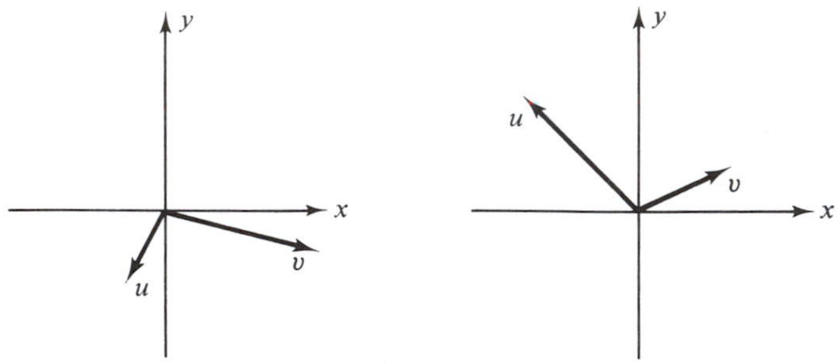

A right-handed coordinate system A left-handed coordinate system

Figure 4.3

In general (see Exercise 11),

$$\mathrm{O}\begin{pmatrix} u \\ v \end{pmatrix} = 1$$

if and only if the ordered basis $\{u, v\}$ forms a right-handed coordinate system, and

$$\mathrm{O}\begin{pmatrix} u \\ v \end{pmatrix} = -1$$

if and only if $\{u, v\}$ forms a left-handed coordinate system. [Recall that a coordinate system $\{u, v\}$ is *right-handed* if u can be rotated to coincide with v by rotating in a counterclockwise direction through an angle θ with measure such that $0 < \theta < \pi$; otherwise, $\{u, v\}$ is a *left-handed* coordinate system (see Figure 4.3).] For convenience, we define

$$\mathrm{O}\begin{pmatrix} u \\ v \end{pmatrix} = 1$$

if $\{u, v\}$ is linearly dependent.

Any ordered set $\{u, v\}$ in R^2 determines a parallelogram in the following manner. Regarding u and v as arrows emanating from the origin of R^2, we call the parallelogram having u and v as adjacent sides the *parallelogram determined by* u *and* v (see Figure 4.4).

Observe that if the set $\{u, v\}$ is linearly dependent, i.e., if u and v are parallel, then the "parallelogram" determined by u and v is actually a line segment, which we consider to be a degenerate parallelogram having area zero.

There is an interesting relationship between

$$\mathrm{A}\begin{pmatrix} u \\ v \end{pmatrix},$$

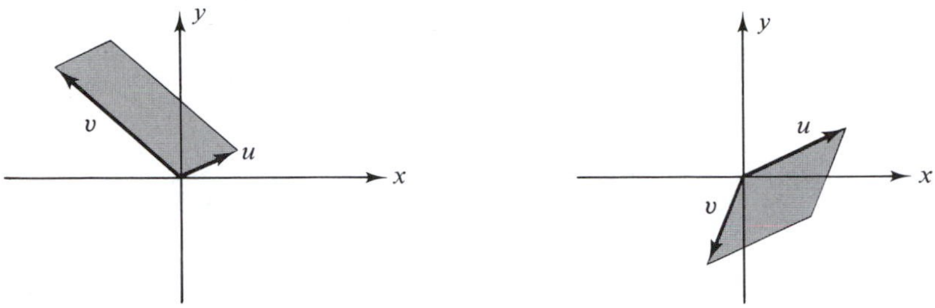

Parallelograms determined by u and v

Figure 4.4

the area of the parallelogram determined by u and v, and

$$\det\begin{pmatrix} u \\ v \end{pmatrix},$$

which we now investigate. Observe first, however, that since

$$\det\begin{pmatrix} u \\ v \end{pmatrix}$$

may be negative, we cannot expect that

$$A\begin{pmatrix} u \\ v \end{pmatrix} = \det\begin{pmatrix} u \\ v \end{pmatrix}.$$

But we can prove that

$$A\begin{pmatrix} u \\ v \end{pmatrix} = O\begin{pmatrix} u \\ v \end{pmatrix} \cdot \det\begin{pmatrix} u \\ v \end{pmatrix},$$

from which it follows that

$$A\begin{pmatrix} u \\ v \end{pmatrix} = \left|\det\begin{pmatrix} u \\ v \end{pmatrix}\right|.$$

In arguing that

$$A\begin{pmatrix} u \\ v \end{pmatrix} = O\begin{pmatrix} u \\ v \end{pmatrix} \cdot \det\begin{pmatrix} u \\ v \end{pmatrix},$$

we will employ a technique that, although somewhat indirect, can be generalized to R^n. First, since

$$O\begin{pmatrix} u \\ v \end{pmatrix} = \pm 1,$$

we may multiply both sides of the desired equation by

$$\mathrm{O}\begin{pmatrix} u \\ v \end{pmatrix}$$

to obtain the equivalent form

$$\mathrm{O}\begin{pmatrix} u \\ v \end{pmatrix} \cdot \mathrm{A}\begin{pmatrix} u \\ v \end{pmatrix} = \det\begin{pmatrix} u \\ v \end{pmatrix}.$$

We will establish this equation by verifying that the three conditions of Theorem 4.2 are satisfied by the function

$$\delta\begin{pmatrix} u \\ v \end{pmatrix} = \mathrm{O}\begin{pmatrix} u \\ v \end{pmatrix} \cdot \mathrm{A}\begin{pmatrix} u \\ v \end{pmatrix}.$$

(a) We begin by showing that

$$\delta\begin{pmatrix} u \\ \lambda v \end{pmatrix} = \lambda \cdot \delta\begin{pmatrix} u \\ v \end{pmatrix}.$$

Observe that this conclusion is immediate if $\lambda = 0$ because

$$\delta\begin{pmatrix} u \\ \lambda v \end{pmatrix} = \mathrm{O}\begin{pmatrix} u \\ 0 \end{pmatrix} \cdot \mathrm{A}\begin{pmatrix} u \\ 0 \end{pmatrix} = 1 \cdot 0 = 0.$$

So assume that $\lambda \neq 0$. Regarding λv as the base of the parallelogram determined by u and λv, we see that

$$\mathrm{A}\begin{pmatrix} u \\ \lambda v \end{pmatrix} = \text{base} \times \text{altitude} = |\lambda|\,(\text{length of } v)(\text{altitude}) = |\lambda| \cdot \mathrm{A}\begin{pmatrix} u \\ v \end{pmatrix}$$

since the altitude h of the parallelogram determined by u and λv is the same as that in the parallelogram determined by u and v (see Figure 4.5). Hence

$$\delta\begin{pmatrix} u \\ \lambda v \end{pmatrix} = \mathrm{O}\begin{pmatrix} u \\ \lambda v \end{pmatrix} \cdot \mathrm{A}\begin{pmatrix} u \\ \lambda v \end{pmatrix} = \left[\frac{\lambda}{|\lambda|} \cdot \mathrm{O}\begin{pmatrix} u \\ v \end{pmatrix}\right]\left[|\lambda| \cdot \mathrm{A}\begin{pmatrix} u \\ v \end{pmatrix}\right]$$

$$= \lambda \cdot \mathrm{O}\begin{pmatrix} u \\ v \end{pmatrix} \cdot \mathrm{A}\begin{pmatrix} u \\ v \end{pmatrix} = \lambda \cdot \delta\begin{pmatrix} u \\ v \end{pmatrix}.$$

A similar argument shows that

$$\delta\begin{pmatrix} \lambda u \\ u \end{pmatrix} = \lambda \cdot \delta\begin{pmatrix} u \\ v \end{pmatrix}.$$

We next prove that

$$\delta\begin{pmatrix} u \\ au + bw \end{pmatrix} = b \cdot \delta\begin{pmatrix} u \\ w \end{pmatrix}$$

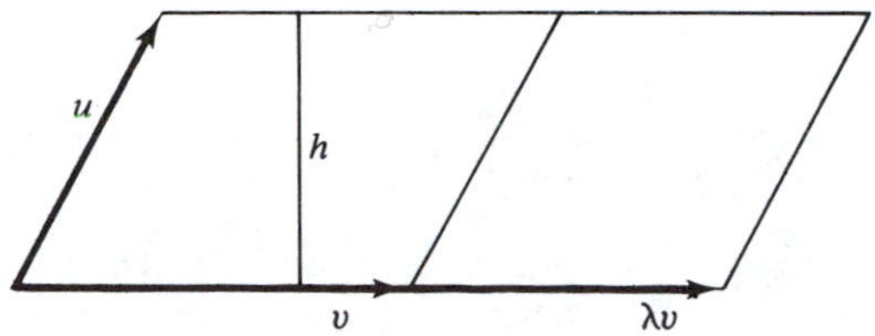

Figure 4.5

for any u, $w \in \mathsf{R}^2$ and any real numbers a and b. Observe that since the parallelograms determined by u and w and by u and $u + w$ have a common base u and the same altitude (see Figure 4.6),

$$\mathrm{A}\begin{pmatrix} u \\ w \end{pmatrix} = \mathrm{A}\begin{pmatrix} u \\ u + w \end{pmatrix}.$$

If $a = 0$, then

$$\delta\begin{pmatrix} u \\ au + bw \end{pmatrix} = \delta\begin{pmatrix} u \\ bw \end{pmatrix} = b \cdot \delta\begin{pmatrix} u \\ w \end{pmatrix}$$

by the first paragraph of part (a). Otherwise, if $a \neq 0$, then

$$\delta\begin{pmatrix} u \\ au + bw \end{pmatrix} = a \cdot \delta\begin{pmatrix} u \\ u + \dfrac{b}{a} w \end{pmatrix} = a \cdot \delta\begin{pmatrix} u \\ \dfrac{b}{a} w \end{pmatrix} = b \cdot \delta\begin{pmatrix} u \\ w \end{pmatrix}.$$

So the desired conclusion is obtained in either case.

We are now able to show that

$$\delta\begin{pmatrix} u \\ v_1 + v_2 \end{pmatrix} = \delta\begin{pmatrix} u \\ v_1 \end{pmatrix} + \delta\begin{pmatrix} u \\ v_2 \end{pmatrix}$$

for all $u, v_1, v_2 \in \mathsf{R}^2$. Since the result is immediate if $u = 0$, we assume that $u \neq 0$. Choose any vector $w \in \mathsf{R}^2$ such that $\{u, w\}$ is linearly independent. Then for any vectors $v_1, v_2 \in \mathsf{R}^2$ there exist scalars a_i and b_i such that $v_i = a_i u + b_i w \, (i = 1, 2)$.

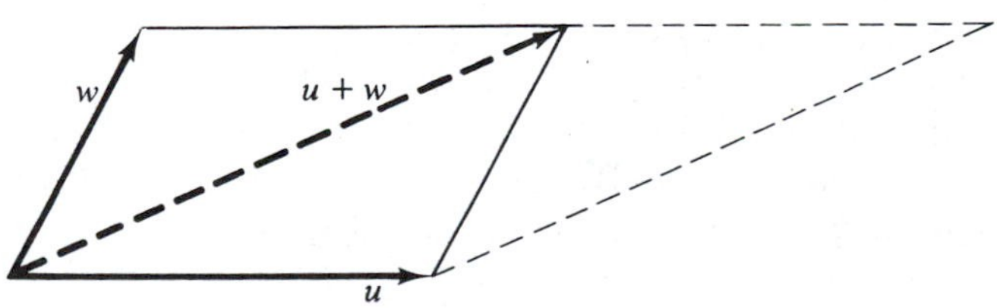

Figure 4.6

Thus

$$\delta\begin{pmatrix} u \\ v_1 + v_2 \end{pmatrix} = \delta\begin{pmatrix} u \\ (a_1 + a_2)u + (b_1 + b_2)w \end{pmatrix} = (b_1 + b_2)\delta\begin{pmatrix} u \\ w \end{pmatrix}$$

$$= \delta\begin{pmatrix} u \\ a_1u + b_1w \end{pmatrix} + \delta\begin{pmatrix} u \\ a_2u + b_2w \end{pmatrix} = \delta\begin{pmatrix} u \\ v_1 \end{pmatrix} + \delta\begin{pmatrix} u \\ v_2 \end{pmatrix}.$$

A similar argument shows that

$$\delta\begin{pmatrix} u_1 + u_2 \\ v \end{pmatrix} = \delta\begin{pmatrix} u_1 \\ v \end{pmatrix} + \delta\begin{pmatrix} u_2 \\ v \end{pmatrix}$$

for all $u_1, u_2, v \in \mathsf{R}^2$.

(b) Since

$$\mathrm{A}\begin{pmatrix} u \\ u \end{pmatrix} = 0, \qquad \delta\begin{pmatrix} u \\ u \end{pmatrix} = \mathrm{O}\begin{pmatrix} u \\ u \end{pmatrix} \cdot \mathrm{A}\begin{pmatrix} u \\ u \end{pmatrix} = 0$$

for any $u \in \mathsf{R}^2$.

(c) Because the parallelogram determined by e_1 and e_2 is the unit square,

$$\delta\begin{pmatrix} e_1 \\ e_2 \end{pmatrix} = \mathrm{O}\begin{pmatrix} e_1 \\ e_2 \end{pmatrix} \cdot \mathrm{A}\begin{pmatrix} e_1 \\ e_2 \end{pmatrix} = 1 \cdot 1 = 1.$$

Therefore, δ satisfies the three conditions of Theorem 4.2, and hence $\delta = \det$. So the area of the parallelogram determined by u and v equals

$$\mathrm{O}\begin{pmatrix} u \\ v \end{pmatrix} \cdot \det\begin{pmatrix} u \\ v \end{pmatrix}.$$

Thus, for example, we see that the area of the parallelogram determined by $u = (-1, 5)$ and $v = (4, -2)$ is

$$\left|\det\begin{pmatrix} u \\ v \end{pmatrix}\right| = \left|\det\begin{pmatrix} -1 & 5 \\ 4 & -2 \end{pmatrix}\right| = 18.$$

EXERCISES

1. Label the following statements as being true or false.
   (a) The determinant of a 2×2 matrix is a linear function of each row of the matrix when the other row is held fixed.
   (b) If I is the 2×2 identity matrix, then $\det(I) = 0$.
   (c) If both rows of a 2×2 matrix A are identical, then $\det(A) = 0$.
   (d) If u and v are vectors in R^2 emanating from the origin, then the area of the parallelogram having u and v as adjacent sides is

$$\det\begin{pmatrix} u \\ v \end{pmatrix}.$$

(e) A coordinate system $\{u, v\}$ is right-handed if and only if its orientation equals 1.

(f) The determinant is a linear transformation from $\mathsf{M}_{2\times 2}(F)$ into F.

2. Compute the determinants of the following elements of $\mathsf{M}_{2\times 2}(R)$.

(a) $\begin{pmatrix} 6 & -3 \\ 2 & 4 \end{pmatrix}$ **(b)** $\begin{pmatrix} -5 & 2 \\ 6 & 1 \end{pmatrix}$ **(c)** $\begin{pmatrix} 8 & 0 \\ 3 & -1 \end{pmatrix}$

3. Compute the determinants of the following elements of $\mathsf{M}_{2\times 2}(C)$.

(a) $\begin{pmatrix} -1+i & 1-4i \\ 3+2i & 2-3i \end{pmatrix}$ **(b)** $\begin{pmatrix} 5-2i & 6+4i \\ -3+i & 7i \end{pmatrix}$ **(c)** $\begin{pmatrix} 2i & 3 \\ 4 & 6i \end{pmatrix}$

4. For each of the following pairs of vectors u and v in R^2, compute the area of the parallelogram determined by u and v.

(a) $u = (3, -2)$ and $v = (2, 5)$

(b) $u = (1, 3)$ and $v = (-3, 1)$

(c) $u = (4, -1)$ and $v = (-6, -2)$

(d) $u = (3, 4)$ and $v = (2, -6)$

5. Prove that if B is the matrix obtained by interchanging the rows of a 2×2 matrix A, then $\det(B) = -\det(A)$.

6. Prove that for any $A \in \mathsf{M}_{2\times 2}(F)$, $\det(A^t) = \det(A)$.

7. Prove that if A is an upper triangular 2×2 matrix, then the determinant of A equals the product of the entries of A lying on the diagonal.

8. Prove that for any $A, B \in \mathsf{M}_{2\times 2}(F)$, $\det(AB) = \det(A)\cdot\det(B)$.

9. The *classical adjoint* of a 2×2 matrix A is the matrix

$$C = \begin{pmatrix} A_{22} & -A_{12} \\ -A_{21} & A_{11} \end{pmatrix}.$$

Prove the following.

(a) $CA = AC = [\det(A)]I$.

(b) $\det(C) = \det(A)$.

(c) The classical adjoint of A^t is C^t.

10. (a) Use Exercise 9(a) to prove that a 2×2 matrix A is invertible if and only if $\det(A) \neq 0$.

(b) Prove that if A is invertible, then $A^{-1} = [\det(A)]^{-1}C$, where C is the classical adjoint of A.

11. Prove that

$$\mathrm{O}\begin{pmatrix} u \\ v \end{pmatrix} = 1$$

if and only if the ordered basis $\{u, v\}$ for R^2 forms a right-handed coordinate system. *Hint:* Recall the definition of a rotation as given in Example 5 of Section 2.1.

4.2 DETERMINANTS OF ORDER *n*

We have seen in Theorem 4.2 that the determinant of a 2×2 matrix is completely characterized by three properties. We will soon define the determinant of an $n \times n$ matrix in terms of these properties, but first we require some preliminary results. To begin, we name the first of the conditions that characterizes the determinant of a 2×2 matrix.

Definition. *A function* $\delta\colon \mathsf{M}_{n\times n}(F) \to F$ *is said to be an* n-linear function *if* δ *is a linear function of each row of an* $n \times n$ *matrix when the remaining* $n-1$ *rows are held fixed, that is, if*

$$\delta\begin{pmatrix} A_{(1)} \\ \vdots \\ cA_{(i)} + A'_{(i)} \\ \vdots \\ A_{(n)} \end{pmatrix} = c\cdot\delta\begin{pmatrix} A_{(1)} \\ \vdots \\ A_{(i)} \\ \vdots \\ A_{(n)} \end{pmatrix} + \delta\begin{pmatrix} A_{(1)} \\ \vdots \\ A'_{(i)} \\ \vdots \\ A_{(n)} \end{pmatrix} \qquad \text{for} \quad i = 1, 2, \ldots, n$$

whenever

$$\begin{pmatrix} A_{(1)} \\ \vdots \\ cA_{(i)} + A'_{(i)} \\ \vdots \\ A_{(n)} \end{pmatrix}$$

is an element of $\mathsf{M}_{n\times n}(F)$.

Example 1

Theorem 4.1 shows that the function det: $\mathsf{M}_{2\times 2}(F) \to F$ defined by $\det(A) = A_{11}A_{22} - A_{12}A_{21}$ is a 2-linear function. ∎

Example 2

The function $\delta\colon \mathsf{M}_{n\times n}(F) \to F$ defined by $\delta(A) = 0$ for each $A \in \mathsf{M}_{n\times n}(F)$ is an n-linear function. ∎

Example 3

Define $\delta\colon \mathsf{M}_{n\times n}(F) \to F$ by $\delta(A) = A_{1j}A_{2j}\cdots A_{nj}$; that is, $\delta(A)$ equals the product of the entries of the jth column of A. Then δ is an n-linear function for each

$j (1 \leqslant j \leqslant n)$ since

$$\delta\begin{pmatrix} A_{(1)} \\ \vdots \\ cA_{(i)} + A'_{(i)} \\ \vdots \\ A_{(n)} \end{pmatrix} = A_{1j} \cdots A_{(i-1)j}(cA_{ij} + A'_{ij})A_{(i+1)j} \cdots A_{nj}$$

$$= c(A_{1j} \cdots A_{ij} \cdots A_{nj}) + (A_{1j} \cdots A_{(i-1)j}A'_{ij}A_{(i+1)j} \cdots A_{nj})$$

$$= c \cdot \delta\begin{pmatrix} A_{(1)} \\ \vdots \\ A_{(i)} \\ \vdots \\ A_{(n)} \end{pmatrix} + \delta\begin{pmatrix} A_{(1)} \\ \vdots \\ A'_{(i)} \\ \vdots \\ A_{(n)} \end{pmatrix}. \quad \blacksquare$$

Example 4

The function $\delta\colon \mathsf{M}_{n \times n}(F) \to F$ defined by $\delta(A) = A_{11}A_{22} \cdots A_{nn}$ (that is, $\delta(A)$ equals the product of the entries of A lying on the diagonal) is an n-linear function. ∎

Example 5

The function $\delta\colon \mathsf{M}_{n \times n}(F) \to F$ defined by $\delta(A) = \operatorname{tr}(A)$ is not an n-linear function. ∎

Our next result shows that n-linear functions may be combined to produce other n-linear functions.

Proposition 4.3. *A linear combination of two* n*-linear functions is an* n*-linear function (where the sum and scalar product are as defined in Example 3 of Section 1.2).*

Proof. Let δ_1 and δ_2 be n-linear functions, and let a and b be scalars. If δ is the linear combination $\delta = a\delta_1 + b\delta_2$, then

$$\delta\begin{pmatrix} A_{(1)} \\ \vdots \\ cA_{(i)} + A'_{(i)} \\ \vdots \\ A_{(n)} \end{pmatrix} = a \cdot \delta_1\begin{pmatrix} A_{(1)} \\ \vdots \\ cA_{(i)} + A'_{(i)} \\ \vdots \\ A_{(n)} \end{pmatrix} + b \cdot \delta_2\begin{pmatrix} A_{(1)} \\ \vdots \\ cA_{(i)} + A'_{(i)} \\ \vdots \\ A_{(n)} \end{pmatrix}$$

$$= a\left[c\cdot\delta_1\begin{pmatrix}A_{(1)}\\ \vdots\\ A_{(i)}\\ \vdots\\ A_{(n)}\end{pmatrix} + \delta_1\begin{pmatrix}A_{(1)}\\ \vdots\\ A'_{(i)}\\ \vdots\\ A_{(n)}\end{pmatrix}\right] + b\left[c\cdot\delta_2\begin{pmatrix}A_{(1)}\\ \vdots\\ A_{(i)}\\ \vdots\\ A_{(n)}\end{pmatrix} + \delta_2\begin{pmatrix}A_{(1)}\\ \vdots\\ A'_{(i)}\\ \vdots\\ A_{(n)}\end{pmatrix}\right]$$

$$= c\left[a\cdot\delta_1\begin{pmatrix}A_{(1)}\\ \vdots\\ A_{(i)}\\ \vdots\\ A_{(n)}\end{pmatrix} + b\cdot\delta_2\begin{pmatrix}A_{(1)}\\ \vdots\\ A_{(i)}\\ \vdots\\ A_{(n)}\end{pmatrix}\right] + \left[a\cdot\delta_1\begin{pmatrix}A_{(1)}\\ \vdots\\ A'_{(i)}\\ \vdots\\ A_{(n)}\end{pmatrix} + b\cdot\delta_2\begin{pmatrix}A_{(1)}\\ \vdots\\ A'_{(i)}\\ \vdots\\ A_{(n)}\end{pmatrix}\right]$$

$$= c\cdot\delta\begin{pmatrix}A_{(1)}\\ \vdots\\ A_{(i)}\\ \vdots\\ A_{(n)}\end{pmatrix} + \delta\begin{pmatrix}A_{(1)}\\ \vdots\\ A'_{(i)}\\ \vdots\\ A_{(n)}\end{pmatrix} \qquad \text{for each } i,\ 1 \leqslant i \leqslant n.$$

Thus δ is an n-linear function. ∎

Corollary. *Any linear combination of* n*-linear functions is an* n*-linear function.*

Proof. Exercise. ∎

The following definition names the second of the three properties that characterizes the determinant of a 2×2 matrix.

Definition. *An* n*-linear function* δ *is said to be* alternating *if* $\delta(A) = 0$ *whenever two adjacent rows are identical.*

Example 6

Of the three n-linear functions given in Examples 2, 3, and 4, only the first is alternating. ∎

The following result shows that the preceding definition is stronger than it first appears. In particular, there is no need for the rows to be assumed adjacent in the definition.

Proposition 4.4. *Let* $\delta\colon \mathsf{M}_{n\times n}(F) \to F$ *be an alternating* n*-linear function. Then both of the following are true:*

(a) *If* B *is obtained by interchanging any two rows of an* $n \times n$ *matrix* A*, then* $\delta(B) = -\delta(A)$.

(b) *If two rows of an* n × n *matrix* A *are identical, then* $\delta(A) = 0$.

Proof. We first prove that if B is obtained by interchanging any two adjacent rows of A, then $\delta(B) = -\delta(A)$. Suppose that B is obtained by interchanging rows i and $i+1$ of

$$A = \begin{pmatrix} A_{(1)} \\ \vdots \\ A_{(i)} \\ A_{(i+1)} \\ \vdots \\ A_{(n)} \end{pmatrix}; \quad \text{thus} \quad B = \begin{pmatrix} A_{(1)} \\ \vdots \\ A_{(i+1)} \\ A_{(i)} \\ \vdots \\ A_{(n)} \end{pmatrix}.$$

Now

$$0 = \delta \begin{pmatrix} A_{(1)} \\ \vdots \\ A_{(i)} + A_{(i+1)} \\ A_{(i)} + A_{(i+1)} \\ \vdots \\ A_{(n)} \end{pmatrix} = \delta \begin{pmatrix} A_{(1)} \\ \vdots \\ A_{(i)} \\ A_{(i)} + A_{(i+1)} \\ \vdots \\ A_{(n)} \end{pmatrix} + \delta \begin{pmatrix} A_{(1)} \\ \vdots \\ A_{(i+1)} \\ A_{(i)} + A_{(i+1)} \\ \vdots \\ A_{(n)} \end{pmatrix}$$

$$= \delta \begin{pmatrix} A_{(1)} \\ \vdots \\ A_{(i)} \\ A_{(i)} \\ \vdots \\ A_{(n)} \end{pmatrix} + \delta \begin{pmatrix} A_{(1)} \\ \vdots \\ A_{(i)} \\ A_{(i+1)} \\ \vdots \\ A_{(n)} \end{pmatrix} + \delta \begin{pmatrix} A_{(1)} \\ \vdots \\ A_{(i+1)} \\ A_{(i)} \\ \vdots \\ A_{(n)} \end{pmatrix} + \delta \begin{pmatrix} A_{(1)} \\ \vdots \\ A_{(i+1)} \\ A_{(i+1)} \\ \vdots \\ A_{(n)} \end{pmatrix}$$

$$= 0 + \delta(A) + \delta(B) + 0$$

since δ is an alternating n-linear function. Thus $\delta(B) = -\delta(A)$.

Now suppose that B is obtained from A by interchanging rows i and j, where $i < j$. Beginning with rows i and $i+1$, successively interchange adjacent rows of A until the rows are in the order

$$A_{(1)}, \ldots, A_{(i-1)}, A_{(i+1)}, \ldots, A_{(j)}, A_{(i)}, A_{(j+1)}, \ldots, A_{(n)}.$$

In all, $j - i$ interchanges are needed to produce this ordering. Now successively interchange $A_{(j)}$ with the preceding row until the rows are in the order

$$A_{(1)}, \ldots, A_{(i-1)}, A_{(j)}, A_{(i+1)}, \ldots, A_{(j-1)}, A_{(i)}, A_{(j+1)}, \ldots, A_{(n)}.$$

This process requires $j - i - 1$ interchanges of adjacent rows and produces the matrix B. Hence by the first paragraph of the proof we see that

$$\delta(B) = (-1)^{j-i}(-1)^{j-i-1}\,\delta(A) = (-1)^{2(j-i)-1}\,\delta(A) = -\delta(A).$$

It remains to show that if two rows of A are identical, say rows i and j $(i < j)$, then $\delta(A) = 0$. If $j = i + 1$, then two adjacent rows of A are identical and $\delta(A) = 0$ by the hypothesis. If $j > i + 1$, interchange rows $i + 1$ and j to obtain a matrix B in which two adjacent rows are equal. Then $\delta(B) = 0$, but since $\delta(B) = -\delta(A)$ by the second paragraph of the proof, it follows that $\delta(A) = 0$. Hence δ satisfies conditions (a) and (b). ∎

We are now prepared to define a determinant on $\mathsf{M}_{n\times n}(F)$. Observe that the determinant is defined in terms of the three properties in Theorem 4.2 that characterize the determinant of a 2×2 matrix.

Definition. *A* determinant *on* $\mathsf{M}_{n\times n}(F)$ *is an alternating* n*-linear function* $\delta\colon \mathsf{M}_{n\times n}(F) \to F$ *such that* $\delta(I) = 1$.

A simple example of a determinant can be given on $\mathsf{M}_{1\times 1}(F)$, for the function $\delta\colon \mathsf{M}_{1\times 1}(F) \to F$ defined by $\delta(A) = A_{11}$ (the only entry of A) clearly satisfies the requirements of this definition. Moreover, Theorem 4.1 shows that by defining the determinant of a 2×2 matrix A as $A_{11}A_{22} - A_{12}A_{21}$, we obtain a determinant on $\mathsf{M}_{2\times 2}(F)$ in the sense of the definition above. Our next result enables us to define a determinant on $\mathsf{M}_{n\times n}(F)$ inductively for any $n \geqslant 3$.

Proposition 4.5. *Let* δ *be an alternating* n*-linear function on* $\mathsf{M}_{n\times n}(F)$. *For each* $(n + 1) \times (n + 1)$ *matrix* A *and each* j $(1 \leqslant j \leqslant n + 1)$, *define*

$$\epsilon_j(A) = \sum_{i=1}^{n+1} (-1)^{i+j} A_{ij} \cdot \delta(\tilde{A}_{ij}),$$

where $\tilde{A}_{ij}$ *is the* $n \times n$ *matrix obtained from* A *by deleting the* ith *row and* jth *column. Then* ϵ_j *is an alternating* $(n + 1)$*-linear function on the* $(n + 1) \times (n + 1)$ *matrices with entries from* F.

Proof. Since $\tilde{A}_{ij}$ is obtained from A by deleting the ith row and jth column, $\delta(\tilde{A}_{ij})$ is independent of the ith row of A. Thus, because δ is an n-linear function, $\delta(\tilde{A}_{ij})$ is a linear function of each row of A except row i. Hence $A_{ij} \cdot \delta(\tilde{A}_{ij})$ is an $(n + 1)$-linear function on the $(n + 1) \times (n + 1)$ matrices with entries from F. Therefore, since

$$\epsilon_j(A) = \sum_{i=1}^{n+1} (-1)^{i+j} A_{ij} \cdot \delta(\tilde{A}_{ij})$$

is a linear combination of the $(n + 1)$-linear functions $A_{ij} \cdot \delta(\tilde{A}_{ij})$, ϵ_j is an $(n + 1)$-linear function by the corollary to Proposition 4.3.

We now prove that ϵ_j is alternating. If A is an $(n + 1) \times (n + 1)$ matrix in which rows k and $k+1$ are identical, then $\tilde{A}_{ij}$ has two identical rows whenever $i \neq k$ and $i \neq k+1$. Thus $\delta(\tilde{A}_{ij}) = 0$ whenever $i \neq k$ and $i \neq k+1$; so

$$\epsilon_j(A) = (-1)^{k+j} A_{kj} \cdot \delta(\tilde{A}_{kj}) + (-1)^{(k+1)+j} A_{(k+1)j} \cdot \delta(\tilde{A}_{(k+1)j}).$$

But because rows k and $k+1$ of A are equal, $A_{kj} = A_{(k+1)j}$ and $\tilde{A}_{kj} = \tilde{A}_{(k+1)j}$. Hence $\epsilon_j(A) = 0$, proving that ϵ_j is alternating. ∎

Corollary 1. *Let δ and ϵ_j be as in the statement of Proposition 4.5. If δ is a determinant on $\mathsf{M}_{n \times n}(F)$, then ϵ_j is a determinant on the $(n+1) \times (n+1)$ matrices with entries from F.*

Proof. Let I denote the $(n+1) \times (n+1)$ identity matrix, and let $\tilde{I}_{ij}$ denote the $n \times n$ matrix obtained from I by deleting row i and column j. Then $\tilde{I}_{jj}$ is the $n \times n$ identity matrix. Since $I_{ij} = 0$ if $i \neq j$ and $I_{jj} = 1$, we have

$$\begin{aligned} \epsilon_j(I) &= \sum_{i=1}^{n+1} (-1)^{i+j} I_{ij} \cdot \delta(\tilde{I}_{ij}) = (-1)^{j+j} \cdot \delta(\tilde{I}_{jj}) \\ &= \delta(\tilde{I}_{jj}) = 1 \end{aligned}$$

because δ is a determinant on $\mathsf{M}_{n \times n}(F)$. Thus ϵ_j is a determinant on the $(n+1) \times (n+1)$ matrices with entries from F. ∎

Corollary 2. *There exists a determinant on $\mathsf{M}_{n \times n}(F)$ for any positive integer n.*

Proof. The proof will be by induction on n. When $n = 1$, the function $\det: \mathsf{M}_{1 \times 1}(F) \to F$ defined by $\det(A) = A_{11}$ is a determinant on $\mathsf{M}_{1 \times 1}(F)$. Assume that there exists a determinant δ on $\mathsf{M}_{n \times n}(F)$. Then for any j $(1 \leqslant j \leqslant n+1)$, the function ϵ_j defined in Proposition 4.5 is a determinant on the $(n+1) \times (n+1)$ matrices with entries from F. This completes the induction. ∎

Definitions. *If δ is a determinant on $\mathsf{M}_{n \times n}(F)$, then the determinant*

$$\epsilon_j(A) = \sum_{i=1}^{n+1} (-1)^{i+j} A_{ij} \cdot \delta(\tilde{A}_{ij})$$

defined in Proposition 4.5 is called the expansion of A along the jth column. *The scalar $(-1)^{i+j} \cdot \delta(\tilde{A}_{ij})$ is called the* cofactor of A_{ij} *(with respect to the determinant δ).*

Example 7

Let A denote the following element of $\mathsf{M}_{3 \times 3}(F)$:

$$\begin{pmatrix} 1 & 2 & 3 \\ 4 & 5 & 6 \\ 7 & 8 & 9 \end{pmatrix}.$$

The cofactors of A_{12}, A_{22}, and A_{32} are

$$(-1)^{1+2}\det\begin{pmatrix}4 & 6\\ 7 & 9\end{pmatrix} = (-1)(4\cdot 9 - 6\cdot 7) = 6,$$

$$(-1)^{2+2}\det\begin{pmatrix}1 & 3\\ 7 & 9\end{pmatrix} = 1(1\cdot 9 - 3\cdot 7) = -12,$$

$$(-1)^{3+2}\det\begin{pmatrix}1 & 3\\ 4 & 6\end{pmatrix} = (-1)(1\cdot 6 - 3\cdot 4) = 6,$$

respectively. Hence the expansion of A along the second column is

$$\begin{aligned}\epsilon_2(A) &= A_{12}(6) + A_{22}(-12) + A_{32}(6)\\ &= 2\cdot 6 + 5(-12) + 8\cdot 6 = 0.\end{aligned}$$

Similarly, the cofactors of A_{13}, A_{23}, and A_{33} are

$$(-1)^{1+3}\det\begin{pmatrix}4 & 5\\ 7 & 8\end{pmatrix} = 1(4\cdot 8 - 5\cdot 7) = -3,$$

$$(-1)^{2+3}\det\begin{pmatrix}1 & 2\\ 7 & 8\end{pmatrix} = (-1)(1\cdot 8 - 2\cdot 7) = 6,$$

$$(-1)^{3+3}\det\begin{pmatrix}1 & 2\\ 4 & 5\end{pmatrix} = 1(1\cdot 5 - 2\cdot 4) = -3,$$

respectively. Hence the expansion of A along the third column is

$$\epsilon_3(A) = A_{13}(-3) + A_{23}(6) + A_{33}(-3) = 3(-3) + 6(6) + 9(-3) = 0. \quad \blacksquare$$

We will see in Theorem 4.9 that the equality of $\epsilon_2(A)$ and $\epsilon_3(A)$ in Example 7 is not coincidental. In fact, we will see that there is exactly one determinant on $\mathsf{M}_{n\times n}(F)$.

EXERCISES

1. Label the following statements as being true or false.

(a) A determinant on $\mathsf{M}_{n\times n}(F)$ is a linear function of each row of an $n \times n$ matrix with entries from F when the other $n - 1$ rows are held fixed.

(b) If δ is a determinant and any two rows of A are identical, then $\delta(A) = 0$.

(c) Let δ be a determinant. If B is a matrix obtained from A by interchanging any two rows, then $\delta(A) = \delta(B)$.

(d) The function $\delta\colon \mathsf{M}_{n\times n}(F) \to F$ defined by $\delta(A) = 0$ for each $A \in \mathsf{M}_{n\times n}(F)$ is a determinant on $\mathsf{M}_{n\times n}(F)$.

(e) For any $n \geqslant 2$ there is a determinant on $\mathsf{M}_{n\times n}(F)$.

(f) Any determinant $\delta\colon \mathsf{M}_{n\times n}(F) \to F$ is a linear transformation.

2. Verify that if A is the 3×3 matrix in Example 7, then the expansion of A along the first column equals zero.

3. Evaluate the determinant of each of the following matrices by expanding along the second column and also along the third column. [Each matrix is an element of $\mathsf{M}_{3\times 3}(C)$.]

(a) $\begin{pmatrix} -3 & 2 & 5 \\ 1 & 0 & -1 \\ 4 & -6 & 7 \end{pmatrix}$ **(b)** $\begin{pmatrix} 8 & -4 & 0 \\ 0 & 6 & -3 \\ -1 & 5 & 2 \end{pmatrix}$

(c) $\begin{pmatrix} 1 & 2 & -5 \\ 6 & -4 & 3 \\ 0 & 1 & 1 \end{pmatrix}$ **(d)** $\begin{pmatrix} 1+i & -1 & 0 \\ 2 & 3i & 4i \\ 0 & 2-i & -1+2i \end{pmatrix}$

4. Which of the following functions $\delta\colon \mathsf{M}_{3\times 3}(F) \to F$ are 3-linear functions? Justify each answer.
(a) $\delta(A) = c$, where c is any nonzero scalar
(b) $\delta(A) = A_{22}$
(c) $\delta(A) = A_{11}A_{23}A_{32}$
(d) $\delta(A) = A_{11}A_{21}A_{32}$
(e) $\delta(A) = A_{11}A_{31}A_{32}$
(f) $\delta(A) = A_{11}^2A_{22}^2A_{33}^2$
(g) $\delta(A) = A_{11}A_{22}A_{33} - A_{11}A_{21}A_{32}$

5. **(a)** Determine all 1-linear functions $\delta\colon \mathsf{M}_{1\times 1}(F) \to F$.
(b) Determine all determinants on $\mathsf{M}_{1\times 1}(F)$.

6. Prove the equality of the three functions $\epsilon_j\colon \mathsf{M}_{3\times 3}(F) \to F$ $(j = 1, 2, 3)$ defined in Proposition 4.5 for each $A \in \mathsf{M}_{3\times 3}(F)$ by

$$\epsilon_j(A) = \sum_{i=1}^{3} (-1)^{i+j} A_{ij} \cdot \det(\tilde{A}_{ij}),$$

where $\tilde{A}_{ij}$ is the 2×2 matrix obtained from A by deleting the ith row and jth column and det denotes the unique determinant on $\mathsf{M}_{2\times 2}(F)$.

7. Prove that the unique determinant on $\mathsf{M}_{2\times 2}(F)$ is a 2-linear function of the columns of a 2×2 matrix and that the determinant of a 2×2 matrix in which both columns are identical is zero.

8. The proof of Theorem 4.2 shows that if δ is a 2-linear function $\delta\colon \mathsf{M}_{2\times 2}(F) \to F$, then

$$\delta(A) = A_{11}A_{22} \cdot \delta(I) + A_{11}A_{21} \cdot \delta(M_1) + A_{12}A_{22} \cdot \delta(M_2) + A_{12}A_{21} \cdot \delta(M_3),$$

where I, M_1, M_2, and M_3 are as in the proof of the theorem. Prove that for any scalars $a, b, c, d \in F$ the function

$$\epsilon(A) = A_{11}A_{22}a + A_{11}A_{21}b + A_{12}A_{22}c + A_{12}A_{21}d$$

is a 2-linear function. Deduce that $\delta'\colon \mathsf{M}_{2\times 2}(F) \to F$ is a 2-linear function if and only if it is of the form above for some scalars a, b, c, and d.

9. Show that if F is a field not of characteristic two (as defined in Appendix D), then condition (a) of Proposition 4.4 implies condition (b) of that proposition. This result is not true in arbitrary fields, however.

10. Prove the corollary to Proposition 4.3.

4.3 PROPERTIES OF DETERMINANTS

There are several important properties that are quite useful in evaluating a determinant of a given matrix. These are summarized in the next theorem.

Theorem 4.6. *Any determinant δ on $\mathsf{M}_{n\times n}(F)$ has the following properties:*

(a) *If B is a matrix obtained from A by multiplying each entry of some row of A by a scalar c, then $\delta(B) = c \cdot \delta(A)$.*

(b) *If two rows of A are identical, then $\delta(A) = 0$.*

(c) *If B is a matrix obtained from A by interchanging two rows, then $\delta(B) = -\delta(A)$.*

(d) *If one row of A consists entirely of zero entries, then $\delta(A) = 0$.*

(e) *If B is a matrix obtained from A by adding a multiple of row i to row j $(i \neq j)$, then $\delta(B) = \delta(A)$.*

Proof. Property (a) is a consequence of the fact that δ is an n-linear function, whereas properties (b) and (c) are consequences of Proposition 4.4.

(d) Suppose that $A_{(i)}$ consists entirely of zero entries. Then

$$\delta(A) = \delta\begin{pmatrix} A_{(1)} \\ \vdots \\ 0A_{(i)} \\ \vdots \\ A_{(n)} \end{pmatrix} = 0 \cdot \delta\begin{pmatrix} A_{(1)} \\ \vdots \\ A_{(i)} \\ \vdots \\ A_{(n)} \end{pmatrix} = 0.$$

(e) Let B be obtained from $A \in \mathsf{M}_{n\times n}(F)$ by adding c times row i to row j. Assume for the sake of argument that $i < j$. Thus if

$$A = \begin{pmatrix} A_{(1)} \\ \vdots \\ A_{(i)} \\ \vdots \\ A_{(j)} \\ \vdots \\ A_{(n)} \end{pmatrix}, \quad \text{then} \quad B = \begin{pmatrix} A_{(1)} \\ \vdots \\ A_{(i)} \\ \vdots \\ cA_{(i)} + A_{(j)} \\ \vdots \\ A_{(n)} \end{pmatrix}.$$

So

$$\delta(B) = c \cdot \delta\begin{pmatrix} A_{(1)} \\ \vdots \\ A_{(i)} \\ \vdots \\ A_{(i)} \\ \vdots \\ A_{(n)} \end{pmatrix} + \delta\begin{pmatrix} A_{(1)} \\ \vdots \\ A_{(i)} \\ \vdots \\ A_{(j)} \\ \vdots \\ A_{(n)} \end{pmatrix} = c \cdot 0 + \delta(A) = \delta(A)$$

by the n-linearity of δ and property (b). ∎

Observe that properties (a), (c), and (e) of Theorem 4.6 show how the determinant of a matrix changes when an elementary row operation is performed on the matrix. We can reformulate these properties in terms of elementary matrices as follows.

Corollary. *Let* E_1, E_2, *and* E_3 *be elementary matrices in* $M_{n \times n}(F)$ *of types 1, 2, and 3, respectively. If* E_2 *is obtained by multiplying a row of* I *by the nonzero scalar* c, *then for any determinant* δ *on* $M_{n \times n}(F)$, $\delta(E_1) = -1$, $\delta(E_2) = c$, *and* $\delta(E_3) = 1$.

This corollary is one of the key ingredients in a proof of the uniqueness of a determinant on $M_{n \times n}(F)$. We now prove the remaining two theorems needed to establish this uniqueness. Our first result computes the determinant of any noninvertible matrix.

Theorem 4.7. *Let* δ *be a determinant on* $M_{n \times n}(F)$, *and let* A *be an element of* $M_{n \times n}(F)$ *having rank less than* n. *Then* $\delta(A) = 0$.

Proof. Since $\text{rank}(A) < n$, the rows of A are linearly dependent (Corollary 2 of Theorem 3.6). Hence there are scalars $c_1, \ldots, c_n$, not all zero, such that $c_1 A_{(1)} + c_2 A_{(2)} + \cdots + c_n A_{(n)} = 0$. Assume for the sake of argument that $c_1 \neq 0$; then

$$A_{(1)} + c_1^{-1} c_2 A_{(2)} + \cdots + c_1^{-1} c_n A_{(n)} = 0.$$

Let B be the matrix obtained from A by adding to the first row the multiple $c_1^{-1} c_i A_{(i)}$ of row i for each i ($i = 2, \ldots, n$). Then the first row of B consists entirely of zero entries, so that $\delta(B) = 0$. But $\delta(B) = \delta(A)$ by property (e) of Theorem 4.6. Therefore, $\delta(A) = 0$. ∎

The next result establishes the final fact needed to prove the uniqueness of a determinant on $M_{n \times n}(F)$—that a determinant behaves well with respect to matrix multiplication. This theorem is of considerable importance in its own right, however. In particular, its second corollary, which provides a determinant test for invertibility of a matrix, will be frequently used in the following chapters.

Lemma. *If* E *is an* n × n *elementary matrix with entries from* F, *and if* δ *is a determinant on* $\mathsf{M}_{n\times n}(F)$, *then* $\delta(EB) = \delta(E)\cdot\delta(B)$ *for any* $B \in \mathsf{M}_{n\times n}(F)$.

Proof. Suppose that multiplication on the left by E interchanges two rows of B. Then $\delta(EB) = -\delta(B)$ by Theorem 4.6(c). But $\delta(E) = -1$ by the corollary to Theorem 4.6; so $\delta(EB) = \delta(E)\cdot\delta(B)$. Similar proofs establish the result for multiplication of a row of B by a nonzero scalar or addition of a multiple of one row to another. ∎

Theorem 4.8. *Let* δ *be a determinant on* $\mathsf{M}_{n\times n}(F)$, *and let* A *and* B *be arbitrary elements of* $\mathsf{M}_{n\times n}(F)$. *Then* $\delta(AB) = \delta(A)\cdot\delta(B)$.

Proof. If $\text{rank}(A) < n$, then by Proposition 3.7 $\text{rank}(AB) \leqslant \text{rank}(A) < n$. Hence by Theorem 4.7 $\delta(AB) = 0$ and $\delta(A) = 0$. Thus in this case

$$\delta(AB) = \delta(A)\cdot\delta(B).$$

If $\text{rank}(A) = n$, then A is invertible and hence is the product of elementary matrices (Corollary 3 of Theorem 3.6). Let $A = E_m \cdots E_1$, where each E_i is an elementary matrix. Then by the lemma we have

$$\begin{aligned}\delta(AB) &= \delta(E_m \cdots E_1 B) = \delta(E_m)\cdot\delta(E_{m-1}\cdots E_1 B) = \cdots \\ &= \delta(E_m)\cdot\cdots\cdot\delta(E_1)\cdot\delta(B) = \delta(E_m\cdots E_1)\cdot\delta(B) = \delta(A)\cdot\delta(B).\end{aligned}$$ ∎

Corollary 1. *Let* δ *be a determinant on* $\mathsf{M}_{n\times n}(F)$, *and let* $A \in \mathsf{M}_{n\times n}(F)$ *be invertible. Then* $\delta(A) \neq 0$, *and* $\delta(A^{-1}) = [\delta(A)]^{-1}$.

Proof. By Theorem 4.8 we have

$$\delta(A)\cdot\delta(A^{-1}) = \delta(AA^{-1}) = \delta(I_n) = 1.$$

So $\delta(A) \neq 0$, and $\delta(A^{-1}) = [\delta(A)]^{-1}$. ∎

Corollary 2. *Let* δ *be a determinant on* $\mathsf{M}_{n\times n}(F)$, *and let* $A \in \mathsf{M}_{n\times n}(F)$. *Then the following conditions are equivalent:*

(a) $\delta(A) = 0$.
(b) A *is not invertible.*
(c) $\text{rank}(A) < n$.

Proof. Corollary 1 shows that if $\delta(A) = 0$, then A is not invertible. Hence condition (a) implies condition (b).

That condition (b) implies condition (c) follows from a remark in Section 3.2.

Finally, Theorem 4.7 shows that condition (c) implies condition (a). ∎

It was proved in Theorems 4.1 and 4.2 that there is exactly one determinant on $\mathsf{M}_{2\times 2}(F)$. We are now able to prove a similar result for $\mathsf{M}_{n\times n}(F)$.

Theorem 4.9. *There is exactly one determinant on* $\mathsf{M}_{n\times n}(F)$.

Proof. The existence of a determinant on $\mathsf{M}_{n\times n}(F)$ was proved in Corollary 2 of Proposition 4.5.

We will complete the proof by showing that if δ_1 and δ_2 are both determinants on $\mathsf{M}_{n\times n}(F)$, then $\delta_1 = \delta_2$. Let A be an arbitrary $n \times n$ matrix with entries from F. If $\text{rank}(A) < n$, then $\delta_1(A) = \delta_2(A) = 0$ by Corollary 2 of Theorem 4.8. If $\text{rank}(A) = n$, then A is invertible and hence is the product of elementary matrices (Corollary 3 of Theorem 3.6). Let $A = E_m \cdots E_1$, where each E_i is an elementary matrix. Since $\delta_1(E_i) = \delta_2(E_i)$ for each i $(1 \leqslant i \leqslant m)$ by the corollary to Theorem 4.6,

$$\begin{aligned}\delta_1(A) &= \delta_1(E_m \cdots E_1) = \delta_1(E_m) \cdots \delta_1(E_1)\\ &= \delta_2(E_m) \cdots \delta_2(E_1) = \delta_2(E_m \cdots E_1) = \delta_2(A)\end{aligned}$$

by Theorem 4.8. Hence $\delta_1 = \delta_2$. ∎

Henceforth we denote the unique determinant on $\mathsf{M}_{n\times n}(F)$ by *det*.

Corollary. *Let* $A \in \mathsf{M}_{n\times n}(F)$. *For any* j $(1 \leqslant j \leqslant n)$

$$det(A) = \sum_{i=1}^{n} (-1)^{i+j} A_{ij} \cdot det(\tilde{A}_{ij}),$$

where $\tilde{A}_{ij}$ *is the* $(n-1) \times (n-1)$ *matrix obtained from* A *by deleting row* i *and column* j.

Evaluating Determinants

The preceding corollary shows that the determinant of an $n \times n$ matrix can be evaluated by expanding along any column; if $n > 2$, the resulting expansion contains n determinants of matrices of size $(n-1) \times (n-1)$. The determinant of each of these $(n-1) \times (n-1)$ matrices can then be expanded along any column, and this process can be continued until an expansion involves only determinants of 2×2 matrices, which can be evaluated by the rule $\det(A) = A_{11}A_{22} - A_{12}A_{21}$.

Observe, however, that the evaluation of $\det(\tilde{A}_{ij})$ can be avoided whenever $A_{ij} = 0$, for the product $A_{ij} \cdot \det(\tilde{A}_{ij})$ is zero regardless of the value of the determinant. Therefore it is beneficial to expand along a column containing as many zero entries as possible. The next two examples illustrate this procedure.

Example 1

Let A denote the following element of $\mathsf{M}_{4\times 4}(F)$:

$$\begin{pmatrix} 1 & 1 & 0 & 1 \\ 1 & 0 & 1 & 1 \\ 0 & 0 & 1 & 1 \\ 1 & 1 & 1 & 1 \end{pmatrix}.$$

To minimize the computation required to evaluate $\det(A)$, we expand along the second column:

$$\det(A) = \sum_{i=1}^{4} (-1)^{i+2} A_{i2} \cdot \det(\tilde{A}_{i2})$$

$$= (-1)^{1+2} \cdot 1 \cdot \det \begin{pmatrix} 1 & 1 & 1 \\ 0 & 1 & 1 \\ 1 & 1 & 1 \end{pmatrix} + (-1)^{2+2} \cdot 0 \cdot \det \begin{pmatrix} 1 & 0 & 1 \\ 0 & 1 & 1 \\ 1 & 1 & 1 \end{pmatrix}$$

$$+ (-1)^{3+2} \cdot 0 \cdot \det \begin{pmatrix} 1 & 0 & 1 \\ 1 & 1 & 1 \\ 1 & 1 & 1 \end{pmatrix} + (-1)^{4+2} \cdot 1 \cdot \det \begin{pmatrix} 1 & 0 & 1 \\ 1 & 1 & 1 \\ 0 & 1 & 1 \end{pmatrix}$$

$$= (-1) \cdot 1 \cdot 0 + 0 + 0 + 1 \cdot 1 \cdot \det \begin{pmatrix} 1 & 0 & 1 \\ 1 & 1 & 1 \\ 0 & 1 & 1 \end{pmatrix} = \det \begin{pmatrix} 1 & 0 & 1 \\ 1 & 1 & 1 \\ 0 & 1 & 1 \end{pmatrix}.$$

(The first of the four 3×3 matrices has two identical rows, so that its determinant is zero.) We now evaluate the remaining determinant by expanding along the first column to obtain

$$\det(A) = \det \begin{pmatrix} 1 & 0 & 1 \\ 1 & 1 & 1 \\ 0 & 1 & 1 \end{pmatrix}$$

$$= (-1)^{1+1} \cdot 1 \cdot \det \begin{pmatrix} 1 & 1 \\ 1 & 1 \end{pmatrix} + (-1)^{2+1} \cdot 1 \cdot \det \begin{pmatrix} 0 & 1 \\ 1 & 1 \end{pmatrix}$$

$$+ (-1)^{3+1} \cdot 0 \cdot \det \begin{pmatrix} 0 & 1 \\ 1 & 1 \end{pmatrix}$$

$$= 1 \cdot 1 \cdot 0 + (-1) \cdot 1 \cdot (-1) + 0 = 1.$$

Now let B denote the following element of $\mathsf{M}_{5\times 5}(R)$:

$$\begin{pmatrix} 1 & -1 & 0 & 0 & 0 \\ -2 & 0 & -3 & 1 & 6 \\ 5 & -4 & 0 & 2 & 0 \\ 0 & 3 & 0 & -1 & 4 \\ -9 & 8 & 0 & 0 & 0 \end{pmatrix}.$$

Expanding successively along the third, fourth, and third columns, we see that

$$\begin{aligned}
\det(B) &= (-1)^{2+3}\cdot(-3)\cdot\det\begin{pmatrix} 1 & -1 & 0 & 0 \\ 5 & -4 & 2 & 0 \\ 0 & 3 & -1 & 4 \\ -9 & 8 & 0 & 0 \end{pmatrix} \\
&= 3(-1)^{3+4}\cdot 4\cdot\det\begin{pmatrix} 1 & -1 & 0 \\ 5 & -4 & 2 \\ -9 & 8 & 0 \end{pmatrix} \\
&= -12\cdot(-1)^{2+3}\cdot 2\cdot\det\begin{pmatrix} 1 & -1 \\ -9 & 8 \end{pmatrix} \\
&= 24[1\cdot 8-(-1)(-9)] = 24(-1) = -24. \quad \blacksquare
\end{aligned}$$

As these examples suggest, the process of evaluating a determinant is often tedious even when there are zero entries present. Without zero entries the evaluation of a determinant by expanding along a column is quite inefficient. Instead of this procedure we can utilize property (e) of Theorem 4.6 to change a matrix A into a matrix B having the same determinant as A and having zero entries in one or more columns. This is essentially the same process that is used to reduce A to row echelon form. Examples of this technique follow.

Example 2

Let A denote the following element of $\mathsf{M}_{4\times 4}(R)$:

$$\begin{pmatrix} 1 & -1 & 2 & -1 \\ -3 & 4 & 1 & -1 \\ 2 & -5 & -3 & 8 \\ -2 & 6 & -4 & 1 \end{pmatrix}.$$

Then

$$\det(A) = \det\begin{pmatrix} A_{(1)} \\ A_{(2)} \\ A_{(3)} \\ A_{(4)} \end{pmatrix} = \det\begin{pmatrix} A_{(1)} \\ 3A_{(1)} + A_{(2)} \\ -2A_{(1)} + A_{(3)} \\ 2A_{(1)} + A_{(4)} \end{pmatrix}$$

$$= \det\begin{pmatrix} 1 & -1 & 2 & -1 \\ 0 & 1 & 7 & -4 \\ 0 & -3 & -7 & 10 \\ 0 & 4 & 0 & -1 \end{pmatrix} = (-1)^{1+1}\cdot 1\cdot\det\begin{pmatrix} 1 & 7 & -4 \\ -3 & -7 & 10 \\ 4 & 0 & -1 \end{pmatrix}$$

$$= \det\begin{pmatrix} 1 & 7 & -4 \\ -2 & 0 & 6 \\ 4 & 0 & -1 \end{pmatrix} = (-1)^{1+2}\cdot 7\cdot\det\begin{pmatrix} -2 & 6 \\ 4 & -1 \end{pmatrix}$$

$$= -7[(-2)(-1) - 6\cdot 4] = -7(-22) = 154. \quad \blacksquare$$

Example 3

Let A denote the following element of $\mathsf{M}_{4\times 4}(R)$:

$$\begin{pmatrix} 1 & -1 & 2 & 1 \\ 2 & -1 & -1 & 4 \\ -4 & 5 & -10 & -6 \\ 3 & -2 & 10 & -1 \end{pmatrix}.$$

We will introduce zero entries by use of Theorem 4.6(e) so that A is transformed into an upper triangular matrix having the same determinant as A. The determinant of the upper triangular matrix will then be evaluated by successive expansions along the first column.

$$\det(A) = \det\begin{pmatrix} 1 & -1 & 2 & 1 \\ 2 & -1 & -1 & 4 \\ -4 & 5 & -10 & -6 \\ 3 & -2 & 10 & -1 \end{pmatrix} = \det\begin{pmatrix} 1 & -1 & 2 & 1 \\ 0 & 1 & -5 & 2 \\ 0 & 1 & -2 & -2 \\ 0 & 1 & 4 & -4 \end{pmatrix}$$

$$= \det\begin{pmatrix} 1 & -1 & 2 & 1 \\ 0 & 1 & -5 & 2 \\ 0 & 0 & 3 & -4 \\ 0 & 0 & 9 & -6 \end{pmatrix} = \det\begin{pmatrix} 1 & -1 & 2 & 1 \\ 0 & 1 & -5 & 2 \\ 0 & 0 & 3 & -4 \\ 0 & 0 & 0 & 6 \end{pmatrix}$$

$$= 1 \cdot \det\begin{pmatrix} 1 & -5 & 2 \\ 0 & 3 & -4 \\ 0 & 0 & 6 \end{pmatrix} = 1 \cdot 1 \cdot \det\begin{pmatrix} 3 & -4 \\ 0 & 6 \end{pmatrix}$$

$$= 1 \cdot 1 \cdot 3 \cdot 6 = 18. \quad \blacksquare$$

Until now, the roles played by the rows and columns of a matrix in the study of determinants have been quite different; a determinant was defined as a function on $\mathsf{M}_{n \times n}(F)$ that satisfies certain properties involving the rows of a matrix, whereas the evaluation of a determinant is accomplished by expanding along columns of a matrix. These roles are reversible, and we now verify this fact by showing that the determinants of A and A^t are equal. (Since the rows of A are columns of A^t, and vice versa, this result will be sufficient to prove that the roles of rows and columns are interchangeable.)

Theorem 4.10. *For any* n × n *matrix* A, *det*$(\mathrm{A}^t) =$ *det*(A).

Proof. If A is not an invertible matrix then $\operatorname{rank}(A) < n$. But since $\operatorname{rank}(A^t) = \operatorname{rank}(A)$ (Corollary 2 of Theorem 3.6), it follows that A^t is not invertible. So $\det(A) = 0 = \det(A^t)$ in this case.

If A is invertible, then $A = E_m \cdots E_1$, where $E_1, \ldots, E_m$ are elementary matrices. Since $\det(E_i^t) = \det(E_i)$ for each i (see Exercise 5), we have

$$\begin{aligned} \det(A^t) &= \det(E_1^t \cdots E_m^t) = \det(E_1^t) \cdots \det(E_m^t) \\ &= \det(E_1) \cdots \det(E_m) = \det(E_m) \cdots \det(E_1) \\ &= \det(E_m \cdots E_1) = \det(A). \quad \blacksquare \end{aligned}$$

Corollary. *Any statement about determinants that involves the rows of a matrix can be restated in terms of the columns of the matrix, and any statement about determinants that involves the columns of a matrix can be restated in terms of the rows of the matrix. In particular, if* A *is an* n × n *matrix,*

$$det(\mathrm{A}) = \sum_{j=1}^{n} (-1)^{i+j} \mathrm{A}_{ij} \cdot det(\tilde{\mathrm{A}}_{ij}),$$

where $\tilde{\mathrm{A}}_{ij}$ *is the* $(n-1) \times (n-1)$ *matrix obtained from* A *by a deleting row* i *and column* j.

Example 4

Let A denote the following element of $\mathsf{M}_{4 \times 4}(R)$:

$$\begin{pmatrix} 4 & -1 & -3 & 6 \\ -2 & 3 & 1 & 4 \\ 0 & 5 & 0 & 0 \\ 1 & 2 & 3 & -1 \end{pmatrix}.$$

In this case the computation required to evaluate det(A) can be minimized by expanding along the third row. Thus

$$\det(A) = -5\det\begin{pmatrix} 4 & -3 & 6 \\ -2 & 1 & 4 \\ 1 & 3 & -1 \end{pmatrix} = -5\det\begin{pmatrix} 0 & -15 & 10 \\ 0 & 7 & 2 \\ 1 & 3 & -1 \end{pmatrix}$$

$$= -5\det\begin{pmatrix} -15 & 10 \\ 7 & 2 \end{pmatrix} = -5[(-15)\cdot 2 - 10\cdot 7] = 500. \quad \blacksquare$$

Our final result allows us to evaluate easily the determinant of an upper triangular matrix. This result makes the technique used in Example 3 a very efficient method for evaluating determinants.

Theorem 4.11. *The determinant of an upper triangular square matrix is the product of its diagonal entries.*

Proof. Let A be an upper triangular $n \times n$ matrix. The proof is by induction on n. If $n = 2$, then A has the form

$$A = \begin{pmatrix} A_{11} & A_{12} \\ 0 & A_{22} \end{pmatrix},$$

and so $\det(A) = A_{11}A_{22} - A_{12}\cdot 0 = A_{11}A_{22}$, proving the theorem if $n = 2$.

Assume that the theorem is true for upper triangular $(n-1) \times (n-1)$ matrices, and let A be an upper triangular $n \times n$ matrix. Then A has the form

$$\begin{pmatrix} A_{11} & A_{12} & \cdots & A_{1(n-1)} & A_{1n} \\ 0 & A_{22} & \cdots & A_{2(n-1)} & A_{2n} \\ \vdots & \vdots & & \vdots & \vdots \\ 0 & 0 & \cdots & 0 & A_{nn} \end{pmatrix}.$$

Expanding along the first column, we see that

$$\det(A) = A_{11}\cdot\det\begin{pmatrix} A_{22} & \cdots & A_{2(n-1)} & A_{2n} \\ \vdots & & \vdots & \vdots \\ 0 & \cdots & 0 & A_{nn} \end{pmatrix}$$

$$= A_{11}\cdot(A_{22}\cdots A_{nn})$$

by the induction hypothesis. This completes the proof. $\blacksquare$

Corollary. *If* $A \in M_{n\times n}(F)$ *is lower triangular (that is,* $A_{ij} = 0$ *whenever* $i < j$*), then the determinant of* A *is the product of its diagonal entries.*

Proof. Exercise. $\blacksquare$

As in Section 4.1, it is possible to interpret the determinant of an element A in $\mathsf{M}_{n\times n}(R)$ geometrically. We can interpret

$$\left|\det\begin{pmatrix} A_{(1)} \\ \vdots \\ A_{(n)} \end{pmatrix}\right|$$

as the n-dimensional volume (the generalization of area in R^2 and volume in R^3) of the parallelepiped having the vectors $A_{(1)}, \ldots, A_{(n)}$ as adjacent sides. (For a proof of this result, see Serge Lang, *Analysis I*, Addison-Wesley, Reading, Mass., 1968, pp. 413–418.)

In our earlier discussion of the geometric significance of the determinant formed from the vectors in an ordered basis for R^2, we also saw that this determinant is positive if and only if the basis induces a right-handed coordinate system. A similar statement is true in R^n. Specifically, if γ is any ordered basis for R^n and β is the standard ordered basis for R^n, then γ induces a right-handed coordinate system if and only if $\det(Q) > 0$, where Q is the change of coordinate matrix changing γ-coordinates into β-coordinates. Thus, for instance,

$$\gamma = \left\{\begin{pmatrix} 1 \\ 1 \\ 0 \end{pmatrix}, \begin{pmatrix} 1 \\ -1 \\ 0 \end{pmatrix}, \begin{pmatrix} 0 \\ 0 \\ 1 \end{pmatrix}\right\}$$

induces a left-handed coordinate system in R^3 since

$$\det\begin{pmatrix} 1 & 1 & 0 \\ 1 & -1 & 0 \\ 0 & 0 & 1 \end{pmatrix} = -2 < 0,$$

whereas

$$\gamma' = \left\{\begin{pmatrix} 1 \\ 2 \\ 0 \end{pmatrix}, \begin{pmatrix} -2 \\ 1 \\ 0 \end{pmatrix}, \begin{pmatrix} 0 \\ 0 \\ 1 \end{pmatrix}\right\}$$

induces a right-handed coordinate system in R^3 since

$$\det\begin{pmatrix} 1 & -2 & 0 \\ 2 & 1 & 0 \\ 0 & 0 & 1 \end{pmatrix} = 5 > 0.$$

More generally, if β and γ are any two ordered bases for R^n, then the coordinate systems induced by β and γ have the same orientation (both right-handed or both left-handed) if and only if $\det(Q) > 0$, where Q is the change of coordinate matrix changing γ-coordinates into β-coordinates.

Cramer's Rule

The next result shows that the solution of a system of equations with an invertible coefficient matrix can be expressed in terms of determinants.

Theorem 4.12 (Cramer's Rule). *Let* $AX = B$ *be the matrix form of a system of* n *linear equations in* n *unknowns, where* $X = (x_1, x_2, \ldots, x_n)^t$ *and* $B = (b_1, b_2, \ldots, b_n)^t$. *If* $det(A) \neq 0$, *then this system has a unique solution, and for each* k $(1 \leqslant k \leqslant n)$

$$x_k = [det(A)]^{-1} \cdot det(M_k),$$

where M_k *is the* $n \times n$ *matrix obtained from* A *by replacing* $A^{(k)}$ *by* B.

Proof. If $\det(A) \neq 0$, then this system has a unique solution by Corollary 2 of Theorem 4.8 and Theorem 3.10. Let k be an integer such that $1 \leqslant k \leqslant n$, and define X_k to be the matrix obtained from the $n \times n$ identity matrix by replacing its kth column by X. Expanding X_k along the kth row produces

$$\det(X_k) = x_k \cdot \det(I_{n-1}) = x_k.$$

Thus by Theorem 2.14 we have

$$\begin{aligned} AX_k &= A(e_1, \ldots, e_{k-1}, X, e_{k+1}, \ldots, e_n) \\ &= (Ae_1, \ldots, Ae_{k-1}, AX, Ae_{k+1}, \ldots, Ae_n) \\ &= (A^{(1)}, \ldots, A^{(k-1)}, B, A^{(k+1)}, \ldots, A^{(n)}) \\ &= M_k. \end{aligned}$$

So, by Theorem 4.8,

$$\det(M_k) = \det(AX_k) = \det(A) \cdot \det(X_k) = \det(A) \cdot x_k.$$

Hence

$$x_k = [\det(A)]^{-1} \cdot \det(M_k). \quad \blacksquare$$

Example 5

We will use Cramer's rule to solve the matrix equation $AX = B$, where

$$A = \begin{pmatrix} 1 & 2 & 3 \\ 1 & 0 & 1 \\ 1 & 1 & -1 \end{pmatrix} \quad \text{and} \quad B = \begin{pmatrix} 2 \\ 3 \\ 1 \end{pmatrix}.$$

It is easily checked that $\det(A) = 6$, so that Cramer's rule applies. Now

$$x_1 = \frac{\det(M_1)}{\det(A)} = \frac{\det\begin{pmatrix} 2 & 2 & 3 \\ 3 & 0 & 1 \\ 1 & 1 & -1 \end{pmatrix}}{\det(A)} = \frac{15}{6} = \frac{5}{2},$$

$$x_2 = \frac{\det(M_2)}{\det(A)} = \frac{\det\begin{pmatrix} 1 & 2 & 3 \\ 1 & 3 & 1 \\ 1 & 1 & -1 \end{pmatrix}}{\det(A)} = \frac{-6}{6} = -1,$$

and

$$x_3 = \frac{\det(M_3)}{\det(A)} = \frac{\det\begin{pmatrix} 1 & 2 & 2 \\ 1 & 0 & 3 \\ 1 & 1 & 1 \end{pmatrix}}{\det(A)} = \frac{3}{6} = \frac{1}{2};$$

so the unique solution of the given system is

$$(x_1, x_2, x_3) = (\tfrac{5}{2}, -1, \tfrac{1}{2}). \quad \blacksquare$$

In applications involving a system of linear equations, we sometimes need to know that the solution consists of integers. In this situation Cramer's rule is useful because it shows that a system of linear equations with integral coefficients has an integral solution if $\det(A) = \pm 1$. On the other hand, Cramer's rule is not useful for computation because the solution of a system of n linear equations in n unknowns requires evaluating $n+1$ determinants of $n \times n$ matrices and hence is much less efficient than the method of Gaussian elimination discussed in Section 3.4. Thus our interest in Cramer's rule is theoretical and aesthetic, rather than computational.

EXERCISES

1. Label the following statements as being true or false.

(a) If two rows of A are identical, then $\det(A) = 0$.

(b) If B is a matrix obtained from A by interchanging two rows, then $\det(B) = -\det(A)$.

(c) If B is a matrix obtained from A by multiplying a row of A by a scalar c, then $\det(A) = \det(B)$.

(d) If B is a matrix obtained from A by adding a scalar multiple of row i to row j ($i \neq j$), then $\det(B) = \det(A)$.

(e) If E is an elementary matrix, then $\det(E) = \pm 1$.

(f) If $A, B \in \mathsf{M}_{n \times n}(F)$, then $\det(AB) = \det(A) \cdot \det(B)$.

(g) A matrix M is invertible if and only if $\det(M) = 0$.

(h) A matrix $M \in \mathsf{M}_{n \times n}(F)$ has rank n if and only if $\det(M) \neq 0$.

(i) The determinant of a matrix may be evaluated by expanding along any row or column.

(j) $\det(A^t) = -\det(A)$.

(k) The determinant of a diagonal matrix is the product of its diagonal entries.

(l) Every system of n linear equations in n unknowns can be solved by Cramer's rule.

(m) Let $AX = B$ be the matrix form of a system of n linear equations in n unknowns, where $X = (x_1, x_2, \ldots, x_n)^t$. If $\det(A) \neq 0$ and if M_k is the matrix obtained from A by replacing the kth row of A by B^t, then for each k $(1 \leqslant k \leqslant n)$,

$$x_k = [\det(A)]^{-1} \cdot \det(M_k).$$

2. Evaluate each determinant in the manner indicated.

(a) Expand

$$\begin{pmatrix} 2 & -1 & 4 \\ 3 & 6 & 1 \\ -1 & 2 & 3 \end{pmatrix}$$

along the second column.

(b) Expand

$$\begin{pmatrix} -3 & 1 & 2 \\ 1 & -1 & 4 \\ 5 & 6 & 1 \end{pmatrix}$$

along the first row.

(c) Expand

$$\begin{pmatrix} 1 & 2 & 1 & 1 \\ -1 & 1 & 0 & 1 \\ 2 & 0 & 0 & -1 \\ 0 & 1 & 1 & 1 \end{pmatrix}$$

along the third column.

(d) Expand

$$\begin{pmatrix} 1 & 2 & 0 & -1 \\ 1 & -1 & 0 & 2 \\ 1 & 0 & 1 & 0 \\ 0 & 1 & 2 & 1 \end{pmatrix}$$

along the fourth row.

3. Evaluate the determinants of the matrices below by any legitimate method. In each case C is the field of scalars.

(a) $\begin{pmatrix} 4 & -7 & 3 \\ 1 & 2 & -1 \\ -3 & 4 & 5 \end{pmatrix}$ **(b)** $\begin{pmatrix} 9 & 0 & 0 \\ 4 & 8 & 0 \\ 3 & 2 & 7 \end{pmatrix}$ **(c)** $\begin{pmatrix} 4 & -5 & 2 \\ 2 & 8 & 1 \\ 6 & -1 & 3 \end{pmatrix}$

(d) $\begin{pmatrix} -2+i & -1 & 5i \\ 3 & 3+2i & -2i \\ 4i & 0 & 1+i \end{pmatrix}$ (e) $\begin{pmatrix} 1 & 2 & -1 & -1 \\ -3 & 0 & 2 & 1 \\ 2 & -1 & 5 & 4 \\ -1 & 6 & 3 & 3 \end{pmatrix}$

(f) $\begin{pmatrix} 2 & 0 & -1 & 3 \\ -4 & 3 & 5 & 1 \\ 1 & 6 & 0 & 2 \\ 0 & -5 & 3 & 7 \end{pmatrix}$ (g) $\begin{pmatrix} -1+3i & 2i & 6 & 0 \\ 4 & 0 & 3+i & 4i \\ 0 & 1-2i & 0 & 2-i \\ 2i & 5 & 0 & 1+i \end{pmatrix}$

4. Prove that an upper or lower triangular $n \times n$ matrix is invertible if and only if all its diagonal entries are nonzero.

5. Complete the proof of Theorem 4.10 by proving that if E is an elementary matrix, then $\det(E^t) = \det(E)$. *Hint:* E^t is an elementary matrix of the same type as E.

6. Prove that if $A \in \mathsf{M}_{n\times n}(F)$, then $\det(cA) = c^n \det(A)$ for any scalar c.

7. **(a)** A matrix B in $\mathsf{M}_{n\times n}(R)$ is called *orthogonal* if $BB^t = I$. Prove that if B is orthogonal, then $\det(B) = \pm 1$.

(b) A matrix B in $\mathsf{M}_{n\times n}(C)$ is called *unitary* if $BB^* = I$, where $(B^*)_{ij} = \overline{B_{ji}}$, the complex conjugate of B_{ji}. Prove that if B is unitary, then $|\det(B)| = 1$. *Hint:* Prove that $\det(\overline{B}) = \overline{\det(B)}$ where $(\overline{B})_{ij} = \overline{B_{ij}}$.

8. A matrix B in $\mathsf{M}_{n\times n}(C)$ is called *skew-symmetric* if $B^t = -B$. Prove that if $B \in \mathsf{M}_{n\times n}(C)$ is skew-symmetric and n is odd, then $\det(B) = 0$.

9.† Suppose that $A \in \mathsf{M}_{n\times n}(F)$ can be written in the form

$$A = \begin{pmatrix} B_1 & B_2 \\ O & B_3 \end{pmatrix},$$

where B_1 and B_3 are square matrices, respectively. Prove that $\det(A) = \det(B_1)\cdot\det(B_3)$.

10. Let $\beta = \{x_1, \ldots, x_n\}$ be a subset of F^n containing n distinct vectors, and let B denote the element of $\mathsf{M}_{n\times n}(F)$ whose jth column is the vector x_j. Prove that β is a basis for F^n if and only if $\det(B) \neq 0$.

11. Complete the proof of the lemma to Theorem 4.8.

12. Recall the linear transformation $\mathsf{T}\colon \mathsf{P}_n(F) \to \mathsf{F}^{n+1}$ defined in Exercise 20 of Section 2.4 by $\mathsf{T}(f) = (f(c_0), f(c_1), \ldots, f(c_n))$, where $c_0, c_1, \ldots, c_n$ are distinct elements of an infinite field F. Let β be the standard ordered basis for $\mathsf{P}_n(F)$ and γ be the standard ordered basis for F^{n+1}.

(a) Show that $M = [\mathsf{T}]_\beta$ has the form

$$\begin{pmatrix} 1 & c_0 & c_0^2 & \cdots & c_0^n \\ 1 & c_1 & c_1^2 & \cdots & c_1^n \\ \vdots & \vdots & \vdots & & \vdots \\ 1 & c_n & c_n^2 & \cdots & c_n^n \end{pmatrix}.$$

A matrix with this form is called a *Vandermonde matrix.*

(b) Show that $\det(M) \neq 0$ by using Exercise 20 of Section 2.4.

(c) Prove that

$$\det(M) = \prod_{0 \leqslant i < j \leqslant n} (c_j - c_i),$$

the product of all terms of the form $c_j - c_i$ for $0 \leqslant i < j \leqslant n$.

13. Let $A \in \mathsf{M}_{n\times n}(F)$ be nonzero. For any m $(1 \leqslant m \leqslant n)$, an $m \times m$ submatrix of A is obtained by deleting any $n - m$ rows and any $n - m$ columns from A. Let k $(1 \leqslant k \leqslant n)$ denote the largest integer such that some $k \times k$ submatrix of A has a nonzero determinant. Prove $\text{rank}(A) = k$.

14. Use the results of this section to prove Exercise 8 of Section 2.4: If A and B are $n \times n$ matrices such that $AB = I_n$, then A is invertible (and hence $B = A^{-1}$).

15. Prove that if A and B are similar matrices, then $\det(A) = \det(B)$.

16. Solve the following systems of equations by Cramer's rule.

(a) $\begin{cases} a_{11}x_1 + a_{12}x_2 = b_1 \\ a_{21}x_1 + a_{22}x_2 = b_2, \end{cases}$

where $a_{11}a_{22} - a_{12}a_{21} \neq 0$.

(b) $\begin{cases} 2x_1 + x_2 - 3x_3 = 5 \\ x_1 - 2x_2 + x_3 = 10 \\ 3x_1 + 4x_2 - 2x_3 = 0 \end{cases}$

(c) $\begin{cases} 2x_1 + x_2 - 3x_3 = 1 \\ x_1 - 2x_2 + x_3 = 0 \\ 3x_1 + 4x_2 - 2x_3 = -5 \end{cases}$

(d) $\begin{cases} x_1 - x_2 + 4x_3 = -4 \\ -8x_1 + 3x_2 + x_3 = 8 \\ 2x_1 - x_2 + x_3 = 0 \end{cases}$

(e) $\begin{cases} x_1 - x_2 + 4x_3 = -2 \\ -8x_1 + 3x_2 + x_3 = 0 \\ 2x_1 - x_2 + x_3 = 6 \end{cases}$

(f) $\begin{cases} 3x_1 + x_2 + x_3 = 4 \\ -2x_1 - x_2 = 12 \\ x_1 + 2x_2 + x_3 = -8 \end{cases}$

17. **(a)** Let c_{jk} denote the cofactor of A_{jk} in the $n \times n$ matrix A. Prove that

$$\det(A^{(1)}, \ldots, A^{(k-1)}, e_j, A^{(k+1)}, \ldots, A^{(n)}) = c_{jk}.$$

(b) Show that for $1 \leqslant j \leqslant n$

$$A \begin{pmatrix} c_{j1} \\ c_{j2} \\ \vdots \\ c_{jn} \end{pmatrix} = \det(A) \cdot e_j.$$

Hint: Apply Cramer's rule to $AX = e_j$.

(c) Deduce that if C is the $n \times n$ matrix such that $C_{ij} = c_{ji}$, then $AC = \det(A)I_n$.

(d) Show that if $\det(A) \neq 0$, then $A^{-1} = [\det(A)]^{-1}C$.

Definition. *The matrix C defined in part (c) of Exercise 17 is called the* classical adjoint *of* A.

18. Find the classical adjoint of the following matrices.

(a) $\begin{pmatrix} A_{11} & A_{12} \\ A_{21} & A_{22} \end{pmatrix}$ **(b)** $\begin{pmatrix} -4 & 0 & 0 \\ 0 & 2 & 0 \\ 0 & 0 & 5 \end{pmatrix}$

(c) $\begin{pmatrix} 1-i & 0 & 0 \\ 4 & 3i & 0 \\ 2i & 1+4i & -1 \end{pmatrix}$ **(d)** $\begin{pmatrix} 3 & 6 & 7 \\ 0 & 4 & 8 \\ 0 & 0 & 5 \end{pmatrix}$

(e) $\begin{pmatrix} 7 & 1 & 4 \\ 6 & -3 & 0 \\ -3 & 5 & -2 \end{pmatrix}$ **(f)** $\begin{pmatrix} 3 & 2+i & 0 \\ -1+i & 0 & i \\ 0 & 1 & 3-2i \end{pmatrix}$

(g) $\begin{pmatrix} -1 & 2 & 5 \\ 8 & 0 & -3 \\ 4 & 6 & 1 \end{pmatrix}$ **(h)** $\begin{pmatrix} 4 & 0 & 0 \\ 0 & 4 & 0 \\ 0 & 0 & 4 \end{pmatrix}$

19. Let C be the classical adjoint of $A \in \mathsf{M}_{n\times n}(F)$. Prove the following.
(a) $\det(C) = [\det(A)]^{n-1}$.
(b) C^t is the classical adjoint of A^t.
(c) If A is an invertible upper triangular matrix, then C and A^{-1} are upper triangular.

20. Let $y_1, y_2, \ldots, y_n$ be linearly independent functions in C^∞, and let $\mathsf{T}\colon \mathsf{C}^\infty \to \mathsf{C}^\infty$ be the linear transformation defined by

$$\mathsf{T}(y) = \det \begin{pmatrix} y & y_1 & y_2 & \cdots & y_n \\ y' & y_1' & y_2' & \cdots & y_n' \\ \vdots & \vdots & \vdots & & \vdots \\ y^{(n)} & y_1^{(n)} & y_2^{(n)} & \cdots & y_n^{(n)} \end{pmatrix}.$$

Prove that $\mathsf{N}(\mathsf{T}) = \mathrm{span}(\{y_1, y_2, \ldots, y_n\})$.

4.4 SUMMARY—IMPORTANT FACTS ABOUT DETERMINANTS

In this section we summarize the important properties of the determinant needed for the remainder of the text. The results contained in this section have been derived in Sections 4.2 and 4.3; consequently the facts presented here are stated without proofs.

The *determinant* of an $n \times n$ matrix A having entries from a field F is an element of F denoted $\det(A)$, which can be computed in the following manner:

1. If A is 1×1, then $\det(A) = A_{11}$, the single entry of A.
2. If A is 2×2, then $\det(A) = A_{11}A_{22} - A_{12}A_{21}$. Thus, for example,

$$\det\begin{pmatrix} -1 & 2 \\ 5 & 3 \end{pmatrix} = (-1)(3) - (2)(5) = -13.$$

3. If A is $n \times n$ for $n > 2$, then the determinant of A can be expressed as the sum of products of each entry of some row or column of A multiplied by ± 1 times the determinant of an $(n-1) \times (n-1)$ matrix obtained by deleting from A the row and column containing the entry in question. The precise formula is

$$\det(A) = \sum_{j=1}^{n} (-1)^{i+j} A_{ij} \cdot \det(\tilde{A}_{ij})$$

(if the determinant is evaluated by the entries of row i of A) or

$$\det(A) = \sum_{i=1}^{n} (-1)^{i+j} A_{ij} \cdot \det(\tilde{A}_{ij})$$

(if the determinant is evaluated by the entries of column j of A), where $\tilde{A}_{ij}$ is the $(n-1) \times (n-1)$ matrix obtained by deleting row i and column j from A.

In the formulas above the scalar $(-1)^{i+j} \det(\tilde{A}_{ij})$ is called the *cofactor* of the entry A_{ij}. In this language the determinant of A is evaluated as the sum of products of each entry of some row or column of A multiplied by the cofactor of that entry. Thus $\det(A)$ is expressed in terms of n determinants of $(n-1) \times (n-1)$ matrices. These determinants are then evaluated in terms of determinants of $(n-2) \times (n-2)$ matrices, and so forth, until 2×2 matrices are obtained. The determinants of the 2×2 matrices are then evaluated as in item 2 above.

Let us consider several examples of this technique in evaluating the determinant of the 4×4 matrix

$$A = \begin{pmatrix} 2 & 1 & 1 & 5 \\ 1 & 1 & -4 & -1 \\ 2 & 0 & -3 & 1 \\ 3 & 6 & 1 & 2 \end{pmatrix}.$$

First we will evaluate the determinant of A by expanding along the fourth row. This requires knowing the cofactors of each entry of that row. The cofactor of

$A_{41} = 3$ is

$$(-1)^{4+1}\det\begin{pmatrix}1 & 1 & 5\\ 1 & -4 & -1\\ 0 & -3 & 1\end{pmatrix}.$$

Let us evaluate the determinant above by expanding along the first column. Then

$$\det\begin{pmatrix}1 & 1 & 5\\ 1 & -4 & -1\\ 0 & -3 & 1\end{pmatrix} = (-1)^{1+1}(1)\det\begin{pmatrix}-4 & -1\\ -3 & 1\end{pmatrix} + (-1)^{2+1}(1)\det\begin{pmatrix}1 & 5\\ -3 & 1\end{pmatrix}$$
$$+ (-1)^{3+1}(0)\det\begin{pmatrix}1 & 5\\ -4 & -1\end{pmatrix}$$
$$= 1(1)[(-4)(1) - (-1)(-3)]$$
$$+ (-1)(1)[(1)(1) - (5)(-3)] + 0$$
$$= -7 - 16 + 0 = -23.$$

Thus the cofactor of A_{41} is $(-1)^5(-23) = 23$. Similarly, the cofactor of $A_{42} = 6$ is

$$(-1)^{4+2}\det\begin{pmatrix}2 & 1 & 5\\ 1 & -4 & -1\\ 2 & -3 & 1\end{pmatrix}.$$

Evaluating this determinant along the second row gives

$$\det\begin{pmatrix}2 & 1 & 5\\ 1 & -4 & -1\\ 2 & -3 & 1\end{pmatrix} = (-1)^{2+1}(1)\det\begin{pmatrix}1 & 5\\ -3 & 1\end{pmatrix} + (-1)^{2+2}(-4)\det\begin{pmatrix}2 & 5\\ 2 & 1\end{pmatrix}$$
$$+ (-1)^{2+3}(-1)\det\begin{pmatrix}2 & 1\\ 2 & -3\end{pmatrix}$$
$$= (-1)(1)[(1)(1) - (5)(-3)] + (1)(-4)[(2)(1) - (5)(2)]$$
$$+ (-1)(-1)[(2)(-3) - (1)(2)]$$
$$= -16 + 32 - 8 = 8.$$

So the cofactor of A_{42} is $(-1)^6(8) = 8$. The cofactor of $A_{43} = 1$ is

$$(-1)^{4+3}\det\begin{pmatrix}2 & 1 & 5\\ 1 & 1 & -1\\ 2 & 0 & 1\end{pmatrix}.$$

If we evaluate this determinant by expanding along the third row, we find

$$\det\begin{pmatrix} 2 & 1 & 5 \\ 1 & 1 & -1 \\ 2 & 0 & 1 \end{pmatrix} = (-1)^{3+1}(2)\det\begin{pmatrix} 1 & 5 \\ 1 & -1 \end{pmatrix} + (-1)^{3+2}(0)\det\begin{pmatrix} 2 & 5 \\ 1 & -1 \end{pmatrix}$$

$$+ (-1)^{3+3}(1)\det\begin{pmatrix} 2 & 1 \\ 1 & 1 \end{pmatrix}$$

$$= 1(2)[(1)(-1) - (5)(1)] + 0 + 1(1)[(2)(1) - (1)(1)]$$

$$= -12 + 0 + 1 = -11.$$

Hence the cofactor of A_{43} is $(-1)^7(-11) = 11$. Finally, the cofactor of $A_{44} = 2$ is

$$(-1)^{4+4}\det\begin{pmatrix} 2 & 1 & 1 \\ 1 & 1 & -4 \\ 2 & 0 & -3 \end{pmatrix}.$$

Computing this determinant by expanding along the second column, we obtain

$$\det\begin{pmatrix} 2 & 1 & 1 \\ 1 & 1 & -4 \\ 2 & 0 & -3 \end{pmatrix} = (-1)^{1+2}(1)\det\begin{pmatrix} 1 & -4 \\ 2 & -3 \end{pmatrix} + (-1)^{2+2}(1)\det\begin{pmatrix} 2 & 1 \\ 2 & -3 \end{pmatrix}$$

$$+ (-1)^{3+2}(0)\det\begin{pmatrix} 2 & 1 \\ 1 & -4 \end{pmatrix}$$

$$= (-1)(1)[(1)(-3) - (-4)(2)] + 1(1)[(2)(-3) - (1)(2)] + 0$$

$$= -5 - 8 + 0 = -13.$$

Therefore, the cofactor of A_{44} is $(-1)^8(-13) = -13$. We can now evaluate the determinant of A by multiplying each entry of the fourth row by its cofactor; this gives

$$\det(A) = 3(23) + 6(8) + 1(11) + 2(-13) = 102.$$

For the sake of comparison we will also compute the determinant of A by expanding along the second column. The reader should verify that the cofactors of A_{12}, A_{22}, and A_{42} are 14, 40, and 8, respectively. Thus

$$\det(A) = (-1)^{1+2}(1)\det\begin{pmatrix} 1 & -4 & -1 \\ 2 & -3 & 1 \\ 3 & 1 & 2 \end{pmatrix} + (-1)^{2+2}(1)\det\begin{pmatrix} 2 & 1 & 5 \\ 2 & -3 & 1 \\ 3 & 1 & 2 \end{pmatrix}$$

$$+(-1)^{3+2}(0)\det\begin{pmatrix}2 & 1 & 5\\ 1 & -4 & -1\\ 3 & 1 & 2\end{pmatrix}+(-1)^{4+2}(6)\det\begin{pmatrix}2 & 1 & 5\\ 1 & -4 & -1\\ 2 & -3 & 1\end{pmatrix}$$

$$= 14 + 40 + 0 + 48 = 102.$$

Of course, the fact that the value 102 is obtained again is no surprise since the value of the determinant of A is independent of the choice of row or column used in the expansion.

Observe that the computation of $\det(A)$ is easier when expanded along the second column than when expanded along the fourth row. The difference is the presence of a zero in the second column, which made it unnecessary to evaluate one of the cofactors (the cofactor of A_{32}). For this reason it is beneficial to evaluate the determinant of a matrix by expanding along a row or column of the matrix that contains the largest number of zero entries. In fact, it is often helpful to introduce zeros into the matrix by means of elementary row operations before computing the determinant. This technique utilizes the first three properties of the determinant.

Properties of the Determinant

1. If B is a matrix obtained by interchanging two rows or two columns of A, then $\det(B) = -\det(A)$.
2. If B is a matrix obtained by multiplying each entry of some row or column of A by a scalar c, then $\det(B) = c\det(A)$.
3. If B is a matrix obtained from A by adding a multiple of row i to row j or a multiple of column i to column j, where $i \neq j$, then $\det(B) = \det(A)$.

To illustrate the use of these three properties in evaluating determinants, we will compute the determinant of the 4×4 matrix A considered previously. Our procedure will be to introduce zeros into the second column of A by employing property 3 and then to expand along that column. (The elementary row operations used here consist of adding multiples of row 1 to rows 2 and 4.) This procedure yields

$$\det(A) = \det\begin{pmatrix}2 & 1 & 1 & 5\\ 1 & 1 & -4 & -1\\ 2 & 0 & -3 & 1\\ 3 & 6 & 1 & 2\end{pmatrix} = \det\begin{pmatrix}2 & 1 & 1 & 5\\ -1 & 0 & -5 & -6\\ 2 & 0 & -3 & 1\\ -9 & 0 & -5 & -28\end{pmatrix}$$

$$= 1(-1)^{1+2}\det\begin{pmatrix}-1 & -5 & -6\\ 2 & -3 & 1\\ -9 & -5 & -28\end{pmatrix}.$$

The resulting determinant of a 3 × 3 matrix can be evaluated in the same manner: Use type 3 elementary row operations to introduce two zeros into the first column and then expand along that column. Continuing from above, we have

$$\det(A) = (-1)\cdot\det\begin{pmatrix} -1 & -5 & -6 \\ 2 & -3 & 1 \\ -9 & -5 & -28 \end{pmatrix} = (-1)\cdot\det\begin{pmatrix} -1 & -5 & -6 \\ 0 & -13 & -11 \\ 0 & 40 & 26 \end{pmatrix}$$

$$= (-1)\left[(-1)^{1+1}(-1)\det\begin{pmatrix} -13 & -11 \\ 40 & 26 \end{pmatrix}\right]$$

$$= (-13)(26) - (-11)(40) = 102.$$

The reader should compare this calculation of $\det(A)$ with the preceding ones to see how much less work is required when properties 1, 2, and 3 are employed.

In the following chapters we often have to evaluate the determinant of matrices having special forms. The next three properties of the determinant are useful in this regard.

4. $\det(I) = 1$.
5. If two rows (or columns) of a matrix are identical, then the determinant of the matrix is zero.
6. The determinant of an upper triangular matrix is the product of its diagonal entries.

As an illustration of property 6, notice that

$$\det\begin{pmatrix} -3 & 1 & 2 \\ 0 & 4 & 5 \\ 0 & 0 & -6 \end{pmatrix} = (-3)(4)(-6) = 72.$$

Property 6 provides an efficient method for evaluating the determinant of a matrix.

1. Use Gaussian elimination and properties 1, 2, and 3 above to reduce to an upper triangular matrix.
2. Compute the product of the diagonal entries.

For instance,

$$\det\begin{pmatrix} 1 & -1 & 2 & 1 \\ 2 & -1 & -1 & 4 \\ -4 & 5 & -10 & -6 \\ 3 & -2 & 10 & -1 \end{pmatrix} = \det\begin{pmatrix} 1 & -1 & 2 & 1 \\ 0 & 1 & -5 & 2 \\ 0 & 1 & -2 & -2 \\ 0 & 1 & 4 & -4 \end{pmatrix}$$

$$= \det \begin{pmatrix} 1 & -1 & 2 & 1 \\ 0 & 1 & -5 & 2 \\ 0 & 0 & 3 & -4 \\ 0 & 0 & 9 & -6 \end{pmatrix} = \det \begin{pmatrix} 1 & -1 & 2 & 1 \\ 0 & 1 & -5 & 2 \\ 0 & 0 & 3 & -4 \\ 0 & 0 & 0 & 6 \end{pmatrix}$$

$$= 1 \cdot 1 \cdot 3 \cdot 6 = 18.$$

The remaining four properties of the determinant are frequently used in later chapters. Indeed, perhaps the most significant property of the determinant is that it provides a simple characterization of invertible matrices (see property 10).

7. For any A, $\det(A) = \det(A^t)$.
8. For any $A, B \in \mathsf{M}_{n \times n}(F)$, $\det(AB) = \det(A) \cdot \det(B)$.
9. If Q is an invertible matrix, then $\det(Q^{-1}) = [\det(Q)]^{-1}$.
10. A matrix Q is invertible if and only if $\det(Q) \neq 0$.

Thus, for example, property 10 guarantees that the matrix A on page 206 is invertible because $\det(A) = 102$. Therefore $\text{rank}(A) = 4$ by a remark on page 132.

EXERCISES

1. Label the following statements as being true or false.

(a) The determinant of a square matrix may be computed by expanding the matrix along any row or column.

(b) In evaluating the determinant of a matrix, it is wise to expand along a row or column containing the largest number of zero entries.

(c) If two rows or columns of A are identical, then $\det(A) = 0$.

(d) If B is a matrix obtained by interchanging two rows or two columns of A, then $\det(B) = \det(A)$.

(e) If B is a matrix obtained by multiplying each entry of some row or column of A by a scalar, then $\det(B) = \det(A)$.

(f) If B is a matrix obtained from A by adding a multiple of some row to a different row (or a multiple of some column to a different column), then $\det(B) = \det(A)$.

(g) The determinant of an upper triangular $n \times n$ matrix is the product of its diagonal entries.

(h) $\det(A^t) = -\det(A)$.

(i) If $A, B \in \mathsf{M}_{n \times n}(F)$, then $\det(AB) = \det(A) \cdot \det(B)$.

(j) If Q is an invertible matrix, then $\det(Q^{-1}) = [\det(Q)]^{-1}$.

(k) A matrix Q is invertible if and only if $\det(Q) \neq 0$.

2. Evaluate the determinant of the following 2×2 matrices.

(a) $\begin{pmatrix} 4 & -5 \\ 2 & 3 \end{pmatrix}$ **(b)** $\begin{pmatrix} -1 & 7 \\ 3 & 8 \end{pmatrix}$

(c) $\begin{pmatrix} 2+i & -1+3i \\ 1-2i & 3-i \end{pmatrix}$ **(d)** $\begin{pmatrix} 3 & 4i \\ -6i & 2i \end{pmatrix}$

3. Evaluate the determinant of the following matrices in the manner indicated.

(a) Expand

$$\begin{pmatrix} 0 & 1 & 2 \\ -1 & 0 & -3 \\ 2 & 3 & 0 \end{pmatrix}$$

along the second column.

(b) Expand

$$\begin{pmatrix} 1 & 0 & 2 \\ 0 & 1 & 5 \\ -1 & 3 & 0 \end{pmatrix}$$

along the third row.

(c) Expand

$$\begin{pmatrix} 0 & 1+i & 2 \\ -2i & 0 & 1-i \\ 3 & 4i & 0 \end{pmatrix}$$

along the first column.

(d) Expand

$$\begin{pmatrix} i & 2+i & 0 \\ -1 & 3 & 2i \\ 0 & -1 & 1-i \end{pmatrix}$$

along the first row.

(e) Expand

$$\begin{pmatrix} 0 & 2 & 1 & 3 \\ 1 & 0 & -2 & 2 \\ 3 & -1 & 0 & 1 \\ -1 & 1 & 2 & 0 \end{pmatrix}$$

along the fourth column.

4. Evaluate the determinant of the following matrices by any legitimate method.

(a) $\begin{pmatrix} 2 & 5 & 0 \\ -6 & 1 & 3 \\ 0 & -4 & 2 \end{pmatrix}$ **(b)** $\begin{pmatrix} -1 & 3 & 2 \\ 4 & -1 & 1 \\ 2 & 2 & 5 \end{pmatrix}$ **(c)** $\begin{pmatrix} 1 & 0 & 3 \\ 2 & -1 & 4 \\ -3 & 2 & 1 \end{pmatrix}$

(d) $\begin{pmatrix} i & 2 & -1 \\ 3 & 1+i & 2 \\ -2i & 1 & 4-i \end{pmatrix}$ **(e)** $\begin{pmatrix} -1 & 2+i & 3 \\ 1-i & i & 1 \\ 3i & 2 & -1+i \end{pmatrix}$

(f) $\begin{pmatrix} 1 & 0 & -2 & 3 \\ -3 & 1 & 1 & 2 \\ 0 & 4 & -1 & 1 \\ 2 & 3 & 0 & 1 \end{pmatrix}$ **(g)** $\begin{pmatrix} 4 & 2+i & 2i & 5+2i \\ 0 & 1-i & 1 & 3-4i \\ 0 & 0 & 3i & 6 \\ 0 & 0 & 0 & 2 \end{pmatrix}$

5.† Work Exercise 9 of Section 4.3.

6. Let $A \in \mathsf{M}_{n \times n}(F)$, and consider the system of linear equations $AX = B$, where

$$X = \begin{pmatrix} x_1 \\ x_2 \\ \vdots \\ x_n \end{pmatrix} \quad \text{and} \quad B = \begin{pmatrix} b_1 \\ b_2 \\ \vdots \\ b_n \end{pmatrix}.$$

(a) Compute $\det(X_k)$, where X_k is the matrix obtained from the $n \times n$ identity matrix by replacing its kth column by X.

(b) Show that $AX_k = M_k$, where M_k is the $n \times n$ matrix obtained from A by replacing its kth column by B.

(c) Prove Cramer's rule: If $\det(A) \neq 0$, the solution of the system $AX = B$ is such that $x_k = [\det(A)]^{-1} \cdot \det(M_k)$ for each k $(1 \leqslant k \leqslant n)$.

INDEX OF DEFINITIONS FOR CHAPTER 4

5

Diagonalization

This chapter is concerned with the so-called "diagonalization problem." Given a linear transformation $\mathsf{T}\colon \mathsf{V} \to \mathsf{V}$, where V is a finite-dimensional vector space, we seek answers to the following questions:

1. Does there exist an ordered basis β for V such that $[\mathsf{T}]_\beta$ is a diagonal matrix?
2. If such a basis exists, how can it be found?

Since computations involving diagonal matrices are simple, an affirmative answer to question 1 leads us to a clearer understanding of how the transformation T acts on V, and an answer to question 2 enables us to obtain easy solutions to many practical problems that can be formulated in a linear algebra context. We will consider some of these problems and their solutions in this chapter; see, for example, Section 5.3.

A solution to the diagonalization problem leads naturally to the concepts of "eigenvalue" and "eigenvector." Aside from the important role that these concepts play in the diagonalization problem, they will also prove to be useful tools in the study of many nondiagonalizable transformations, as we will see in Chapter 7.

5.1 EIGENVALUES AND EIGENVECTORS

Since the diagonalization problem involves the study of a linear transformation that maps a vector space into itself, it is useful to name such a transformation. Accordingly, we call a linear transformation $\mathsf{T}\colon \mathsf{V} \to \mathsf{V}$ on a vector space V a *linear operator* on V.

For a given linear operator T on a finite-dimensional vector space V, we are concerned with the matrices that represent T relative to various ordered bases for V.

Throughout this chapter, we usually omit the adjective "ordered" from the expression "ordered basis."

Consider a linear operator T on a finite-dimensional vector space V and any two bases β and β' for V. Recall from Theorem 2.24 that the matrices $[\mathsf{T}]_\beta$ and $[\mathsf{T}]_{\beta'}$ are related by

$$[\mathsf{T}]_{\beta'} = Q^{-1}[\mathsf{T}]_\beta Q,$$

where Q is the change of coordinate matrix changing β'-coordinates into β-coordinates. In Section 2.5 we defined such matrices to be *similar*. A useful special case of this relationship is proved in the following theorem.

Theorem 5.1. *Let* $A \in \mathsf{M}_{n\times n}(F)$, *and let* $\gamma = \{x_1, x_2, \ldots, x_n\}$ *be any basis for* F^n. *Then* $[\mathsf{L}_A]_\gamma = Q^{-1}AQ$, *where* Q *is the* $n \times n$ *matrix in which the* jth *column is* x_j $(j = 1, 2, \ldots, n)$.

Proof. Let β be the standard basis for F^n. It is easily seen that the matrix Q is the change of coordinate matrix changing γ-coordinates into β-coordinates. Hence

$$[\mathsf{L}_A]_\gamma = Q^{-1}[\mathsf{L}_A]_\beta Q = Q^{-1}AQ. \quad \blacksquare$$

Example 1

To illustrate Theorem 5.1, let

$$A = \begin{pmatrix} 1 & 2 \\ 0 & 1 \end{pmatrix} \in \mathsf{M}_{2\times 2}(R) \quad \text{and} \quad \gamma = \left\{ \begin{pmatrix} 1 \\ 3 \end{pmatrix}, \begin{pmatrix} 1 \\ 2 \end{pmatrix} \right\}.$$

It is a simple matter to check that if

$$Q = \begin{pmatrix} 1 & 1 \\ 3 & 2 \end{pmatrix},$$

then

$$Q^{-1} = \begin{pmatrix} -2 & 1 \\ 3 & -1 \end{pmatrix}$$

and

$$[\mathsf{L}_A]_\gamma = Q^{-1}AQ = \begin{pmatrix} -2 & 1 \\ 3 & -1 \end{pmatrix}\begin{pmatrix} 1 & 2 \\ 0 & 1 \end{pmatrix}\begin{pmatrix} 1 & 1 \\ 3 & 2 \end{pmatrix} = \begin{pmatrix} -11 & -8 \\ 18 & 13 \end{pmatrix}. \quad \blacksquare$$

As mentioned above, matrices that represent the same linear operator relative to different bases are similar. We now establish the converse of this result.

Theorem 5.2. *Let* T *be a linear operator on an* n*-dimensional vector space* V, *and let* β *be a basis for* V. *If* B *is any* $n \times n$ *matrix similar to* $[\mathsf{T}]_\beta$, *then there exists a basis* β' *for* V *such that* $B = [\mathsf{T}]_{\beta'}$.

Proof. If B is similar to $[\mathsf{T}]_\beta$, then there exists an invertible matrix Q such that $B = Q^{-1}[\mathsf{T}]_\beta Q$. Suppose that $\beta = \{x_1, x_2, \ldots, x_n\}$, and define

$$x'_j = \sum_{i=1}^{n} Q_{ij} x_i \qquad \text{for } 1 \leqslant j \leqslant n.$$

Then $\beta' = \{x'_1, x'_2, \ldots, x'_n\}$ is a basis for V such that Q is the change of coordinate matrix changing β'-coordinates into β-coordinates (Exercise 12 of Section 2.5). Hence

$$[\mathsf{T}]_{\beta'} = Q^{-1}[\mathsf{T}]_\beta Q = B$$

by Theorem 2.24. ∎

The concept of similarity is useful in studying the diagonalization problem since it can be used to reformulate the problem in the context of matrices. We now introduce the definitions of diagonalizability.

Definitions. *A linear operator* T *on a finite-dimensional vector space* V *is said to be* diagonalizable *if there exists a basis* β *for* V *such that* $[\mathsf{T}]_\beta$ *is a diagonal matrix.*

A square matrix A *is said to be* diagonalizable *if* A *is similar to a diagonal matrix.*

The following theorem relates these two concepts and leads to a reformulation of the diagonalization problem in the context of matrices.

Theorem 5.3. *Let* T *be a linear operator on a finite-dimensional vector space* V *and let* β *be a basis for* V. *Then* T *is diagonalizable if and only if* $[\mathsf{T}]_\beta$ *is a diagonalizable matrix.*

Proof. If T is diagonalizable, then there exists a basis γ of V such that $[\mathsf{T}]_\gamma$ is a diagonal matrix. By Theorem 2.24, the matrices $[\mathsf{T}]_\beta$ and $[\mathsf{T}]_\gamma$ are similar. Therefore, $[\mathsf{T}]_\beta$ is diagonalizable.

Now suppose that $[\mathsf{T}]_\beta$ is diagonalizable. Then $[\mathsf{T}]_\beta$ is similar to a diagonal matrix B. By Theorem 5.2, there exists a basis β' for V such that $B = [\mathsf{T}]_{\beta'}$. Therefore, T is diagonalizable. ∎

As an immediate consequence of this theorem we have the following useful result.

Corollary. *A matrix* A *is diagonalizable if and only if* L_A *is diagonalizable.*

Because of Theorem 5.3, we can reformulate the diagonalization problem as follows.

1. Is a given square matrix A diagonalizable?

2. If A is diagonalizable, how can an invertible matrix Q be determined so that $Q^{-1}AQ$ is a diagonal matrix?

We now present the first of several results leading to a solution of the diagonalization problem.

Theorem 5.4. *A linear operator* T *on a finite-dimensional vector space* V *is diagonalizable if and only if there exists a basis* $\beta = \{x_1, \ldots, x_n\}$ *for* V *and scalars* $\lambda_1, \ldots, \lambda_n$ *(not necessarily distinct) such that* $T(x_j) = \lambda_j x_j$, *for* $1 \leqslant j \leqslant n$. *Under these circumstances*

$$[T]_\beta = \begin{pmatrix} \lambda_1 & 0 & \cdots & 0 \\ 0 & \lambda_2 & \cdots & 0 \\ \vdots & \vdots & & \vdots \\ 0 & 0 & \cdots & \lambda_n \end{pmatrix}.$$

Proof. Suppose that T is diagonalizable. Then there is a basis β for V such that $[T]_\beta = D$ is a diagonal matrix. Let $\lambda_j = D_{jj}$ and $\beta = \{x_1, \ldots, x_n\}$. Then for each j,

$$T(x_j) = \sum_{i=1}^{n} D_{ij}x_i = D_{jj}x_j = \lambda_j x_j.$$

Conversely, suppose there exists a basis $\beta = \{x_1, \ldots, x_n\}$ and scalars $\lambda_1, \ldots, \lambda_n$ such that $T(x_j) = \lambda_j x_j$. Then clearly

$$[T]_\beta = \begin{pmatrix} \lambda_1 & 0 & \cdots & 0 \\ 0 & \lambda_2 & \cdots & 0 \\ \vdots & \vdots & & \vdots \\ 0 & 0 & \cdots & \lambda_n \end{pmatrix}. \quad \blacksquare$$

Theorem 5.4 motivates the following definitions.

Definitions. *Let* T *be a linear operator on a vector space* V. *A nonzero element* $x \in V$ *is called an* eigenvector *of* T *if there exists a scalar* λ *such that* $T(x) = \lambda x$. *The scalar* λ *is called the* eigenvalue *corresponding to the eigenvector* x.

Similarly, if A *is an* $n \times n$ *matrix over a field* F, *a nonzero element* $x \in F^n$ *is called an* eigenvector *of the matrix* A *if* x *is an eigenvector of* L_A; *that is,* $Ax = \lambda x$ *for some scalar* λ. *The scalar* λ *is called the* eigenvalue *of* A *corresponding to the eigenvector* x.

The words *characteristic vector* and *proper vector* are often used in place of *eigenvector*. The corresponding terms for an eigenvalue are *characteristic value* and *proper value*.

In this terminology we see that in Theorem 5.4 the basis β consists of eigenvectors of T and that the diagonal entries of $[\mathsf{T}]_\beta$ are eigenvalues of T. Thus Theorem 5.4 can be restated as follows: *A linear operator* T *on a finite-dimensional vector space* V *is diagonalizable if and only if there exists a basis* β *for* V *consisting of eigenvectors of* T. *Furthermore, if* T *is diagonalizable,* $\beta = \{x_1, x_2, \ldots, x_n\}$ *is a basis of eigenvectors of* T, *and if* $D = [\mathsf{T}]_\beta$, *then* D *is a diagonal matrix and* D_{ii} *is the eigenvalue corresponding to* x_i $(i = 1, 2, \ldots, n)$.

Before continuing our examination of the diagonalization problem, we consider two examples involving eigenvectors and eigenvalues.

Example 2

Let $\mathsf{C}^\infty(R)$ denote the set of all functions $f\colon R \to R$ having derivatives of all orders. (Thus $\mathsf{C}^\infty(R)$ includes all polynomial functions, the sine and cosine functions, the exponential functions, etc.) It is easy to see that $\mathsf{C}^\infty(R)$ is a subspace of the vector space $\mathscr{F}(R, R)$ of all functions from R to R as defined in Section 1.2. Define $\mathsf{T}\colon \mathsf{C}^\infty(R) \to \mathsf{C}^\infty(R)$ by $\mathsf{T}(y) = y'$, where y' denotes the derivative of y. It is easily verified that T is a linear operator on $\mathsf{C}^\infty(R)$. We determine the eigenvalues and eigenvectors of T.

If λ is an eigenvalue of T, then there is an eigenvector $y \in \mathsf{C}^\infty(R)$ such that $y' = \mathsf{T}(y) = \lambda y$. This is a first-order differential equation whose solutions are of the form $y(t) = ce^{\lambda t}$ for some constant c. Consequently every real number λ is an eigenvalue of T, and the corresponding eigenvectors are of the form $ce^{\lambda t}$ for $c \neq 0$. (Note that if $\lambda = 0$, the eigenvectors are the nonzero constant functions.) ∎

Example 3

Let

$$A = \begin{pmatrix} 1 & 3 \\ 4 & 2 \end{pmatrix}, \qquad x_1 = \begin{pmatrix} 1 \\ -1 \end{pmatrix}, \quad \text{and} \quad x_2 = \begin{pmatrix} 3 \\ 4 \end{pmatrix}.$$

Since

$$\mathsf{L}_A(x_1) = \begin{pmatrix} 1 & 3 \\ 4 & 2 \end{pmatrix}\begin{pmatrix} 1 \\ -1 \end{pmatrix} = \begin{pmatrix} -2 \\ 2 \end{pmatrix} = -2\begin{pmatrix} 1 \\ -1 \end{pmatrix} = -2x_1,$$

x_1 is an eigenvector of L_A (and hence of A). Also $\lambda_1 = -2$ is the eigenvalue corresponding to x_1. Moreover,

$$\mathsf{L}_A(x_2) = \begin{pmatrix} 1 & 3 \\ 4 & 2 \end{pmatrix}\begin{pmatrix} 3 \\ 4 \end{pmatrix} = \begin{pmatrix} 15 \\ 20 \end{pmatrix} = 5\begin{pmatrix} 3 \\ 4 \end{pmatrix} = 5x_2.$$

Thus x_2 is an eigenvector of L_A (and of A) with $\lambda_2 = 5$ as the associated eigenvalue. Note that $\beta = \{x_1, x_2\}$ is a basis for R^2, and hence by Theorem 5.4

$$[\mathsf{L}_A]_\beta = \begin{pmatrix} -2 & 0 \\ 0 & 5 \end{pmatrix}.$$

Finally, if

$$Q = \begin{pmatrix} 1 & 3 \\ -1 & 4 \end{pmatrix},$$

then

$$Q^{-1}AQ = [\mathsf{L}_A]_\beta = \begin{pmatrix} -2 & 0 \\ 0 & 5 \end{pmatrix}$$

by Theorem 5.1. ∎

Example 3 demonstrates a technique for diagonalizing an $n \times n$ matrix A: If $\beta = \{x_1, x_2, \ldots, x_n\}$ is a basis for F^n consisting of eigenvectors of A and Q is the $n \times n$ matrix whose jth column is the eigenvector x_j ($j = 1, 2, \ldots, n$), then $Q^{-1}AQ$ is a diagonal matrix. In order to use this procedure, we need a method for determining the eigenvectors of a matrix or operator. As we will see, eigenvectors are easily determined once the eigenvalues are known. For this reason we begin by discussing a method for computing eigenvalues. As an aid in this computation we will utilize the following theorem to introduce the "determinant" of a linear operator.

Theorem 5.5. *Let* T *be a linear operator on a finite-dimensional vector space* V, *and let* β *and* β' *be any two bases for* V. *Then* $\det([\mathsf{T}]_\beta) = \det([\mathsf{T}]_{\beta'})$.

Proof. Let $A = [\mathsf{T}]_\beta$ and $B = [\mathsf{T}]_{\beta'}$. Since A and B are similar, there exists an invertible matrix Q such that $B = Q^{-1}AQ$. Thus

$$\begin{aligned}\det(B) &= \det(Q^{-1}AQ) \\ &= \det(Q^{-1})\cdot\det(A)\cdot\det(Q) \\ &= [\det(Q)]^{-1}\cdot[\det(A)]\cdot[\det(Q)] = \det(A). \quad ∎\end{aligned}$$

This result motivates the following definition. Note that by Theorem 5.5, $\det(\mathsf{T})$ is well-defined, that is, independent of the choice of basis.

Definition. *Let* T *be a linear operator on a finite-dimensional vector space* V. *We define the* determinant *of* T, *denoted* $\det(\mathsf{T})$, *as follows: Choose any basis* β *for* V, *and define* $\det(\mathsf{T}) = \det([\mathsf{T}]_\beta)$.

Example 4

Let $\mathsf{T}\colon \mathsf{P}_2(R) \to \mathsf{P}_2(R)$ be defined by $\mathsf{T}(f) = f'$, the derivative of f. To compute $\det(\mathsf{T})$, let $\beta = \{1, x, x^2\}$. Then β is a basis for $\mathsf{P}_2(R)$ and

$$[\mathsf{T}]_\beta = \begin{pmatrix} 0 & 1 & 0 \\ 0 & 0 & 2 \\ 0 & 0 & 0 \end{pmatrix}.$$

Thus $\det(\mathsf{T}) = \det([\mathsf{T}]_\beta) = 0$. ∎

Our next result establishes some important properties of the determinant of a linear operator. Note the similarity of these properties to those proved for the determinant of a matrix in Chapter 4.

Theorem 5.6. *Let* T *be a linear operator on a finite-dimensional vector space* V. *Then*

(a) T *is invertible if and only if* $\det(\mathsf{T}) \neq 0$.

(b) *If* T *is invertible, then* $\det(\mathsf{T}^{-1}) = [\det(\mathsf{T})]^{-1}$.

(c) *If* $\mathsf{U}: \mathsf{V} \to \mathsf{V}$ *is linear, then* $\det(\mathsf{TU}) = \det(\mathsf{T}) \cdot \det(\mathsf{U})$.

(d) *If* λ *is any scalar and* β *any basis for* V, *then*

$$\det(\mathsf{T} - \lambda \mathsf{I}_\mathsf{V}) = \det(A - \lambda I),$$

where $A = [\mathsf{T}]_\beta$.

Proof. The proofs of (a), (b), and (c) are exercises. To prove (d), suppose that λ is a scalar, β is a basis for V, and $A = [\mathsf{T}]_\beta$. Then $[\mathsf{I}_\mathsf{V}]_\beta = I$, and hence $[\mathsf{T} - \lambda \mathsf{I}_\mathsf{V}]_\beta = A - \lambda I$. Thus by definition we have $\det(\mathsf{T} - \lambda \mathsf{I}_\mathsf{V}) = \det(A - \lambda I)$. ∎

The following theorem provides us with a method for computing eigenvalues.

Theorem 5.7. *Let* T *be a linear operator on a finite-dimensional vector space* V *over a field* F. *A scalar* $\lambda \in F$ *is an eigenvalue of* T *if and only if* $\det(\mathsf{T} - \lambda \mathsf{I}) = 0$.

Proof. A scalar λ is an eigenvalue of T if and only if there exists a nonzero vector x in V such that $\mathsf{T}(x) = \lambda x$, or $(\mathsf{T} - \lambda \mathsf{I})(x) = 0$. By Theorem 2.5, this is true if and only if $\mathsf{T} - \lambda \mathsf{I}$ is not invertible. However, by Theorem 5.6, this result is equivalent to the statement that $\det(\mathsf{T} - \lambda \mathsf{I}) = 0$. ∎

Corollary 1. *Let* A *be an* $n \times n$ *matrix over a field* F. *Then a scalar* $\lambda \in F$ *is an eigenvalue of* A *if and only if* $\det(A - \lambda I) = 0$.

Proof. Exercise. ∎

Example 5

Let

$$A = \begin{pmatrix} 1 & 1 \\ 4 & 1 \end{pmatrix} \in \mathsf{M}_{2 \times 2}(R).$$

Since

$$\det(A - \lambda I) = \det \begin{pmatrix} 1 - \lambda & 1 \\ 4 & 1 - \lambda \end{pmatrix} = \lambda^2 - 2\lambda - 3 = (\lambda - 3)(\lambda + 1),$$

the only eigenvalues of A are 3 and -1. ∎

Example 6

Let $\mathsf{T}: \mathsf{P}_2(R) \to \mathsf{P}_2(R)$ be the linear operator that is defined by $\mathsf{T}(f(x)) = f(x) + xf'(x) + f'(x)$, and let β be the standard basis for $\mathsf{P}_2(R)$. Then

$$[\mathsf{T}]_\beta = \begin{pmatrix} 1 & 1 & 0 \\ 0 & 2 & 2 \\ 0 & 0 & 3 \end{pmatrix}.$$

Since

$$\begin{aligned} \det(\mathsf{T} - \lambda\mathsf{I}) = \det([\mathsf{T}]_\beta - \lambda I) &= \det \begin{pmatrix} 1-\lambda & 1 & 0 \\ 0 & 2-\lambda & 2 \\ 0 & 0 & 3-\lambda \end{pmatrix} \\ &= (1-\lambda)(2-\lambda)(3-\lambda) \\ &= -(\lambda - 1)(\lambda - 2)(\lambda - 3), \end{aligned}$$

λ is an eigenvalue of T if and only if $\lambda = 1$, 2, or 3. ∎

Example 6 makes use of the following obvious consequence of Theorem 5.6.

Corollary 2. *Let* T *be a linear operator on a finite-dimensional vector space* V, *and let* β *be a basis for* V. *Then* λ *is an eigenvalue of* T *if and only if* λ *is an eigenvalue of* $[\mathsf{T}]_\beta$.

In Examples 5 and 6 the reader may have observed that if A is an $n \times n$ matrix, then $\det(A - \lambda I_n)$ is a polynomial in λ of degree n with leading coefficient $(-1)^n$. The eigenvalues of A are simply the zeros of this polynomial. Thus the following definition is appropriate.

Definition. *If* $A \in \mathsf{M}_{n \times n}(F)$, *the polynomial* $\det(A - tI_n)$ *in the indeterminate* t *is called the* characteristic polynomial *of* A.†

It is easily shown that similar matrices have the same characteristic polynomial (see Exercise 12). This fact permits the following definition.

Definition. *Let* T *be a linear operator on a finite-dimensional vector space* V *with basis* β. *We define the* characteristic polynomial $f(t)$ *of* T *to be the*

†The observant reader may have noticed that the entries of the matrix $A - tI_n$ are not elements of the field F. They are, however, elements of another field $F(t)$. (The field $F(t)$ is the field of quotients of the polynomial ring $F[t]$. It is usually studied in abstract algebra courses.) Consequently, the results proved about determinants in Chapter 4 remain true in this context.

characteristic polynomial of $A = [T]_\beta$*; that is,*

$$f(t) = det(A - tI).$$

The remark preceding the definition shows that this definition is independent of the choice of the basis β. We often denote the characteristic polynomial of an operator T by $\det(T - tI)$.

The next result confirms our observations about Examples 5 and 6. It can be proved by a straightforward induction argument.

Theorem 5.8. *The characteristic polynomial of* $A \in M_{n \times n}(F)$ *is a polynomial of degree* n *with leading coefficient* $(-1)^n$.

The following consequences of Theorem 5.8 are immediate (see also Corollary 2 of Theorem E.1).

Corollary 1. *Let* A *be any* $n \times n$ *matrix, and let* f(t) *be the characteristic polynomial of* A. *Then*

(a) *A scalar* λ *is an eigenvalue of* A *if and only if* λ *is a zero of the polynomial* f(t) [*i.e., if and only if* $f(\lambda) = 0$].

(b) A *has at most* n *distinct eigenvalues.*

Corollary 2. *Let* T *be a linear operator on an* n*-dimensional vector space* V *with characteristic polynomial* f(t). *Then*

(a) *A scalar* λ *is an eigenvalue of* T *if and only if* λ *is a zero of the polynomial* f(t) [*i.e., if and only if* $f(\lambda) = 0$].

(b) T *has at most* n *distinct eigenvalues.*

The two corollaries above provide us with a method for determining all the eigenvalues of a matrix or an operator. Our next result gives us a procedure for determining the eigenvectors corresponding to a given eigenvalue.

Theorem 5.9. *Let* T *be a linear operator on a vector space* V, *and let* λ *be an eigenvalue of* T. *A vector* $x \in V$ *is an eigenvector of* T *corresponding to* λ *if and only if* $x \neq 0$ *and* $x \in N(T - \lambda I)$.

Proof. Exercise. ∎

Example 7

To find all the eigenvectors of the matrix

$$A = \begin{pmatrix} 1 & 1 \\ 4 & 1 \end{pmatrix}$$

in Example 5, recall that A has two eigenvalues $\lambda_1 = 3$ and $\lambda_2 = -1$. We begin by finding all the eigenvectors corresponding to $\lambda_1 = 3$. Let

$$B = A - \lambda_1 I = \begin{pmatrix} 1 & 1 \\ 4 & 1 \end{pmatrix} - \begin{pmatrix} 3 & 0 \\ 0 & 3 \end{pmatrix} = \begin{pmatrix} -2 & 1 \\ 4 & -2 \end{pmatrix}.$$

Then

$$x = \begin{pmatrix} x_1 \\ x_2 \end{pmatrix} \in \mathsf{R}^2$$

is an eigenvector corresponding to $\lambda_1 = 3$ if and only if $x \neq \mathit{0}$ and $x \in \mathsf{N}(\mathsf{L}_B)$, that is, $x \neq \mathit{0}$ and

$$\begin{pmatrix} -2 & 1 \\ 4 & -2 \end{pmatrix}\begin{pmatrix} x_1 \\ x_2 \end{pmatrix} = \begin{pmatrix} -2x_1 + x_2 \\ 4x_1 - 2x_2 \end{pmatrix} = \begin{pmatrix} 0 \\ 0 \end{pmatrix}.$$

Clearly the set of all solutions to the equation above is

$$\left\{t\begin{pmatrix} 1 \\ 2 \end{pmatrix}: \quad t \in R\right\}.$$

Hence x is an eigenvector corresponding to $\lambda_1 = 3$ if and only if

$$x = t\begin{pmatrix} 1 \\ 2 \end{pmatrix} \qquad \text{for some } t \neq 0.$$

Now suppose that x is an eigenvector of A corresponding to $\lambda_2 = -1$. Let

$$B = A - \lambda_2 I = \begin{pmatrix} 1 & 1 \\ 4 & 1 \end{pmatrix} - \begin{pmatrix} -1 & 0 \\ 0 & -1 \end{pmatrix} = \begin{pmatrix} 2 & 1 \\ 4 & 2 \end{pmatrix};$$

then

$$x = \begin{pmatrix} x_1 \\ x_2 \end{pmatrix} \in \mathsf{N}(\mathsf{L}_B)$$

if and only if x is a solution to the system

$$\begin{cases} 2x_1 + x_2 = 0 \\ 4x_1 + 2x_2 = 0. \end{cases}$$

Hence

$$\mathsf{N}(\mathsf{L}_B) = \left\{t\begin{pmatrix} 1 \\ -2 \end{pmatrix}: \quad t \in R\right\}.$$

Thus x is an eigenvector corresponding to $\lambda_2 = -1$ if and only if

$$x = t\begin{pmatrix} 1 \\ -2 \end{pmatrix} \qquad \text{for some } t \neq 0.$$

Observe that

$$\left\{\begin{pmatrix}1\\2\end{pmatrix}, \begin{pmatrix}1\\-2\end{pmatrix}\right\}$$

is a basis for R^2 consisting of eigenvectors of A. Thus, by Theorem 5.4, L_A, and hence A, is diagonalizable. In fact, if

$$Q = \begin{pmatrix}1 & 1\\2 & -2\end{pmatrix},$$

then Theorem 5.1 implies that

$$Q^{-1}AQ = \begin{pmatrix}3 & 0\\0 & -1\end{pmatrix}. \quad \blacksquare$$

Example 8

To find the eigenvectors of the linear operator T on $\mathsf{P}_2(R)$ defined in Example 6, first recall that T has eigenvalues 1, 2, and 3. Now consider the diagram shown in Figure 5.1, which is a special case of Figure 2.2 of Section 2.4 as applied to T. Here $\mathsf{V} = \mathsf{W} = \mathsf{P}_2(R)$, $\beta = \gamma$, and

$$A = [\mathsf{T}]_\beta = \begin{pmatrix}1 & 1 & 0\\0 & 2 & 2\\0 & 0 & 3\end{pmatrix}.$$

We will show that $v \in \mathsf{P}_2(R)$ is an eigenvector of T corresponding to λ if and only if $\phi_\beta(v)$ is an eigenvector of A corresponding to λ. (This argument is valid for any operator on a finite-dimensional vector space.) If v is an eigenvector of T corresponding to λ, then $\mathsf{T}(v) = \lambda v$. Hence

$$\mathsf{L}_A\phi_\beta(v) = \phi_\beta\mathsf{T}(v) = \phi_\beta(\lambda v) = \lambda\phi_\beta(v).$$

Now $\phi_\beta(v) \neq 0$ since ϕ_β is an isomorphism. Thus $\phi_\beta(v)$ is an eigenvector of L_A (and hence of A) corresponding to λ. Since the argument above is reversible, we can establish similarly that if $\phi_\beta(v)$ is an eigenvector of A corresponding to λ, then v is an eigenvector of T corresponding to λ (see Exercise 13).

An equivalent formulation of the result proved in the preceding paragraph is that for any eigenvalue λ of A (and hence of T), a vector $y \in \mathsf{R}^3$ is an eigenvector of A corresponding to λ if and only if $\phi_\beta^{-1}(y)$ is an eigenvector of T

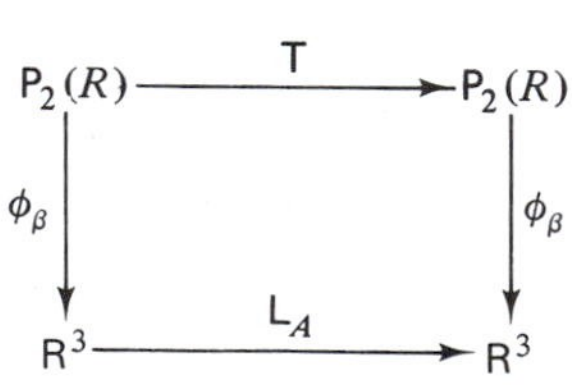

Figure 5.1

corresponding to λ. This fact allows us to compute eigenvectors of T as we did in Example 7.

Let $\lambda_1 = 1$, and define

$$B = A - \lambda_1 I = \begin{pmatrix} 0 & 1 & 0 \\ 0 & 1 & 2 \\ 0 & 0 & 2 \end{pmatrix}.$$

It is easily shown that

$$\mathsf{N}(\mathsf{L}_B) = \left\{ a\begin{pmatrix} 1 \\ 0 \\ 0 \end{pmatrix} : \quad a \in R \right\}.$$

Thus the eigenvectors of A corresponding to λ_1 are of the form

$$a\begin{pmatrix} 1 \\ 0 \\ 0 \end{pmatrix}$$

for any $a \neq 0$. Consequently, the eigenvectors of T corresponding to $\lambda_1 = 1$ are of the form

$$\phi_\beta^{-1}\left(a\begin{pmatrix} 1 \\ 0 \\ 0 \end{pmatrix} \right) = a\phi_\beta^{-1}(e_1) = a$$

for any $a \neq 0$. Hence the nonzero constant polynomials are the eigenvectors of T corresponding to λ_1.

Next let $\lambda_2 = 2$, and define

$$B = A - \lambda_2 I = \begin{pmatrix} -1 & 1 & 0 \\ 0 & 0 & 2 \\ 0 & 0 & 1 \end{pmatrix}.$$

Again it is easily verified that

$$\mathsf{N}(\mathsf{L}_B) = \left\{ a\begin{pmatrix} 1 \\ 1 \\ 0 \end{pmatrix} : \quad a \in R \right\}.$$

Thus the eigenvectors of T corresponding to λ_2 are of the form

$$\phi_\beta^{-1}\left(a\begin{pmatrix} 1 \\ 1 \\ 0 \end{pmatrix} \right) = a\phi_\beta^{-1}(e_1 + e_2) = a(1 + x) = a + ax$$

for any $a \neq 0$.

Finally, consider $\lambda_3 = 3$ and

$$B = A - \lambda_3 I = \begin{pmatrix} -2 & 1 & 0 \\ 0 & -1 & 2 \\ 0 & 0 & 0 \end{pmatrix}.$$

Since

$$\mathsf{N}(\mathsf{L}_B) = \left\{ a \begin{pmatrix} 1 \\ 2 \\ 1 \end{pmatrix} : \quad a \in R \right\},$$

any eigenvector of T corresponding to $\lambda_3 = 3$ is of the form

$$\phi_\beta^{-1}\left(a \begin{pmatrix} 1 \\ 2 \\ 1 \end{pmatrix} \right) = a\phi_\beta^{-1}(e_1 + 2e_2 + e_3) = a(1 + 2x + x^2) = a + 2ax + ax^2$$

for any $a \neq 0$.

Note also that $\gamma = \{1, 1 + x, 1 + 2x + x^2\}$ is a basis for $\mathsf{P}_2(R)$ consisting of eigenvectors of T. Thus T is diagonalizable and

$$[\mathsf{T}]_\gamma = \begin{pmatrix} 1 & 0 & 0 \\ 0 & 2 & 0 \\ 0 & 0 & 3 \end{pmatrix}. \quad \blacksquare$$

We close this section by analyzing eigenvectors and eigenvalues from a geometric viewpoint. If x is an eigenvector of the linear operator T on V, then $\mathsf{T}(x) = \lambda x$ for some scalar λ. Let $\mathsf{W} = \text{span}(\{x\})$ be the one-dimensional subspace of V spanned by x. If $y \in \mathsf{W}$, then $y = cx$ for some scalar c. So

$$\mathsf{T}(y) = \mathsf{T}(cx) = c\mathsf{T}(x) = c\lambda x = \lambda y \in \mathsf{W}.$$

Thus T maps W into itself. If V is a vector space over the field of real numbers, then W can be regarded as a line passing through the origin of V (i.e., through $\mathit{0}$). The operator T acts on elements of W by multiplying each element by the scalar λ. There are several possibilities for the action of T depending on the value of λ (see Figure 5.2).

CASE 1. If $\lambda > 1$, then T moves elements of W to points farther from $\mathit{0}$ by a factor of λ.

CASE 2. If $\lambda = 1$, then T acts as the identity transformation on W.

CASE 3. If $0 < \lambda < 1$, T moves elements of W to points closer to $\mathit{0}$ by a factor of λ.

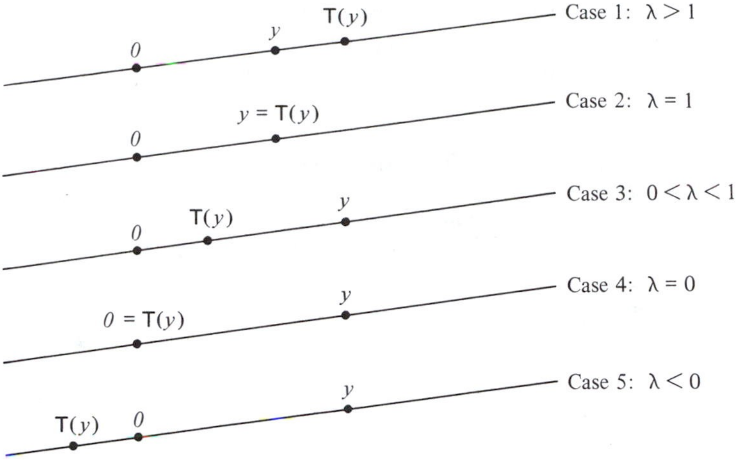

The action of T on W = span({x}) when x is an eigenvector of T.

Figure 5.2

CASE 4. If $\lambda = 0$, then T acts as the zero transformation on W.

CASE 5. If $\lambda < 0$, then T reverses the orientation of W; that is, T moves points of W from one side of *0* to the other.

To illustrate these ideas, consider the linear operators introduced in Examples 5, 6, and 7 of Section 2.1. Recall that the operator T: $\mathsf{R}^2 \to \mathsf{R}^2$ defined by $\mathsf{T}(x_1, x_2) = (x_1, -x_2)$ is a reflection about the x-axis. It is easily seen that T maps both axes onto themselves; thus e_1 and e_2 are eigenvectors of T (corresponding to the eigenvalues 1 and -1, respectively). Observe that T acts as the identity on the x-axis and reverses the orientation of the y-axis. Next consider the projection on the x-axis defined by $\mathsf{U}(x_1, x_2) = (x_1, 0)$. Again it is geometrically clear that U acts as the identity on the x-axis and acts as the zero transformation on the y-axis. This behavior is a consequence of the fact that e_1 and e_2 are eigenvectors of U corresponding to the eigenvalues 1 and 0, respectively. Finally, recall that the rotation through the angle θ is the operator T_θ: $\mathsf{R}^2 \to \mathsf{R}^2$ defined by $\mathsf{T}_\theta(x_1, x_2) = (x_1 \cos\theta - x_2 \sin\theta,\ x_1 \sin\theta + x_2 \cos\theta)$. If $0 < \theta < \pi$, then it is geometrically clear that T_θ maps no one-dimensional subspace of R^2 into itself. This observation implies that T_θ has no eigenvectors (and hence no eigenvalues). To confirm this conclusion using Corollary 2 of Theorem 5.8, we note that the characteristic polynomial of T_θ is

$$\det(\mathsf{T}_\theta - t\mathsf{I}) = \det\begin{pmatrix} \cos\theta - t & -\sin\theta \\ \sin\theta & \cos\theta - t \end{pmatrix} = t^2 - (2\cos\theta)t + 1,$$

which has no real zeros since the discriminant $4\cos^2\theta - 4$ is negative for $0 < \theta < \pi$. Thus there exist operators (and hence matrices) with no eigenvalues or eigenvectors. Of course, such operators and matrices are not diagonalizable.

EXERCISES

1. Label the following statements as being true or false.
   (a) Every linear operator on an n-dimensional vector space has n distinct eigenvalues.
   (b) If a real matrix has one eigenvector, then it has an infinite number of eigenvectors.
   (c) There exists a square matrix with no eigenvectors.
   (d) Eigenvalues must be nonzero scalars.
   (e) Any two eigenvectors are linearly independent.
   (f) The sum of two eigenvalues of a linear operator T is also an eigenvalue of T.
   (g) Linear operators on infinite-dimensional vector spaces never have eigenvalues.
   (h) An $n \times n$ matrix A with entries from a field F is similar to a diagonal matrix if and only if there is a basis for F^n consisting of eigenvectors of A.
   (i) Similar matrices always have the same eigenvalues.
   (j) Similar matrices always have the same eigenvectors.
   (k) The sum of two eigenvectors of an operator T is always an eigenvector of T.
2. For each matrix A and basis β find $[\mathsf{L}_A]_\beta$. Also find an invertible matrix Q such that $[\mathsf{L}_A]_\beta = Q^{-1}AQ$.

   (a) $$A = \begin{pmatrix} 1 & 3 \\ 1 & 1 \end{pmatrix} \quad \text{and} \quad \beta = \left\{ \begin{pmatrix} 1 \\ 1 \end{pmatrix}, \begin{pmatrix} 1 \\ 2 \end{pmatrix} \right\}$$

   (b) $$A = \begin{pmatrix} 1 & 2 \\ 2 & 1 \end{pmatrix} \quad \text{and} \quad \beta = \left\{ \begin{pmatrix} 1 \\ 1 \end{pmatrix}, \begin{pmatrix} 1 \\ -1 \end{pmatrix} \right\}$$

   (c) $$A = \begin{pmatrix} 1 & 1 & -1 \\ 2 & 0 & 1 \\ 1 & 1 & 0 \end{pmatrix} \quad \text{and} \quad \beta = \left\{ \begin{pmatrix} 1 \\ 1 \\ 1 \end{pmatrix}, \begin{pmatrix} 1 \\ 0 \\ 1 \end{pmatrix}, \begin{pmatrix} 1 \\ 1 \\ 2 \end{pmatrix} \right\}$$

   (d) $$A = \begin{pmatrix} 13 & 1 & 4 \\ 1 & 13 & 4 \\ 4 & 4 & 10 \end{pmatrix} \quad \text{and} \quad \beta = \left\{ \begin{pmatrix} 1 \\ 1 \\ -2 \end{pmatrix}, \begin{pmatrix} 1 \\ -1 \\ 0 \end{pmatrix}, \begin{pmatrix} 1 \\ 1 \\ 1 \end{pmatrix} \right\}$$
3. For each of the following matrices $A \in \mathsf{M}_{n \times n}(F)$
   (1) Determine all the eigenvalues of A.

(2) For each eigenvalue λ of A, find the set of eigenvectors corresponding to λ.
(3) If possible, find a basis for F^n consisting of eigenvectors of A.
(4) If successful in finding a basis, determine an invertible matrix Q and a diagonal matrix D such that $Q^{-1}AQ = D$.

(a) $A = \begin{pmatrix} 1 & 2 \\ 3 & 2 \end{pmatrix}$ for $F = R$

(b) $A = \begin{pmatrix} 0 & -2 & -3 \\ -1 & 1 & -1 \\ 2 & 2 & 5 \end{pmatrix}$ for $F = R$

(c) $A = \begin{pmatrix} i & 1 \\ 2 & -i \end{pmatrix}$ for $F = C$

(d) $A = \begin{pmatrix} 2 & 0 & -1 \\ 4 & 1 & -4 \\ 2 & 0 & -1 \end{pmatrix}$ for $F = R$

4. Let $\mathsf{T}\colon \mathsf{P}_2(R) \to \mathsf{P}_2(R)$ be defined by $\mathsf{T}(f(x)) = f(x) + xf'(x)$. Find all the eigenvalues of T, and find a basis β for $\mathsf{P}_2(R)$ such that $[\mathsf{T}]_\beta$ is a diagonal matrix.

5. Prove parts (a), (b), and (c) of Theorem 5.6.

6. Prove Corollaries 1 and 2 of Theorem 5.7.

7. Prove Theorem 5.9.

8. **(a)** Prove that a linear operator T on a finite-dimensional vector space is invertible if and only if zero is not an eigenvalue of T.
(b) Let T be an invertible linear operator. Prove that a scalar λ is an eigenvalue of T if and only if λ^{-1} is an eigenvalue of T^{-1}.

9. Prove that the eigenvalues of an upper triangular matrix M are the diagonal entries of M.

10. Let V be a finite-dimensional vector space and λ be any scalar.
(a) For any basis β for V prove that $[\lambda \mathsf{I}_\mathsf{V}]_\beta = \lambda I$.
(b) Compute the characteristic polynomial of $\lambda \mathsf{I}_\mathsf{V}$.
(c) Show that $\lambda \mathsf{I}_\mathsf{V}$ is diagonalizable and has only one eigenvalue.

11. A *scalar matrix* is a square matrix of the form λI for some scalar λ; i.e., a scalar matrix is a diagonal matrix in which all the diagonal entries are equal.
(a) Prove that if A is similar to a scalar matrix λI, then $A = \lambda I$.
(b) Show that a diagonalizable matrix having only one eigenvalue is a scalar matrix.

(c) Conclude that the matrix

$$\begin{pmatrix} 1 & 1 \\ 0 & 1 \end{pmatrix}$$

is not diagonalizable.

12. (a) Prove that similar matrices have the same characteristic polynomial.
(b) Show that the definition of the characteristic polynomial of a linear operator on a finite-dimensional vector space V is independent of the choice of basis for V.

13. Prove the following assertions made in Example 8.
(a) If $v \in \mathsf{P}_2(R)$ and $\phi_\beta(v)$ is an eigenvector of A corresponding to the eigenvalue λ, then v is an eigenvector of T corresponding to the eigenvalue λ.
(b) If λ is an eigenvalue of A (and hence of T), then a vector $y \in \mathsf{R}^3$ is an eigenvector of A corresponding to λ if and only if $\phi_\beta^{-1}(y)$ is an eigenvector of T corresponding to λ.

14.† For any square matrix A, prove that A and A^t have the same characteristic polynomial (and hence the same eigenvalues).

15.† (a) Let T be a linear operator on a vector space V, and let x be an eigenvector of T corresponding to the eigenvalue λ. For any positive integer m, prove that x is an eigenvector of T^m corresponding to the eigenvalue λ^m.
(b) State and prove the result for matrices that is analogous to that in part (a).

16. (a) Prove that similar matrices have the same trace. *Hint:* Use Exercise 12 of Section 2.3.
(b) How would you define the trace of a linear operator on a finite-dimensional vector space? Justify that your definition is well-defined.

17. Let $\mathsf{T}: \mathsf{M}_{n \times n}(F) \to \mathsf{M}_{n \times n}(F)$ be the mapping defined by $\mathsf{T}(A) = A^t$, the transpose of A.
(a) Verify that T is a linear operator on $\mathsf{M}_{n \times n}(F)$.
(b) Show that ± 1 are the only eigenvalues of T.
(c) Describe the matrices that are eigenvectors corresponding to the eigenvalues 1 and -1, respectively.

18. Show that for any $A, B \in \mathsf{M}_{n \times n}(C)$ such that B is invertible, there exists a scalar $c \in C$ such that $A + cB$ is not invertible. *Hint:* Examine $\det(A + cB)$.

19.† Let A and B be similar $n \times n$ matrices. Prove that there exists an n-dimensional vector space V, a linear operator T on V, and bases β and γ for V such that $A = [\mathsf{T}]_\beta$ and $B = [\mathsf{T}]_\gamma$. *Hint:* Use Exercise 13 of Section 2.5.

20. Let A be an $n \times n$ matrix with characteristic polynomial

$$f(t) = (-1)^n t^n + a_{n-1} t^{n-1} + \cdots + a_1 t + a_0.$$

Prove that $f(0) = a_0 = \det(A)$. Deduce that A is invertible if and only if $a_0 \neq 0$.

21. Let A and $f(t)$ be as in Exercise 20.

(a) Prove that $f(t) = (A_{11} - t)(A_{22} - t)\cdots(A_{nn} - t) + q(t)$, where $q(t)$ is a polynomial in t of degree at most $n - 2$.

(b) Show that $\text{tr}(A) = (-1)^{n-1}a_{n-1}$.

22.† Let T be a linear operator on a finite-dimensional vector space V over the field F. Prove that if $g(t) \in \mathsf{P}(F)$ and x is an eigenvector of T corresponding to eigenvalue λ, then $g(\mathsf{T})(x) = g(\lambda)x$.

23. Use Exercise 22 to prove that if $f(t)$ is the characteristic polynomial of a diagonalizable linear operator T, then $f(\mathsf{T}) = \mathsf{T}_0$, the zero operator. (As we will see in Section 5.4, this result does not depend on the diagonalizability of T.)

24. Prove Theorem 5.8.

25. Find the number of characteristic polynomials for matrices in $\mathsf{M}_{2\times 2}(Z_2)$.

5.2 DIAGONALIZABILITY

In Section 5.1 we presented the diagonalization problem and observed that not all linear operators or matrices are diagonalizable. Although we were able to diagonalize certain operators and matrices and even obtained a necessary and sufficient condition for diagonalizability (Theorem 5.4), we have not solved the diagonalization problem. What is still needed is a simple test to determine if an operator or a matrix can be diagonalized and, if so, an algorithm for obtaining a basis of eigenvectors. In this section we develop such a test and an algorithm.

In Example 7 of Section 5.1 we obtained a basis of eigenvectors by choosing one eigenvector corresponding to each eigenvalue. In general such a procedure will not yield a basis, but the following theorem shows that any set constructed in this manner must be linearly independent.

Theorem 5.10. *Let* T *be a linear operator on* V, *and let* $\lambda_1, \lambda_2, \ldots, \lambda_k$ *be distinct eigenvalues of* T. *If* $x_1, x_2, \ldots, x_k$ *are eigenvectors of* T *such that* λ_j *corresponds to* x_j $(1 \leqslant j \leqslant k)$, *then* $\{x_1, x_2, \ldots, x_k\}$ *is linearly independent.*

Proof. The proof is by mathematical induction on the integer k. Suppose that $k = 1$. Then $x_1 \neq 0$ since x_1 is an eigenvector, and hence $\{x_1\}$ is linearly independent. Assume that the theorem always holds for $k - 1$ eigenvectors, where $k - 1 \geqslant 1$, and that we have k eigenvectors $x_1, \ldots, x_k$ corresponding to distinct eigenvalues $\lambda_1, \ldots, \lambda_k$. We wish to show that $\{x_1, \ldots, x_k\}$ is linearly independent. Suppose that there are scalars $a_1, \ldots, a_k$ such that

$$a_1x_1 + \cdots + a_kx_k = 0. \tag{1}$$

Applying $\mathsf{T} - \lambda_k \mathsf{I}$ to both sides of (1), we obtain

$$a_1(\lambda_1 - \lambda_k)x_1 + \cdots + a_{k-1}(\lambda_{k-1} - \lambda_k)x_{k-1} = 0.$$

By the induction hypothesis $\{x_1, \ldots, x_{k-1}\}$ is linearly independent; hence

$$a_1(\lambda_1 - \lambda_k) = \cdots = a_{k-1}(\lambda_{k-1} - \lambda_k) = 0.$$

Since $\lambda_1, \ldots, \lambda_k$ are distinct, it follows that $\lambda_i - \lambda_k \neq 0$ for $1 \leqslant i \leqslant k-1$. So $a_1 = \cdots = a_{k-1} = 0$. Thus (1) reduces to $a_k x_k = 0$. Since $x_k \neq 0$, $a_k = 0$. Therefore $a_1 = \cdots = a_k = 0$, and hence $\{x_1, \ldots, x_k\}$ is linearly independent. ∎

Corollary. *Let* T *be a linear operator on* V, *a finite-dimensional vector space of dimension* n. *If* T *has* n *distinct eigenvalues, then* T *is diagonalizable.*

Proof. Let T have n distinct eigenvalues $\lambda_1, \ldots, \lambda_n$, and let $x_1, \ldots, x_n$ be eigenvectors of T such that λ_j corresponds to x_j for $1 \leqslant j \leqslant n$. By Theorem 5.10 $\{x_1, \ldots, x_n\}$ is linearly independent, and since $\dim(\mathsf{V}) = n$, this set is a basis for V. Thus, by Theorem 5.4, T is diagonalizable. ∎

Example 1

Let

$$A = \begin{pmatrix} 1 & 1 \\ 1 & 1 \end{pmatrix} \in \mathsf{M}_{2 \times 2}(R).$$

The characteristic polynomial of A (and hence of L_A) is

$$\det(A - tI) = \det\begin{pmatrix} 1-t & 1 \\ 1 & 1-t \end{pmatrix} = t(t-2),$$

and thus the eigenvalues of L_A are 0 and 2. Since L_A is a linear operator on the two-dimensional vector space R^2, we conclude from the corollary above that L_A (and hence A) is diagonalizable. ∎

Although the corollary to Theorem 5.10 provides a sufficient condition for diagonalizability, this condition is not necessary. In fact the identity operator is diagonalizable but has only one eigenvalue, namely $\lambda = 1$.

We have seen that diagonalizability requires the existence of eigenvalues. Actually, diagonalizability imposes a much stronger condition on the characteristic polynomial.

Definition. *A polynomial* f(x) *in* P(F) splits (*in* F) *if there are scalars* $a_0, a_1, \ldots, a_n$ (*not necessarily distinct*) *in* F *such that*

$$f(x) = a_0(x - a_1)(x - a_2)\cdots(x - a_n).$$

For example, $x^2 - 1 = (x+1)(x-1)$ splits in R, but $(x^2 + 1)(x - 2)$ does not split in R because $x^2 + 1$ cannot be factored into linear factors. However, $(x^2 + 1)(x - 2)$ does split in C because it factors into the product $(x + i)(x - i)(x - 2)$. If $f(t)$ is the characteristic polynomial of a linear operator or a matrix, then the field under consideration will be the field associated with the operator or the matrix.

Theorem 5.11. *The characteristic polynomial of any diagonalizable linear operator* T *splits.*

Proof. Suppose that T is diagonalizable. Then there exists a basis β for V such that $[\mathsf{T}]_\beta = D$ is a diagonal matrix. If

$$D = \begin{pmatrix} \lambda_1 & 0 & \cdots & 0 \\ 0 & \lambda_2 & \cdots & 0 \\ \vdots & \vdots & & \vdots \\ 0 & 0 & \cdots & \lambda_n \end{pmatrix},$$

and $f(t)$ is the characteristic polynomial of T, then

$$f(t) = \det(D - tI) = \det \begin{pmatrix} \lambda_1 - t & 0 & \cdots & 0 \\ 0 & \lambda_2 - t & \cdots & 0 \\ \vdots & \vdots & & \vdots \\ 0 & 0 & \cdots & \lambda_n - t \end{pmatrix}$$

$$= (\lambda_1 - t)(\lambda_2 - t)\cdots(\lambda_n - t) = (-1)^n(t - \lambda_1)(t - \lambda_2)\cdots(t - \lambda_n). \quad \blacksquare$$

From this theorem it is clear that if T is a diagonalizable linear operator on an n-dimensional vector space that fails to have n distinct eigenvalues, then the characteristic polynomial of T must have repeated zeros. This observation leads us to the following definition.

Definition. *Let λ be an eigenvalue of a linear operator or matrix with characteristic polynomial* f(t). *The* (algebraic) multiplicity *of λ is the largest positive integer* k *for which* $(t - \lambda)^k$ *is a factor of* f(t).

Example 2

Let

$$A = \begin{pmatrix} 1 & 1 & 0 \\ 0 & 1 & 3 \\ 0 & 0 & 2 \end{pmatrix}.$$

If $f(t)$ is the characteristic polynomial of A, then $f(t) = -(t - 1)^2(t - 2)$. Hence $\lambda = 1$ is an eigenvalue of A with multiplicity 2, and $\lambda = 2$ is an eigenvalue of A with multiplicity 1. $\blacksquare$

If T is a diagonalizable linear operator on a finite-dimensional vector space V, then there is a basis β for V consisting of eigenvectors of T. We know from Theorem 5.4 that $[\mathsf{T}]_\beta$ is a diagonal matrix in which the diagonal entries are the eigenvalues of T. Since the characteristic polynomial of T is $\det([\mathsf{T}]_\beta - tI)$, it is easily seen that each eigenvalue of T must occur as a diagonal entry of $[\mathsf{T}]_\beta$ exactly as many times as its multiplicity. Hence β contains as many (linearly independent) eigenvectors corresponding to an eigenvalue as the multiplicity of that eigenvalue. Thus we see that the number of linearly independent eigenvectors corresponding to a given eigenvalue is of great interest in determining when an operator can be diagonalized. Recalling from Theorem 5.9 that the eigenvectors of T corresponding to the eigenvalue λ are the nonzero vectors in the null space of $\mathsf{T} - \lambda\mathsf{I}$, we are led naturally to the study of this set.

Definition. *Let* T *be a linear operator on a vector space* V, *and let* λ *be an eigenvalue of* T. *Define* $\mathsf{E}_\lambda = \{x \in \mathsf{V}: \mathsf{T}(x) = \lambda x\} = \mathsf{N}(\mathsf{T} - \lambda\mathsf{I}_\mathsf{V})$. *The set* E_λ *is called the* eigenspace of T corresponding to the eigenvalue λ. *As expected, by an* eigenspace *of a matrix* A, *we mean the corresponding eigenspace of the operator* L_A.

Clearly, E_λ is a subspace of V consisting of the zero vector and the eigenvectors of T corresponding to the eigenvalue λ. The number of linearly independent eigenvectors of T corresponding to the eigenvalue λ is therefore the dimension of E_λ. Our next result relates this dimension to the multiplicity of λ.

Theorem 5.12. *Let* T *be a linear operator on a finite-dimensional vector space* V. *If* λ *is an eigenvalue of* T *having multiplicity* m, *then* $1 \leqslant \dim(\mathsf{E}_\lambda) \leqslant m$.

Proof. Pick a basis $\{x_1, \ldots, x_p\}$ for E_λ, and extend it to a basis $\beta = \{x_1, \ldots, x_p, x_{p+1}, \ldots, x_n\}$ for V. Observe that x_i $(1 \leqslant i \leqslant p)$ is an eigenvector of T corresponding to λ, and let $A = [\mathsf{T}]_\beta$. Then A may be written in the form

$$A = \begin{pmatrix} \lambda I_p & B \\ O & C \end{pmatrix}.$$

By Exercise 9 of Section 4.3 the characteristic polynomial of T is

$$\begin{aligned} f(t) = \det(A - tI_n) &= \det\begin{pmatrix} (\lambda - t)I_p & B \\ O & C - tI_{n-p} \end{pmatrix} \\ &= \det((\lambda - t)I_p)\det(C - tI_{n-p}) \\ &= (-1)^p(\lambda - t)^p g(t), \end{aligned}$$

where $g(t)$ is a polynomial. Hence $(\lambda - t)^p$ is a factor of $f(t)$, and the multiplicity of λ is at least p. But $\dim(\mathsf{E}_\lambda) = p$; so $\dim(\mathsf{E}_\lambda) \leqslant m$. ∎

Example 3

Let $\mathsf{T}: \mathsf{P}_2(R) \to \mathsf{P}_2(R)$ be the linear operator defined by $\mathsf{T}(f) = f'$, the derivative of f. The matrix of T with respect to the standard basis for $\mathsf{P}_2(R)$ is

$$[\mathsf{T}]_\beta = \begin{pmatrix} 0 & 1 & 0 \\ 0 & 0 & 2 \\ 0 & 0 & 0 \end{pmatrix}.$$

Consequently the characteristic polynomial of T is

$$\det([\mathsf{T}]_\beta - tI) = \det \begin{pmatrix} -t & 1 & 0 \\ 0 & -t & 2 \\ 0 & 0 & -t \end{pmatrix} = -t^3.$$

Thus T has only one eigenvalue ($\lambda = 0$) with multiplicity 3. So $\mathsf{E}_\lambda = \mathsf{N}(\mathsf{T} - \lambda\mathsf{I}) = \mathsf{N}(\mathsf{T})$. Hence E_λ is the subspace of $\mathsf{P}_2(R)$ containing the constant polynomials. So in this case $\{1\}$ is a basis for E_λ, and $\dim(\mathsf{E}_\lambda) = 1$. Consequently there is no basis for $\mathsf{P}_2(R)$ consisting of eigenvectors of T, so that T is not diagonalizable. ▮

Example 4

Let T be the linear operator on R^3 defined by

$$\mathsf{T}\begin{pmatrix} a_1 \\ a_2 \\ a_3 \end{pmatrix} = \begin{pmatrix} 4a_1 \qquad\quad + a_3 \\ 2a_1 + 3a_2 + 2a_3 \\ a_1 \qquad\quad + 4a_3 \end{pmatrix}.$$

We will determine the eigenspace of T corresponding to each eigenvalue. If β is the standard basis for R^3, then

$$[\mathsf{T}]_\beta = \begin{pmatrix} 4 & 0 & 1 \\ 2 & 3 & 2 \\ 1 & 0 & 4 \end{pmatrix}.$$

Hence the characteristic polynomial of T is

$$\det([\mathsf{T}]_\beta - tI) = \det \begin{pmatrix} 4-t & 0 & 1 \\ 2 & 3-t & 2 \\ 1 & 0 & 4-t \end{pmatrix} = -(t-5)(t-3)^2.$$

So the eigenvalues of T are $\lambda_1 = 5$ and $\lambda_2 = 3$ with multiplicities 1 and 2, respectively.

Since

$$\mathsf{E}_{\lambda_1} = \mathsf{N}(\mathsf{T} - \lambda_1\mathsf{I}) = \left\{ \begin{pmatrix} x_1 \\ x_2 \\ x_3 \end{pmatrix} \in \mathsf{R}^3 \colon \begin{pmatrix} -1 & 0 & 1 \\ 2 & -2 & 2 \\ 1 & 0 & -1 \end{pmatrix} \begin{pmatrix} x_1 \\ x_2 \\ x_3 \end{pmatrix} = \begin{pmatrix} 0 \\ 0 \\ 0 \end{pmatrix} \right\},$$

E_{λ_1} is the solution space of the system of equations

$$\begin{cases} -x_1 \qquad\quad + \ x_3 = 0 \\ 2x_1 - 2x_2 + 2x_3 = 0 \\ \ \ x_1 \qquad\quad - \ x_3 = 0. \end{cases}$$

It is easily seen (using the techniques of Chapter 3) that

$$\left\{ \begin{pmatrix} 1 \\ 2 \\ 1 \end{pmatrix} \right\}$$

is a basis for E_{λ_1}. Hence $\dim(\mathsf{E}_{\lambda_1}) = 1$.

Similarly, $\mathsf{E}_{\lambda_2} = \mathsf{N}(\mathsf{T} - \lambda_2\mathsf{I})$ is the solution space of the system

$$\begin{cases} x_1 + \ x_3 = 0 \\ 2x_1 + 2x_3 = 0 \\ x_1 + \ x_3 = 0. \end{cases}$$

So

$$\left\{ \begin{pmatrix} 1 \\ 0 \\ -1 \end{pmatrix}, \begin{pmatrix} 0 \\ 1 \\ 0 \end{pmatrix} \right\}$$

is a basis for E_{λ_2}, and $\dim(\mathsf{E}_{\lambda_2}) = 2$.

In this case the multiplicity of each eigenvalue λ_i equals the dimension of the corresponding eigenspace E_{λ_i}. Observe that

$$\left\{ \begin{pmatrix} 1 \\ 2 \\ 1 \end{pmatrix}, \begin{pmatrix} 1 \\ 0 \\ -1 \end{pmatrix}, \begin{pmatrix} 0 \\ 1 \\ 0 \end{pmatrix} \right\}$$

is a basis for R^3 consisting of eigenvectors of T. Consequently, T is diagonalizable. ∎

Examples 3 and 4 suggest that for a linear operator T whose characteristic polynomial splits, diagonalizability of T is equivalent to equality of the dimension of the eigenspace and multiplicity of the eigenvalue for each eigenvalue of T. This is indeed true as we now show. We begin with the following lemma, which is a slight variation of Theorem 5.10.

Lemma. *Let* T *be a linear operator on a vector space* V, *and let* $\lambda_1, \lambda_2, \ldots, \lambda_k$ *be distinct eigenvalues of* T. *For each* $i = 1, 2, \ldots, k$, *let* $x_i \in E_{\lambda_i}$, *the eigenspace corresponding to* λ_i. *If*

$$x_1 + x_2 + \cdots + x_k = 0,$$

then $x_i = 0$ *for all* i.

Proof. Suppose otherwise. By renumbering if necessary suppose that for some integer m ($1 \leqslant m \leqslant k$) we have that $x_i \neq 0$ for $1 \leqslant i \leqslant m$, and $x_i = 0$ for $i > m$. Then for each $i \leqslant m$ we have that x_i is an eigenvector of T corresponding to λ_i and

$$x_1 + x_2 + \cdots + x_m = 0.$$

But this contradicts Theorem 5.10, which states that these x_i's form a linearly independent set. We therefore conclude that $x_i = 0$ for all i. ∎

The next two theorems give us the information for recognizing when a linear operator is diagonalizable and for choosing a basis for the vector space consisting of eigenvectors of the operator.

Theorem 5.13. *Let* T *be a linear operator on a vector space* V, *and let* $\lambda_1, \lambda_2, \ldots, \lambda_k$ *be distinct eigenvalues of* T. *For each* $i = 1, 2, \ldots, k$ *let* S_i *be a finite linearly independent subset of the eigenspace* E_{λ_i}. *Then* $S = S_1 \cup S_2 \cup \cdots \cup S_k$ *is a linearly independent subset of* V.

Proof. Suppose that for each i,

$$S_i = \{x_{i1}, x_{i2}, \ldots, x_{in_i}\}.$$

Then $S = \{x_{ij}\colon 1 \leqslant j \leqslant n_i, \text{ and } 1 \leqslant i \leqslant k\}$. Consider any scalars $\{a_{ij}\}$ such that

$$\sum_{i=1}^{k} \sum_{j=1}^{n_i} a_{ij} x_{ij} = 0.$$

For each i let

$$y_i = \sum_{j=1}^{n_i} a_{ij} x_{ij}.$$

Then $y_i \in E_{\lambda_i}$ for each i and $y_1 + \cdots + y_k = 0$. Therefore, by the lemma, $y_i = 0$ for all i. But S_i is linearly independent for all i. Thus, for each i, it now follows that $a_{ij} = 0$ for all j. We conclude that S is linearly independent. ∎

Theorem 5.13 tells us how to build up a linearly independent subset of eigenvectors, namely by collecting bases for the individual eigenspaces. The next theorem tells us when the result will be a basis for V.

Theorem 5.14. *Let T be a linear operator on a finite-dimensional vector space V such that the characteristic polynomial of T splits. Let $\lambda_1, \lambda_2, \ldots, \lambda_k$ be the distinct eigenvalues of T. Then*

(a) *T is diagonalizable if and only if the multiplicity of λ_i is equal to* $\dim(\mathsf{E}_{\lambda_i})$ *for all* i.

(b) *If T is diagonalizable and S_i is a basis for E_{λ_i} for each* i, *then $\beta = S_1 \cup S_2 \cup \cdots \cup S_k$ is a basis for V consisting of eigenvectors of T.*

Proof. For each i, let m_i denote the multiplicity of λ_i, $d_i = \dim(\mathsf{E}_{\lambda_i})$, and $n = \dim(\mathsf{V})$.

First, suppose that T is diagonalizable. Let β be a basis for V consisting of eigenvectors of T. For each i let $\beta_i = \beta \cap \mathsf{E}_{\lambda_i}$, the set of vectors in β that are eigenvectors corresponding to λ_i, and let n_i denote the number of vectors in β_i. Then $n_i \leqslant d_i$ for each i because β_i is a linearly independent subset of a space of dimension d_i, and $d_i \leqslant m_i$ for all i by Theorem 5.12. Thus

$$n = \sum_{i=1}^{k} n_i \leqslant \sum_{i=1}^{k} d_i \leqslant \sum_{i=1}^{k} m_i = n.$$

The n_i's sum to n because β contains n elements. The m_i's sum to n because the degree of the characteristic polynomial of T is equal to the sum of the multiplicities of the eigenvalues. It follows that

$$\sum_{i=1}^{k} (m_i - d_i) = 0 \quad \text{and} \quad (m_i - d_i) \geqslant 0 \qquad \text{for each } i.$$

From this we conclude that $m_i = d_i$ for all i.

Conversely, suppose that $m_i = d_i$ for all i. We will show that T is diagonalizable, and in the process we also prove (b). For each i, let S_i be a basis for E_{λ_i}, and let $\beta = S_1 \cup S_2 \cup \cdots \cup S_k$. Then by Theorem 5.13 β is a linearly independent subset of V. Since $d_i = m_i$ for all i, we have that β contains

$$\sum_{i=1}^{k} d_i = \sum_{i=1}^{k} m_i = n$$

elements. Therefore, β is a basis for V consisting of eigenvectors of V. We conclude that T is diagonalizable. ∎

This theorem completes our study of the diagonalization problem. We summarize some of our previous results in the following test and algorithm.

A Test for Diagonalizability

Let T be a linear operator on an n-dimensional vector space. Then T is diagonalizable if and only if both of the following conditions hold.

1. The characteristic polynomial of T splits.
2. The multiplicity of λ equals $n - \text{rank}(\mathsf{T} - \lambda \mathsf{I})$ for each eigenvalue λ of T.

Observe that condition 2 makes use of the dimension theorem. Also observe that condition 2 is automatically satisfied for eigenvalues having multiplicity 1 (Theorem 5.12). Thus condition 2 need only be checked for those eigenvalues having multiplicity greater than 1.

Since diagonalizability of a matrix A is equivalent to diagonalizability of the operator L_A, a similar test holds for matrices.

An Algorithm for Diagonalization

Let T be a diagonalizable linear operator on a finite-dimensional vector space V, and let $\lambda_1, \ldots, \lambda_k$ denote the distinct eigenvalues of T. For each j, let β_j be a basis for $\mathsf{E}_{\lambda_j} = \mathsf{N}(\mathsf{T} - \lambda_j \mathsf{I})$ and $\beta = \beta_1 \cup \beta_2 \cup \cdots \cup \beta_k$. Then β is a basis for V, and $[\mathsf{T}]_\beta$ is a diagonal matrix.

Example 5

We will test the matrix

$$A = \begin{pmatrix} 3 & 1 & 0 \\ 0 & 3 & 0 \\ 0 & 0 & 4 \end{pmatrix} \in \mathsf{M}_{3 \times 3}(R)$$

for diagonalizability.

The characteristic polynomial of A is $\det(A - tI) = -(t-4)(t-3)^2$. Hence A has eigenvalues $\lambda_1 = 4$ and $\lambda_2 = 3$ with multiplicities 1 and 2, respectively. Clearly condition 1 of the test for diagonalizability is satisfied, and since λ_1 has multiplicity 1, condition 2 is satisfied for λ_1. Thus we need only check condition 2 of the test for λ_2. Since

$$B = A - \lambda_2 I = \begin{pmatrix} 0 & 1 & 0 \\ 0 & 0 & 0 \\ 0 & 0 & 1 \end{pmatrix}$$

has rank 2, $3 - \text{rank}(B) = 1$. Thus condition 2 of the test fails for λ_2, and consequently A is not diagonalizable. ▮

Example 6

Let $\mathsf{T}\colon \mathsf{R}^3 \to \mathsf{R}^3$ be defined by

$$\mathsf{T}\begin{pmatrix} a \\ b \\ c \end{pmatrix} = \begin{pmatrix} -2b - 3c \\ a + 3b + 3c \\ c \end{pmatrix}.$$

We will test T for diagonalizability. Letting γ denote the standard basis for R^3, we have

$$[\mathsf{T}]_\gamma = \begin{pmatrix} 0 & -2 & -3 \\ 1 & 3 & 3 \\ 0 & 0 & 1 \end{pmatrix}.$$

The characteristic polynomial of T is $-(t-1)^2(t-2)$. Thus T has two eigenvalues: $\lambda_1 = 1$ with multiplicity 2 and $\lambda_2 = 2$ with multiplicity 1. Note that condition 1 of the test for diagonalizability is satisfied. We now consider condition 2.

For $\lambda_1 = 1$,

$$3 - \text{rank}(\mathsf{T} - \lambda_1 \mathsf{I}) = 3 - \text{rank} \begin{pmatrix} -1 & -2 & -3 \\ 1 & 2 & 3 \\ 0 & 0 & 0 \end{pmatrix} = 3 - 1 = 2.$$

Thus the dimension of E_{λ_1} is the same as the multiplicity of λ_1. Since λ_2 has multiplicity 1, the dimension of E_{λ_2} is equal to the multiplicity of λ_2. Hence T is diagonalizable.

We now find a basis β for R^3 such that $[\mathsf{T}]_\beta$ is a diagonal matrix. Since

$$\mathsf{E}_{\lambda_1} = \mathsf{N}(\mathsf{T} - \lambda_1 \mathsf{I}) = \left\{ \begin{pmatrix} x_1 \\ x_2 \\ x_3 \end{pmatrix} \in \mathsf{R}^3 \colon \begin{pmatrix} -1 & -2 & -3 \\ 1 & 2 & 3 \\ 0 & 0 & 0 \end{pmatrix} \begin{pmatrix} x_1 \\ x_2 \\ x_3 \end{pmatrix} = 0 \right\},$$

E_{λ_1} is the solution set of

$$\begin{cases} -x_1 - 2x_2 - 3x_3 = 0 \\ x_1 + 2x_2 + 3x_3 = 0, \end{cases}$$

which has

$$\beta_1 = \left\{ \begin{pmatrix} -2 \\ 1 \\ 0 \end{pmatrix}, \begin{pmatrix} -3 \\ 0 \\ 1 \end{pmatrix} \right\}$$

as a basis. Also,

$$\mathsf{E}_{\lambda_2} = \mathsf{N}(\mathsf{T} - \lambda_2 \mathsf{I}) = \left\{ \begin{pmatrix} x_1 \\ x_2 \\ x_3 \end{pmatrix} \in \mathsf{R}^3 \colon \begin{pmatrix} -2 & -2 & -3 \\ 1 & 1 & 3 \\ 0 & 0 & -1 \end{pmatrix} \begin{pmatrix} x_1 \\ x_2 \\ x_3 \end{pmatrix} = 0 \right\}.$$

Thus E_{λ_2} is the solution set of

$$\begin{cases} -2x_1 - 2x_2 - 3x_3 = 0 \\ x_1 + x_2 + 3x_3 = 0 \\ - x_3 = 0, \end{cases}$$

which has

$$\beta_2 = \left\{ \begin{pmatrix} -1 \\ 1 \\ 0 \end{pmatrix} \right\}$$

as a basis. Let $\beta = \beta_1 \cup \beta_2$; then β is a basis for V and

$$[\mathsf{T}]_\beta = \begin{pmatrix} 1 & 0 & 0 \\ 0 & 1 & 0 \\ 0 & 0 & 2 \end{pmatrix}. \quad \blacksquare$$

Our next example is an application of diagonalization that will be of interest in Section 5.3.

Example 7

Let

$$A = \begin{pmatrix} 0 & -2 \\ 1 & 3 \end{pmatrix}.$$

We will show that A is diagonalizable and find a 2×2 matrix Q such that $Q^{-1}AQ$ is a diagonal matrix. This information will then be used to compute A^n for any positive integer n.

Recall that A is diagonalizable if and only if L_A is diagonalizable. Now the characteristic polynomial of A is $(t-1)(t-2)$. Thus A has two distinct eigenvalues, and so A is diagonalizable. To find a basis β for R^2 such that $[\mathsf{L}_A]_\beta$ is a diagonal matrix, note that L_A has eigenvalues $\lambda_1 = 1$ and $\lambda_2 = 2$. It is easily seen that

$$\left\{ \begin{pmatrix} 2 \\ -1 \end{pmatrix} \right\}$$

is a basis for E_{λ_1} and that

$$\left\{ \begin{pmatrix} 1 \\ -1 \end{pmatrix} \right\}$$

is a basis for E_{λ_2}. Thus for the basis

$$\beta = \left\{ \begin{pmatrix} 2 \\ -1 \end{pmatrix}, \begin{pmatrix} 1 \\ -1 \end{pmatrix} \right\}$$

we have

$$[\mathsf{L}_A]_\beta = \begin{pmatrix} 1 & 0 \\ 0 & 2 \end{pmatrix}.$$

Moreover, if

$$Q = \begin{pmatrix} 2 & 1 \\ -1 & -1 \end{pmatrix},$$

then

$$Q^{-1}AQ = \begin{pmatrix} 1 & 0 \\ 0 & 2 \end{pmatrix}$$

by Theorem 5.1.

Finally, since

$$Q^{-1}AQ = \begin{pmatrix} 1 & 0 \\ 0 & 2 \end{pmatrix}, \qquad A = Q\begin{pmatrix} 1 & 0 \\ 0 & 2 \end{pmatrix}Q^{-1}.$$

Thus

$$\begin{aligned} A^n &= \left[Q\begin{pmatrix} 1 & 0 \\ 0 & 2 \end{pmatrix}Q^{-1}\right]^n \\ &= Q\begin{pmatrix} 1 & 0 \\ 0 & 2 \end{pmatrix}Q^{-1}Q\begin{pmatrix} 1 & 0 \\ 0 & 2 \end{pmatrix}Q^{-1}\cdots Q\begin{pmatrix} 1 & 0 \\ 0 & 2 \end{pmatrix}Q^{-1} \\ &= Q\begin{pmatrix} 1 & 0 \\ 0 & 2 \end{pmatrix}^n Q^{-1} \\ &= Q\begin{pmatrix} 1^n & 0 \\ 0 & 2^n \end{pmatrix}Q^{-1} \\ &= \begin{pmatrix} 2 & 1 \\ -1 & -1 \end{pmatrix}\begin{pmatrix} 1 & 0 \\ 0 & 2^n \end{pmatrix}\begin{pmatrix} 1 & 1 \\ -1 & -2 \end{pmatrix} = \begin{pmatrix} 2-2^n & 2-2^{n+1} \\ -1+2^n & -1+2^{n+1} \end{pmatrix}. \quad \blacksquare \end{aligned}$$

We will now discuss an application that uses diagonalization to solve a system of differential equations.

Systems of Differential Equations

Consider the system of differential equations

$$\begin{cases} x_1' = 3x_1 + x_2 + x_3 \\ x_2' = 2x_1 + 4x_2 + 2x_3 \\ x_3' = -x_1 - x_2 + x_3, \end{cases}$$

where, for each i, $x_i = x_i(t)$ is a differentiable real-valued function of the real variable t. Clearly, this system has a solution, namely the solution in which each $x_i(t)$ is the zero function. We will determine all the solutions of this system.

Let $X: R \to \mathsf{R}^3$ be the function defined by

$$X(t) = \begin{pmatrix} x_1(t) \\ x_2(t) \\ x_3(t) \end{pmatrix}.$$

The derivative of X is defined as the function X', where

$$X'(t) = \begin{pmatrix} x_1'(t) \\ x_2'(t) \\ x_3'(t) \end{pmatrix}.$$

Letting

$$A = \begin{pmatrix} 3 & 1 & 1 \\ 2 & 4 & 2 \\ -1 & -1 & 1 \end{pmatrix}$$

be the coefficient matrix of the given system, we can rewrite the system in the matrix form $X' = AX$, where AX is the matrix product of A and X.

The reader should verify that A is diagonalizable and that if

$$Q = \begin{pmatrix} 1 & 0 & 1 \\ 0 & 1 & 2 \\ -1 & -1 & -1 \end{pmatrix},$$

then

$$Q^{-1}AQ = \begin{pmatrix} 2 & 0 & 0 \\ 0 & 2 & 0 \\ 0 & 0 & 4 \end{pmatrix}.$$

Set

$$D = \begin{pmatrix} 2 & 0 & 0 \\ 0 & 2 & 0 \\ 0 & 0 & 4 \end{pmatrix},$$

and substitute $A = QDQ^{-1}$ into $X' = AX$ to find $X' = QDQ^{-1}X$ or, equivalently, $Q^{-1}X' = DQ^{-1}X$. Define $Y: R \to \mathsf{R}^3$ by $Y(t) = Q^{-1}X(t)$. It can be shown that Y is a differentiable function and, in fact, $Y' = Q^{-1}X'$. Hence the original system can be written as $Y' = DY$.

Since D is a diagonal matrix, the system $Y' = DY$ is easy to solve. For if

$$Y(t) = \begin{pmatrix} y_1(t) \\ y_2(t) \\ y_3(t) \end{pmatrix},$$

then $Y' = DY$ can be written

$$\begin{pmatrix} y_1'(t) \\ y_2'(t) \\ y_3'(t) \end{pmatrix} = \begin{pmatrix} 2 & 0 & 0 \\ 0 & 2 & 0 \\ 0 & 0 & 4 \end{pmatrix} \begin{pmatrix} y_1(t) \\ y_2(t) \\ y_3(t) \end{pmatrix} = \begin{pmatrix} 2y_1(t) \\ 2y_2(t) \\ 4y_3(t) \end{pmatrix}.$$

The three equations

$$y_1'(t) = 2y_1(t)$$
$$y_2'(t) = 2y_2(t)$$
$$y_3'(t) = 4y_3(t)$$

are independent of each other and thus can be solved individually. It is easily seen (as in Example 2 of Section 5.1) that the general solution of these equations is $y_1(t) = c_1e^{2t}$, $y_2(t) = c_2e^{2t}$, and $y_3(t) = c_3e^{4t}$, where c_1, c_2, and c_3 are arbitrary scalars. Finally,

$$\begin{pmatrix} x_1(t) \\ x_2(t) \\ x_3(t) \end{pmatrix} = X(t) = QY(t) = \begin{pmatrix} 1 & 0 & 1 \\ 0 & 1 & 2 \\ -1 & -1 & -1 \end{pmatrix} \begin{pmatrix} c_1e^{2t} \\ c_2e^{2t} \\ c_3e^{4t} \end{pmatrix}$$
$$= \begin{pmatrix} c_1e^{2t} + c_3e^{4t} \\ c_2e^{2t} + 2c_3e^{4t} \\ -c_1e^{2t} - c_2e^{2t} - c_3e^{4t} \end{pmatrix}$$

yields the general solution of the original system. Note that this solution can be written as

$$X(t) = e^{2t}\left[c_1\begin{pmatrix} 1 \\ 0 \\ -1 \end{pmatrix} + c_2\begin{pmatrix} 0 \\ 1 \\ -1 \end{pmatrix}\right] + e^{4t}\left[c_3\begin{pmatrix} 1 \\ 2 \\ -1 \end{pmatrix}\right].$$

The expressions in brackets are simply arbitrary elements of E_{λ_1} and E_{λ_2}, respectively, where $\lambda_1 = 2$ and $\lambda_2 = 4$. Thus the general solution of the original system is $X(t) = e^{2t}z_1 + e^{4t}z_2$, where $z_1 \in \mathsf{E}_{\lambda_1}$ and $z_2 \in \mathsf{E}_{\lambda_2}$.

Direct Sums*

Let T be a linear operator on a finite-dimensional vector space V. There is a way of decomposing V into simple subspaces that offers some insight into the behavior of T. This approach is especially useful in Chapter 7, where we study nondiagonalizable linear operators. In the case of a diagonalizable operator the simple subspaces are the eigenspaces of the operator.

Definition. *Let* V *be a vector space and let* $W_1, W_2, \ldots, W_k$ *be subspaces of* V. *We define the* sum *of the subspaces,* $W_1 + W_2 + \cdots + W_k = \sum_{i=1}^{k} W_i$ *by*

$$W_1 + W_2 + \cdots + W_k = \sum_{i=1}^{k} W_i = \{x_1 + x_2 + \cdots + x_k\colon\ x_i \in W_i,\ 1 \leqslant i \leqslant k\}.$$

It is a simple exercise to show that the sum of subspaces of a vector space is a subspace.

Example 8

Let $V = R^3$, let W_1 denote the *xy*-plane, and let W_2 denote the *yz*-plane. Then $R^3 = W_1 + W_2$ because for any vector (a, b, c) in R^3, we have that

$$(a, b, c) = (a, 0, 0) + (0, b, c)$$

and $(a, 0, 0) \in W_1$, and $(0, b, c) \in W_2$. ∎

Notice that in Example 8 the representation of (a, b, c) as a sum of vectors in W_1 and W_2 is not unique. For example, $(a, b, c) = (a, b, 0) + (0, 0, c)$ is another representation. We are interested in sums for which representations are unique. We now impose a constraint on sums to assure us of this outcome.

The definition of *direct sum* that follows is a generalization of the definition given in the exercises of Section 1.3.

Definition. *Let* $W_1, W_2, \ldots, W_k$ *be subspaces of a vector space* V. *We write* $V = W_1 \oplus W_2 \oplus \cdots \oplus W_k$ *and call* V *the* direct sum *of* $W_1, W_2, \ldots, W_k$ *if*

$$V = \sum_{i=1}^{k} W_i$$

and

$$W_i \cap \left(\sum_{j \neq i} W_j \right) = \{0\} \qquad \textit{for each } i\ (1 \leqslant i \leqslant k).$$

Example 9

Let $V = R^4$, and let $W_1 = \{(a, b, 0, 0)\colon a, b \in R\}$, $W_2 = \{(0, 0, c, 0)\ c \in R\}$, and $W_3 = \{(0, 0, 0, d)\colon d \in R\}$. For any element (a, b, c, d) of V,

$$(a, b, c, d) = (a, b, 0, 0) + (0, 0, c, 0) + (0, 0, 0, d) \in W_1 + W_2 + W_3.$$

Thus

$$V = \sum_{i=1}^{3} W_i.$$

To show that V is the direct sum of W_1, W_2, and W_3, we must prove that $W_1 \cap (W_2 + W_3) = \{0\}$, $W_2 \cap (W_1 + W_3) = \{0\}$, and $W_3 \cap (W_1 + W_2) = \{0\}$. But these equalities are obvious; so $V = W_1 \oplus W_2 \oplus W_3$. ∎

Our next result contains several conditions that are equivalent to the definition of a direct sum.

Theorem 5.15. *Let* $\mathsf{W}_1, \mathsf{W}_2, \ldots, \mathsf{W}_k$ *be subspaces of a finite-dimensional vector space* V. *The following conditions are equivalent:*

(a) $\mathsf{V} = \mathsf{W}_1 \oplus \mathsf{W}_2 \oplus \cdots \oplus \mathsf{W}_k$.

(b) $\mathsf{V} = \sum_{i=1}^{k} \mathsf{W}_i$ *and, for any vectors* $x_1, x_2, \ldots, x_k$ *such that* $x_i \in \mathsf{W}_i$ $(i = 1, 2, \ldots, k)$, *if* $x_1 + x_2 \cdots + x_k = 0$, *then* $x_i = 0$ $(i = 1, 2, \ldots, k)$.

(c) *Each vector* v *in* V *can be uniquely written in the form* $v = x_1 + x_2 + \cdots + x_k$, *where* $x_i \in \mathsf{W}_i$ $(i = 1, 2, \ldots, k)$.

(d) *If, for each* $i = 1, 2, \ldots, k$, γ_i *is any ordered basis for* W_i, *then* $\gamma_1 \cup \gamma_2 \cup \cdots \cup \gamma_k$† *is an ordered basis for* V.

(e) *For each* $i = 1, 2, \ldots, k$ *there exists an ordered basis* γ_i *for* W_i $(i = 1, 2, \ldots, k)$ *such that* $\gamma_1 \cup \gamma_2 \cup \cdots \cup \gamma_k$ *is an ordered basis for* V.

Proof. If (a) is true, then

$$\mathsf{V} = \sum_{i=1}^{k} \mathsf{W}_i$$

by definition. Suppose that $x_1, x_2, \ldots, x_k$ are vectors such that $x_i \in \mathsf{W}_i$ $(i = 1, 2, \ldots, k)$ and $x_1 + x_2 + \cdots + x_k = 0$. Then for any i

$$-x_i = \sum_{j \neq i} x_j \in \sum_{j \neq i} \mathsf{W}_j.$$

But also

$$-x_i \in \mathsf{W}_i, \text{ and so } -x_i \in \mathsf{W}_i \cap \left(\sum_{j \neq i} \mathsf{W}_j \right) = \{0\}.$$

Hence $x_i = 0$, proving (b).

We next prove that (b) implies (c). Since

$$\mathsf{V} = \sum_{i=1}^{k} \mathsf{W}_i$$

by (b), any vector $v \in \mathsf{V}$ can be represented in the form $v = x_1 + x_2 + \cdots + x_k$ for some elements $x_i \in \mathsf{W}_i$ $(i = 1, 2, \ldots, k)$. We must show that this representation is unique. Suppose therefore that $v = y_1 + y_2 + \cdots + y_k$, where $y_i \in \mathsf{W}_i$ $(i = 1, 2, \ldots, k)$. Then

$$(x_1 - y_1) + (x_2 - y_2) + \cdots + (x_k - y_k) = 0.$$

But since $x_i - y_i \in \mathsf{W}_i$, it follows from (b) that $x_i - y_i = 0$ $(i = 1, 2, \ldots, k)$. Thus $x_i = y_i$ for each i, proving the uniqueness of the representation.

†We regard $\gamma_1 \cup \gamma_2 \cup \cdots \cup \gamma_k$ as an ordered basis in the natural way—the vectors in γ_1 are listed first (in the same order as in γ_1), then the vectors in γ_2 (in the same order as in γ_2), etc.

To show that (c) implies (d), let γ_i be a basis for W_i ($i = 1, 2, \ldots, k$). Since

$$\mathsf{V} = \sum_{i=1}^{k} \mathsf{W}_i$$

by (c), it is clear that $\gamma_1 \cup \gamma_2 \cup \cdots \cup \gamma_k$ generates V. Suppose that there are vectors $x_{ij} \in \gamma_i$ ($j = 1, 2, \ldots, m_i$ and $i = 1, 2, \ldots, k$) and scalars a_{ij} such that

$$\sum_{i,j} a_{ij} x_{ij} = 0.$$

Set

$$y_i = \sum_{j=1}^{m_i} a_{ij} x_{ij};$$

then $y_i \in \text{span}(\gamma_i) = \mathsf{W}_i$ and

$$y_1 + y_2 + \cdots + y_k = \sum_{i,j} a_{ij} x_{ij} = 0.$$

Since $0 \in \mathsf{W}_i$ for each i and $0 + 0 + \cdots + 0 = y_1 + y_2 + \cdots + y_k$, condition (c) implies that $y_i = 0$ for each i. Thus

$$0 = y_i = \sum_{j=1}^{m_i} a_{ij} x_{ij}$$

for each i. But since γ_i is linearly independent, it follows that $a_{ij} = 0$ for $j = 1, 2, \ldots, m_i$ and each i. Hence $\gamma_1 \cup \gamma_2 \cup \cdots \cup \gamma_k$ is linearly independent and therefore is a basis for V.

It is immediate that (d) implies (e).

Finally, we shall show that (e) implies (a). If γ_i is a basis for W_i ($i = 1, 2, \ldots, k$) such that $\gamma_1 \cup \gamma_2 \cup \cdots \cup \gamma_k$ is a basis for V, then

$$\begin{aligned} \mathsf{V} &= \text{span}(\gamma_1 \cup \gamma_2 \cup \cdots \cup \gamma_k) \\ &= \text{span}(\gamma_1) + \text{span}(\gamma_2) + \cdots + \text{span}(\gamma_k) = \sum_{i=1}^{k} \mathsf{W}_i \end{aligned}$$

by repeated applications of Exercise 12 of Section 1.4. Fix an index i and suppose that

$$0 \neq v \in \mathsf{W}_i \cap \left(\sum_{j \neq i} \mathsf{W}_j \right).$$

Then

$$v \in \mathsf{W}_i = \text{span}(\gamma_i) \quad \text{and} \quad v \in \sum_{j \neq i} \mathsf{W}_j = \text{span}\left(\bigcup_{j \neq i} \gamma_j \right).$$

Hence v is a nontrivial linear combination of both γ_i and $\bigcup_{j \neq i} \gamma_j$, so that v can be expressed as a linear combination of $\gamma_1 \cup \gamma_2 \cup \cdots \cup \gamma_k$ in more than one way.

But these representations contradict Theorem 1.7, so we conclude that

$$\mathsf{W}_i \cap \left(\sum_{j \neq i} \mathsf{W}_j \right) = \{0\},$$

proving (a). ∎

With the aid of Theorem 5.15 we are now able to characterize diagonalizability in terms of direct sums.

Theorem 5.16. *Let* T *be a linear operator on a finite-dimensional vector space* V. *Then* T *is diagonalizable if and only if* V *is a direct sum of the eigenspaces of* T.

Proof. First, suppose that T is diagonalizable. Then the characteristic polynomial of T splits. Let $\lambda_1, \lambda_2, \ldots, \lambda_k$ be the distinct eigenvalues of T. For each i, choose an ordered basis S_i for the eigenspace E_{λ_i}. It follows from Theorem 5.14 that $S_1 \cup S_2 \cup \cdots \cup S_k$ is a basis for V. Therefore, by Theorem 5.15 we conclude that V is a direct sum of the eigenspaces.

Conversely, suppose that V is a direct sum of the eigenspaces of T: $\mathsf{E}_{\lambda_1}, \ldots, \mathsf{E}_{\lambda_k}$. For each i, choose a basis S_i of E_{λ_i}. By Theorem 5.15 we have that $S_1 \cup S_2 \cup \cdots \cup S_k$ is a basis for V. Since this basis consists of eigenvectors of T, we conclude that T is diagonalizable. ∎

Example 10

Let $\mathsf{T}\colon \mathsf{R}^4 \to \mathsf{R}^4$ be defined by

$$\mathsf{T}(a, b, c, d) = (a, b, 2c, 3d).$$

Then it is easily seen that T is diagonalizable, with eigenvalues $\lambda_1 = 1$, $\lambda_2 = 2$, and $\lambda_3 = 3$. Furthermore, the corresponding eigenspaces coincide with the subspaces W_1, W_2, and W_3 of Example 9. Theorem 5.16 provides us with another proof that $\mathsf{R}^4 = \mathsf{W}_1 \oplus \mathsf{W}_2 \oplus \mathsf{W}_3$. ∎

EXERCISES

1. Label the following statements as being true or false.
- **(a)** Any linear operator on an n-dimensional vector space that has fewer than n distinct eigenvalues is not diagonalizable.
- **(b)** Eigenvectors corresponding to the same eigenvalue are always linearly dependent.
- **(c)** If λ is an eigenvalue of a linear operator T, then each element of E_λ is an eigenvector of T.
- **(d)** If λ_1 and λ_2 are distinct eigenvalues of a linear operator T, then $\mathsf{E}_{\lambda_1} \cap \mathsf{E}_{\lambda_2} = \{0\}$.

(e) Let $A \in \mathsf{M}_{n \times n}(F)$ and $\beta = \{x_1, \dots, x_n\}$ be a basis for F^n consisting of eigenvectors of A. If Q is the $n \times n$ matrix whose ith column is x_i $(i = 1, 2, \dots, n)$, then $Q^{-1}AQ$ is a diagonal matrix.

(f) A linear operator T on a finite-dimensional vector space is diagonalizable if and only if the multiplicity of each eigenvalue λ equals the dimension of E_λ.

(g) Every diagonalizable linear operator has at least one eigenvalue.

(h) If a vector space is the direct sum of subspaces $\mathsf{W}_1, \mathsf{W}_2, \dots, \mathsf{W}_k$, then $\mathsf{W}_i \cap \mathsf{W}_j = \{0\}$ for $i \neq j$.

(i) If

$$\mathsf{V} = \sum_{i=1}^{k} \mathsf{W}_i \quad \text{and} \quad \mathsf{W}_i \cap \mathsf{W}_j = \{0\} \qquad \text{for } i \neq j,$$

then $\mathsf{V} = \mathsf{W}_1 \oplus \mathsf{W}_2 \oplus \cdots \oplus \mathsf{W}_k$.

2. For each of the following matrices A in $\mathsf{M}_{n \times n}(R)$, test A for diagonalizability, and if A is diagonalizable, find a matrix Q such that $Q^{-1}AQ$ is a diagonal matrix.

(a) $\begin{pmatrix} 1 & 2 \\ 0 & 1 \end{pmatrix}$ **(b)** $\begin{pmatrix} 1 & 3 \\ 3 & 1 \end{pmatrix}$ **(c)** $\begin{pmatrix} 1 & 4 \\ 3 & 2 \end{pmatrix}$

(d) $\begin{pmatrix} 7 & -4 & 0 \\ 8 & -5 & 0 \\ 6 & -6 & 3 \end{pmatrix}$ **(e)** $\begin{pmatrix} 0 & 0 & 1 \\ 1 & 0 & -1 \\ 0 & 1 & 1 \end{pmatrix}$ **(f)** $\begin{pmatrix} 1 & 1 & 0 \\ 0 & 1 & 2 \\ 0 & 0 & 3 \end{pmatrix}$

(g) $\begin{pmatrix} 3 & 1 & 1 \\ 2 & 4 & 2 \\ -1 & -1 & 1 \end{pmatrix}$

3. For each of the following linear operators T, test T for diagonalizability. If T is diagonalizable, find a basis β such that $[\mathsf{T}]_\beta$ is a diagonal matrix.

(a) $\mathsf{T}\colon \mathsf{P}_3(R) \to \mathsf{P}_3(R)$ defined by $\mathsf{T}(f) = f' + f''$, where f' and f'' are the first and second derivatives of f, respectively.

(b) $\mathsf{T}\colon \mathsf{P}_2(R) \to \mathsf{P}_2(R)$ defined by $\mathsf{T}(ax^2 + bx + c) = cx^2 + bx + a$.

(c) $\mathsf{T}\colon \mathsf{R}^3 \to \mathsf{R}^3$ defined by

$$\mathsf{T}\begin{pmatrix} a_1 \\ a_2 \\ a_3 \end{pmatrix} = \begin{pmatrix} a_2 \\ -a_1 \\ 2a_3 \end{pmatrix}.$$

(d) $\mathsf{T}\colon \mathsf{P}_2(R) \to \mathsf{P}_2(R)$ defined by $\mathsf{T}(f) = f(0) + f(1)(x + x^2)$.

(e) $\mathsf{T}\colon \mathsf{C}^2 \to \mathsf{C}^2$ defined by $\mathsf{T}(z, w) = (z + iw, iz + w)$.

(f) $\mathsf{T}\colon \mathsf{M}_{2 \times 2}(R) \to \mathsf{M}_{2 \times 2}(R)$ defined by $\mathsf{T}(A) = A^t$.

4. Prove the matrix version of the corollary to Theorem 5.10: If $A \in \mathsf{M}_{n \times n}(F)$ has n distinct eigenvalues, then A is diagonalizable.

5. State and prove the matrix version of Theorem 5.11.

6. **(a)** Justify the test for diagonalizability and the algorithm for diagonalization stated in this section.
   (b) Formulate part (a) for matrices.

7. If

$$A = \begin{pmatrix} 1 & 4 \\ 2 & 3 \end{pmatrix} \in \mathsf{M}_{2\times 2}(R),$$

find A^n for any positive integer n.

8. Let $A \in \mathsf{M}_{n\times n}(F)$ have two distinct eigenvalues λ_1 and λ_2. If $\dim(\mathsf{E}_{\lambda_1}) = n - 1$, prove that A is diagonalizable.

9. Let T be a linear operator on a finite-dimensional vector space V for which the distinct eigenvalues $\lambda_1, \lambda_2, \ldots, \lambda_k$ occur with multiplicities $m_1, m_2, \ldots, m_k$, respectively. If β is a basis for V such that $[\mathsf{T}]_\beta$ is a triangular matrix, prove that the diagonal entries of $[\mathsf{T}]_\beta$ are $\lambda_1, \lambda_2, \ldots, \lambda_k$ and that each λ_j occurs m_j times $(j = 1, 2, \ldots, k)$.

10. Suppose that A is an $n \times n$ matrix whose characteristic polynomial splits and that the distinct eigenvalues of A are $\lambda_1, \ldots, \lambda_k$. For each j, let m_j denote the multiplicity of λ_j. Prove that
    (a) $\operatorname{tr}(A) = \sum_{j=1}^{k} m_j\lambda_j$.
    (b) $\det(A) = (\lambda_1)^{m_1}(\lambda_2)^{m_2}\cdots(\lambda_k)^{m_k}$.

11. Let T be an invertible linear operator on a finite-dimensional vector space. Prove that T is diagonalizable if and only if T^{-1} is diagonalizable.

12. Let $A \in \mathsf{M}_{n\times n}(F)$. Show that A is diagonalizable if and only if A^t is diagonalizable.

13. Find the general solution of each system of differential equations.

    (a) $\begin{cases} x' = x + y \\ y' = 3x - y \end{cases}$ **(b)** $\begin{cases} x_1' = 8x_1 + 10x_2 \\ x_2' = -5x_1 - 7x_2 \end{cases}$

    (c) $\begin{cases} x_1' = x_1 + x_3 \\ x_2' = x_2 + x_3 \\ x_3' = 2x_3 \end{cases}$

14. Let

$$A = \begin{pmatrix} a_{11} & a_{12} & \cdots & a_{1n} \\ a_{21} & a_{22} & \cdots & a_{2n} \\ \vdots & \vdots & & \vdots \\ a_{n1} & a_{n2} & \cdots & a_{nn} \end{pmatrix}$$

be the coefficient matrix of the system of differential equations

$$\begin{cases} x_1' = a_{11}x_1 + a_{12}x_2 + \cdots + a_{1n}x_n \\ x_2' = a_{21}x_1 + a_{22}x_2 + \cdots + a_{2n}x_n \\ \vdots \\ x_n' = a_{n1}x_1 + a_{n2}x_2 + \cdots + a_{nn}x_n. \end{cases}$$

Suppose that A is diagonalizable and that the distinct eigenvalues of A are $\lambda_1, \lambda_2, \ldots, \lambda_k$. Prove that a differentiable function $X\colon R \to \mathsf{R}^n$ is a solution to the system if and only if X is of the form

$$X(t) = e^{\lambda_1 t}z_1 + e^{\lambda_2 t}z_2 + \cdots + e^{\lambda_k t}z_k,$$

where $z_i \in \mathsf{E}_{\lambda_i}$ for $i = 1, 2, \ldots, k$. Conclude that the set of solutions to the system is an n-dimensional real vector space.

Exercises 15 through 17 are concerned with simultaneous diagonalization.

Definitions. *Two linear operators* T *and* U *on the same finite-dimensional vector space* V *are called* simultaneously diagonalizable *if there exists a basis* β *for* V *such that both* $[\mathsf{T}]_\beta$ *and* $[\mathsf{U}]_\beta$ *are diagonal matrices. Similarly,* $A, B \in \mathsf{M}_{n\times n}(F)$ *are called* simultaneously diagonalizable *if there exists an invertible matrix* $Q \in \mathsf{M}_{n\times n}(F)$ *such that both* $Q^{-1}AQ$ *and* $Q^{-1}BQ$ *are diagonal matrices.*

15. (a) If T and U are simultaneously diagonalizable linear operators on a finite-dimensional vector space V, prove that $[\mathsf{T}]_\beta$ and $[\mathsf{U}]_\beta$ are simultaneously diagonalizable matrices for any basis β.

(b) Show that if A and B are simultaneously diagonalizable matrices, then L_A and L_B are simultaneously diagonalizable operators.

16. (a) Show that if T and U are simultaneously diagonalizable operators, then T and U commute (i.e., TU = UT).

(b) Show that if A and B are simultaneously diagonalizable matrices, then A and B commute.

The converse of (a) and (b) will be established in Exercise 25 of Section 5.4.

17. Let T be a diagonalizable linear operator on a finite-dimensional vector space, and let m be any positive integer. Prove that T and T^m are simultaneously diagonalizable.

Exercises 18 through 21 are concerned with direct sums.

18. Let $\mathsf{W}_1, \mathsf{W}_2, \ldots, \mathsf{W}_k$ be subspaces of a finite-dimensional vector space V such that

$$\sum_{i=1}^{k} \mathsf{W}_i = \mathsf{V}.$$

Prove that V is the direct sum of $W_1, W_2, \ldots, W_k$ if and only if

$$\dim(V) = \sum_{i=1}^{k} \dim(W_i).$$

19. Let V be a finite-dimensional vector space with basis $\beta = \{x_1, x_2, \ldots, x_n\}$, and let $\beta_1, \beta_2, \ldots, \beta_k$ be a partition of β (that is, $\beta_1, \beta_2, \ldots, \beta_k$ are subsets of β such that $\beta = \beta_1 \cup \beta_2 \cup \cdots \cup \beta_k$ and $\beta_i \cap \beta_j = \varnothing$ if $i \neq j$). Prove that $V = \text{span}(\beta_1) \oplus \text{span}(\beta_2) \oplus \cdots \oplus \text{span}(\beta_k)$.

20. Let T be a linear operator on a finite-dimensional vector space V for which the distinct eigenvalues of T are $\lambda_1, \lambda_2, \ldots, \lambda_k$. Prove that

$$\text{span}(\{x \in V\colon \ x \text{ is an eigenvector of } T\}) = E_{\lambda_1} \oplus E_{\lambda_2} \oplus \cdots \oplus E_{\lambda_k}.$$

21. Let $W_1, W_2, K_1, K_2, \ldots, K_p, M_1, M_2, \ldots, M_q$ be subspaces of a vector space V such that $W_1 = K_1 \oplus K_2 \oplus \cdots \oplus K_p$ and $W_2 = M_1 \oplus M_2 \oplus \cdots \oplus M_q$. Prove that if $W_1 \cap W_2 = \{0\}$, then $W_1 + W_2 = W_1 \oplus W_2 = K_1 \oplus K_2 \oplus \cdots \oplus K_p \oplus M_1 \oplus M_2 \oplus \cdots \oplus M_q$.

5.3* MATRIX LIMITS AND MARKOV CHAINS

If A is a square matrix having complex entries, then for any positive integer m, A^m is a square matrix of the same size that also has complex entries. In many of the life and natural sciences there are important practical applications that require determining the "limit" (if one exists) of the sequence of matrices A, A^2, $A^3, \ldots$.

In this section we consider such limits and examine one important situation in which this type of limit exists. We assume familiarity with limits of sequences of real numbers. The limit of a sequence of complex numbers $\{z_m\colon m = 1, 2, \ldots\}$ can be defined in terms of the limits of the sequences of real and imaginary parts: If $z_m = r_m + is_m$, where r_m and s_m are real numbers, then

$$\lim_{m\to\infty} z_m = \lim_{m\to\infty} r_m + i \lim_{m\to\infty} s_m.$$

Definition. *Let* L, $A_1, A_2, A_3, \ldots$ *be* $n \times p$ *matrices having complex entries. The sequence* $A_1, A_2, A_3, \ldots$ *is said to* converge *to the matrix* L, *called the* limit *of the sequence, if, for all* i *and* j,

$$\lim_{m\to\infty} (A_m)_{ij} = L_{ij}.$$

To denote that the sequence $A_1, A_2, A_3, \ldots$ *converges to* L, *we write*

$$\lim_{m\to\infty} A_m = L.$$

Example 1

If

$$A_m = \begin{pmatrix} 1 - \dfrac{1}{m} & \left(-\dfrac{3}{4}\right)^m & \dfrac{3m^2}{m^2+1} + i\left(\dfrac{2m+1}{m-1}\right) \\ \left(\dfrac{i}{2}\right)^m & 2 & \left(1 + \dfrac{1}{m}\right)^m \end{pmatrix},$$

then

$$\lim_{m\to\infty} A_m = \begin{pmatrix} 1 & 0 & 3+2i \\ 0 & 2 & e \end{pmatrix},$$

where e is the base of the natural logarithm. ∎

A simple but important property of matrix limits is contained in the next theorem. Note the analogy with the familiar property of limits of sequences of real numbers which asserts that if $\lim_{m\to\infty} a_m$ exists, then

$$\lim_{m\to\infty} ca_m = c\left(\lim_{m\to\infty} a_m\right).$$

Theorem 5.17. *Let* $A_1, A_2, A_3, \ldots$ *be a sequence of* $n \times p$ *matrices having complex entries such that*

$$\lim_{m\to\infty} A_m = L \in M_{n\times p}(C).$$

Then for any $P \in M_{r\times n}(C)$ *and* $Q \in M_{p\times s}(C)$,

$$\lim_{m\to\infty} PA_m = PL \quad \textit{and} \quad \lim_{m\to\infty} A_mQ = LQ.$$

Proof. For any i $(1 \leqslant i \leqslant r)$ and j $(1 \leqslant j \leqslant p)$,

$$\lim_{m\to\infty} [(PA_m)_{ij}] = \lim_{m\to\infty} \left[\sum_{k=1}^{n} P_{ik}(A_m)_{kj}\right]$$

$$= \sum_{k=1}^{n} P_{ik}\left\{\lim_{m\to\infty} [(A_m)_{kj}]\right\} = \sum_{k=1}^{n} P_{ik}L_{kj} = (PL)_{ij}.$$

Hence $\lim_{m\to\infty} PA_m = PL$. The proof that $\lim_{m\to\infty} A_mQ = LQ$ is similar. ∎

Corollary. *Let* $A \in M_{n\times n}(C)$, *and let* $\lim_{m\to\infty} A^m = L$. *Then for any invertible matrix* $Q \in M_{n\times n}(C)$,

$$\lim_{m\to\infty} (QAQ^{-1})^m = QLQ^{-1}.$$

Proof. Since

$$(QAQ^{-1})^m = (QAQ^{-1})(QAQ^{-1})\cdots(QAQ^{-1}) = QA^mQ^{-1},$$

we have

$$\lim_{m\to\infty} [(QAQ^{-1})^m] = \lim_{m\to\infty} (QA^mQ^{-1}) = Q\left(\lim_{m\to\infty} A^m\right)Q^{-1} = QLQ^{-1}$$

by applying Theorem 5.17 twice. ∎

In the discussion that follows, we frequently encounter the set

$$S = \{\lambda \in C\colon |\lambda| < 1 \text{ or } \lambda = 1\}.$$

Geometrically, this set consists of the complex number 1 and the interior of the unit disk (the disk of radius 1 centered at the origin). This set is of interest because if λ is a complex number, $\lim_{m\to\infty} \lambda^m$ exists if and only if $\lambda \in S$. (This fact, which is obviously true if λ is a real number, can be shown to be true for complex numbers also.)

The following important result gives necessary and sufficient conditions for the existence of the type of limit under consideration.

Theorem 5.18. *Let* A *be a square matrix having complex entries. Then* $\lim_{m\to\infty} \mathrm{A}^m$ *exists if and only if the following conditions hold:*

(a) *If* λ *is an eigenvalue of* A, *then* $\lambda \in$ S.
(b) *If* 1 *is an eigenvalue of* A, *then the dimension of the eigenspace corresponding to* 1 *equals the multiplicity of* 1 *as an eigenvalue of* A.

Unfortunately, it will not be possible to prove this theorem until we study the Jordan canonical form in Section 7.2; for this reason the proof will be deferred until Exercise 18 of that section. The necessity of condition (a) is easy to see, however. For suppose that λ is an eigenvalue of A such that $\lambda \notin S$. Let x be an eigenvector of A corresponding to λ. Regarding x as an $n \times 1$ matrix, we see that

$$\lim_{m\to\infty} (A^m x) = \left(\lim_{m\to\infty} A^m\right)x = Lx$$

by Theorem 5.17, where $L = \lim_{m\to\infty} A^m$. But $\lim_{m\to\infty} (A^m x) = \lim_{m\to\infty} (\lambda^m x)$ diverges since $\lim_{m\to\infty} \lambda^m$ does not exist. Hence if $\lim_{m\to\infty} A^m$ exists, then condition (a) of Theorem 5.18 must hold. Although we are unable to prove the necessity of condition (b) at this time, let us consider an example for which this condition fails. Observe that for the matrix

$$B = \begin{pmatrix} 1 & 1 \\ 0 & 1 \end{pmatrix},$$

the eigenvalue $\lambda = 1$ has multiplicity 2, whereas $\dim(\mathsf{E}_\lambda) = 1$. But

$$B^m = \begin{pmatrix} 1 & m \\ 0 & 1 \end{pmatrix}$$

by an easy induction, and hence $\lim_{m \to \infty} B^m$ does not exist. (We will see later that if A is a matrix for which condition (b) fails, then the Jordan canonical form of A can be chosen so that its upper left 2×2 submatrix is precisely this matrix B.)

In most of the applications involving this type of limit, however, the matrix A is diagonalizable. When condition (b) of Theorem 5.18 is replaced by the stronger condition that A is diagonalizable (see Theorem 5.14), then the existence of the limit is easily shown.

Theorem 5.19. *Let* $\mathsf{A} \in \mathsf{M}_{n \times n}(C)$ *be such that the following conditions hold:*

(a) *If* λ *is an eigenvalue of* A, *then* $\lambda \in \mathsf{S}$.
(b) A *is diagonalizable.*

Then $\lim_{m \to \infty} \mathsf{A}^m$ *exists.*

Proof. Since A is diagonalizable, there exists an invertible matrix Q such that $Q^{-1}AQ = D$, a diagonal matrix. Let

$$D = \begin{pmatrix} \lambda_1 & 0 & \cdots & 0 \\ 0 & \lambda_2 & \cdots & 0 \\ \vdots & \vdots & & \vdots \\ 0 & 0 & \cdots & \lambda_n \end{pmatrix}.$$

Because $\lambda_1, \lambda_2, \ldots, \lambda_n$ are the eigenvalues of A, condition (a) shows that either $\lambda_i = 1$ or $|\lambda_i| < 1$ for $1 \leqslant i \leqslant n$. Thus

$$\lim_{m \to \infty} \lambda_i^m = \begin{cases} 1 & \text{if } \lambda_i = 1 \\ 0 & \text{otherwise.} \end{cases}$$

But since

$$D^m = \begin{pmatrix} \lambda_1^m & 0 & \cdots & 0 \\ 0 & \lambda_2^m & \cdots & 0 \\ \vdots & \vdots & & \vdots \\ 0 & 0 & \cdots & \lambda_n^m \end{pmatrix},$$

the sequence $D, D^2, D^3, \ldots$ converges to a limit L. Hence

$$\lim_{m \to \infty} A^m = \lim_{m \to \infty} (QDQ^{-1})^m = QLQ^{-1}$$

by the corollary to Theorem 5.17. ∎

The technique for computing $\lim_{m\to\infty} A^m$ that is used in the proof of Theorem 5.19 is quite useful. We now employ this method to compute $\lim_{m\to\infty} A^m$ for the matrix

$$A = \begin{pmatrix} \frac{7}{4} & -\frac{9}{4} & -\frac{15}{4} \\ \frac{3}{4} & \frac{7}{4} & \frac{3}{4} \\ \frac{3}{4} & -\frac{9}{4} & -\frac{11}{4} \end{pmatrix}.$$

If

$$Q = \begin{pmatrix} 1 & 3 & -1 \\ -3 & -2 & 1 \\ 2 & 3 & -1 \end{pmatrix},$$

then

$$D = Q^{-1}(AQ) = \begin{pmatrix} -1 & 0 & 1 \\ -1 & 1 & 2 \\ -5 & 3 & 7 \end{pmatrix} \begin{pmatrix} 1 & -\frac{3}{2} & -\frac{1}{4} \\ -3 & 1 & \frac{1}{4} \\ 2 & -\frac{3}{2} & -\frac{1}{4} \end{pmatrix} = \begin{pmatrix} 1 & 0 & 0 \\ 0 & -\frac{1}{2} & 0 \\ 0 & 0 & \frac{1}{4} \end{pmatrix}.$$

Hence

$$\lim_{m\to\infty} A^m = \lim_{m\to\infty} (QDQ^{-1})^m = \lim_{m\to\infty} (QD^mQ^{-1}) = Q\left(\lim_{m\to\infty} D^m\right)Q^{-1}$$

$$= \begin{pmatrix} 1 & 3 & -1 \\ -3 & -2 & 1 \\ 2 & 3 & -1 \end{pmatrix} \left[\lim_{m\to\infty} \begin{pmatrix} 1 & 0 & 0 \\ 0 & (-\frac{1}{2})^m & 0 \\ 0 & 0 & (\frac{1}{4})^m \end{pmatrix} \right] \begin{pmatrix} -1 & 0 & 1 \\ -1 & 1 & 2 \\ -5 & 3 & 7 \end{pmatrix}$$

$$= \begin{pmatrix} 1 & 3 & -1 \\ -3 & -2 & 1 \\ 2 & 3 & -1 \end{pmatrix} \begin{pmatrix} 1 & 0 & 0 \\ 0 & 0 & 0 \\ 0 & 0 & 0 \end{pmatrix} \begin{pmatrix} -1 & 0 & 1 \\ -1 & 1 & 2 \\ -5 & 3 & 7 \end{pmatrix} = \begin{pmatrix} -1 & 0 & 1 \\ 3 & 0 & -3 \\ -2 & 0 & 2 \end{pmatrix}.$$

Let us now consider a simple example in which the limit of powers of a matrix occurs. Suppose that the population of a certain metropolitan area remains constant but that there is a continual movement of people between the city and the suburbs. Specifically, let the entries of the matrix A below represent the probabilities that someone living in the city or in the suburbs on January 1 will be living in each region on January 1 of the next year.

	Presently living in the city	Presently living in the suburbs
Living next year in the city	0.90	0.02
Living next year in the suburbs	0.10	0.98

$$\begin{pmatrix} 0.90 & 0.02 \\ 0.10 & 0.98 \end{pmatrix} = A$$

For instance, the probability that someone living in the city (on January 1) will be living in the suburbs next year (on January 1) is 0.10. Notice that since the entries of each column of A represent probabilities of residing in each of the two locations, the entries of A are nonnegative. Moreover, the assumption of a constant population in the metropolitan area requires that the sum of the entries of each column of A be 1. Any matrix having these two properties (that the entries are nonnegative and that the sum of the entries in each column is 1) is called a *transition matrix* (or a *stochastic matrix*). For an arbitrary $n \times n$ transition matrix M, the rows and columns correspond to n *states*, and the entry M_{ij} represents the probability of moving from state j into state i in one *stage*. In our example, there are two states (residing in the city and residing in the suburbs), and A_{21} represents the probability of moving from the city to the suburbs in one stage (year).

Let us now determine the probability that a city resident will be living in the suburbs after 2 years. Observe first that there are two different ways in which such a move can be made—either by remaining in the city for 1 year and then moving to the suburbs or by moving to the suburbs during the first year and remaining there the second (see Figure 5.3). The probability that a city dweller stays in the city during the next year is 0.90, and the probability that a city dweller moves to the suburbs during the following year is 0.10. Hence the probability that a city resident stays in the city for 1 year and moves to the suburbs during the next is 0.90(0.10). Likewise the probability of a city dweller moving to the suburbs during the first year and remaining there the next is 0.10(0.98). Thus the probability that a city resident will be living in the suburbs after 2 years is $0.90(0.10) + 0.10(0.98) = 0.188$. Observe that this number is obtained by the same calculation as that which produces $(A^2)_{21}$—hence $(A^2)_{21}$

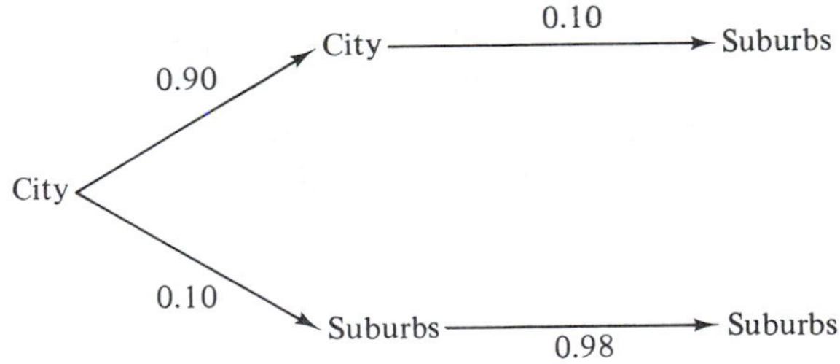

Figure 5.3

represents the probability that a city dweller will be residing in the suburbs after 2 years. In general, for any transition matrix M, $(M^m)_{ij}$ represents the probability of moving from state j to state i in m stages.

Suppose additionally that 70% of the 1970 population of the metropolitan area lived in the city and 30% lived in the suburbs. Let us record this data as a column vector:

$$\begin{matrix}\text{Proportion of city dwellers} \\ \text{Proportion of suburb residents}\end{matrix} \quad \begin{pmatrix}0.70\\0.30\end{pmatrix} = P.$$

Notice that the rows of P correspond to the states of residing in the city and residing in the suburbs, respectively—the same order as the states are listed in the transition matrix A. Observe also that P is a column vector containing nonnegative entries whose sum is 1; such a vector is called a *probability vector*. In this terminology each column of a transition matrix is a probability vector.

Let us now consider the significance of the vector AP. The first coordinate of this vector is formed by the calculation $0.90(0.70) + 0.02(0.30)$. The term $0.90(0.70)$ represents the proportion of the 1970 metropolitan population that remained in the city during the next year, and the term $0.02(0.30)$ represents the proportion of the 1970 metropolitan population that moved into the city during the next year. Hence the first coordinate of AP represents the proportion of the metropolitan population that was living in the city in 1971. Similarly, the second coordinate of

$$AP = \begin{pmatrix}0.636\\0.364\end{pmatrix}$$

represents the proportion of the metropolitan population that was living in the suburbs in 1971. This argument can be easily extended to show that the coordinates of

$$A^2P = A(AP) = \begin{pmatrix}0.57968\\0.42032\end{pmatrix}$$

represent the proportions of the metropolitan population that were living in each location in 1972. In general, the coordinates of A^mP represent the proportion of the metropolitan population that will be living in the city and suburbs, respectively, after m stages (m years after 1970).

Will the city eventually be depleted if this trend continues? In view of the preceding discussion it is natural to define the eventual proportion of city dwellers and suburbanites to be the first and second coordinates, respectively, of $\lim_{m\to\infty} A^mP$. Let us now compute this limit. We can construct Q and D as in earlier computations to obtain

$$D = Q^{-1}AQ = \begin{pmatrix}\frac{1}{6} & \frac{1}{6}\\ \frac{5}{6} & -\frac{1}{6}\end{pmatrix}\begin{pmatrix}0.90 & 0.02\\0.10 & 0.98\end{pmatrix}\begin{pmatrix}1 & 1\\5 & -1\end{pmatrix} = \begin{pmatrix}1 & 0\\0 & 0.88\end{pmatrix};$$

thus

$$L = \lim_{m \to \infty} A^m = \lim_{m \to \infty} (QD^mQ^{-1}) = Q\begin{pmatrix} 1 & 0 \\ 0 & 0 \end{pmatrix}Q^{-1} = \begin{pmatrix} \frac{1}{6} & \frac{1}{6} \\ \frac{5}{6} & \frac{5}{6} \end{pmatrix}.$$

Hence

$$\lim_{m \to \infty} A^m P = LP = \begin{pmatrix} \frac{1}{6} \\ \frac{5}{6} \end{pmatrix};$$

so eventually $\frac{1}{6}$ of the population will live in the city and $\frac{5}{6}$ will live in the suburbs. It is easy to show that

$$LP = \begin{pmatrix} \frac{1}{6} \\ \frac{5}{6} \end{pmatrix}$$

for any probability vector P. Hence in this example the eventual proportions of city dwellers and suburbanites are independent of the initial proportions (as given by the vector P)!

In analyzing the city–suburb problem we gave probabilistic interpretations of A^2 and AP, showing that A^2 is a transition matrix and AP is a probability vector. Analogous arguments can be used to show that the product of two transition matrices is a transition matrix and that the product of a transition matrix and a probability vector is a probability vector. Another proof of these results can be based on the following theorem, which characterizes transition matrices and probability vectors.

Theorem 5.20. *Let* M *be an* $n \times n$ *matrix having (real) nonnegative entries,* x *be a column vector in* R^n *having nonnegative coordinates, and* $u \in R^n$ *be the column vector in which each coordinate equals* 1. *Then*

(a) M *is a transition matrix if and only if* $M^t u = u$.

(b) x *is a probability vector if and only if* $u^t x = (1)$.

Proof. Exercise. ∎

Corollary.

(a) *The product of two* $n \times n$ *transition matrices is an* $n \times n$ *transition matrix. In particular, any power of a transition matrix is a transition matrix.*

(b) *The product of a transition matrix and a probability vector is a probability vector.*

Proof. Exercise. ∎

A *stochastic process* is concerned with predicting the state of an object that is constrained to be in exactly one of a number of possible states at any given time but that changes states in some random manner. Normally, the probability that the object is in some particular state at a given time will depend on such factors as

1. The state in question
2. The time in question
3. Some or all of the previous states in which the object has been
4. The states that other objects are in or have been in

For instance, the object could be an American voter and the state of the object could be his or her preference of political party, or the object could be a molecule of H_2O and the states could be the physical states in which H_2O can exist (the solid, liquid, and gaseous states). In these examples all four of the factors mentioned above will influence the probability that the objects are in a particular state at a particular time.

If, however, the probability that an object in one state will change to a different state depends only on the two states (and not on the time, earlier states, or other factors), then the stochastic process is called a *Markov process*. If, in addition, the number of possible states is finite, then the Markov process is called a *Markov chain*. The preceding example of the movement of population between the city and suburbs is a two-state Markov chain.

Let us consider another Markov chain. A certain junior college would like to obtain information about the likelihood that various categories of presently enrolled students will graduate. The school classifies a student as a sophomore or a freshman depending on the number of credits that the student has earned. Data from the school indicate that from one fall semester to the next 40% of the sophomores will graduate, 30% will remain sophomores, and 30% will quit permanently. For freshmen the data show that 10% will graduate by next fall, 50% will become sophomores, 20% will remain freshmen, and 20% will quit permanently. During the present year 50% of the students at the school are sophomores and 50% are freshmen. Assuming that the trend indicated by the data continues indefinitely, the school would like to know

1. The percentage of the present students who will graduate, the percentage who will be sophomores, the percentage who will be freshmen, and the percentage who will quit school permanently by next fall
2. The same percentages as in item 1 for the fall semester 2 years hence
3. The percentage of its present students who will eventually graduate

The preceding paragraph describes a four-state Markov chain with states

1. Having graduated
2. Being a sophomore
3. Being a freshman
4. Having quit permanently

The data cited above provide us with the transition matrix

$$A = \begin{pmatrix} 1 & 0.4 & 0.1 & 0 \\ 0 & 0.3 & 0.5 & 0 \\ 0 & 0 & 0.2 & 0 \\ 0 & 0.3 & 0.2 & 1 \end{pmatrix}$$

of the Markov chain. (Notice that students who have graduated or have quit permanently are assumed to remain indefinitely in those respective states. Thus a freshman who quits the school and returns during a later semester is not regarded as having changed states—the student is assumed to have remained in the state of being a freshman during the time he or she was not enrolled.) Moreover, we are told that the present distribution of students is half in each of states 2 and 3 and none in states 1 and 4. The vector

$$P = \begin{pmatrix} 0 \\ 0.5 \\ 0.5 \\ 0 \end{pmatrix}$$

that describes the initial probability of being in each state is called the *initial probability vector* for the Markov chain.

To answer question 1, we must determine the probability that a present student will be in each state by next fall. As we have seen, these probabilities are the coordinates of the vector

$$AP = \begin{pmatrix} 0.25 \\ 0.40 \\ 0.10 \\ 0.25 \end{pmatrix}.$$

Hence by next fall 25% of the present students will graduate, 40% will be sophomores, 10% will be freshmen, and 25% will quit the school. Similarly,

$$A^2P = A(AP) = \begin{pmatrix} 0.42 \\ 0.17 \\ 0.02 \\ 0.39 \end{pmatrix}$$

provides the information needed to answer question 2: within 2 years 42% of the present students will graduate, 17% will be sophomores, 2% will be freshmen, and 39% will quit the school.

Finally, the answer to question 3 is provided by the vector LP, where

$L = \lim_{m \to \infty} A^m$. The reader should verify that if

$$Q = \begin{pmatrix} 1 & -4 & 19 & 0 \\ 0 & 7 & -40 & 0 \\ 0 & 0 & 8 & 0 \\ 0 & -3 & 13 & 1 \end{pmatrix}.$$

then

$$D = Q^{-1}AQ = \begin{pmatrix} 1 & \frac{4}{7} & \frac{27}{56} & 0 \\ 0 & \frac{1}{7} & \frac{5}{7} & 0 \\ 0 & 0 & \frac{1}{8} & 0 \\ 0 & \frac{3}{7} & \frac{29}{56} & 1 \end{pmatrix} \begin{pmatrix} 1 & 0.4 & 0.1 & 0 \\ 0 & 0.3 & 0.5 & 0 \\ 0 & 0 & 0.2 & 0 \\ 0 & 0.3 & 0.2 & 1 \end{pmatrix} \begin{pmatrix} 1 & -4 & 19 & 0 \\ 0 & 7 & -40 & 0 \\ 0 & 0 & 8 & 0 \\ 0 & -3 & 13 & 1 \end{pmatrix}$$

$$= \begin{pmatrix} 1 & 0 & 0 & 0 \\ 0 & 0.3 & 0 & 0 \\ 0 & 0 & 0.2 & 0 \\ 0 & 0 & 0 & 1 \end{pmatrix}.$$

Thus

$$L = \lim_{m \to \infty} A^m = Q\left(\lim_{m \to \infty} D^m\right)Q^{-1}$$

$$= \begin{pmatrix} 1 & -4 & 19 & 0 \\ 0 & 7 & -40 & 0 \\ 0 & 0 & 8 & 0 \\ 0 & -3 & 13 & 1 \end{pmatrix} \begin{pmatrix} 1 & 0 & 0 & 0 \\ 0 & 0 & 0 & 0 \\ 0 & 0 & 0 & 0 \\ 0 & 0 & 0 & 1 \end{pmatrix} \begin{pmatrix} 1 & \frac{4}{7} & \frac{27}{56} & 0 \\ 0 & \frac{1}{7} & \frac{5}{7} & 0 \\ 0 & 0 & \frac{1}{8} & 0 \\ 0 & \frac{3}{7} & \frac{29}{56} & 1 \end{pmatrix} = \begin{pmatrix} 1 & \frac{4}{7} & \frac{27}{56} & 0 \\ 0 & 0 & 0 & 0 \\ 0 & 0 & 0 & 0 \\ 0 & \frac{3}{7} & \frac{29}{56} & 1 \end{pmatrix}.$$

So

$$LP = \begin{pmatrix} \frac{59}{112} \\ 0 \\ 0 \\ \frac{53}{112} \end{pmatrix},$$

and hence the probability that one of the present students will graduate is $\frac{59}{112}$.

In the two preceding examples we have seen that $\lim_{m \to \infty} A^m P$, where A is the transition matrix and P is the initial probability vector of the Markov chain, gives the eventual proportions in each state. In general, however, the limit of powers of a transition matrix need not exist. For example, if

$$M = \begin{pmatrix} 0 & 1 \\ 1 & 0 \end{pmatrix},$$

then $\lim_{m\to\infty} M^m$ clearly does not exist. (Odd powers of M equal M and even powers of M equal I.) The reason that the limit fails to exist is that condition (a) of Theorem 5.18 does not hold for M (-1 is an eigenvalue). In fact, it can be shown (see Exercise 20 of Section 7.2) that the only transition matrices A such that $\lim_{m\to\infty} A^m$ does not exist are precisely those matrices for which condition (a) of Theorem 5.18 fails to hold.

But even if the limit of powers of the transition matrix exists, the computation of the limit may be quite difficult. (The reader is encouraged to work Exercise 6 to appreciate the truth of the last sentence.) Fortunately, there is a large and important class of transition matrices for which this limit exists and is easily computed—this is the class of "regular" transition matrices.

Definition. *If some power of a transition matrix contains only positive entries, then the matrix is called a* regular *transition matrix.*

Example 2

The transition matrix

$$\begin{pmatrix} 0.90 & 0.02 \\ 0.10 & 0.98 \end{pmatrix}$$

of the Markov chain describing the movement of population between the city and suburbs is clearly regular since each entry is positive. On the other hand, the transition matrix

$$A = \begin{pmatrix} 1 & 0.4 & 0.1 & 0 \\ 0 & 0.3 & 0.5 & 0 \\ 0 & 0 & 0.2 & 0 \\ 0 & 0.3 & 0.2 & 1 \end{pmatrix}$$

of the Markov chain describing the junior college enrollments is not regular. [It is easy to show that the first column of A^m is

$$\begin{pmatrix} 1 \\ 0 \\ 0 \\ 0 \end{pmatrix}$$

for any m. Hence $(A^m)_{41}$, for instance, is never positive.]

Observe that a regular transition matrix may contain zero entries; for example,

$$M = \begin{pmatrix} 0.9 & 0.5 & 0 \\ 0 & 0.5 & 0.4 \\ 0.1 & 0 & 0.6 \end{pmatrix}$$

is regular since every entry of M^2 is positive. ∎

In the remainder of this section we are concerned primarily with proving that if A is a regular transition matrix, then $L = \lim_{m \to \infty} A^m$ exists and the columns of L are identical. (Recall the appearance of L in the city–suburb problem.) From this fact it is easy to compute the limit. In the course of proving this result we obtain some interesting bounds for the magnitudes of the eigenvalues of any square matrix. These bounds are given in terms of the sum of the absolute values of the entries of the rows and columns of the matrix. The necessary terminology is introduced in the definition below.

Definitions. *Let* $A \in M_{n \times n}(C)$. *Define* $\rho_i(A)$ *to be the sum of the absolute values of the entries of row* i *of* A and $v_j(A)$ *to be the sum of the absolute values of the entries of column* j *of* A. *Thus*

$$\rho_i(A) = \sum_{j=1}^{n} |A_{ij}| \qquad \textit{for } i = 1, 2, \ldots, n$$

and

$$v_j(A) = \sum_{i=1}^{n} |A_{ij}| \qquad \textit{for } j = 1, 2, \ldots, n.$$

The row sum *of* A, *denoted* $\rho(A)$, *and the* column sum *of* A, *denoted* $v(A)$, *are defined as*

$$\rho(A) = \max\{\rho_i(A)\colon 1 \leqslant i \leqslant n\} \quad \textit{and} \quad v(A) = \max\{v_j(A)\colon 1 \leqslant j \leqslant n\}.$$

Example 3

For the matrix

$$A = \begin{pmatrix} 1 & -1 & 5 \\ -4 & 0 & 6 \\ 3 & 2 & -1 \end{pmatrix},$$

$\rho_1(A) = 7$, $\rho_2(A) = 10$, $\rho_3(A) = 6$, $v_1(A) = 8$, $v_2(A) = 3$, and $v_3(A) = 12$. Hence $\rho(A) = 10$ and $v(A) = 12$. ∎

Our next results show that the smaller of $\rho(A)$ and $v(A)$ is an upper bound for the absolute values of the eigenvalues of A. In the preceding example, for instance, A has no eigenvalue with absolute value greater than 10.

Theorem 5.21 (Gerschgorin's Disk Theorem). *Let* $A \in M_{n \times n}(C)$. *For* $i = 1, 2, \ldots, n$ *define*

$$r_i = \rho_i(A) - |A_{ii}|,$$

and let C_i *denote the disk centered at* A_{ii} *of radius* r_i. *Then each eigenvalue of* A *lies in some* C_i.

Proof. Let λ be an eigenvalue of A with corresponding eigenvector

$$x = \begin{pmatrix} x_1 \\ \vdots \\ x_n \end{pmatrix}.$$

Then x satisfies the matrix equation $Ax = \lambda x$, which can be written as

$$\sum_{j=1}^{n} A_{ij}x_j = \lambda x_i \qquad (i = 1, 2, \ldots, n). \tag{4}$$

Suppose that x_k is the coordinate of x having the largest absolute value, and note that $x_k \neq 0$ because x is an eigenvector of A.

We will show that λ lies in C_k, that is, $|\lambda - A_{kk}| \leqslant r_k$. It follows from the kth equation in (4) that

$$\begin{aligned} |\lambda x_k - A_{kk}x_k| &= \left|\sum_{j=1}^{n} A_{kj}x_j - A_{kk}x_k\right| = \left|\sum_{j \neq k} A_{kj}x_j\right| \\ &\leqslant \sum_{j \neq k} |A_{kj}|\,|x_j| \leqslant \sum_{j \neq k} |A_{kj}|\,|x_k| \\ &= |x_k| \sum_{j \neq k} |A_{kj}| = |x_k| r_k. \end{aligned}$$

Thus

$$|x_k|\,|\lambda - A_{kk}| \leqslant |x_k| r_k;$$

so

$$|\lambda - A_{kk}| \leqslant r_k$$

because $|x_k| > 0$. ∎

Corollary 1. *Let λ be any eigenvalue of $A \in M_{n \times n}(C)$. Then $|\lambda| \leqslant \rho(A)$.*

Proof. By Gerschgorin's disk theorem, $|\lambda - A_{kk}| \leqslant r_k$ for some k. Hence

$$\begin{aligned} |\lambda| &= |(\lambda - A_{kk}) + A_{kk}| \leqslant |\lambda - A_{kk}| + |A_{kk}| \\ &\leqslant r_k + |A_{kk}| = \rho_k(A) \leqslant \rho(A). \end{aligned}$$ ∎

Corollary 2. *Let λ be any eigenvalue of $A \in M_{n \times n}(C)$. Then*

$$|\lambda| \leqslant \min\{\rho(A), \nu(A)\}.$$

Proof. Since $|\lambda| \leqslant \rho(A)$ by Corollary 1, it suffices to show that $|\lambda| \leqslant \nu(A)$. Exercise 14 of Section 5.1 shows that λ is an eigenvalue of A^t. Hence $|\lambda| \leqslant \rho(A^t)$. But the rows of A^t are the columns of A. Thus $\rho(A^t) = \nu(A)$; so $|\lambda| \leqslant \nu(A)$. ∎

The following conclusion is immediate from Corollary 2.

Corollary 3. *If λ is an eigenvalue of a transition matrix, then $|\lambda| \leqslant 1$.*

The next result shows that the upper bound in Corollary 3 is attained.

Theorem 5.22. *Every transition matrix has* 1 *as an eigenvalue.*

Proof. Let A be an $n \times n$ transition matrix, and let $u \in \mathsf{R}^n$ be the column vector in which each coordinate is 1. Then $A^t u = u$ by Theorem 5.20, and hence u is an eigenvector of A^t corresponding to the eigenvalue 1. But since A and A^t have the same eigenvalues, it follows that 1 is also an eigenvalue of A. ∎

Suppose now that A is a transition matrix for which some eigenvector corresponding to the eigenvalue 1 has only nonnegative coordinates. Then some multiple of this vector will be a probability vector P as well as an eigenvector of A corresponding to the eigenvalue 1. It is interesting to observe that if P is the initial probability vector of a Markov chain having A as its transition matrix, then the Markov chain is completely static. For in this situation $A^m P = P$ for every positive integer m, and hence the probability of being in each state never changes. Consider, for instance, the city–suburb problem with

$$P = \begin{pmatrix} \frac{1}{6} \\ \frac{5}{6} \end{pmatrix}.$$

Theorem 5.23. *Let* $\mathrm{A} \in \mathsf{M}_{n \times n}(\mathrm{C})$ *be a matrix in which each entry is positive, and let* λ *be an eigenvalue of* A *such that* $|\lambda| = \rho(\mathrm{A})$. *Then* $\lambda = \rho(\mathrm{A})$, *and* $\{\mathrm{u}\}$ *is a basis for* E_λ, *where*

$$\mathrm{u} = \begin{pmatrix} 1 \\ \vdots \\ 1 \end{pmatrix}.$$

Proof. Let

$$x = \begin{pmatrix} x_1 \\ \vdots \\ x_n \end{pmatrix}$$

be an eigenvector of A corresponding to the eigenvalue λ, let x_k be the coordinate of x having the largest absolute value, and let $b = |x_k|$. Then

$$|\lambda| b = |\lambda|\,|x_k| = |\lambda x_k| = \left| \sum_{j=1}^{n} A_{kj} x_j \right| \leqslant \sum_{j=1}^{n} |A_{kj} x_j|$$

$$= \sum_{j=1}^{n} |A_{kj}|\,|x_j| \leqslant \sum_{j=1}^{n} |A_{kj}| b = \rho_k(A) b \leqslant \rho(A) b. \tag{5}$$

Since $|\lambda| = \rho(A)$, the three inequalities in (5) are actually equalities; that is,

(a) $\left| \sum_{j=1}^{n} A_{kj} x_j \right| = \sum_{j=1}^{n} |A_{kj} x_j|,$

(b) $\sum_{j=1}^{n} |A_{kj}||x_j| = \sum_{j=1}^{n} |A_{kj}|b$, and

(c) $\rho_k(A) = \rho(A)$.

We will see in Exercise 15(b) of Section 6.1 that (a) holds if and only if all the terms $A_{kj}x_j$ $(j = 1, 2, \ldots, n)$ are nonnegative multiples of some nonzero complex number z. Without loss of generality we assume that $|z| = 1$. Thus there exist nonnegative real numbers $c_1, \ldots, c_n$ such that

$$A_{kj}x_j = c_j z. \tag{6}$$

Clearly, (b) holds if and only if for each j we have $A_{kj} = 0$ or $|x_j| = b$. Since each entry of A is assumed to be positive, we conclude that (b) holds if and only if

$$|x_j| = b \qquad \text{for } j = 1, 2, \ldots, n. \tag{7}$$

Thus (5), and hence (c) above, is valid for $k = 1, 2, \ldots, n$.

From (6) we see that

$$x_j = \frac{c_j}{A_{kj}} z \qquad (j = 1, 2, \ldots, n),$$

and hence

$$b = |x_j| = \left|\frac{c_j}{A_{kj}} z\right| = \frac{c_j}{A_{kj}} \qquad (j = 1, 2, \ldots, n)$$

by (7). Therefore, $x_j = bz$ for $j = 1, 2, \ldots, n$. So

$$x = \begin{pmatrix} x_1 \\ \vdots \\ x_n \end{pmatrix} = \begin{pmatrix} bz \\ \vdots \\ bz \end{pmatrix} = bz \begin{pmatrix} 1 \\ \vdots \\ 1 \end{pmatrix},$$

and hence $\{u\}$ is a basis for E_λ.

Clearly, u is an eigenvector of A corresponding to $\rho(A)$ since

$$Au = A\begin{pmatrix} 1 \\ \vdots \\ 1 \end{pmatrix} = \begin{pmatrix} \sum_{j=1}^{n} A_{1j} \\ \vdots \\ \sum_{j=1}^{n} A_{nj} \end{pmatrix} = \begin{pmatrix} \rho_1(A) \\ \vdots \\ \rho_n(A) \end{pmatrix} = \begin{pmatrix} \rho(A) \\ \vdots \\ \rho(A) \end{pmatrix} = \rho(A)\begin{pmatrix} 1 \\ \vdots \\ 1 \end{pmatrix} = \rho(A)u$$

by (c) above. But the preceding paragraph shows that if λ is any eigenvalue of A such that $|\lambda| = \rho(A)$, then λ corresponds to the eigenvector u. Hence $\lambda = \rho(A)$. ∎

Corollary 1. *Let $A \in M_{n\times n}(C)$ be a matrix in which each entry is positive, and let λ be an eigenvalue of A such that $|\lambda| = \nu(A)$. Then $\lambda = \nu(A)$, and the dimension of E_λ is 1.*

Proof. Exercise. ∎

Corollary 2. *Let $A \in M_{n\times n}(C)$ be a transition matrix in which each entry is positive, and let λ denote any eigenvalue of A other than 1. Then $|\lambda| < 1$. Moreover, the dimension of the eigenspace corresponding to the eigenvalue 1 is 1.*

Proof. Exercise. ∎

Our next result extends Corollary 2 to regular transition matrices and thus shows that regular transition matrices satisfy condition (a) of Theorems 5.18 and 5.19.

Theorem 5.24. *Let A be a regular transition matrix.*
(a) *If λ is an eigenvalue of A, then $\lambda \in S$.*
(b) *Furthermore, if $\lambda = 1$, then* $\dim(E_\lambda) = 1$.

Proof. Statement (a) was proved as Corollary 2 of Theorem 5.23.
Since A is regular, there exists a positive integer s such that A^s has only positive entries. Because A is a transition matrix and the entries of A^s are positive, the entries of $A^{s+1} = A^s(A)$ are positive. Let λ be an eigenvalue of A having absolute value 1. Then λ^s and λ^{s+1} are eigenvalues of A^s and A^{s+1}, respectively, having absolute value 1. So by Corollary 2 of Theorem 5.23, $\lambda^s = \lambda^{s+1} = 1$. Thus $\lambda = 1$. Let E_λ and E'_λ denote the eigenspaces of A and A^s, respectively, corresponding to $\lambda = 1$. Then $E_\lambda \subseteq E'_\lambda$, but E'_λ has dimension 1 (Corollary 2 of Theorem 5.23). Hence $E_\lambda = E'_\lambda$, and $\dim(E_\lambda) = 1$. ∎

Corollary. *Let A be a regular transition matrix that is diagonalizable. Then $\lim_{m\to\infty} A^m$ exists.*

The preceding corollary, which follows immediately from Theorems 5.24 and 5.19, is not the best possible result. In fact, it can be shown that if A is a regular transition matrix, then the multiplicity of 1 as an eigenvalue of A is 1. Thus by Theorem 5.12 condition (b) of Theorem 5.18 is satisfied. So if A is a regular transition matrix, $\lim_{m\to\infty} A^m$ exists whether A is diagonalizable or not. As with Theorem 5.18, however, the fact that the multiplicity of 1 as an eigenvalue of A is 1 cannot be proved at this time. Nevertheless, we will state this result here (leaving the proof until Exercise 20 of Section 7.2) and deduce further facts about $\lim_{m\to\infty} A^m$ when A is a regular transition matrix.

Theorem 5.25. *Let* A *be an* n × n *regular transition matrix. Then*

(a) *The multiplicity of* 1 *as an eigenvalue of* A *is* 1.

(b) $\lim_{m\to\infty} A^m$ *exists.*

(c) $L = \lim_{m\to\infty} A^m$ *is a transition matrix.*

(d) AL = LA = L.

(e) *The columns of* L *are identical. In fact, each column of* L *is equal to the unique probability vector* v *that is also an eigenvector corresponding to the eigenvalue* 1 *of* A.

(f) *For any probability vector* x, $\lim_{m\to\infty} (A^m x) = v$.

Proof. (a) See Exercise 20 of Section 7.2.

(b) The proof that $\lim_{m\to\infty} A^m$ exists follows from part (a) and Theorems 5.24 and 5.18.

(c) Since A^m is a transition matrix by the corollary to Theorem 5.20, each entry of A^m is nonnegative ($m = 1, 2, 3, \ldots$). Hence

$$L_{ij} = \lim_{m\to\infty} (A^m)_{ij} \geqslant 0 \qquad \text{for } 1 \leqslant i, j \leqslant n.$$

Moreover,

$$\sum_{i=1}^{n} L_{ij} = \sum_{i=1}^{n} \left[\lim_{m\to\infty} (A^m)_{ij} \right] = \lim_{m\to\infty} \left[\sum_{i=1}^{n} (A^m)_{ij} \right] = \lim_{m\to\infty} (1) = 1 \qquad \text{for } 1 \leqslant j \leqslant n.$$

Thus L is a transition matrix.

(d) By Theorem 5.17

$$AL = A\left(\lim_{m\to\infty} A^m \right) = \lim_{m\to\infty} (AA^m) = \lim_{m\to\infty} A^{m+1} = L.$$

Similarly, $LA = L$.

(e) Since $AL = L$ by part (d), each column of L is an eigenvector of A corresponding to the eigenvalue 1. Moreover, by part (c), each column of L is a probability vector. Thus by part (a) each column of L is equal to the unique probability vector v corresponding to the eigenvalue 1 of A.

(f) Let x be any probability vector, and set $y = Lx$. Then y is a probability vector (corollary to Theorem 5.20), and $Ay = ALx = Lx = y$ by part (d). Hence y is also an eigenvector corresponding to the eigenvalue 1 of A. So $y = v$ by part (e). ∎

Definition. *The vector* v *in part (e) of Theorem 5.25 is called the* fixed probability vector (*or* stationary vector) *of the regular transition matrix* A.

Theorem 5.25 can be used to deduce information about the eventual percentage in each state of a Markov chain having a regular transition matrix.

Example 4

A survey in ancient Persia showed that on a particular day 50% of the Persians preferred a loaf of bread, 30% preferred a jug of wine, and 20% preferred thou beside me in the wilderness. A subsequent survey 1 month later yielded the following data: Of those who preferred a loaf of bread on the first survey, 40% continued to prefer a loaf of bread, 10% now preferred a jug of wine, and 50% preferred thou; of those who preferred a jug of wine on the first survey, 20% now preferred a loaf of bread, 70% continued to prefer a jug of wine, and 10% now preferred thou; of those who preferred thou on the first survey, 20% now preferred a loaf of bread, 20% now preferred a jug of wine, and 60% continued to prefer thou.

The situation described in the preceding paragraph is a three-state Markov chain in which the states are the three possible preferences. Assuming that the trend described above continues, we can predict the percentage of Persians in each state for each month following the original survey. Letting the first, second, and third states be preference for bread, wine, and thou, respectively, we see that the probability vector that gives the initial probability of being in each state is

$$P = \begin{pmatrix} 0.50 \\ 0.30 \\ 0.20 \end{pmatrix},$$

and the transition matrix is

$$A = \begin{pmatrix} 0.40 & 0.20 & 0.20 \\ 0.10 & 0.70 & 0.20 \\ 0.50 & 0.10 & 0.60 \end{pmatrix}.$$

The probabilities of being in each state m months after the original survey are the coordinates of the vector A^mP. The reader may check that

$$AP = \begin{pmatrix} 0.30 \\ 0.30 \\ 0.40 \end{pmatrix}, \qquad A^2P = A(AP) = \begin{pmatrix} 0.26 \\ 0.32 \\ 0.42 \end{pmatrix}, \qquad A^3P = A(A^2P) = \begin{pmatrix} 0.252 \\ 0.334 \\ 0.414 \end{pmatrix},$$

and

$$A^4P = A(A^3P) = \begin{pmatrix} 0.2504 \\ 0.3418 \\ 0.4078 \end{pmatrix}.$$

Note the seeming convergence of A^mP.

Since A is regular, the long-range prediction concerning the Persians' preferences can be found by computing the fixed probability vector for A. This vector is the unique probability vector v such that $(A - I)v = 0$. Letting

$$v = \begin{pmatrix} v_1 \\ v_2 \\ v_3 \end{pmatrix},$$

we see that the matrix equation $(A - I)v = 0$ yields the following system of linear equations:

$$\begin{cases} -0.60v_1 + 0.20v_2 + 0.20v_3 = 0 \\ 0.10v_1 - 0.30v_2 + 0.20v_3 = 0 \\ 0.50v_1 + 0.10v_2 - 0.40v_3 = 0. \end{cases}$$

It is easily shown that

$$\begin{pmatrix} 5 \\ 7 \\ 8 \end{pmatrix}$$

is a basis for the solution space of this system. Hence the unique fixed probability vector for A is

$$\begin{pmatrix} \dfrac{5}{5+7+8} \\ \dfrac{7}{5+7+8} \\ \dfrac{8}{5+7+8} \end{pmatrix} = \begin{pmatrix} 0.25 \\ 0.35 \\ 0.40 \end{pmatrix}.$$

Thus in the long run 25% of the Persians will prefer a loaf of bread, 35% will prefer a jug of wine, and 40% will prefer thou beside me in the wilderness.

Note that if

$$Q = \begin{pmatrix} 5 & 0 & 3 \\ 7 & 1 & 1 \\ 8 & -1 & -4 \end{pmatrix},$$

then

$$Q^{-1}AQ = \begin{pmatrix} 1 & 0 & 0 \\ 0 & 0.5 & 0 \\ 0 & 0 & 0.2 \end{pmatrix}.$$

So

$$\lim_{m\to\infty} A^m = Q\left[\lim_{m\to\infty}\begin{pmatrix}1 & 0 & 0\\ 0 & 0.5 & 0\\ 0 & 0 & 0.2\end{pmatrix}^m\right]Q^{-1} = Q\begin{pmatrix}1 & 0 & 0\\ 0 & 0 & 0\\ 0 & 0 & 0\end{pmatrix}Q^{-1}$$

$$= \begin{pmatrix}0.25 & 0.25 & 0.25\\ 0.35 & 0.35 & 0.35\\ 0.40 & 0.40 & 0.40\end{pmatrix}. \quad \blacksquare$$

Example 5

Farmers in Lamron plant one crop per year—either corn, soybeans, or wheat. Because they believe in the necessity of rotating their crops, these farmers will not plant the same crop in successive years. In fact, of the total acreage on which a particular crop is planted, exactly half will be planted with each of the other two crops during the succeeding year. This year 300 acres of corn were planted, 200 acres of soybeans were planted, and 100 acres of wheat were planted.

The situation described in the paragraph above is another three-state Markov chain in which the three states correspond to the planting of corn, soybeans, and wheat, respectively. In this problem, however, the amount of land devoted to each crop, rather than the percentage of the total acreage (600 acres), was given. By converting these amounts into fractions of the total acreage, we see that the transition matrix A and the initial probability vector P of the Markov chain are

$$A = \begin{pmatrix}0 & \frac{1}{2} & \frac{1}{2}\\ \frac{1}{2} & 0 & \frac{1}{2}\\ \frac{1}{2} & \frac{1}{2} & 0\end{pmatrix} \quad\text{and}\quad P = \begin{pmatrix}\frac{300}{600}\\ \frac{200}{600}\\ \frac{100}{600}\end{pmatrix} = \begin{pmatrix}\frac{1}{2}\\ \frac{1}{3}\\ \frac{1}{6}\end{pmatrix}.$$

The fraction of the total acreage devoted to each crop in m years is given by the coordinates of A^mP, and the eventual proportions of the total acreage to be used for each crop are the coordinates of $\lim_{m\to\infty} A^mP$. Thus the eventual amounts of land to be devoted to each crop are found by multiplying this limit by the total acreage; i.e., the eventual amounts of land to be used for each crop are the coordinates of $600(\lim_{m\to\infty} A^mP)$.

Since A is a regular transition matrix, Theorem 5.25 shows that $\lim_{m\to\infty} A^m$ is a matrix L in which each column equals the unique fixed probability vector for

A. It is easily seen that the fixed probability vector for A is

$$\begin{pmatrix} \frac{1}{3} \\ \frac{1}{3} \\ \frac{1}{3} \end{pmatrix}.$$

Hence

$$L = \begin{pmatrix} \frac{1}{3} & \frac{1}{3} & \frac{1}{3} \\ \frac{1}{3} & \frac{1}{3} & \frac{1}{3} \\ \frac{1}{3} & \frac{1}{3} & \frac{1}{3} \end{pmatrix};$$

so

$$600\left(\lim_{m\to\infty} A^m P\right) = 600LP = \begin{pmatrix} 200 \\ 200 \\ 200 \end{pmatrix}.$$

Thus in the long run we expect 200 acres of each crop to be planted each year. (For a direct computation of $600\,(\lim_{m\to\infty} A^m P)$, see Exercise 14.) ∎

In this section we have concentrated primarily on the theory of regular transition matrices. There is another interesting class of transition matrices that can be represented in the form

$$\begin{pmatrix} I & B \\ O & C \end{pmatrix},$$

where I is an identity matrix and O is a zero matrix. (Such transition matrices are not regular since the lower left block remains O in any power of the matrix.) The states corresponding to the identity submatrix are called *absorbing states* because such a state is never left once it is entered. A Markov chain is called an *absorbing Markov chain* if it is possible to go from an arbitrary state into an absorbing state in a finite number of stages. Observe that the Markov chain that described the enrollment pattern in a junior college is an absorbing Markov chain with states 1 and 4 as its absorbing states. Readers interested in learning more about absorbing Markov chains are referred to *Introduction to Finite Mathematics* (third edition) by J. Kemeny, J. Snell, and G. Thompson (Prentice-Hall, Inc., Englewood Cliffs, N.J., 1974) or *Discrete Mathematical Models* by Fred S. Roberts (Prentice-Hall, Inc., Englewood Cliffs, N.J., 1976).

An Application

In species that reproduce sexually, the characteristics of an offspring with respect to a particular genetic trait are determined by a pair of genes, one inherited from each parent. The genes for a particular trait are of two types, which we denote by G and g. The gene G represents the dominant characteristic and g represents the recessive characteristic. Offspring with genotypes GG or Gg exhibit the dominant characteristic, whereas offspring with genotype gg exhibit the recessive characteristic. For example, in humans, brown eyes are a dominant characteristic and blue eyes are the corresponding recessive characteristic; thus offspring with genotypes GG or Gg will be brown-eyed, whereas those of type gg will be blue-eyed.

Let us consider the probability of offspring of each genotype for a male parent of genotype Gg. (We assume that the population under consideration is large, that mating is random with respect to genotype, and that the distribution of each genotype within the population is independent of sex and life expectancy.) Let

$$P = \begin{pmatrix} p \\ q \\ r \end{pmatrix}$$

denote the proportion of the adult population with genotypes GG, Gg, and gg, respectively, at the start of the experiment. This experiment describes a three-state Markov chain with transition matrix

$$\begin{array}{ll} & \text{Genotype of female parent} \\ & \begin{array}{ccc} \text{GG} & \text{Gg} & \text{gg} \end{array} \\ \begin{array}{ll} \text{Genotype of offspring} & \begin{array}{l} \text{GG} \\ \text{Gg} \\ \text{gg} \end{array} \end{array} & \begin{pmatrix} \frac{1}{2} & \frac{1}{4} & 0 \\ \frac{1}{2} & \frac{1}{2} & \frac{1}{2} \\ 0 & \frac{1}{4} & \frac{1}{2} \end{pmatrix} = B. \end{array}$$

It is easily checked that B^2 contains only positive entries; so B is regular. Thus by permitting only males of genotype Gg to reproduce, the proportion of offspring in the population having a certain genotype will stabilize at the fixed probability vector for B, which is

$$\begin{pmatrix} \frac{1}{4} \\ \frac{1}{2} \\ \frac{1}{4} \end{pmatrix}.$$

Now suppose that similar experiments are to be performed with males of genotypes GG and gg. As above, these experiments are three-state Markov chains with transition matrices

$$A = \begin{pmatrix} 1 & \frac{1}{2} & 0 \\ 0 & \frac{1}{2} & 1 \\ 0 & 0 & 0 \end{pmatrix} \quad \text{and} \quad C = \begin{pmatrix} 0 & 0 & 0 \\ 1 & \frac{1}{2} & 0 \\ 0 & \frac{1}{2} & 1 \end{pmatrix},$$

respectively. In order to consider the case where all male genotypes are permitted to reproduce, we must form the transition matrix $M = pA + qB + rC$, which is a linear combination of A, B, and C weighted by the proportion of males of each genotype. Thus

$$M = \begin{pmatrix} p + \frac{1}{2}q & \frac{1}{2}p + \frac{1}{4}q & 0 \\ \frac{1}{2}q + r & \frac{1}{2}p + \frac{1}{2}q + \frac{1}{2}r & p + \frac{1}{2}q \\ 0 & \frac{1}{4}q + \frac{1}{2}r & \frac{1}{2}q + r \end{pmatrix}.$$

To simplify the notation, let $a = p + \frac{1}{2}q$ and $b = \frac{1}{2}q + r$. (The numbers a and b represent the proportion of G and g genes, respectively, in the population.) Then

$$M = \begin{pmatrix} a & \frac{1}{2}a & 0 \\ b & \frac{1}{2} & a \\ 0 & \frac{1}{2}b & b \end{pmatrix},$$

where $a + b = p + q + r = 1$.

Let p', q', and r' denote the proportion of the first-generation offspring having genotypes GG, Gg, and gg, respectively. Then

$$\begin{pmatrix} p' \\ q' \\ r' \end{pmatrix} = MP = \begin{pmatrix} ap + \frac{1}{2}aq \\ bp + \frac{1}{2}q + ar \\ \frac{1}{2}bq + br \end{pmatrix} = \begin{pmatrix} a^2 \\ 2ab \\ b^2 \end{pmatrix}.$$

In order to consider the effects of unrestricted matings among the first-generation offspring, a new transition matrix $\tilde{M}$ must be determined based upon the distribution of first-generation genotypes. As before, we find that

$$\tilde{M} = \begin{pmatrix} p' + \frac{1}{2}q' & \frac{1}{2}p' + \frac{1}{4}q' & 0 \\ \frac{1}{2}q' + r' & \frac{1}{2}p' + \frac{1}{2}q' + \frac{1}{2}r' & p' + \frac{1}{2}q' \\ 0 & \frac{1}{4}q' + \frac{1}{2}r' & \frac{1}{2}q' + r' \end{pmatrix} = \begin{pmatrix} a' & \frac{1}{2}a' & 0 \\ b' & \frac{1}{2} & a' \\ 0 & \frac{1}{2}b' & b' \end{pmatrix},$$

where $a' = p' + \frac{1}{2}q'$ and $b' = \frac{1}{2}q' + r'$. But

$$a' = a^2 + \tfrac{1}{2}(2ab) = a(a + b) = a$$

and

$$b' = \tfrac{1}{2}(2ab) + b^2 = b(a + b) = b.$$

Thus $\tilde{M} = M$; so the distribution of second-generation offspring among the three genotypes is

$$\tilde{M}(MP) = M^2P = \begin{pmatrix} a^3 + a^2b \\ a^2b + ab + ab^2 \\ ab^2 + b^3 \end{pmatrix} = \begin{pmatrix} a^2(a + b) \\ ab(a + 1 + b) \\ b^2(a + b) \end{pmatrix} = \begin{pmatrix} a^2 \\ 2ab \\ b^2 \end{pmatrix}$$

$$= MP,$$

the same as the first-generation offspring. In other words, MP is the fixed probability vector for M, and genetic equilibrium is achieved in the population after only one generation. (This result is called the *Hardy–Weinberg law.*) Notice that in the important special case where $a = b$ (or equivalently, where $p = r$), the distribution at equilibrium is

$$MP = \begin{pmatrix} a^2 \\ 2ab \\ b^2 \end{pmatrix} = \begin{pmatrix} a^2 \\ 2a^2 \\ a^2 \end{pmatrix} = \begin{pmatrix} \frac{1}{4} \\ \frac{1}{2} \\ \frac{1}{4} \end{pmatrix}.$$

EXERCISES

1. Label the following statements as being true or false.

(a) If $A \in \mathsf{M}_{n \times n}(C)$ and $\lim_{m \to \infty} A^m = L$, then, for any invertible matrix $Q \in \mathsf{M}_{n \times n}(C)$, $\lim_{m \to \infty} QA^mQ^{-1} = QLQ^{-1}$.

(b) If 2 is an eigenvalue of $A \in \mathsf{M}_{n \times n}(C)$, then $\lim_{m \to \infty} A^m$ does not exist.

(c) A vector

$$\begin{pmatrix} x_1 \\ \vdots \\ x_n \end{pmatrix} \in \mathsf{R}^n$$

such that $x_1 + \cdots + x_n = 1$ is a probability vector.

(d) The sum of the entries in each row of a transition matrix equals 1.

(e) The product of a transition matrix and a probability vector is a probability vector.

(f) The matrix

$$\begin{pmatrix} 1 & 0 & -1 \\ 0 & 1 & 1 \\ -1 & 1 & 0 \end{pmatrix}$$

does not have 3 as an eigenvalue.

(g) Every transition matrix has 1 as an eigenvalue.

(h) No transition matrix can have -1 as an eigenvalue.

(i) If A is a transition matrix, then $\lim_{m \to \infty} A^m$ exists.

(j) If A is a regular transition matrix, then $\lim_{m \to \infty} A^m$ exists and has rank 1.

2. Determine whether or not $\lim_{m \to \infty} A^m$ exists for the following matrices A. If the limit exists, compute it.

(a) $\begin{pmatrix} 0.1 & 0.7 \\ 0.7 & 0.1 \end{pmatrix}$ **(b)** $\begin{pmatrix} 0.50 & 2 \\ 0.75 & 0 \end{pmatrix}$ **(c)** $\begin{pmatrix} 0.4 & 0.7 \\ 0.6 & 0.3 \end{pmatrix}$

(d) $\begin{pmatrix} 2 & 1 \\ -2 & 0 \end{pmatrix}$ **(e)** $\begin{pmatrix} -2 & -1 \\ 4 & 3 \end{pmatrix}$ **(f)** $\begin{pmatrix} -0.5 & 0.5 \\ 0.9 & -0.1 \end{pmatrix}$

(g) $\begin{pmatrix} -1.8 & 0 & -1.4 \\ -5.6 & 1 & -2.8 \\ 2.8 & 0 & 2.4 \end{pmatrix}$ **(h)** $\begin{pmatrix} -2.5 & 4.5 & 7.5 \\ -1.5 & -2.5 & -1.5 \\ -1.5 & 4.5 & 6.5 \end{pmatrix}$

(i) $\begin{pmatrix} -\frac{1}{2} - 2i & 4i & \frac{1}{2} + 5i \\ 1 + 2i & -3i & -1 - 4i \\ -1 - 2i & 4i & 1 + 5i \end{pmatrix}$

(j) $\begin{pmatrix} \dfrac{-26}{3} + \dfrac{i}{3} & \dfrac{-28}{3} - \dfrac{4i}{3} & 28 \\ \dfrac{-7}{3} + \dfrac{2i}{3} & \dfrac{-5}{3} + \dfrac{i}{3} & 7 - 2i \\ \dfrac{-13}{6} + i & \dfrac{-5}{6} + i & \dfrac{35}{6} - \dfrac{10i}{3} \end{pmatrix}$

3. Prove that if $A_1, A_2, A_3, \ldots$ is a sequence of $n \times p$ matrices with complex number entries such that $\lim_{m \to \infty} A_m = L$, then $\lim_{m \to \infty} A_m^t = L^t$.

4. Prove that if $A \in \mathsf{M}_{n \times n}(C)$ is diagonalizable and $L = \lim_{m \to \infty} A^m$ exists, then either $L = I_n$ or $\operatorname{rank}(L) < n$.

5. Find 2×2 matrices A and B having real number entries such that $\lim_{m \to \infty} A^m$, $\lim_{m \to \infty} B^m$, and $\lim_{m \to \infty} (AB)^m$ all exist but $\lim_{m \to \infty} (AB)^m \neq (\lim_{m \to \infty} A^m)(\lim_{m \to \infty} B^m)$.

6. A hospital trauma unit has determined that 30% of its patients are ambulatory and 70% are bedridden at the time of arrival at the hospital. A month after arrival, 60% of the ambulatory patients have recovered, 20% remain ambulatory, and 20% have become bedridden. After the same time, 10% of the bedridden patients have recovered, 20% have become ambulatory, 50% remain bedridden, and 20% have died. Determine the percentage of patients who have recovered, are ambulatory, are bedridden, and have died 1 month after arrival. Also determine the eventual percentage of patients of each type.

7. A player begins a game of chance by placing a marker in box 2 (marked *start*) (see Figure 5.4). A die is rolled, and the marker is moved one square to

Win	Start		Lose
1	2	3	4

Figure 5.4

the left if a 1 or 2 is rolled and one square to the right if a 3, 4, 5, or 6 is rolled. This process continues until the marker lands in square 1 (in which case the player wins the game) or square 4 (in which case the player loses the game). What is the probability of winning this game?

8. Which of the following are regular transition matrices?

(a) $\begin{pmatrix} 0.2 & 0.3 & 0.5 \\ 0.3 & 0.2 & 0.5 \\ 0.5 & 0.5 & 0 \end{pmatrix}$ **(b)** $\begin{pmatrix} 0.5 & 0 & 1 \\ 0.5 & 0 & 0 \\ 0 & 1 & 0 \end{pmatrix}$ **(c)** $\begin{pmatrix} 0.5 & 0 & 0 \\ 0.5 & 0 & 1 \\ 0 & 1 & 0 \end{pmatrix}$

(d) $\begin{pmatrix} 0.5 & 0 & 1 \\ 0.5 & 1 & 0 \\ 0 & 0 & 0 \end{pmatrix}$ **(e)** $\begin{pmatrix} \frac{1}{3} & 0 & 0 \\ \frac{1}{3} & 1 & 0 \\ \frac{1}{3} & 0 & 1 \end{pmatrix}$ **(f)** $\begin{pmatrix} 1 & 0 & 0 \\ 0 & 0.7 & 0.2 \\ 0 & 0.3 & 0.8 \end{pmatrix}$

(g) $\begin{pmatrix} 0 & \frac{1}{2} & 0 & 0 \\ \frac{1}{2} & 0 & 0 & 0 \\ \frac{1}{4} & \frac{1}{4} & 1 & 0 \\ \frac{1}{4} & \frac{1}{4} & 0 & 1 \end{pmatrix}$ **(h)** $\begin{pmatrix} \frac{1}{4} & \frac{1}{4} & 0 & 0 \\ \frac{1}{4} & \frac{1}{4} & 0 & 0 \\ \frac{1}{4} & \frac{1}{4} & 1 & 0 \\ \frac{1}{4} & \frac{1}{4} & 0 & 1 \end{pmatrix}$

9. Compute $\lim_{m \to \infty} A^m$, if it exists, for each of the matrices A in Exercise 8.

10. Each of the following matrices is a regular transition matrix for a three-state Markov chain. In each case the initial probability vector is

$$P = \begin{pmatrix} 0.3 \\ 0.3 \\ 0.4 \end{pmatrix}.$$

For each transition matrix, compute the proportion of objects in each state after two stages and the eventual proportion of objects in each state by determining the fixed probability vector.

(a) $\begin{pmatrix} 0.6 & 0.1 & 0.1 \\ 0.1 & 0.9 & 0.2 \\ 0.3 & 0 & 0.7 \end{pmatrix}$ **(b)** $\begin{pmatrix} 0.8 & 0.1 & 0.2 \\ 0.1 & 0.8 & 0.2 \\ 0.1 & 0.1 & 0.6 \end{pmatrix}$ **(c)** $\begin{pmatrix} 0.9 & 0.1 & 0.1 \\ 0.1 & 0.6 & 0.1 \\ 0 & 0.3 & 0.8 \end{pmatrix}$

(d) $\begin{pmatrix} 0.4 & 0.2 & 0.2 \\ 0.1 & 0.7 & 0.2 \\ 0.5 & 0.1 & 0.6 \end{pmatrix}$ **(e)** $\begin{pmatrix} 0.5 & 0.3 & 0.2 \\ 0.2 & 0.5 & 0.3 \\ 0.3 & 0.2 & 0.5 \end{pmatrix}$ **(f)** $\begin{pmatrix} 0.6 & 0 & 0.4 \\ 0.2 & 0.8 & 0.2 \\ 0.2 & 0.2 & 0.4 \end{pmatrix}$

11. In 1940 a county land-use survey showed that 10% of the county land was urban, 50% was unused, and 40% was agricultural. Five years later a follow-up survey revealed that 70% of the urban land had remained urban, 10% had become unused, and 20% had become agricultural. Likewise, 20% of the unused land had become urban, 60% had remained unused, and 20% had become agricultural. Finally, the 1945 survey showed that 20% of the agricultural land had become unused while 80% remained agricultural. Assuming that the trends indicated by the 1945 survey continue, compute the percentage of urban, unused, and agricultural land in the county in 1950 and the corresponding eventual percentages.

12. A diaper liner is placed in each diaper worn by a baby. If, after a diaper change, the liner is soiled, then it is discarded. Otherwise the liner is washed with the diapers and reused, except that each liner is discarded after its third use (even if it has never been soiled). The probability that the baby will soil any diaper liner is one-third. If there are only new diaper liners at first, eventually what proportion of the diaper liners being used will be new, once-used, and twice-used?

13. In 1975 the automobile industry determined that 40% of American car owners drove large cars, 20% drove intermediate-sized cars, and 40% drove small cars. A second survey in 1985 showed that 70% of the large-car owners in 1975 still owned large cars in 1985, but 30% had changed to an intermediate-sized car. Of those who owned intermediate-sized cars in 1975, 10% had switched to large cars, 70% continued to drive intermediate-sized cars, and 20% had changed to small cars in 1985. Finally, of the small-car owners in 1975, 10% owned intermediate-sized cars and 90% owned small cars in 1985. Assuming that these trends continue, determine the percentage of Americans who will own cars of each size in 1995 and the corresponding eventual percentages.

14. Show that if A and P are as in Example 5, then

$$A^m = \begin{pmatrix} r_m & r_{m+1} & r_{m+1} \\ r_{m+1} & r_m & r_{m+1} \\ r_{m+1} & r_{m+1} & r_m \end{pmatrix},$$

where

$$r_m = \frac{1}{3}\left[1 + \frac{(-1)^m}{2^{m-1}}\right].$$

Deduce that

$$600(A^m P) = A^m \begin{pmatrix} 300 \\ 200 \\ 100 \end{pmatrix} = \begin{pmatrix} 200 + \dfrac{(-1)^m}{2^m}(100) \\ 200 \\ 200 + \dfrac{(-1)^{m+1}}{2^m}(100) \end{pmatrix}$$

15. Prove Theorem 5.20 and its corollary.

16. Prove the two corollaries of Theorem 5.23.

17. Prove the corollary to Theorem 5.24.

Definition. *If* $A \in M_{n \times n}(C)$, *define* $e^A = \lim_{m \to \infty} B_m$, *where*

$$B_m = I + A + \frac{A^2}{2!} + \cdots + \frac{A^m}{m!}$$

(see Exercise 20). Thus e^A *is the sum of the infinite series*

$$I + A + \frac{A^2}{2!} + \frac{A^3}{3!} + \cdots,$$

and B_m *is the* m*th partial sum of this series. (Note the analogy with the power series*

$$e^a = 1 + a + \frac{a^2}{2!} + \frac{a^3}{3!} + \cdots,$$

which is valid for all complex numbers a.)

18. Compute e^O and e^I, where O and I denote the $n \times n$ zero and identity matrices, respectively.

19. Suppose that $P^{-1}AP$ is a diagonal matrix D. Prove that $e^A = Pe^DP^{-1}$.

20. Let $A \in M_{n \times n}(C)$ be diagonalizable. Use the result of Exercise 19 to show that e^A exists. (Exercise 21 of Section 7.2 will show that e^B exists for each $B \in M_{n \times n}(C)$.)

21. Find $A, B \in M_{2 \times 2}(R)$ such that $e^Ae^B \neq e^{A+B}$.

22. Prove that a differentiable function $X: R \to R^n$ is a solution to the system of differential equations defined in Exercise 14 of Section 5.2 if and only if X is of the form $X(t) = e^{tA}v$ for some $v \in R^n$, where A is as defined in that exercise.

5.4 INVARIANT SUBSPACES AND THE CAYLEY–HAMILTON THEOREM

In Section 5.1 we observed that if x is an eigenvector of a linear operator T, then T maps the span of $\{x\}$ into itself. Subspaces that are mapped into themselves are of great importance in the study of linear operators (see, e.g., Exercise 26 of Section 2.1).

Definition. *Let* T *be a linear operator on a vector space* V. *A subspace* W *of* V *is called a* T-invariant *subspace of* V *if* $T(W) \subseteq W$, *that is, if* $T(x) \in W$ *for all* $x \in W$.

Example 1

Suppose that T is a linear operator on a vector space V. Then the following subspaces of V are T-invariant.

(a) $\{0\}$
(b) V
(c) R(T)
(d) N(T)
(e) E_λ, for any eigenvalue λ of T

The proofs that these subspaces are T-invariant are left as exercises. ∎

Example 2

Let $T: R^3 \to R^3$ be defined by

$$T(a, b, c) = (a + b, b + c, 0).$$

Then the xy-plane $= \{(x, y, 0): x, y \in R\}$ and the x-axis $= \{(x, 0, 0): x \in R\}$ are T-invariant subspaces of R^3. ∎

Let T be a linear operator on a vector space V, and let x be a nonzero element of V. The subspace

$$W = \text{span}(\{x, T(x), T^2(x), \ldots\})$$

is called the T-*cyclic subspace* of V *generated by* x. It is a simple matter to show that W is T-invariant. In fact, W is the "smallest" T-invariant subspace of V containing x. That is, any T-invariant subspace of V containing x must also contain W (see the exercises). Cyclic subspaces have various uses. We apply them in this section to establish the Cayley–Hamilton theorem. In the exercises we outline a method for using cyclic subspaces to compute the characteristic polynomial of a linear operator without resorting to determinants. Cyclic subspaces play an important role in Chapter 7, where we study matrix representations of nondiagonalizable linear operators.

Example 3

Let $T: R^3 \to R^3$ be defined by

$$T(a, b, c) = (-b + c, a + c, 3c).$$

We will determine the T-cyclic subspace W generated by $e_1 = (1, 0, 0)$. Since $T(e_1) = T(1, 0, 0) = (0, 1, 0) = e_2$ and $T^2(e_1) = T(T(e_1)) = T(e_2) = (-1, 0, 0) = -e_1$, $W_{e_1} = \text{span}(\{e_1, T(e_1), T^2(e_1), \ldots\}) = \text{span}(\{e_1, e_2\}) = \{(s, t, 0): s, t \in R\}$. ∎

Example 4

Let T be the linear operator on $P(R)$ defined by $T(f) = f'$. Then the T-cyclic subspace generated by x^2 is $\text{span}(\{x^2, 2x, 2\}) = P_2(R)$. ∎

The existence of a T-invariant subspace provides the opportunity to define a new linear operator whose domain is the subspace. If T is a linear operator on V and W is a T-invariant subspace of V, then the restriction T_W of T to W (see Appendix B) is a mapping from W to W, and it is a simple matter to argue that T_W is a linear operator on W (see the exercises). As a linear operator T_W inherits certain properties from its parent operator T. The following result illustrates how the two operators are linked.

Theorem 5.26. *Let T be a linear operator on a finite-dimensional vector space V, and let W be a T-invariant subspace of V. Then the characteristic polynomial of T_W divides the characteristic polynomial of T.*

Proof. Extend a basis $\gamma = \{x_1, \ldots, x_k\}$ for W to a basis $\beta = \{x_1, \ldots, x_k, \ldots, x_n\}$ for V. Let $A = [T]_\beta$ and $B_1 = [T_W]_\beta$. Then, by Exercise 10,

$$A = \begin{pmatrix} B_1 & B_2 \\ O & B_3 \end{pmatrix},$$

where O is an $(n-k) \times k$ zero matrix. If $f(t)$ is the characteristic polynomial of T and $g(t)$ is the characteristic polynomial of T_W, then

$$f(t) = \det(A - tI_n) = \det\begin{pmatrix} B_1 - tI_k & B_2 \\ O & B_3 - tI_{n-k} \end{pmatrix} = g(t)\cdot\det(B_3 - tI_{n-k})$$

by Exercise 9 of Section 4.3. Thus $g(t)$ divides $f(t)$. ∎

Example 5

Let $T: R^4 \to R^4$ be defined by

$$T(a, b, c, d) = (a + b + 2c - d, b + d, 2c - d, c + d),$$

and let $W = \{(t, s, 0, 0): t, s \in R\}$. Observe that W is a T-invariant subspace of R^4, for

$$T(a, b, 0, 0) = (a + b, b, 0, 0) \in W.$$

Let $\gamma = \{e_1, e_2\}$, and note that γ is a basis for W. Extend γ to the standard basis β for R^4. Then

$$B_1 = [T_W]_\gamma = \begin{pmatrix} 1 & 1 \\ 0 & 1 \end{pmatrix} \quad \text{and} \quad A = [T]_\beta = \begin{pmatrix} 1 & 1 & 2 & -1 \\ 0 & 1 & 0 & 1 \\ 0 & 0 & 2 & -1 \\ 0 & 0 & 1 & 1 \end{pmatrix}$$

in the notation of Theorem 5.26. Thus if $f(t)$ is the characteristic polynomial of T and $g(t)$ is the characteristic polynomial of T_W, then

$$f(t) = \det(A - tI_4) = \det\begin{pmatrix} 1-t & 1 & 2 & -1 \\ 0 & 1-t & 0 & 1 \\ 0 & 0 & 2-t & -1 \\ 0 & 0 & 1 & 1-t \end{pmatrix}$$

$$= \det\begin{pmatrix} 1-t & 1 \\ 0 & 1-t \end{pmatrix} \cdot \det\begin{pmatrix} 2-t & -1 \\ 1 & 1-t \end{pmatrix} = g(t) \cdot \det\begin{pmatrix} 2-t & -1 \\ 1 & 1-t \end{pmatrix}. \quad \blacksquare$$

In view of Theorem 5.26 we may use the characteristic polynomial of T_W to gain information about the characteristic polynomial of T itself. In this regard, cyclic subspaces are useful because the characteristic polynomial of the restriction of T to a cyclic subspace is readily computable.

Theorem 5.27. *Let* T *be a linear operator on a finite-dimensional vector space* V, *and let* W *denote the* T*-cyclic subspace of* V *generated by* $x \in V$. *Suppose that* $\dim(W) = k \geqslant 1$ *(and hence* $x \neq 0$*). Then*

(a) $\{x, T(x), T^2(x), \ldots, T^{k-1}(x)\}$ *is a basis for* W.

(b) *If* $T^k(x) = -a_0x - a_1T(x) - \cdots - a_{k-1}T^{k-1}(x)$, *then the characteristic polynomial of* T_W *is* $f(t) = (-1)^k(a_0 + a_1t + \cdots + a_{k-1}t^{k-1} + t^k)$.

Proof. Let j be the smallest integer for which $\{x, T(x), \ldots, T^j(x)\}$ is linearly dependent. (Such a j must exist since W is finite-dimensional.) Since $x \neq 0$, $j \geqslant 1$. Thus $\{x, T(x), \ldots, T^{j-1}(x)\}$ is linearly independent and $T^j(x) \in \text{span}(\{x, T(x), \ldots, T^{j-1}(x)\})$ by Theorem 1.8. We will show by mathematical induction that $T^s(x)$ lies in this span for any nonnegative integer s. This is clear for $0 \leqslant s \leqslant j$. Suppose that $T^m(x) \in \text{span}(\{x, T(x), \ldots, T^{j-1}(x)\})$ for some $m \geqslant j$. Then there exist scalars $b_0, b_1, \ldots, b_{j-1}$ such that

$$T^m(x) = b_0x + b_1T(x) + \cdots + b_{j-1}T^{j-1}(x).$$

Applying T to both sides of the preceding equality, we obtain

$$T^{m+1}(x) = b_0T(x) + b_1T^2(x) + \cdots + b_{j-1}T^j(x).$$

So $T^{m+1}(x)$ is a linear combination of $T(x), T^2(x), \ldots, T^j(x)$, each of which lies in $\text{span}(\{x, T(x), \ldots, T^{j-1}(x)\})$. Thus $T^{m+1}(x)$ lies in this span, completing the induction. Hence

$$W = \text{span}(\{x, T(x), T^2(x), \ldots\}) \subseteq \text{span}(\{x, T(x), \ldots, T^{j-1}(x)\}).$$

But clearly, the reverse inclusion is also true, and so $\{x, T(x), \ldots, T^{j-1}(x)\}$ spans W. Since this set is also linearly independent, it is a basis for W.

But $\dim(\mathsf{W}) = k$; so this set must contain k elements. Therefore, $j = k$, and thus $\{x, \mathsf{T}(x), \ldots, \mathsf{T}^{k-1}(x)\}$ is a basis for W, proving (a).

To prove part (b), let $\beta = \{x, \mathsf{T}(x), \ldots, \mathsf{T}^{k-1}(x)\}$ be the basis from part (a), and let $-a_0, -a_1, \ldots, -a_{k-1}$ be the scalars such that $\mathsf{T}^k(x) = -a_0 x - a_1\mathsf{T}(x) - \cdots - a_{k-1}\mathsf{T}^{k-1}(x)$. Observe that

$$[\mathsf{T}_\mathsf{W}]_\beta = \begin{pmatrix} 0 & 0 & \cdots & 0 & -a_0 \\ 1 & 0 & \cdots & 0 & -a_1 \\ 0 & 1 & \cdots & 0 & -a_2 \\ \vdots & \vdots & & \vdots & \vdots \\ 0 & 0 & \cdots & 0 & -a_{k-2} \\ 0 & 0 & \cdots & 1 & -a_{k-1} \end{pmatrix},$$

so the characteristic polynomial of $[\mathsf{T}_\mathsf{W}]_\beta$ is

$$f(t) = (-1)^k(a_0 + a_1 t + \cdots + a_{k-1}t^{k-1} + t^k)$$

by Exercise 19. Thus $f(t)$ is the characteristic polynomial of T_W, proving (b). ∎

Example 6

Let T be the linear operator of Example 3, and let $\mathsf{W} = \text{span}\{e_1, e_2\}$, the T-cyclic subspace generated by e_1. We can compute the characteristic polynomial $f(t)$ of T_W in two ways: by means of Theorem 5.27 and by means of determinants.

(a) *By means of Theorem 5.27.* From Example 3 we have that $\{e_1, e_2\}$ generates W and that $\mathsf{T}^2(e_1) = -e_1$. Therefore, $f(t) = t^2 + 1$ by part (b) of Theorem 2.27.

(b) *By means of determinants.* Let $\beta = \{e_1, e_2\}$, which is a basis for W. Since $\mathsf{T}(e_1) = e_2$ and $\mathsf{T}(e_2) = -e_1$, we have that

$$[\mathsf{T}]_\beta = \begin{pmatrix} 0 & -1 \\ 1 & 0 \end{pmatrix},$$

and therefore,

$$f(t) = \det\begin{pmatrix} -t & -1 \\ 1 & -t \end{pmatrix} = t^2 + 1. \quad ∎$$

The Cayley–Hamilton Theorem

As an illustration of the usefulness of Theorem 5.27, we prove a well-known result that will be useful in Chapter 7. The reader should refer to Appendix E for the definition of $f(\mathsf{T})$, where T is a linear operator and $f(x)$ is a polynomial.

Theorem 5.28 (Cayley–Hamilton). *Let* T *be a linear operator on a finite-dimensional vector space* V, *and let* f(t) *be the characteristic polynomial of* T. *Then*

$\mathsf{f(T) = T_0}$ *(the zero transformation); that is,* T *satisfies its characteristic polynomial.*

Proof. We must show that $f(\mathsf{T})(x) = 0$ for all $x \in \mathsf{V}$. If $x = 0$, then $f(\mathsf{T})(x) = 0$ since $f(\mathsf{T})$ is a linear transformation. Suppose then that $x \neq 0$, and let W denote the T-cyclic subspace of V generated by x. If $\dim(\mathsf{W}) = k$, then by Theorem 5.27 there exist scalars $-a_0, -a_1, \ldots, -a_{k-1}$ such that

$$\mathsf{T}^k(x) = -a_0 x - a_1\mathsf{T}(x) - \cdots - a_{k-1}\mathsf{T}^{k-1}(x).$$

Hence Theorem 5.27 implies that

$$g(t) = (-1)^k(a_0 + a_1 t + \cdots + a_{k-1}t^{k-1} + t^k)$$

is the characteristic polynomial of $\mathsf{T_W}$. Combining these two equations yields

$$g(\mathsf{T})(x) = (-1)^k(a_0\mathsf{I} + a_1\mathsf{T} + \cdots + a_{k-1}\mathsf{T}^{k-1} + \mathsf{T}^k)(x) = 0.$$

By Theorem 5.26, $g(t)$ divides $f(t)$; hence there exists a polynomial $q(t)$ such that $f(t) = q(t)g(t)$. So

$$f(\mathsf{T})(x) = q(\mathsf{T})g(\mathsf{T})(x) = q(\mathsf{T})(g(\mathsf{T})(x)) = q(\mathsf{T})(0) = 0. \quad \blacksquare$$

Example 7

Define $\mathsf{T}: \mathsf{R}^2 \to \mathsf{R}^2$ by $\mathsf{T}(a, b) = (a + 2b, -2a + b)$, and let $\beta = \{e_1, e_2\}$. Then

$$A = \begin{pmatrix} 1 & 2 \\ -2 & 1 \end{pmatrix},$$

where $A = [\mathsf{T}]_\beta$. The characteristic polynomial of T is therefore

$$f(t) = \det(A - tI) = \det\begin{pmatrix} 1-t & 2 \\ -2 & 1-t \end{pmatrix} = t^2 - 2t + 5.$$

It is easily verified that $\mathsf{T}_0 = f(\mathsf{T}) = \mathsf{T}^2 - 2\mathsf{T} + 5\mathsf{I}$. Similarly,

$$f(A) = A^2 - 2A + 5I = \begin{pmatrix} -3 & 4 \\ -4 & -3 \end{pmatrix} + \begin{pmatrix} -2 & -4 \\ 4 & -2 \end{pmatrix} + \begin{pmatrix} 5 & 0 \\ 0 & 5 \end{pmatrix}$$

$$= \begin{pmatrix} 0 & 0 \\ 0 & 0 \end{pmatrix}. \quad \blacksquare$$

Example 7 suggests the following.

Corollary (Cayley–Hamilton Theorem for Matrices). *Let* A *be an* n × n *matrix and let* f(t) *be the characteristic polynomial of* A. *Then* f(A) = O, *the* n × n *zero matrix.*

Proof. Exercise. ∎

Invariant Subspaces and Direct Sums*

It is useful to decompose a finite-dimensional vector space V into a direct sum of as many T-invariant subspaces as possible because the behavior of T on V can be inferred from its behavior on the direct summands. For example, T is diagonalizable if and only if V can be decomposed into a direct sum of one-dimensional T-invariant subspaces (see Exercise 33). In Chapter 7 we will consider alternative ways of decomposing V into direct sums of T-invariant subspaces if T is not diagonalizable. We now proceed to gather a few facts about direct sums of T-invariant subspaces for use in Section 7.4. The first of these facts is about characteristic polynomials.

Theorem 5.29. *Let* T *be a linear operator on a finite-dimensional vector space* V, *and suppose that* $V = W_1 \oplus \cdots \oplus W_k$, *where* W_i *is a* T*-invariant subspace of* V *for each* i $(1 \leqslant i \leqslant k)$. *If* f(t) *denotes the characteristic polynomial of* T *and* $f_i(t)$ *denotes the characteristic polynomial of* T_{W_i} $(1 \leqslant i \leqslant k)$, *then*

$$f(t) = f_1(t) \cdot f_2(t) \cdot \cdots \cdot f_k(t).$$

Proof. The proof is by induction on k. Suppose first that $k = 2$. Let β_1 be a basis for W_1, β_2 be a basis for W_2, and $\beta = \beta_1 \cup \beta_2$. Then β is a basis for V by Theorem 5.15(d). Let $A = [T]_\beta$, $B_1 = [T_{W_1}]_{\beta_1}$, and $B_2 = [T_{W_2}]_{\beta_2}$. It is easily seen by Exercise 32 that

$$A = \begin{pmatrix} B_1 & O \\ O' & B_2 \end{pmatrix},$$

where O and O' are zero matrices. Thus

$$f(t) = \det(A - tI) = \det(B_1 - tI) \cdot \det(B_2 - tI) = f_1(t) \cdot f_2(t)$$

by Exercise 9 of Section 4.3, proving the result if $k = 2$.

Now assume that the theorem is true for $k - 1$ summands, where $k - 1$ is some integer greater than or equal to 1. Suppose that V is a direct sum of k summands,

$$V = W_1 \oplus W_2 \oplus \cdots \oplus W_k,$$

and define $W = W_1 + W_2 + \cdots + W_{k-1}$. It is easily verified that $V = W \oplus W_k$. So by the case for $k = 2$, $f(t) = g(t) \cdot f_k(t)$, where $g(t)$ is the characteristic polynomial of T_W. Clearly $W = W_1 \oplus W_2 \oplus \cdots \oplus W_{k-1}$. Hence by the induction hypothesis we have $g(t) = f_1(t) \cdot f_2(t) \cdot \cdots \cdot f_{k-1}(t)$. Thus $f(t) = g(t) \cdot f_k(t) = f_1(t) \cdot f_2(t) \cdot \cdots \cdot f_k(t)$. ∎

As an illustration of this result, suppose that T is a diagonalizable linear operator on a finite-dimensional vector space V with distinct eigenvalues $\lambda_1, \lambda_2, \ldots, \lambda_k$. By Theorem 5.16 we have that V is a direct sum of the eigenspaces of T.

Since each eigenspace is T-invariant, we may view this situation in the context of Theorem 5.29. For any eigenvalue λ_i, the restriction of T to E_{λ_i} has characteristic polynomial $(\lambda_i - t)^{m_i}$, where m_i is the dimension of E_{λ_i}. By Theorem 5.29 the characteristic polynomial $f(t)$ of T is the product

$$f(t) = (\lambda_1 - t)^{m_1}(\lambda_2 - t)^{m_2} \cdots (\lambda_k - t)^{m_k}.$$

It follows that the multiplicity of each eigenvalue is equal to the dimension of the corresponding eigenspace, as is expected.

Example 8

Let $\mathsf{T}: \mathsf{R}^4 \to \mathsf{R}^4$ be defined by

$$\mathsf{T}(a, b, c, d) = (2a - b, a + b, c - d, c + d),$$

and let $\mathsf{W}_1 = \{(s, t, 0, 0): s, t \in R\}$ and $\mathsf{W}_2 = \{(0, 0, s, t): s, t \in R\}$. Notice that W_1 and W_2 are each T-invariant and that $\mathsf{R}^4 = \mathsf{W}_1 \oplus \mathsf{W}_2$. Let $\beta_1 = \{e_1, e_2\}$, $\beta_2 = \{e_3, e_4\}$, and $\beta = \{e_1, e_2, e_3, e_4\}$. Then β_1 is a basis for W_1, β_2 is a basis for W_2, and β is a basis for R^4. If $B_1 = [\mathsf{T}_{\mathsf{W}_1}]_{\beta_1}$, $B_2 = [\mathsf{T}_{\mathsf{W}_2}]_{\beta_2}$, and $A = [\mathsf{T}]_\beta$, then

$$B_1 = \begin{pmatrix} 2 & -1 \\ 1 & 1 \end{pmatrix}, \qquad B_2 = \begin{pmatrix} 1 & -1 \\ 1 & 1 \end{pmatrix},$$

and

$$A = \begin{pmatrix} B_1 & O \\ O & B_2 \end{pmatrix} = \begin{pmatrix} 2 & -1 & 0 & 0 \\ 1 & 1 & 0 & 0 \\ 0 & 0 & 1 & -1 \\ 0 & 0 & 1 & 1 \end{pmatrix}.$$

Moreover, if $f(t)$ denotes the characteristic polynomial of T, $f_1(t)$ the characteristic polynomial of $\mathsf{T}_{\mathsf{W}_1}$, and $f_2(t)$ the characteristic polynomial of $\mathsf{T}_{\mathsf{W}_2}$, then

$$f(t) = \det(A - tI) = \det(B_1 - tI)\cdot\det(B_2 - tI) = f_1(t)\cdot f_2(t). \quad \blacksquare$$

The matrix A defined in Example 8 can be obtained by joining the matrices B_1 and B_2 in the manner explained in the following definition.

Definition. *Let* B_1 *and* B_2 *be square matrices (not necessarily of the same size) having entries from the same field. If* B_1 *is an* $\mathrm{m} \times \mathrm{m}$ *matrix and* B_2 *is an* $\mathrm{n} \times \mathrm{n}$ *matrix, then the* direct sum *of* B_1 *and* B_2, *denoted* $\mathrm{B}_1 \oplus \mathrm{B}_2$, *is the* $(\mathrm{m} + \mathrm{n}) \times (\mathrm{m} + \mathrm{n})$ *matrix* A *such that*

$$\mathrm{A}_{ij} = \begin{cases} (\mathrm{B}_1)_{ij} & \text{for } 1 \leqslant \mathrm{i}, \mathrm{j} \leqslant \mathrm{m} \\ (\mathrm{B}_2)_{(i-m),(j-m)} & \text{for } \mathrm{m} + 1 \leqslant \mathrm{i}, \mathrm{j} \leqslant \mathrm{n} + \mathrm{m} \\ 0 & \text{otherwise.} \end{cases}$$

If $B_1, B_2, \ldots, B_k$ *are square matrices with entries from the same field, then we define the direct sum of* $B_1, B_2, \ldots, B_k$ *recursively by*

$$B_1 \oplus B_2 \oplus \cdots \oplus B_k = (B_1 \oplus B_2 \oplus \cdots \oplus B_{k-1}) \oplus B_k.$$

If $A = B_1 \oplus B_2 \oplus \cdots \oplus B_k$, *then we will often write*

$$A = \begin{pmatrix} B_1 & O & \cdots & O \\ O & B_2 & \cdots & O \\ \vdots & \vdots & & \vdots \\ O & O & \cdots & B_k \end{pmatrix}.$$

Example 9

Let

$$B_1 = \begin{pmatrix} 1 & 2 \\ 1 & 1 \end{pmatrix}, \qquad B_2 = (3), \quad \text{and} \quad B_3 = \begin{pmatrix} 1 & 2 & 1 \\ 1 & 2 & 3 \\ 1 & 1 & 1 \end{pmatrix}.$$

Then

$$B_1 \oplus B_2 \oplus B_3 = \left(\begin{array}{cc|c|ccc} 1 & 2 & 0 & 0 & 0 & 0 \\ 1 & 1 & 0 & 0 & 0 & 0 \\ \hline 0 & 0 & 3 & 0 & 0 & 0 \\ \hline 0 & 0 & 0 & 1 & 2 & 1 \\ 0 & 0 & 0 & 1 & 2 & 3 \\ 0 & 0 & 0 & 1 & 1 & 1 \end{array}\right). \quad \blacksquare$$

The final result of this section relates direct sums of matrices to direct sums of invariant subspaces. It states the general case of the relationship among the matrices A, B_1, and B_2 in Example 8.

Theorem 5.30. *Let* T *be a linear operator on a finite-dimensional vector space* V, *and let* $W_1, W_2, \ldots, W_k$ *be* T*-invariant subspaces of* V *such that* $V = W_1 \oplus W_2 \oplus \cdots \oplus W_k$. *For each* i, *let* β_i *be a basis for* W_i *and* $\beta = \beta_1 \cup \beta_2 \cup \cdots \cup \beta_k$. *If* $A = [T]_\beta$ *and* $A_i = [T_{W_i}]_{\beta_i}$ *for* $i = 1, 2, \ldots, k$, *then* $A = A_1 \oplus A_2 \oplus \cdots \oplus A_k$.

Proof. Exercise. ∎

EXERCISES

1. Label the following statements as being true or false.
   (a) There exists a linear operator T with no T-invariant subspace.
   (b) If T is a linear operator on a finite-dimensional vector space V, and W is a T-invariant subspace of V, then the characteristic polynomial of T_W

divides the characteristic polynomial of T.

(c) Let T be a linear operator on a finite-dimensional vector space V, and let x and y be elements of V. If W is the T-cyclic subspace generated by x, W′ is the T-cyclic subspace generated by y, and W = W′, then $x = y$.

(d) If T is a linear operator on a finite-dimensional vector space V, then for any $x \in$ V the T-cyclic subspace generated by x is the same as the T-cyclic subspace generated by T(x).

(e) Let T be a linear operator on an n-dimensional vector space. Then there exists a polynomial $g(t)$ of degree n such that $g(\mathsf{T}) = \mathsf{T}_0$.

(f) Any polynomial of the form

$$(-1)^n(a_0 + a_1 t + \cdots + a_{n-1}t^{n-1} + t^n)$$

is the characteristic polynomial of some linear operator.

(g) If T is a linear operator on a finite-dimensional vector space V, and if V is a direct sum of k T-invariant subspaces, then there is a basis β for V such that $[\mathsf{T}]_\beta$ is a direct sum of k matrices.

2. For each of the following linear operators T, determine if the given subspace W is a T-invariant subspace of V.

(a) $\mathsf{V} = \mathsf{P}_3(R)$, $\mathsf{T}(f) = f'$, and $\mathsf{W} = \mathsf{P}_2(R)$

(b) $\mathsf{V} = \mathsf{P}(R)$, $\mathsf{T}(f)(x) = xf(x)$, and $\mathsf{W} = \mathsf{P}_2(R)$

(c) $\mathsf{V} = \mathsf{R}^3$, $\mathsf{T}(a, b, c) = (a + b + c, a + b + c, a + b + c)$, and $\mathsf{W} = \{(t, t, t): t \in \mathsf{R}\}$

(d) $\mathsf{V} = C([0, 1])$, $\mathsf{T}(f)(t) = \left[\int_0^1 f(x)\,dx\right]t$, and $\mathsf{W} = \{f \in \mathsf{V}: f(t) = at + b \text{ for some } a \text{ and } b\}$

(e) $\mathsf{V} = \mathsf{M}_{2\times 2}(R)$, $\mathsf{T}(A) = \begin{pmatrix} 0 & 1 \\ 1 & 0 \end{pmatrix} A$, and $\mathsf{W} = \{A \in \mathsf{V}: A^t = A\}$

3. Let T be a linear operator on a finite-dimensional vector space V. Prove that the following subspaces are T-invariant.

(a) $\{0\}$ and V

(b) N(T) and R(T)

(c) E_λ, for any eigenvalue λ of T

4. Let T be a linear operator on a vector space V, and let W be a T-invariant subspace of V. Prove that W is $g(\mathsf{T})$-invariant for any polynomial $g(t)$.

5. Let T be a linear operator on a vector space V. Prove that the intersection of any collection of T-invariant subspaces of V is a T-invariant subspace of V.

6. For each linear operator T on the vector space V find a basis for the T-cyclic subspace generated by the vector z.

(a) $\mathsf{V} = \mathsf{R}^4$, $\mathsf{T}(a, b, c, d) = (a + b, b - c, a + c, a + d)$, and $z = e_1$

(b) $\mathsf{V} = \mathsf{P}_3(R)$, $\mathsf{T}(f) = f''$, and $z = x^3$

(c) $\mathsf{V} = \mathsf{M}_{2\times 2}(R)$, $\mathsf{T}(A) = A^t$, and $z = \begin{pmatrix} 0 & 1 \\ 1 & 0 \end{pmatrix}$

(d) $\mathsf{V} = \mathsf{M}_{2\times 2}(R)$, $\mathsf{T}(A) = \begin{pmatrix} 1 & 1 \\ 2 & 2 \end{pmatrix} A$, and $z = \begin{pmatrix} 0 & 1 \\ 1 & 0 \end{pmatrix}$

7. Prove that the restriction of a linear operator T to a T-invariant subspace is a linear operator on that subspace.

8. Let T be a linear operator on a vector space with a T-invariant subspace W. Prove that if x is an eigenvector of T_W with corresponding eigenvalue λ, then the same is true for T.

9. For each linear operator T and cyclic subspace W of Exercise 6, compute the characteristic polynomial of T_W in two ways as in Example 6.

10. Verify that $A = \begin{pmatrix} B_1 & B_2 \\ O & B_3 \end{pmatrix}$ in the proof of Theorem 5.26.

11. Let T be a linear operator on a vector space V, let x be a nonzero element of V, and let W be the T-cyclic subspace of V generated by x. Prove:
    (a) W is T-invariant.
    (b) Any T-invariant subspace of V containing x also contains W.

12. For each linear operator of Exercise 6, find the characteristic polynomial $f(t)$ of T, and verify that the characteristic polynomial of T_W (computed in Exercise 9) divides $f(t)$.

13. Let T be a linear operator on a vector space V, let x be a nonzero element of V, and let W be the T-cyclic subspace of V generated by x. For any $y \in \mathsf{V}$, prove that $y \in \mathsf{W}$ if and only if there exists a polynomial $g(t)$ such that $y = g(\mathsf{T})x$.

14. Prove that the polynomial $g(t)$ of Exercise 13 can always be chosen so that its degree is less than or equal to dim(W).

15. Use the Cayley–Hamilton theorem (Theorem 5.28) to prove its corollary for matrices.

16. Let T be a linear operator on a finite-dimensional vector space V.
    (a) Prove that if the characteristic polynomial of T splits, then so does the characteristic polynomial of the restriction of T to any T-invariant subspace of V.
    (b) Deduce that if the characteristic polynomial of T splits, then any nontrivial T-invariant subspace of V contains an eigenvector of T.

17. Let A be an $n \times n$ matrix. Prove that

$$\dim(\text{span}(\{I_n, A, A^2, \ldots,\})) \leq n.$$

18. Let A be an $n \times n$ matrix with characteristic polynomial

$$f(t) = (-1)^n t^n + a_{n-1}t^{n-1} + \cdots + a_1 t + a_0.$$

    (a) Prove that A is invertible if and only if $a_0 \neq 0$.
    (b) Prove that if A is invertible, then

$$A^{-1} = (-1/a_0)[(-1)^n A^{n-1} + a_{n-1}A^{n-2} + \cdots + a_1 I_n].$$

    (c) Use part (b) to compute A^{-1} for

$$A = \begin{pmatrix} 1 & 2 & 1 \\ 0 & 2 & 3 \\ 0 & 0 & -1 \end{pmatrix}.$$

19. Let A denote the $k \times k$ matrix

$$\begin{pmatrix} 0 & 0 & \cdots & 0 & -a_0 \\ 1 & 0 & \cdots & 0 & -a_1 \\ 0 & 1 & \cdots & 0 & -a_2 \\ \vdots & \vdots & & \vdots & \vdots \\ 0 & 0 & \cdots & 0 & -a_{k-2} \\ 0 & 0 & \cdots & 1 & -a_{k-1} \end{pmatrix}.$$

where $a_0, a_1, \ldots, a_{k-1}$ are arbitrary scalars. Prove that the characteristic polynomial of A is

$$(-1)^k(a_0 + a_1 t + \cdots + a_{k-1}t^{k-1} + t^k).$$

Hint: Use mathematical induction on k, expanding the determinant along the first row.

20. Let T be a linear operator on a vector space V, and suppose that V is a T-cyclic subspace of itself. Prove that if U is a linear operator on V, then UT = TU if and only if U = g(T) for some polynomial $g(t)$. *Hint:* Suppose that V is generated by x. Choose $g(t)$ according to Exercise 13 so that g(T)(x) = U(x).

21. Let T be a linear operator on a two-dimensional vector space V. Prove that either V is a T-cyclic subspace of itself or T = cI for some scalar c.

22. Let T be a linear operator on a two-dimensional vector space V and suppose that T $\neq$ cI for any scalar c. Show that if U is any linear operator on V such that UT = TU, then U = g(T) for some polynomial $g(t)$.

23. Let T be a diagonalizable linear operator on a finite-dimensional vector space V, and let W be a T-invariant subspace of V. Suppose that $x_1, x_2, \ldots, x_k$ are eigenvectors of T corresponding to distinct eigenvalues. Prove that if $x_1 + x_2 + \cdots + x_k$ is in W, then $x_i \in$ W for all i. *Hint:* Use mathematical induction on k.

24. Prove that the restriction of a diagonalizable linear operator T to any nontrivial T-invariant subspace is also diagonalizable. *Hint:* Use the result of Exercise 23.

25. **(a)** Prove a converse of Exercise 16(a) of Section 5.2: If T and U are diagonalizable linear operators on a finite-dimensional vector space V such that UT = TU, then T and U are *simultaneously diagonalizable.* (See the definition in the exercises of Section 5.2.) *Hint:* For any eigenvalue λ of T show that E_λ is U-invariant, and apply Exercise 24 to obtain a basis for E_λ of eigenvectors of T.

(b) State and prove a matrix version of (a).

Exercises 26 through 30 require familiarity with Exercise 29 of Section 1.3 and Exercise 30 of Section 2.1. It is also advisable to review Exercise 26 of Section 1.6 and Exercise 22 of Section 2.4.

26. Let T be a linear operator on a vector space V, and let W be a T-invariant subspace of V. Define $\bar{\mathsf{T}}: \mathsf{V}/\mathsf{W} \to \mathsf{V}/\mathsf{W}$ by

$$\bar{\mathsf{T}}(v + \mathsf{W}) = \mathsf{T}(v) + \mathsf{W} \qquad \text{for any } v + \mathsf{W} \text{ in } \mathsf{V}/\mathsf{W}.$$

(a) Show that $\bar{\mathsf{T}}$ is well-defined. That is, show that $\bar{\mathsf{T}}(v + \mathsf{W}) = \bar{\mathsf{T}}(v' + \mathsf{W})$ whenever $v + \mathsf{W} = v' + \mathsf{W}$.
(b) Prove that $\bar{\mathsf{T}}$ is a linear operator on V/W.
(c) Let $\eta: \mathsf{V} \to \mathsf{V}/\mathsf{W}$ be the linear transformation as defined in Exercise 30 of Section 2.1 by $\eta(v) = v + \mathsf{W}$. Show that the diagram of Figure 5.5 is commutative, that is, $\eta\bar{\mathsf{T}} = \bar{\mathsf{T}}\eta$.

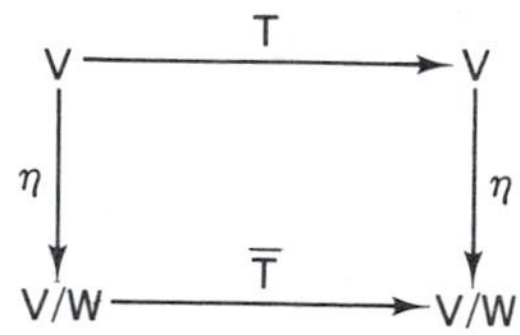

Figure 5.5

In Exercises 27 through 29, T is a linear operator on a finite-dimensional vector space V and W is a nontrivial T-invariant subspace.

27. Let $f(t)$, $g(t)$, and $h(t)$ be the characteristic polynomials of T, T_W, and $\bar{\mathsf{T}}$, respectively. Prove that $f(t) = g(t)h(t)$. *Hint:* Extend a basis $\gamma = \{x_1, x_2, \ldots, x_k\}$ for W to a basis $\beta = \{x_1, x_2, \ldots, x_n\}$ for V, show that $\alpha = \{x_{k+1} + \mathsf{W}, \ldots, x_n + \mathsf{W}\}$ is a basis for V/W, and

$$[\mathsf{T}]_\beta = \begin{pmatrix} B_1 & B_2 \\ O & B_3 \end{pmatrix},$$

where $B_1 = [\mathsf{T}]_\gamma$ and $B_3 = [\bar{\mathsf{T}}]_\alpha$.

28. Use the hint in Exercise 27 to prove that if T is diagonalizable, then so is $\bar{\mathsf{T}}$.

29. Prove the converse to Exercises 24 and 28: If both T_W and $\bar{\mathsf{T}}$ are diagonalizable, then so is T.

The results of Theorem 5.27 and Exercise 27 are useful in devising algorithms for computing characteristic polynomials without the use of determinants. This is illustrated by the next exercise.

30. Let $A = \begin{pmatrix} 1 & 1 & -3 \\ 2 & 3 & 4 \\ 1 & 2 & 1 \end{pmatrix}$, let $\mathsf{T} = \mathsf{L}_A$, and let W be the cyclic subspace of R^3 generated by e_1.

(a) Use Theorem 5.27 to compute the characteristic polynomial of T_W.
(b) Show that $\{e_2 + \mathsf{W}\}$ is a basis for R^3/W, and use that fact to compute the characteristic polynomial of T.
(c) Use the results of parts (a) and (b) to find the characteristic polynomial of A.

Exercises 31 through 38 are concerned with direct sums.

31. Let $\mathsf{W}_1, \mathsf{W}_2, \ldots, W_k$ be T-invariant subspaces of a vector space V. Prove that $\sum_{i=1}^{k} \mathsf{W}_i$ is also a T-invariant subspace of V.

32. Verify that $A = \begin{pmatrix} B_1 & O \\ O' & B_2 \end{pmatrix}$ in the proof of Theorem 5.29.

33. Let T be a linear operator on a finite-dimensional vector space V. Prove that T is diagonalizable if and only if V is the direct sum of one-dimensional T-invariant subspaces.

34. Prove Theorem 5.30.

35. Let T be a linear operator on a finite-dimensional vector space V, and suppose that $\mathsf{W}_1, \mathsf{W}_2, \ldots, \mathsf{W}_k$ are nontrivial T-invariant subspaces of V such that $\mathsf{V} = \mathsf{W}_1 \oplus \mathsf{W}_2 \oplus \cdots \oplus \mathsf{W}_k$. Prove that

$$\det(\mathsf{T}) = \det(\mathsf{T}_{\mathsf{W}_1}) \det(\mathsf{T}_{\mathsf{W}_2}) \cdots \det(\mathsf{T}_{\mathsf{W}_k}).$$

36. Let T be a linear operator on a finite-dimensional vector space V, and suppose that $\mathsf{W}_1, \mathsf{W}_2, \ldots, \mathsf{W}_k$ are nontrivial T-invariant subspaces of V such that $\mathsf{V} = \mathsf{W}_1 \oplus \mathsf{W}_2 \oplus \cdots \oplus \mathsf{W}_k$. Prove that T is diagonalizable if and only if $\mathsf{T}_{\mathsf{W}_i}$ is diagonalizable for all i.

37. Prove that a collection $\mathscr{C}$ of diagonalizable linear operators on a finite-dimensional vector space V is simultaneously diagonalizable (see Exercise 25) if and only if UT = TU for all T and U in $\mathscr{C}$. *Hint:* In the case that UT = TU for all T and U in $\mathscr{C}$, first establish the result if each operator in $\mathscr{C}$ has only one eigenvalue. Then establish the general result by induction on dim(V) using the fact that V is the direct sum of the eigenspaces of some operator in $\mathscr{C}$.

38. Let $B_1, B_2, \ldots, B_k$ be square matrices with entries in the same field, and let $A = B_1 \oplus B_2 \oplus \cdots \oplus B_k$. Prove that the characteristic polynomial of A is the product of the characteristic polynomials of the B_i's.

INDEX OF DEFINITIONS FOR CHAPTER 5

6

Inner Product Spaces

Most applications of mathematics are involved with the concept of measurement and hence of the magnitude or relative size of various quantities. So it is not surprising that the fields of real and complex numbers that have a built-in notion of distance should play a special role. Except for Section 6.7 we assume that all vector spaces are over the field F, where F denotes either R or C.

We will introduce the idea of distance or length into vector spaces, obtaining a much richer structure, the so-called "inner product space" structure. This added structure will provide applications to geometry (Sections 6.5 and 6.10), physics (Section 6.8), conditioning in systems of equations (Section 6.9), least squares applications (Section 6.3), and quadratic forms (Section 6.7).

6.1 INNER PRODUCTS AND NORMS

Many of the geometric notions such as angle, length, and perpendicularity in R^2 and R^3 may be extended to more general real and complex vector spaces. All of these ideas are related to the concept of "inner product."

Definition. *Let* V *be a vector space over* F. *An* inner product *on* V *is a function that assigns to every ordered pair of vectors* x *and* y *in* V *a scalar in* F, *denoted* $\langle \mathsf{x}, \mathsf{y} \rangle$, *such that for all* x, y, *and* z *in* V *and all* c *in* F *we have*

(a) $\langle \mathsf{x} + \mathsf{z}, \mathsf{y} \rangle = \langle \mathsf{x}, \mathsf{y} \rangle + \langle \mathsf{z}, \mathsf{y} \rangle$.

(b) $\langle \mathsf{cx}, \mathsf{y} \rangle = \mathsf{c}\langle \mathsf{x}, \mathsf{y} \rangle$.

(c) $\overline{\langle \mathsf{x}, \mathsf{y} \rangle} = \langle \mathsf{y}, \mathsf{x} \rangle$, *where the bar denotes complex conjugation.*

(d) $\langle \mathsf{x}, \mathsf{x} \rangle > 0$ *if* $\mathsf{x} \neq 0$.

Note that (c) reduces to $\langle x, y \rangle = \langle y, x \rangle$ if $F = R$. Conditions (a) and (b) simply require that the inner product be linear in the first component.

It is easily shown that if $a_1, \ldots, a_n \in F$ and $y, x_1, x_2, \ldots, x_n \in \mathsf{V}$, then

$$\left\langle \sum_{i=1}^{n} a_i x_i, y \right\rangle = \sum_{i=1}^{n} a_i \langle x_i, y \rangle.$$

Example 1

Let $\mathsf{V} = \mathsf{F}^n$. For $x = (a_1, \ldots, a_n)$ and $y = (b_1, \ldots, b_n)$, define

$$\langle x, y \rangle = \sum_{i=1}^{n} a_i \overline{b_i}.$$

$\langle \cdot, \cdot \rangle$ satisfies conditions (a) through (d) and is called the *standard inner product* on F^n. (In elementary courses in linear algebra, this is called the *dot product.*)

The verification of (a) through (d) is easy. For example, if $z = (c_1, \ldots, c_n)$, we have for (a)

$$\begin{aligned}\langle x + z, y \rangle &= \sum_{i=1}^{n} (a_i + c_i)\overline{b_i} = \sum_{i=1}^{n} a_i \overline{b_i} + \sum_{i=1}^{n} c_i \overline{b_i} \\ &= \langle x, y \rangle + \langle z, y \rangle.\end{aligned}$$

Thus for $x = (1 + i, 4)$ and $y = (2 - 3i,\ 4 + 5i)$ in C^2 we have $\langle x, y \rangle = (1 + i)(2 + 3i) + 4(4 - 5i) = 15 - 15i$. ∎

Example 2

If $\langle x, y \rangle$ is any inner product on a vector space V and $r > 0$, we may define another inner product by the rule $\langle x, y \rangle' = r\langle x, y \rangle$. If $r < 0$, then (d) would not hold. ∎

Example 3

Let $\mathsf{V} = \mathsf{C}([0, 1])$, the vector space of real-valued continuous functions on $[0, 1]$. For $f, g \in \mathsf{V}$, define $\langle f, g \rangle = \int_0^1 f(t)g(t)\,dt$. Since the integral above is linear in f, (a) and (b) are immediate, and (c) is trivial. If $f \neq 0$, then f^2 is bounded away from zero on some subinterval of $[0, 1]$ (continuity is used here), and hence $\langle f, f \rangle = \int_0^1 [f(t)]^2\,dt > 0$. ∎

Definition. *Let* A *be an* m × n *matrix with entries from* F. *We define the* conjugate transpose (*or* adjoint) *of* A *to be the* n × m *matrix* A^* *such that* $(\mathsf{A}^*)_{ij} = \overline{\mathsf{A}_{ji}}$.

Example 4

Let

$$A = \begin{pmatrix} i & 1 + 2i \\ 2 & 3 + 4i \end{pmatrix}.$$

Then

$$A^* = \begin{pmatrix} -i & 2 \\ 1-2i & 3-4i \end{pmatrix}. \quad \blacksquare$$

The conjugate transpose of a matrix plays a very important role in the remainder of this chapter. Note that if A has real entries, then A^* is simply the transpose of A.

Example 5

Let $\mathsf{V} = \mathsf{M}_{n \times n}(F)$, and define $\langle A, B \rangle = \operatorname{tr}(B^*A)$ for $A, B \in \mathsf{V}$. (Recall that the trace of a matrix A is defined by $\operatorname{tr}(A) = \sum_{i=1}^{n} A_{ii}$.) We will verify that (a) and (d) of the definition of inner product hold and leave (b) and (c) to the reader. For this purpose, let $A, B, C \in \mathsf{V}$. Then (using Exercise 6 of Section 1.3)

$$\begin{aligned}\langle A + B, C \rangle &= \operatorname{tr}(C^*(A + B)) = \operatorname{tr}(C^*A + C^*B) \\ &= \operatorname{tr}(C^*A) + \operatorname{tr}(C^*B) = \langle A, C \rangle + \langle B, C \rangle.\end{aligned}$$

Also,

$$\begin{aligned}\langle A, A \rangle = \operatorname{tr}(A^*A) &= \sum_{i=1}^{n} (A^*A)_{ii} = \sum_{i=1}^{n} \sum_{k=1}^{n} (A^*)_{ik} A_{ki} \\ &= \sum_{i=1}^{n} \sum_{k=1}^{n} \overline{A_{ki}}\, A_{ki} = \sum_{i=1}^{n} \sum_{k=1}^{n} |A_{ki}|^2.\end{aligned}$$

Now if $A \neq O$, then $A_{ki} \neq 0$ for some k and i. So $\langle A, A \rangle > 0$. $\blacksquare$

A vector space V over F endowed with a specific inner product is called an *inner product space*. If $F = C$, we call V a *complex inner product space*, whereas if $F = R$, we call V a *real inner product space*.

Thus Examples 1, 3, and 5 also provide examples of inner product spaces. *For the remainder of this chapter* F^n *will denote the inner product space with the inner product given in Example 1.*

The reader should be cautioned that two distinct inner products on a given vector space yield two distinct inner product spaces.

A very important inner product space that resembles $\mathsf{C}([0, 1])$ is the space H of continuous complex-valued functions defined on the interval $[0, 2\pi]$ with the inner product

$$\langle f, g \rangle = \frac{1}{2\pi} \int_0^{2\pi} f(t)\overline{g(t)}\, dt.$$

The reason for the constant $1/2\pi$ will become evident later. This inner product space, which arises often in the context of physical situations, will be examined more closely in later sections.

At this point we mention a few facts about integration of complex-valued functions. First, the imaginary number i can be treated as a constant under the integral sign. Second, every complex-valued function f may be written as $f = f_1 + if_2$, where f_1 and f_2 are real-valued functions. Thus we have that

$$\int f = \int f_1 + i \int f_2 \quad \text{and} \quad \overline{\int f} = \int \bar{f}.$$

From these properties, as well as the assumption of continuity, it follows that H is an inner product space [see Exercise 16(a)].

Some properties that follow easily from the definition of an inner product are contained in the next theorem.

Theorem 6.1. *Let* V *be an inner product space. Then for* x, y, z ∈ V *and* c ∈ F

(a) $\langle x, y + z\rangle = \langle x, y\rangle + \langle x, z\rangle$.
(b) $\langle x, cy\rangle = \bar{c}\langle x, y\rangle$.
(c) $\langle x, x\rangle = 0$ *if and only if* $x = 0$.
(d) *If* $\langle x, y\rangle = \langle x, z\rangle$ *for all* $x \in V$, *then* $y = z$.

Proof. (a) $\langle x, y + z\rangle = \overline{\langle y + z, x\rangle} = \overline{\langle y, x\rangle + \langle z, x\rangle}$

$$= \overline{\langle y, x\rangle} + \overline{\langle z, x\rangle} = \langle x, y\rangle + \langle x, z\rangle.$$

The proofs of (b), (c), and (d) are left as exercises. ∎

The reader should observe that (a) and (b) of Theorem 6.1 show that the inner product is *conjugate linear* in the second component.

In order to generalize the notion of length in R^3 to arbitrary inner product spaces, we need only observe that the length of $x = (a, b, c) \in R^3$ is given by $\sqrt{a^2 + b^2 + c^2} = \sqrt{\langle x, x\rangle}$. Hence we make the following definition.

Definition. *Let* V *be an inner product space. For* x ∈ V *we define the* norm *(or* length*) of* x *by* $\|x\| = \sqrt{\langle x, x\rangle}$.

Example 6

Let $V = F^n$. Then

$$\|(a_1, \ldots, a_n)\| = \left[\sum_{i=1}^{n} |a_i|^2\right]^{1/2}$$

is the Euclidean definition of length. Note that if $n = 1$, we have $\|a\| = |a|$. ∎

As we might expect, the well-known properties of length in R^3 hold in general, as shown below.

Theorem 6.2. *Let* V *be an inner product space. Then for all* x, y ∈ V *and* c ∈ F *we have*

(a) $\|cx\| = |c| \cdot \|x\|$.

(b) $\|x\| = 0$ *if and only if* $x = 0$. *In any case,* $\|x\| \geq 0$.

(c) (*Cauchy–Schwarz Inequality*) $|\langle x, y\rangle| \leq \|x\| \cdot \|y\|$.

(d) (*Triangle Inequality*) $\|x + y\| \leq \|x\| + \|y\|$.

Proof. We leave the proofs of (a) and (b) as exercises.

(c) If $y = 0$, then the result is immediate. So assume that $y \neq 0$. Then for any $c \in F$, we have

$$\begin{aligned} 0 \leq \|x - cy\|^2 &= \langle x - cy, x - cy\rangle = \langle x, x - cy\rangle - c\langle y, x - cy\rangle \\ &= \langle x, x\rangle - \bar{c}\langle x, y\rangle - c\langle y, x\rangle + c\bar{c}\langle y, y\rangle. \end{aligned}$$

Setting

$$c = \frac{\langle x, y\rangle}{\langle y, y\rangle},$$

the inequality above becomes

$$0 \leq \langle x, x\rangle - \frac{|\langle x, y\rangle|^2}{\langle y, y\rangle} = \|x\|^2 - \frac{|\langle x, y\rangle|^2}{\|y\|^2},$$

from which (c) follows.

(d)
$$\begin{aligned} \|x + y\|^2 &= \langle x + y, x + y\rangle = \langle x, x\rangle + \langle y, x\rangle + \langle x, y\rangle + \langle y, y\rangle \\ &= \|x\|^2 + 2\,\mathrm{Re}\langle x, y\rangle + \|y\|^2 \\ &\leq \|x\|^2 + 2|\langle x, y\rangle| + \|y\|^2 \\ &\leq \|x\|^2 + 2\|x\| \cdot \|y\| + \|y\|^2 \\ &= (\|x\| + \|y\|)^2, \end{aligned}$$

where $\mathrm{Re}\langle x, y\rangle$ denotes the real part of the complex number $\langle x, y\rangle$. Note that we used (c) to prove (d). ∎

The case when equality results in (c) and (d) is considered in Exercise 15.

Example 7

For $V = F^n$ we may apply (c) and (d) of Theorem 6.2 to the standard inner product to obtain the following well-known inequalities:

$$\left|\sum_{i=1}^{n} a_i \bar{b}_i\right| \leq \left[\sum_{i=1}^{n} |a_i|^2\right]^{1/2} \left[\sum_{i=1}^{n} |b_i|^2\right]^{1/2}$$

and

$$\left[\sum_{i=1}^{n} |a_i + b_i|^2\right]^{1/2} \leq \left[\sum_{i=1}^{n} |a_i|^2\right]^{1/2} + \left[\sum_{i=1}^{n} |b_i|^2\right]^{1/2}. \quad ∎$$

The reader may recall from earlier courses that for $\mathsf{V} = \mathsf{R}^3$ or R^2 we have that $\langle x, y\rangle = \|x\| \cdot \|y\| \cos\theta$, where θ denotes the angle ($0 \leqslant \theta \leqslant \pi$) between x and y. This equation implies (c) immediately since $|\cos\theta| \leqslant 1$. Notice also that x and y are perpendicular if and only if $\cos\theta = 0$, that is, if and only if $\langle x, y\rangle = 0$.

We are now at the point where we can generalize the notion of perpendicularity to arbitrary inner product spaces.

Definitions. *Let* V *be an inner product space. A vector* x *in* V *is a* unit vector *if* $\|x\| = 1$. *Vectors* x *and* y *in* V *are* orthogonal (perpendicular) *if* $\langle x, y\rangle = 0$. *A subset* S *of* V *is* orthogonal *if any two distinct elements of* S *are orthogonal. Finally, a subset* S *of* V *is* orthonormal *if* S *is orthogonal and consists entirely of unit vectors.*

Note that if $S = \{x_1, x_2, \ldots, x_n\}$, then S is orthonormal if and only if $\langle x_i, x_j\rangle = \delta_{ij}$, where δ_{ij} denotes the Kronecker delta. Also observe that for any nonzero vector x, $(1/\|x\|)x$ is a unit vector.

Example 8

In F^3 the set $\{(1, 1, 0), (1, -1, 1), (-1, 1, 2)\}$ is orthogonal, but not orthonormal; however, if we divide each vector by its length, we obtain the orthonormal set

$$\left\{\frac{1}{\sqrt{2}}(1, 1, 0), \frac{1}{\sqrt{3}}(1, -1, 1), \frac{1}{\sqrt{6}}(-1, 1, 2)\right\}. \quad \blacksquare$$

Example 9

Recall the inner product space H (defined on page 297). We will produce an important example of an orthonormal subset of H. Define $S = \{e^{ijx}: j \text{ is an integer}\}$, where i is the imaginary number $\sqrt{-1}$. Clearly, S is a subset of H. (Recall that $e^{ijx} = \cos jx + i \sin jx$.) Using the property that $\overline{e^{it}} = e^{-it}$ for every real number t, we have for $j \neq k$ that

$$\langle e^{ijx}, e^{ikx}\rangle = \frac{1}{2\pi}\int_0^{2\pi} e^{ijt}\overline{e^{ikt}}\, dt = \frac{1}{2\pi}\int_0^{2\pi} e^{i(j-k)t}\, dt$$

$$= \frac{1}{2\pi i(j-k)} e^{i(j-k)t}\Big|_0^{2\pi} = 0.$$

Also,

$$\langle e^{ijx}, e^{ijx}\rangle = \frac{1}{2\pi}\int_0^{2\pi} e^{i(j-j)t}\, dt = \frac{1}{2\pi}\int_0^{2\pi} dt = 1.$$

In other words, $\langle e^{ijx}, e^{ikx}\rangle = \delta_{jk}$. $\blacksquare$

We will use the orthonormal set of Example 9 in later examples.

EXERCISES

1. Label the following statements as being true or false.
   (a) An inner product is a scalar-valued function on the set of ordered pairs of vectors.
   (b) An inner product space must be over the field of real or complex numbers.
   (c) An inner product is linear in both components.
   (d) There is exactly one inner product on the vector space R^n.
   (e) The triangle inequality only holds in finite-dimensional inner product spaces.
   (f) Only square matrices have a conjugate-transpose.
   (g) If $\langle x, y\rangle = 0$ for all x in an inner product space, then $y = 0$.
2. Let $\mathsf{V} = \mathsf{C}^3$ with the standard inner product. Let $x = (2, 1 + i, i)$ and $y = (2 - i, 2, 1 + 2i)$. Compute $\langle x, y\rangle$, $\|x\|$, $\|y\|$, and $\|x + y\|^2$. Then verify both Cauchy's inequality and the triangle inequality.
3. In $\mathsf{C}([0, 1])$, let $f(t) = t$ and $g(t) = e^t$. Compute $\langle f, g\rangle$ (as defined in Example 3), $\|f\|$, $\|g\|$, and $\|f + g\|$. Then verify both Cauchy's inequality and the triangle inequality.
4. Let $\mathsf{V} = \mathsf{M}_{n\times n}(F)$ with $\langle A, B\rangle = \operatorname{tr}(B^*A)$. Complete the proof in Example 5 that $\langle\cdot,\cdot\rangle$ is an inner product. If $n = 2$ and
$$A = \begin{pmatrix} 1 & 2+i \\ 3 & i \end{pmatrix} \quad\text{and}\quad B = \begin{pmatrix} 1+i & 0 \\ i & -i \end{pmatrix},$$
compute $\|A\|$, $\|B\|$, and $\langle A, B\rangle$.
5. On C^2, show that $\langle x, y\rangle = xAy^*$ is an inner product, where
$$A = \begin{pmatrix} 1 & i \\ -i & 2 \end{pmatrix}.$$
Compute $\langle x, y\rangle$ for $x = (1 - i, 2 + 3i)$ and $y = (2 + i, 3 - 2i)$.
6. Complete the proof of Theorem 6.1.
7. Complete the proof of Theorem 6.2.
8. Provide reasons why each of the following is not an inner product on the given vector spaces.
   (a) $\langle (a, b), (c, d)\rangle = ac - bd$ on R^2
   (b) $\langle A, B\rangle = \operatorname{tr}(A + B)$ on $\mathsf{M}_{2\times 2}(R)$
   (c) $\langle f, g\rangle = \int_0^1 f'(t)g(t)\,dt$ on $\mathsf{P}(R)$, where $'$ denotes differentiation
9. Let β be a basis for a finite-dimensional inner product space. Prove that if $\langle x, y\rangle = 0$ for all $x \in \beta$, then $y = 0$.
10.† Let V be an inner product space, and suppose that x and y are orthogonal elements of V. Prove that $\|x + y\|^2 = \|x\|^2 + \|y\|^2$. Deduce the Pythagorean theorem in R^2.

11. Prove the *parallelogram law* on an inner product space V; that is, show that

$$\|x + y\|^2 + \|x - y\|^2 = 2\|x\|^2 + 2\|y\|^2 \qquad \text{for all } x, y \in \mathsf{V}.$$

What does this equation state about parallelograms in R^2?

12.† Let $\{x_1, \ldots, x_k\}$ be an orthogonal set in V, and let $a_1, \ldots, a_k \in F$. Prove that

$$\left\|\sum_{i=1}^{k} a_i x_i\right\|^2 = \sum_{i=1}^{k} |a_i|^2 \|x_i\|^2.$$

13. Suppose that $\langle\cdot,\cdot\rangle_1$ and $\langle\cdot,\cdot\rangle_2$ are two inner products on a vector space V. Prove that $\langle\cdot,\cdot\rangle = \langle\cdot,\cdot\rangle_1 + \langle\cdot,\cdot\rangle_2$ is another inner product on V.

14. Let A and B be $n \times n$ matrices, and let $c \in F$. Prove that $(A + cB)^* = A^* + \bar{c}B^*$.

15. **(a)** Prove that if V is an inner product space, then $|\langle x, y\rangle| = \|x\| \cdot \|y\|$ if and only if one of the vectors x or y is a multiple of the other. *Hint:* If $y \neq 0$, let

$$a = \frac{\langle x, y\rangle}{\|y\|^2}.$$

Then $x = ay + z$, where $\langle y, z\rangle = 0$. By assumption

$$|a| = \frac{\|x\|}{\|y\|}.$$

Apply Exercise 10 to $\|x\|^2 = \|ay + z\|^2$ and obtain $\|z\| = 0$.

(b) Derive a similar result for the equality $\|x + y\| = \|x\| + \|y\|$, and generalize it to the case of n vectors.

16. **(a)** Show that the vector space H defined in this section is an inner product space.

(b) Let $\mathsf{V} = \mathsf{C}([0, 1])$, and define

$$\langle f, g\rangle = \int_0^{1/2} f(t)g(t)\,dt.$$

Is this an inner product on V?

17. Let V be an inner product space, and suppose that $\mathsf{T}: \mathsf{V} \to \mathsf{V}$ is linear and that $\|\mathsf{T}(x)\| = \|x\|$ for all x. Prove that T is one-to-one.

18. Let V be a vector space over F, where $F = R$ or $F = C$, and let W be an inner product space over F with inner product $\langle\cdot,\cdot\rangle$. If $\mathsf{T}: \mathsf{V} \to \mathsf{W}$ is linear, prove that $\langle x, y\rangle' = \langle \mathsf{T}(x), \mathsf{T}(y)\rangle$ defines an inner product on V if and only if T is one-to-one.

19. Let V be an inner product space; prove that

(a) $\|x \pm y\|^2 = \|x\|^2 \pm 2\,\mathrm{Re}\langle x, y\rangle + \|y\|^2$ for all $x, y \in \mathsf{V}$, where $\mathrm{Re}\langle x, y\rangle$ denotes the real part of the complex number $\langle x, y\rangle$.

(b) $\big|\|x\| - \|y\|\big| \leqslant \|x - y\|$ for all $x, y \in \mathsf{V}$.

20. Let V be an inner product space over F. Verify the *polar identities*: For all $x, y \in V$

(a) $\langle x, y\rangle = \frac{1}{4}\|x + y\|^2 - \frac{1}{4}\|x - y\|^2$ if $F = R$.

(b) $\langle x, y\rangle = \frac{1}{4}\sum_{k=1}^{4} i^k\|x + i^k y\|^2$ if $F = C$, where $i = \sqrt{-1}$.

21. Let A be an $n \times n$ matrix. Define

$$A_1 = \frac{1}{2}(A + A^*) \quad \text{and} \quad A_2 = \frac{1}{2i}(A - A^*).$$

(a) Prove that $A_1^* = A_1$, $A_2^* = A_2$, and $A = A_1 + iA_2$. Would it be reasonable to define A_1 and A_2 to be the real and imaginary parts, respectively, of the matrix A?

(b) Let A be an $n \times n$ matrix. Prove that if $A = B_1 + iB_2$, where $B_1^* = B_1$ and $B_2^* = B_2$, then $B_1 = A_1$ and $B_2 = A_2$.

22. Let V be a vector space over F, where F is either R or C. Whether or not V is an inner product space, we may still define a "norm" $\|\cdot\|$ as a real-valued function on V satisfying the following conditions for all $x, y \in V$ and $a \in F$:

(1) $\|x\| \geqslant 0$, and $\|x\| = 0$ if and only if $x = 0$.

(2) $\|ax\| = |a|\cdot\|x\|$.

(3) $\|x + y\| \leqslant \|x\| + \|y\|$.

Prove that the following are norms on the given vector spaces V.

(a) $V = M_{m\times n}(F)$; $\quad \|A\| = \max_{i,j} |A_{ij}| \quad$ for all $A \in V$

(b) $V = C([0, 1])$; $\quad \|f\| = \max_{t\in[0,1]} |f(t)| \quad$ for all $f \in V$

(c) $V = C([0, 1])$; $\quad \|f\| = \int_0^1 |f(t)|\,dt \quad$ for all $f \in V$

(d) $V = R^2$; $\quad \|(a, b)\| = \max\{|a|, |b|\} \quad$ for all (a, b) in V

Use Exercise 20 to show that there is no inner product $\langle\cdot,\cdot\rangle$ on R^2 such that $\|x\|^2 = \langle x, x\rangle$ for all $x \in R^2$ if $\langle\cdot,\cdot\rangle$ is defined as in (d).

23. Let V be an inner product space, and define for each ordered pair of vectors the scalar $d(x, y) = \|x - y\|$, called the *distance* between x and y. Prove for all $x, y, z \in V$ that

(a) $d(x, y) \geqslant 0$.

(b) $d(x, y) = d(y, x)$.

(c) $d(x, y) \leqslant d(x, z) + d(z, y)$.

(d) $d(x, x) = 0$.

(e) $d(x, y) \neq 0$ if $x \neq y$.

24. Let V be a real or complex vector space (possibly infinite-dimensional), and let β be a basis for V. For $x, y \in V$ there exist $x_1, \ldots, x_n \in \beta$ such that

$$x = \sum_{i=1}^{n} a_i x_i \quad \text{and} \quad y = \sum_{i=1}^{n} b_i x_i.$$

Define

$$\langle x, y \rangle = \sum_{i=1}^{n} a_i \bar{b}_i.$$

(a) Prove that $\langle \cdot, \cdot \rangle$ is an inner product on V and that β is an orthonormal basis. Thus every real or complex vector space may be regarded as an inner product space.

(b) Prove that if $V = R^n$ or $V = C^n$ and β is the standard ordered basis, then the inner product defined above is the standard inner product.

25. Let $\|\cdot\|$ be a norm (as defined in Exercise 22) on a real vector space V satisfying the parallelogram law given in Exercise 11. Define

$$\langle x, y \rangle = \tfrac{1}{4}[\|x + y\|^2 - \|x - y\|^2].$$

Prove that $\langle \cdot, \cdot \rangle$ defines an inner product on V such that $\|x\|^2 = \langle x, x \rangle$ for all $x \in V$.

26. Let $\|\cdot\|$ be a norm (as defined in Exercise 22) on a complex vector space V satisfying

$$\sum_{k=1}^{4} \|x + i^k y\|^2 = 4[\|x\|^2 + \|y\|^2],$$

where $i = \sqrt{-1}$. Define

$$\langle x, y \rangle = \frac{1}{4} \sum_{k=1}^{4} i^k \|x + i^k y\|^2.$$

Prove that $\langle \cdot, \cdot \rangle$ defines an inner product on V such that $\|x\|^2 = \langle x, x \rangle$ for all $x \in V$.

6.2 THE GRAM–SCHMIDT ORTHOGONALIZATION PROCESS AND ORTHOGONAL COMPLEMENTS

In previous chapters we have seen the special role of the standard ordered basis in R^n. The special properties of this basis stem from the fact that the basis vectors form an orthonormal set. Just as bases are the building blocks of vector spaces, bases that are also orthonormal sets are the building blocks of inner product spaces. We now name such bases.

Definition. *Let V be an inner product space. A subset β of V is an* orthonormal basis *for V if β is an ordered basis that is orthonormal.*

Example 1

If $V = F^n$, then the standard ordered basis is an orthonormal basis for V. ∎

Example 2

$$\left\{\left(\frac{1}{\sqrt{5}}, \frac{2}{\sqrt{5}}\right), \left(\frac{2}{\sqrt{5}}, \frac{-1}{\sqrt{5}}\right)\right\}$$

is an orthonormal basis for R^2. ∎

The next theorem and its corollaries illustrate why orthonormal sets and, in particular, orthonormal bases are so important.

Theorem 6.3. *Let* V *be an inner product space and let* $S = \{x_1, \ldots, x_k\}$ *be an orthogonal set of nonzero vectors. If*

$$y = \sum_{i=1}^{k} a_i x_i,$$

then $a_j = \langle y, x_j\rangle / \|x_j\|^2$ *for all* j.

Proof. For $1 \leqslant j \leqslant k$, we have

$$\langle y, x_j\rangle = \left\langle \sum_{i=1}^{k} a_i x_i, x_j \right\rangle = \sum_{i=1}^{k} a_i\langle x_i, x_j\rangle = a_j\langle x_j, x_j\rangle$$
$$= a_j\|x_j\|^2. \quad ∎$$

The next corollary follows immediately from Theorem 6.3.

Corollary 1. *If, in addition to the hypotheses of Theorem 6.3,* S *is orthonormal, then*

$$y = \sum_{i=1}^{k} \langle y, x_i\rangle x_i.$$

If V possesses a finite orthonormal basis, then Corollary 1 allows us to compute the coefficients in a linear combination very easily (see Example 3).

Corollary 2. *Let* V *be an inner product space, and let* S *be an orthogonal set of nonzero vectors. Then* S *is linearly independent.*

Proof. Suppose that $x_1, \ldots, x_k \in S$ and

$$\sum_{i=1}^{k} a_i x_i = 0.$$

By Theorem 6.3 $a_i = \langle 0, x_i\rangle / \|x_i\|^2 = 0$ for all i. So S is linearly independent. ∎

Example 3

Using the orthogonal set obtained in Example 8 of Section 6.1 and Corollary 2, we obtain an orthonormal basis for R^3

$$\left\{\frac{1}{\sqrt{2}}(1, 1, 0), \frac{1}{\sqrt{3}}(1, -1, 1), \frac{1}{\sqrt{6}}(-1, 1, 2)\right\}.$$

Let $x = (2, 1, 3)$. The "coefficients" given by Theorem 6.3 that express x as a linear combination of the basis vectors are

$$a_1 = \frac{1}{\sqrt{2}}(2 + 1) = \frac{3}{\sqrt{2}}, \quad a_2 = \frac{1}{\sqrt{3}}(2 - 1 + 3) = \frac{4}{\sqrt{3}},$$

and

$$a_3 = \frac{1}{\sqrt{6}}(-2 + 1 + 6) = \frac{5}{\sqrt{6}}.$$

As a check, we have

$$(2, 1, 3) = \frac{3}{2}(1, 1, 0) + \frac{4}{3}(1, -1, 1) + \frac{5}{6}(-1, 1, 2). \quad \blacksquare$$

Corollary 2 tells us, for example, that the vector space H in Example 9 of Section 6.1 contains an infinite linearly independent set and hence is not a finite-dimensional vector space.

Of course, we have not yet shown that every finite-dimensional inner product space possesses an orthonormal basis. The next theorem takes us most of the way in obtaining this result. It tells us how to construct an orthogonal set from a linearly independent set of vectors in such a way that both sets generate the same subspace.

Before stating this theorem, let us consider a simple case. Suppose that $\{y_1, y_2\}$ is a linearly independent subset of an inner product space (and hence a basis for some two-dimensional subspace). We would like to construct an orthogonal set from $\{y_1, y_2\}$ that spans the same subspace. Figure 6.1 suggests that the set $\{x_1, x_2\}$, where $x_1 = y_1$ and $x_2 = y_2 - cy_1$, will work if c is properly chosen.

To find c, we need only solve the following equation.

$$0 = \langle x_2, y_1 \rangle = \langle y_2 - cy_1, y_1 \rangle = \langle y_2, y_1 \rangle - c\langle y_1, y_1 \rangle.$$

So

$$c = \frac{\langle y_2, y_1 \rangle}{\|y_1\|^2}.$$

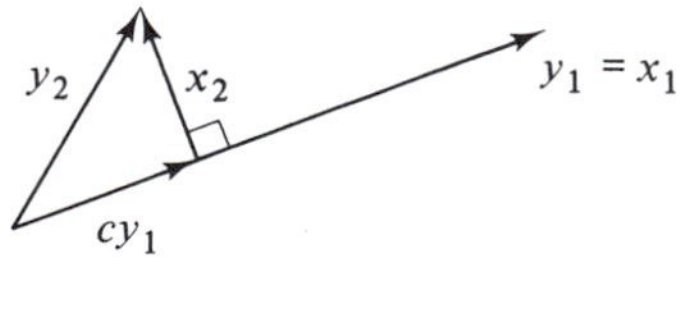

Figure 6.1

Thus

$$x_2 = y_2 - \frac{\langle y_2, y_1\rangle}{\|y_1\|^2} y_1.$$

This process can be extended to any finite linearly independent subset.

Theorem 6.4. *Let* V *be an inner product space, and let* $\mathsf{S} = \{\mathsf{y}_1, \ldots, \mathsf{y}_n\}$ *be a linearly independent subset of* V. *Define* $\mathsf{S}' = \{\mathsf{x}_1, \ldots, \mathsf{x}_n\}$, *where* $\mathsf{x}_1 = \mathsf{y}_1$, *and*

$$\mathsf{x}_k = \mathsf{y}_k - \sum_{j=1}^{k-1} \frac{\langle \mathsf{y}_k, \mathsf{x}_j\rangle}{\|\mathsf{x}_j\|^2} \mathsf{x}_j \qquad \text{for } 2 \leqslant k \leqslant n. \tag{1}$$

Then S' *is an orthogonal set of nonzero vectors such that* $\operatorname{span}(\mathsf{S}') = \operatorname{span}(\mathsf{S})$.

Proof. The proof is by induction on n. Let $S_n = \{y_1, \ldots, y_n\}$. If $n = 1$, then the theorem is proved by taking $S_1' = S_1$; i.e., $x_1 = y_1 \neq 0$. Assume then that the set $S_k' = \{x_1, \ldots, x_k\}$ with the desired properties has been constructed by the use of (1). We will show that the set $S_{k+1}' = \{x_1, \ldots, x_k, x_{k+1}\}$ also has the desired properties, where

$$x_{k+1} = y_{k+1} - \sum_{j=1}^{k} \frac{\langle y_{k+1}, x_j\rangle}{\|x_j\|^2} x_j. \tag{2}$$

If $x_{k+1} = 0$, then (2) would imply that $y_{k+1} \in \operatorname{span}(S_k') = \operatorname{span}(S_k)$, which contradicts the assumption that S_{k+1} is linearly independent.

For $1 \leqslant i \leqslant k$ we have from (2) that

$$\begin{aligned}\langle x_{k+1}, x_i\rangle &= \langle y_{k+1}, x_i\rangle - \sum_{j=1}^{k} \frac{\langle y_{k+1}, x_j\rangle}{\|x_j\|^2} \langle x_j, x_i\rangle \\ &= \langle y_{k+1}, x_i\rangle - \frac{\langle y_{k+1}, x_i\rangle}{\|x_i\|^2} \|x_i\|^2 = 0,\end{aligned}$$

since $\langle x_j, x_i\rangle = 0$ if $i \neq j$ by the inductive assumption that S_k' is orthogonal. Hence S_{k+1}' is orthogonal. Now by (2) we have that $\operatorname{span}(S_{k+1}') \subseteq \operatorname{span}(S_{k+1})$. But by Corollary 2 of Theorem 6.3, S_{k+1}' is linearly independent; so $\dim(\operatorname{span}(S_{k+1}')) = k + 1 = \dim(\operatorname{span}(S_{k+1}))$. Hence $\operatorname{span}(S_{k+1}') = \operatorname{span}(S_{k+1})$. ∎

The construction of $\{x_1, \ldots, x_n\}$ by the use of (1) is called the *Gram–Schmidt orthogonalization process.*

Example 4

Let $\mathsf{V} = \mathsf{R}^3$, and let $y_1 = (1, 1, 0)$, $y_2 = (2, 0, 1)$, and $y_3 = (2, 2, 1)$. Then $\{y_1, y_2, y_3\}$ is linearly independent. We shall use (1) above to compute the orthogonal vectors x_1, x_2, and x_3. Take $x_1 = (1, 1, 0)$. Then $\|x_1\|^2 = 2$, so

$$\begin{aligned} x_2 &= y_2 - \frac{\langle y_2, x_1\rangle}{\|x_1\|^2} x_1 \\ &= (2, 0, 1) - \frac{2}{2}(1, 1, 0) \\ &= (1, -1, 1). \end{aligned}$$

Finally,

$$\begin{aligned} x_3 &= y_3 - \frac{\langle y_3, x_1\rangle}{\|x_1\|^2} x_1 - \frac{\langle y_3, x_2\rangle}{\|x_2\|^2} x_2 \\ &= (2, 2, 1) - \frac{4}{2}(1, 1, 0) - \frac{1}{3}(1, -1, 1) \\ &= \left(-\frac{1}{3}, \frac{1}{3}, \frac{2}{3}\right). \quad \blacksquare \end{aligned}$$

Theorem 6.5. *Let* V *be a finite-dimensional inner product space. Then* V *has an orthonormal basis* β. *Furthermore, if* $\beta = \{x_1, x_2, \ldots, x_n\}$ *and* $x \in V$, *then*

$$x = \sum_{i=1}^{n} \langle x, x_i\rangle x_i.$$

Proof. Let β_0 be an ordered basis for V. Apply Theorem 6.4 to obtain an orthogonal set β' of nonzero vectors with $\text{span}(\beta') = \text{span}(\beta_0) = V$. By dividing each vector in β' by its length, we obtain an orthonormal set β that generates V. By Corollary 2 of Theorem 6.3, β is linearly independent, and therefore β is an orthonormal basis for V. The remainder of the theorem follows from Corollary 1 to Theorem 6.3. ∎

We now have an easy method for computing the matrix representation of a linear operator.

Corollary. *Let* V *be a finite-dimensional inner product space with an orthonormal basis* $\beta = \{x_1, \ldots, x_n\}$. *Let* T *be a linear operator on* V, *and let* $A = [T]_\beta$. *Then* $A_{ij} = \langle T(x_j), x_i\rangle$.

Proof. From Theorem 6.5 we have

$$T(x_j) = \sum_{i=1}^{n} \langle T(x_j), x_i\rangle x_i.$$

Hence $A_{ij} = \langle T(x_j), x_i\rangle$. ∎

The scalars $\langle x, x_i \rangle$ associated with x have been studied extensively for special vector spaces. Although the vectors $x_1, \ldots, x_n$ were chosen from an orthonormal basis, we will consider more general sets β for the definition of the scalars $\langle x, x_i \rangle$.

Definition. *Let β be an orthonormal subset (possibly infinite) of an inner product space* V*, and let* $x \in$ V*. We define the* Fourier coefficients of x relative to β *to be the scalars* $\langle x, y \rangle$*, where* $y \in \beta$.

In the nineteenth century the French mathematician Jean Baptiste Fourier was associated with the study of the coefficients

$$\int_0^{2\pi} f(t) \sin nt \, dt \quad \text{and} \quad \int_0^{2\pi} f(t) \cos nt \, dt,$$

or more generally,

$$c_n = \frac{1}{2\pi} \int_0^{2\pi} f(t) e^{-int} \, dt,$$

of a function f. In the context of Example 9 of Section 6.1, we see that $c_n = \langle f, e^{inx} \rangle$; that is, c_n is the nth Fourier coefficient of a continuous function $f \in$ H relative to S. These coefficients are the "classical" Fourier coefficients of a function, and the literature concerning the behavior of these coefficients is extensive. We will learn more about these Fourier coefficients in the remainder of this chapter.

Example 5

In H define $f(x) = x$. We will compute the Fourier coefficients of f relative to the orthonormal set S of Example 9 of Section 6.1. Using integration by parts, we have, for $n \neq 0$,

$$\langle f, e^{inx} \rangle = \frac{1}{2\pi} \int_0^{2\pi} t \overline{e^{int}} \, dt = \frac{1}{2\pi} \int_0^{2\pi} t e^{-int} \, dt = \frac{-1}{in}.$$

And, for $n = 0$,

$$\langle f, 1 \rangle = \frac{1}{2\pi} \int_0^{2\pi} t(1) \, dt = \pi.$$

Now by Exercise 14 we have that

$$\|f\|^2 \geqslant \sum_{n=1}^{k} |\langle f, e^{inx} \rangle|^2$$

for every k. Thus, using the fact that $\|f\|^2 = \frac{4}{3}\pi^2$, we have

$$\frac{4}{3}\pi^2 \geqslant \sum_{n=1}^{k} \left| \frac{-1}{in} \right|^2 = \sum_{n=1}^{k} \frac{1}{n^2}.$$

Since this inequality is true for all k, we have by the appropriate use of limits that

$$\frac{4}{3}\pi^2 \geqslant \sum_{n=1}^{\infty} \frac{1}{n^2}.$$

Other results may be obtained similarly by using other functions. ∎

We are now ready to proceed with the concept of an "orthogonal complement."

Definition. *Let* V *be an inner product space, and let* S *be a subset of* V. *We define* $S^\perp$ *(read* "S perp"*) to be the set of all those vectors in* V *that are orthogonal to every vector in* S; *that is,* $S^\perp = \{x \in V\colon \langle x, y\rangle = 0$ *for all* $y \in S\}$. $S^\perp$ *is called the* orthogonal complement of *S*.

It is easy to show that $S^\perp$ is a subspace of V for any subset S of V.

Example 6

The reader should verify that $\{0\}^\perp = V$ and $V^\perp = \{0\}$. ∎

Example 7

If $V = R^3$ and $S = \{x\}$, then $S^\perp$ is simply the set of all vectors that are perpendicular to x (see Exercise 5). ∎

Exercise 16 provides an interesting example of an orthogonal complement in an infinite-dimensional inner product space.

Proposition 6.6. *Let* W *be a finite-dimensional subspace of an inner product space* V. *If* $\{x_1, \ldots, x_k\}$ *is an orthonormal basis of* W *and* $y \in V$, *then*

$$y = \sum_{i=1}^{k} \langle y, x_i\rangle x_i + z,$$

where $z \in W^\perp$. *Furthermore, this representation of* y *is unique. That is, if also* $y = u + z'$, *where* $u \in W$ *and* $z' \in W^\perp$, *then*

$$u = \sum_{i=1}^{k} \langle y, x_i\rangle x_i$$

and $z' = z$.

Proof. For $y \in V$, let

$$z = y - \sum_{i=1}^{k} \langle y, x_i\rangle x_i.$$

To show that $z \in W^\perp$, it suffices to show that z is orthogonal to each x_j. We have

$$\langle z, x_j\rangle = \left\langle \left(y - \sum_{i=1}^{k} \langle y, x_i\rangle x_i\right), x_j \right\rangle = \langle y, x_j\rangle - \sum_{i=1}^{k} \langle y, x_i\rangle\langle x_i, x_j\rangle$$

$$= \langle y, x_j\rangle - \sum_{i=1}^{k} \langle y, x_i\rangle\delta_{ij} = \langle y, x_j\rangle - \langle y, x_j\rangle = 0.$$

To show uniqueness, suppose that $y = x_1 + x_2 = x_1' + x_2'$, where $x_1, x_1' \in \mathsf{W}$ and $x_2, x_2' \in \mathsf{W}^\perp$. Then $x_1 - x_1' = x_2' - x_2 \in \mathsf{W} \cap \mathsf{W}^\perp = \{0\}$. Therefore, $x_1 = x_1'$ and $x_2 = x_2'$. ∎

Corollary. *In the notation of Proposition 6.6, the vector*

$$y_1 = \sum_{i=1}^{k} \langle y, x_i\rangle x_i$$

is the unique vector in W *that is "closest" to* y; *that is, if* $u \in \mathsf{W}$, *then* $\|y - u\| \geqslant \|y - y_1\|$. *In addition,* $y - y_1 \in \mathsf{W}^\perp$.

Proof. As in Proposition 6.6, we have that $y = y_1 + z$, where $z \in \mathsf{W}^\perp$. Let $u \in \mathsf{W}$. By Exercise 10 of Section 6.1 we have

$$\|y - u\|^2 = \|y_1 + z - u\|^2 = \|(y_1 - u) + z\|^2 = \|(y_1 - u)\|^2 + \|z\|^2$$

$$\geqslant \|z\|^2 = \|y - y_1\|^2.$$

Now suppose that $\|y - u\|^2 = \|y - y_1\|^2$. Then the inequality above becomes an equality, so $\|(y_1 - u)\|^2 = 0$. That is, $y_1 = u$. Clearly, $y - y_1 = z \in \mathsf{W}^\perp$. ∎

The vector y_1 in the corollary is called the *orthogonal projection of y on* W. We will see the importance of orthogonal projections of vectors in the application to least squares in Section 6.3.

Example 8

Let $\mathsf{V} = \mathsf{P}_3(R)$ with the inner product

$$\langle f, g\rangle = \int_0^1 f(x)g(x)\,dx \qquad \text{for all } f, g \in \mathsf{V}.$$

Let $\mathsf{W} = \text{span}(\{1, x\})$ and let $f(x) = x^2$. We will compute the orthogonal projection f_1 of f on W. We first apply the Gram–Schmidt process to $\{1, x\}$ and obtain the orthonormal basis $\{g_1, g_2\}$ of W, where

$$g_1(x) = 1 \quad \text{and} \quad g_2(x) = 2\sqrt{3}(x - \tfrac{1}{2}).$$

It is easy to compute $\langle f, g_1\rangle = \frac{1}{3}$ and $\langle f, g_2\rangle = \sqrt{3}/6$. Therefore,

$$f_1(x) = \left(\frac{1}{3}\right)1 + \frac{\sqrt{3}}{6}\left(2\sqrt{3}\left(x - \frac{1}{2}\right)\right) = -\frac{1}{6} + x. \quad ∎$$

It was shown (Corollary 5 to Theorem 1.10) that any linearly independent set in a finite-dimensional vector space can be extended to a basis. The next theorem provides an interesting analog for an orthonormal subset of an inner product space.

Theorem 6.7. *Suppose that* $\mathsf{S} = \{\mathsf{x}_1, \ldots, \mathsf{x}_k\}$ *is an orthonormal set in an* n-*dimensional inner product space* V. *Then*

(a) S *can be extended to an orthonormal basis* $\{\mathsf{x}_1, \ldots, \mathsf{x}_k, \mathsf{x}_{k+1}, \ldots, \mathsf{x}_n\}$ *for* V.

(b) *If* W = span(S), *then (in the notation above)* $\mathsf{S}_1 = \{\mathsf{x}_{k+1}, \ldots, \mathsf{x}_n\}$ *is an orthonormal basis for* $\mathsf{W}^\perp$.

(c) *If* W *is any subspace of* V, *then* $\dim(\mathsf{V}) = \dim(\mathsf{W}) + \dim(\mathsf{W}^\perp)$.

Proof. (a) By Corollary 5 to Theorem 1.10 S can be extended to a basis $S' = \{x_1, \ldots, x_k, y_{k+1}, \ldots, y_n\}$ for V. Now apply the Gram–Schmidt process to S' and use Exercise 7 to obtain an orthogonal set. The result now follows.

(b) Because S_1 is orthonormal, it is linearly independent by Corollary 2 to Theorem 6.3. Since S_1 is clearly a subset of $\mathsf{W}^\perp$, we need only show that it spans $\mathsf{W}^\perp$. Note that for any x in V, we have

$$x = \sum_{i=1}^{n} \langle x, x_i \rangle x_i.$$

Now if $x \in \mathsf{W}^\perp$, then $\langle x, x_i \rangle = 0$ for $1 \leqslant i \leqslant k$. Therefore,

$$x = \sum_{i=k+1}^{n} \langle x, x_i \rangle x_i \in \operatorname{span}(S_1).$$

(c) Let W be a subspace of V with an orthonormal basis $\{x_1, \ldots, x_k\}$. By parts (a) and (b), we have

$$\dim(\mathsf{V}) = n = k + (n - k) = \dim(\mathsf{W}) + \dim(\mathsf{W}^\perp). \quad \blacksquare$$

Example 9

Let $\mathsf{V} = \mathsf{F}^3$ and $\mathsf{W} = \operatorname{span}(\{e_1, e_2\})$. Then $x = (a, b, c) \in \mathsf{W}^\perp$ if and only if $0 = \langle x, e_1 \rangle = a$ and $0 = \langle x, e_2 \rangle = b$. So $x = (0, 0, c)$, and therefore $\mathsf{W}^\perp = \operatorname{span}(\{e_3\})$. One can deduce the same result by noting that $e_3 \in \mathsf{W}^\perp$ and from part (c) above that $\dim(\mathsf{W}^\perp) = 3 - 2 = 1$. $\blacksquare$

EXERCISES

1. Label the following statements as being true or false.
   (a) The Gram–Schmidt orthogonalization process allows us to construct an orthonormal set from an arbitrary set of vectors.
   (b) Every finite-dimensional inner product space possesses an orthonormal basis.
   (c) The orthogonal complement of any set is a subspace.

(d) If $\beta = \{x_1, \ldots, x_n\}$ is a basis for an inner product space V, then for any $x \in \mathsf{V}$ the scalars $\langle x, x_i \rangle$ $(i = 1, \ldots, n)$ are the Fourier coefficients of x.
(e) An orthonormal basis must be an ordered basis.
(f) Every orthogonal set is linearly independent.
(g) Every orthonormal set is linearly independent.

2. In each of the following parts, apply the Gram–Schmidt process to the given subset S of the inner product space V. Then find an orthonormal basis β for V and compute the Fourier coefficients of the given vector relative to β. Finally, use Theorem 6.5 to verify your result.
(a) $\mathsf{V} = \mathsf{R}^3$, $S = \{(1, 0, 1), (0, 1, 1), (1, 3, 3)\}$, and $x = (1, 1, 2)$
(b) $\mathsf{V} = \mathsf{R}^3$, $S = \{(1, 1, 1), (0, 1, 1), (0, 0, 1)\}$, and $x = (1, 0, 1)$
(c) $\mathsf{V} = \mathsf{P}_2(R)$ with the inner product $\langle f, g \rangle = \int_0^1 f(t)g(t)\,dt$, $S = \{1, x, x^2\}$, and $f(x) = 1 + x$.
(d) $\mathsf{V} = \operatorname{span}(S)$, where $S = \{(1, i, 0), (1 - i, 2, 4i)\}$ and $x = (3 + i, 4i, -4)$

3. Let $\mathsf{V} = \mathsf{R}^2$ and let

$$\beta = \left\{\left(\frac{1}{\sqrt{2}}, \frac{1}{\sqrt{2}}\right), \left(\frac{1}{\sqrt{2}}, \frac{-1}{\sqrt{2}}\right)\right\}.$$

Find the Fourier coefficients of $(3, 4)$ relative to β.

4. Let $\mathsf{V} = \mathsf{C}^3$, and let $S = \{(1, 0, i), (1, 2, 1)\}$. Compute $S^\perp$.

5. Let $\mathsf{V} = \mathsf{R}^3$, and let $S = \{x_0\}$, where $x_0 \neq 0$. Describe $S^\perp$ geometrically. If $\{x_1, x_2\} = S_0$ is linearly independent, describe $S_0^\perp$ geometrically.

6. Let V be an inner product space, and let W be a finite-dimensional subspace of V. If $x \notin \mathsf{W}$, prove that there exists $y \in \mathsf{V}$ such that $y \in \mathsf{W}^\perp$ but $\langle x, y \rangle \neq 0$. *Hint:* Use Proposition 6.6.

7. Prove that if $\{y_1, \ldots, y_n\}$ is an orthogonal set of nonzero vectors, then the vectors $\{x_1, \ldots, x_n\}$ derived from the Gram–Schmidt process satisfy $x_i = y_i$ for $i = 1, \ldots, n$. *Hint:* Use induction.

8. Let $\mathsf{V} = \mathsf{C}^3$ with the standard inner product, and let $\mathsf{W} = \operatorname{span}(\{(i, 0, 1)\})$. Find orthonormal bases for W and $\mathsf{W}^\perp$.

9. Let W be a finite-dimensional subspace of an inner product space V. Prove that there exists a projection T on W such that $\mathsf{N}(\mathsf{T}) = \mathsf{W}^\perp$. In addition, prove that $\|\mathsf{T}(x)\| \leqslant \|x\|$ for all $x \in \mathsf{V}$. *Hint:* Use Exercise 10 of Section 6.1. (Projections are defined in the exercises of Section 2.1.)

10. Let A be an $n \times n$ matrix with complex entries such that the rows of A form an orthonormal set. Prove that $AA^* = I$.

11. Let W_1 and W_2 be subspaces of a finite-dimensional inner product space. Prove that $(\mathsf{W}_1 + \mathsf{W}_2)^\perp = \mathsf{W}_1^\perp \cap \mathsf{W}_2^\perp$ and $(\mathsf{W}_1 \cap \mathsf{W}_2)^\perp = \mathsf{W}_1^\perp + \mathsf{W}_2^\perp$.

12. Let V be an inner product space, S and S_0 be subsets of V, and W be a finite-dimensional subspace of V. Prove the following:
(a) $S_0 \subseteq S$ implies that $S^\perp \subseteq S_0^\perp$.
(b) $S \subseteq (S^\perp)^\perp$, so $\operatorname{span}(S) \subseteq (S^\perp)^\perp$.

(c) $W = (W^\perp)^\perp$. *Hint:* Use Exercise 6.
(d) $V = W \oplus W^\perp$ (see the exercises of Section 1.3).

13. (a) *Parseval's Identity.* Let $\{x_1, \ldots, x_n\}$ be an orthonormal basis for V. For any $x, y \in V$ prove that

$$\langle x, y\rangle = \sum_{i=1}^{n} \langle x, x_i\rangle \overline{\langle y, x_i\rangle}.$$

(b) Use part (a) to prove that if β is an orthonormal basis of a finite-dimensional inner product space V over F with inner product $\langle \cdot, \cdot \rangle$, then for any x and y in V

$$\langle \phi_\beta(x), \phi_\beta(y)\rangle' = \langle [x]_\beta, [y]_\beta\rangle' = \langle x, y\rangle,$$

where $\langle \cdot, \cdot \rangle'$ is the standard inner product on F^n.

14. *Bessel's Inequality.* Let V be an inner product space, and let $S = \{x_1, \ldots, x_n\}$ be any orthonormal subset of V. Prove that for any x in V we have

$$\|x\|^2 \geqslant \sum_{i=1}^{n} |\langle x, x_i\rangle|^2.$$

Hint: Apply Proposition 6.6 to $x \in V$ and $W = \text{span}(S)$. Then use Exercise 10 of Section 6.1.

15. Let T be a linear operator on a finite-dimensional inner product space V. If $\langle T(x), y\rangle = 0$ for all $x, y \in V$, prove that $T = T_0$. In fact, prove this result if the equality holds for all x and y in some basis for V.

16. Let $V = C([-1, 1])$. Suppose that W_e and W_o denote the subspaces of V consisting of the even and odd functions, respectively. Prove that $W_e = W_o$ if the inner product on V is

$$\langle f, g\rangle = \int_{-1}^{1} f(t)g(t)\, dt.$$

17. In each of the following parts, find the orthogonal projection of the given vector on the given subspace W of the inner product space V.
(a) $V = R^2$, $y = (2, 6)$, $W = \{(x, y): y = 4x\}$
(b) $V = R^3$, $y = (2, 1, 3)$, $W = \{(x, y, z): x + 3y - 2z = 0\}$
(c) $V = P(R)$ with the inner product $\langle f, g\rangle = \int_0^1 f(t)g(t)\, dt$, $h(x) = 4 + 3x - 2x^2$, $W = P_1(R)$

18. In Exercise 17 find the distance from the given vector to the subspace W.

6.3 THE ADJOINT OF A LINEAR OPERATOR

In Section 6.1 we defined the conjugate transpose A^* of a matrix A. For a linear operator T on an inner product space V, we now define a related linear operator on V called the "adjoint" of T, whose matrix representation with respect to any orthonormal basis β of V is $[\mathsf{T}]_\beta^*$. The analogy between conjugation of complex numbers and adjoints of linear operators will become apparent. We first need a preliminary result, however.

Let V be an inner product space, and let $y \in \mathsf{V}$. The function $\mathsf{g}\colon \mathsf{V} \to F$ defined by $\mathsf{g}(x) = \langle x, y\rangle$ for all $x \in \mathsf{V}$ is clearly linear. More interesting is the fact that if V is finite-dimensional, every linear transformation from V into F is of this form.

Theorem 6.8. *Let* V *be a finite-dimensional inner product space over* F, *and let* $\mathsf{g}\colon \mathsf{V} \to \mathrm{F}$ *be a linear transformation. Then there exists a unique vector* $\mathrm{y} \in \mathsf{V}$ *such that* $\mathsf{g}(\mathrm{x}) = \langle \mathrm{x}, \mathrm{y}\rangle$ *for all* $\mathrm{x} \in \mathsf{V}$.

Proof. Let β be an orthonormal basis for V, say $\beta = \{x_1, \ldots, x_n\}$, and let

$$y = \sum_{i=1}^{n} \overline{\mathsf{g}(x_i)}\, x_i.$$

If we define $\mathsf{h}\colon \mathsf{V} \to F$ by $\mathsf{h}(x) = \langle x, y\rangle$, then h is clearly linear. Now for $1 \leqslant j \leqslant n$ we have

$$\mathsf{h}(x_j) = \langle x_j, y\rangle = \left\langle x_j, \sum_{i=1}^{n} \overline{\mathsf{g}(x_i)}\, x_i \right\rangle = \sum_{i=1}^{n} \mathsf{g}(x_i)\langle x_j, x_i\rangle$$
$$= \sum_{i=1}^{n} \mathsf{g}(x_i)\delta_{ji} = \mathsf{g}(x_j).$$

Since g and h both agree on β, we have $\mathsf{g} = \mathsf{h}$ by the corollary to Theorem 2.6.

To show that y is unique, suppose that $\mathsf{g}(x) = \langle x, y'\rangle$ for all x. Then $\langle x, y\rangle = \langle x, y'\rangle$ for all x, so by Theorem 6.1(d) we have $y = y'$. ∎

Example 1

Define $\mathsf{g}\colon \mathsf{R}^2 \to R$ by $\mathsf{g}(a_1, a_2) = 2a_1 + a_2$; clearly g is a linear transformation. Let $\beta = \{e_1, e_2\}$, and let $y = \mathsf{g}(e_1)e_1 + \mathsf{g}(e_2)e_2 = 2e_1 + e_2 = (2, 1)$ as in the proof of Theorem 6.8. Then $\mathsf{g}(a_1, a_2) = \langle (a_1, a_2), (2, 1)\rangle = 2a_1 + a_2$. ∎

Theorem 6.9. *Let* V *be a finite-dimensional inner product space, and let* T *be a linear operator on* V. *Then there exists a unique function* $\mathsf{T}^*\colon \mathsf{V} \to \mathsf{V}$ *such that* $\langle \mathsf{T}(\mathrm{x}), \mathrm{y}\rangle = \langle \mathrm{x}, \mathsf{T}^*(\mathrm{y})\rangle$ *for all* $\mathrm{x}, \mathrm{y} \in \mathsf{V}$. *Moreover,* T^* *is linear.*

Proof. Let $y \in \mathsf{V}$. Define $\mathsf{g}\colon \mathsf{V} \to F$ by $\mathsf{g}(x) = \langle \mathsf{T}(x), y\rangle$ for all $x \in \mathsf{V}$. We first show that g is linear. Let $x_1, x_2 \in \mathsf{V}$ and $c \in F$. Then

$$\mathsf{g}(cx_1 + x_2) = \langle \mathsf{T}(cx_1 + x_2), y\rangle = \langle c\mathsf{T}(x_1) + \mathsf{T}(x_2), y\rangle$$
$$= c\langle \mathsf{T}(x_1), y\rangle + \langle \mathsf{T}(x_2), y\rangle = c\mathsf{g}(x_1) + \mathsf{g}(x_2).$$

Hence g is linear.

We now may apply Theorem 6.8 to obtain a unique vector $y' \in \mathsf{V}$ such that

$g(x) = \langle x, y' \rangle$; i.e., $\langle T(x), y \rangle = \langle x, y' \rangle$ for all $x \in V$. Defining $T^*: V \to V$ by $T^*(y) = y'$, we have $\langle T(x), y \rangle = \langle x, T^*(y) \rangle$.

To show that T^* is linear, let $y_1, y_2 \in V$ and $c \in F$. Then for any $x \in V$, we have

$$\begin{aligned} \langle x, T^*(cy_1 + y_2) \rangle &= \langle T(x), cy_1 + y_2 \rangle \\ &= \bar{c}\langle T(x), y_1 \rangle + \langle T(x), y_2 \rangle \\ &= \bar{c}\langle x, T^*(y_1) \rangle + \langle x, T^*(y_2) \rangle \\ &= \langle x, cT^*(y_1) + T^*(y_2) \rangle. \end{aligned}$$

Since x is arbitrary, we have $T^*(cy_1 + y_2) = cT^*(y_1) + T^*(y_2)$ by Theorem 6.1(d).

Finally, we need only show that T^* is unique. Suppose that $U: V \to V$ is linear and that it satisfies $\langle T(x), y \rangle = \langle x, U(y) \rangle$ for all $x, y \in V$. Then $\langle x, T^*(y) \rangle = \langle x, U(y) \rangle$ for all $x, y \in V$, so $T^* = U$. ■

The linear operator T^* described in Theorem 6.9 is called the *adjoint* of the operator T. The symbol T^* is read "T star."

Thus T^* is the unique operator on V satisfying $(T(x), y) = (x, T^*(y))$ for all $x, y \in V$. Note that we also have

$$\langle x, T(y) \rangle = \overline{\langle T(y), x \rangle} = \overline{\langle y, T^*(x) \rangle} = \langle T^*(x), y \rangle,$$

so $\langle x, T(y) \rangle = \langle T^*(x), y \rangle$ for all $x, y \in V$. We may view these equations symbolically as adding a * to T when we shift its position inside the inner product symbol.

In the infinite-dimensional case the adjoint of a linear operator T may be defined to be the function T^* such that $\langle T(x), y \rangle = \langle x, T^*(y) \rangle$ for all $x, y \in V$. The uniqueness and linearity of T^* will follow as before. However, the existence of an adjoint is not guaranteed. The reader should observe the necessity of the hypothesis of finite-dimensionality in the proof of Theorem 6.8. Many of the theorems we prove about adjoints, nevertheless, are independent of the dimension of V. *Thus for the remainder of this chapter we adopt the convention for the exercises that a reference to the adjoint of a linear operator on an infinite-dimensional inner product space assumes its existence.*

A useful result for computing adjoints is Theorem 6.10 below.

Theorem 6.10. *Let V be a finite-dimensional inner product space, and let β be an orthonormal basis for V. If T is a linear operator on V, then*

$$[T^*]_\beta = [T]_\beta^*.$$

Proof. Let $A = [T]_\beta$, $B = [T^*]_\beta$, and $\beta = \{x_1, \ldots, x_n\}$. Then from the corollary to Theorem 6.5 we have

$$\begin{aligned} B_{ij} &= \langle T^*(x_j), x_i \rangle = \overline{\langle x_i, T^*(x_j) \rangle} \\ &= \overline{\langle T(x_i), x_j \rangle} = \overline{A_{ji}} = (A^*)_{ij}. \end{aligned}$$

Hence $B = A^*$. ∎

Corollary. *Let* A *be an* n × n *matrix. Then* $\mathsf{L}_{A^*} = (\mathsf{L}_A)^*$.

Proof. If β is the standard ordered basis for F^n, then by Theorem 2.16 we have that $[\mathsf{L}_A]_\beta = A$. Hence $[(\mathsf{L}_A)^*]_\beta = [\mathsf{L}_A]_\beta^* = A^* = [\mathsf{L}_{A^*}]_\beta$, and so $(\mathsf{L}_A)^* = \mathsf{L}_{A^*}$. ∎

As an application of Theorem 6.10, we will compute the adjoint of a specific linear operator.

Example 2

Define $\mathsf{T}: \mathsf{C}^2 \to \mathsf{C}^2$ by $\mathsf{T}(a_1, a_2) = (2ia_1 + 3a_2, a_1 - a_2)$. If β is the standard ordered basis for C^2, then

$$[\mathsf{T}]_\beta = \begin{pmatrix} 2i & 3 \\ 1 & -1 \end{pmatrix}.$$

So

$$[\mathsf{T}^*]_\beta = [\mathsf{T}]_\beta^* = \begin{pmatrix} -2i & 1 \\ 3 & -1 \end{pmatrix}.$$

Hence

$$\mathsf{T}^*(a_1, a_2) = (-2ia_1 + a_2, 3a_1 - a_2). \quad ∎$$

The following theorem demonstrates the analogy between the conjugates of complex numbers and the adjoints of linear operators.

Theorem 6.11. *Let* V *be a finite-dimensional inner product space, and let* T *and* U *be linear operators on* V. *Then*

(a) $(\mathsf{T} + \mathsf{U})^* = \mathsf{T}^* + \mathsf{U}^*$.
(b) $(c\mathsf{T})^* = \bar{c}\mathsf{T}^*$ *for any* $c \in \mathsf{F}$.
(c) $(\mathsf{TU})^* = \mathsf{U}^*\mathsf{T}^*$.
(d) $\mathsf{T}^{**} = \mathsf{T}$.
(e) $\mathsf{I}^* = \mathsf{I}$.

Proof. We will prove parts (a) and (d); the rest are proved similarly. Let $x, y \in \mathsf{V}$.

Since

$$\begin{aligned}\langle x, (\mathsf{T} + \mathsf{U})^*(y)\rangle &= \langle (\mathsf{T} + \mathsf{U})(x), y\rangle = \langle \mathsf{T}(x) + \mathsf{U}(x), y\rangle \\ &= \langle \mathsf{T}(x), y\rangle + \langle \mathsf{U}(x), y\rangle = \langle x, \mathsf{T}^*(y)\rangle + \langle x, \mathsf{U}^*(y)\rangle \\ &= \langle x, \mathsf{T}^*(y) + \mathsf{U}^*(y)\rangle = \langle x, (\mathsf{T}^* + \mathsf{U}^*)(y)\rangle,\end{aligned}$$

(a) follows.

Similarly, since

$$\begin{aligned}\langle x, \mathsf{T}(y)\rangle &= \langle \mathsf{T}^*(x), y\rangle \\ &= \langle x, \mathsf{T}^{**}(y)\rangle,\end{aligned}$$

(d) follows. ∎

The same proof works in the infinite-dimensional case provided that the existence of T^* and U^* is assumed.

Corollary. *Let* A *and* B *be* n × n *matrices. Then*
(a) $(A + B)^* = A^* + B^*$.
(b) $(cA)^* = \bar{c}A^*$ *for all* $c \in F$.
(c) $(AB)^* = B^*A^*$.
(d) $A^{**} = A$.
(e) $I^* = I$.

Proof. We will prove only (c); the remaining parts can be proved similarly. Since $L_{(AB)^*} = (L_{AB})^* = (L_A L_B)^* = (L_B)^*(L_A)^* = L_{B^*}L_{A^*} = L_{B^*A^*}$, we have $(AB)^* = B^*A^*$. ∎

In the proof above we relied on the corollary to Theorem 6.10. An alternative proof can be given by appealing directly to the definition of the conjugate transposes of the matrices A and B (see Exercise 5).

Least Squares Approximation

Consider the following problem: An experimenter collects data by taking measurements $y_1, y_2, \ldots, y_m$ at times $t_1, t_2, \ldots, t_m$, respectively. For example, he or she may be measuring unemployment at various times during some period. Suppose that the data $(t_1, y_1), \ldots, (t_m, y_m)$ are plotted as points in the plane (see Figure 6.2). From this distribution, he or she feels that there exists an essentially linear relationship between y and t, say, $y = ct + d$, and would like to find the constants c and d so that the line $y = ct + d$ represents the best possible "fit" to the data collected. One such estimate of fit is to calculate the error E that represents the sum of the squares of the vertical distances from the points to the line; i.e.,

$$E = \sum_{i=1}^{m} (y_i - ct_i - d)^2.$$

Thus the problem is to find the constants c and d that minimize E. (For this reason the line $y = ct + d$ is called the *least squares line.*) If we let

$$A = \begin{pmatrix} t_1 & 1 \\ t_2 & 1 \\ \vdots & \vdots \\ t_m & 1 \end{pmatrix}, \qquad X = \begin{pmatrix} c \\ d \end{pmatrix}, \quad \text{and} \quad y = \begin{pmatrix} y_1 \\ y_2 \\ \vdots \\ y_m \end{pmatrix},$$

then it follows that $E = \|y - Ax\|^2$.

We now develop a general method to find an explicit vector $x_0 \in F^n$ that minimizes E; that is, given an $m \times n$ matrix A, we will find $x_0 \in F^n$ such that

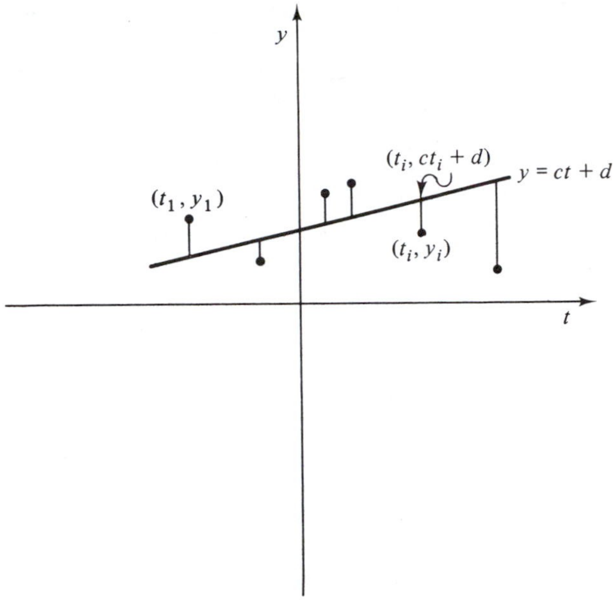

Figure 6.2

$\|y - Ax_0\| \leqslant \|y - Ax\|$ for all vectors $x \in \mathsf{F}^n$. This method will not only allow us to find the linear function which best fits the data but also the polynomial of any fixed degree which best fits the data.

We first need some notation and two simple lemmas. For $x, y \in \mathsf{F}^n$, we let $\langle x, y\rangle_n$ denote the standard inner product of x and y in F^n. Notice that $\langle x, y\rangle_n = y^*x$.

Lemma 1. *Let* A *be an* m × n *matrix over* F, $x \in \mathsf{F}^n$, *and* $y \in \mathsf{F}^m$. *Then*

$$\langle Ax, y\rangle_m = \langle x, A^*y\rangle_n.$$

Proof. By the corollary to Theorem 6.11 we have

$$\langle Ax, y\rangle_m = y^*(Ax) = (A^*y)^*x = \langle x, A^*y\rangle_n. \quad \blacksquare$$

Lemma 2. *Let* A *be an* m × n *matrix over* F. *Then* rank(A^*A) = rank(A).

Proof. By the dimension theorem we need only show that for $x \in \mathsf{F}^n$ we have $A^*Ax = 0$ if and only if $Ax = 0$. Clearly, $Ax = 0$ implies that $A^*Ax = 0$. So assume that $A^*Ax = 0$. Then $0 = \langle A^*Ax, x\rangle_n = \langle Ax, A^{**}x\rangle_m = \langle Ax, Ax\rangle_m$, so that $Ax = 0$. $\blacksquare$

Corollary. *If* A *is an* m × n *matrix such that* rank(A) = n (*i.e.,* A *has "full rank"*), *then* A^*A *is invertible.*

Now consider the system $AX = y$, where A is an $m \times n$ matrix and $y \in \mathsf{F}^m$. Define $\mathsf{W} = \{Ax : x \in \mathsf{F}^n\}$; that is, $\mathsf{W} = \mathsf{R}(\mathsf{L}_A)$. By the corollary to Proposition 6.6 there exists a unique vector in W, say Ax_0 where $x_0 \in \mathsf{F}^n$, that is closest to y. So $\|Ax_0 - y\| \leqslant \|Ax - y\|$ for all $x \in \mathsf{F}^n$.

To develop a practical method for finding such an x_0, we note from the corollary to Proposition 6.6 that $Ax_0 - y \in \mathsf{W}^\perp$, so $\langle Ax, Ax_0 - y\rangle_m = 0$ for all $x \in \mathsf{F}^n$. Thus by Lemma 1 we have that $\langle x, A^*(Ax_0 - y)\rangle_n = 0$ for all $x \in \mathsf{F}^n$; that is, $A^*(Ax_0 - y) = 0$. So we need only find a solution to $A^*AX = A^*y$. If, in addition, we assume that $\text{rank}(A) = n$, then by Lemma 2 we have $x_0 = (A^*A)^{-1}A^*y$. We may summarize this discussion in the following theorem.

Theorem 6.12. *Let* $A \in \mathsf{M}_{m\times n}(F)$ *and* $y \in \mathsf{F}^m$. *Then there exists* $x_0 \in \mathsf{F}^n$ *such that* $(A^*A)x_0 = A^*y$ *and* $\|Ax_0 - y\| \leqslant \|Ax - y\|$ *for all* $x \in \mathsf{F}^n$. *Furthermore, if* $\text{rank}(A) = n$, *then* $x_0 = (A^*A)^{-1}A^*y$.

To return to our experimenter, let us suppose that the data collected is (1, 2), (2, 3), (3, 5), and (4, 7). Then

$$A = \begin{pmatrix} 1 & 1 \\ 2 & 1 \\ 3 & 1 \\ 4 & 1 \end{pmatrix} \quad \text{and} \quad y = \begin{pmatrix} 2 \\ 3 \\ 5 \\ 7 \end{pmatrix};$$

hence

$$A^*A = \begin{pmatrix} 1 & 2 & 3 & 4 \\ 1 & 1 & 1 & 1 \end{pmatrix}\begin{pmatrix} 1 & 1 \\ 2 & 1 \\ 3 & 1 \\ 4 & 1 \end{pmatrix} = \begin{pmatrix} 30 & 10 \\ 10 & 4 \end{pmatrix},$$

so

$$(A^*A)^{-1} = \frac{1}{20}\begin{pmatrix} 4 & -10 \\ -10 & 30 \end{pmatrix}.$$

Therefore,

$$\begin{pmatrix} c \\ d \end{pmatrix} = x_0 = \frac{1}{20}\begin{pmatrix} 4 & -10 \\ -10 & 30 \end{pmatrix}\begin{pmatrix} 1 & 2 & 3 & 4 \\ 1 & 1 & 1 & 1 \end{pmatrix}\begin{pmatrix} 2 \\ 3 \\ 5 \\ 7 \end{pmatrix} = \begin{pmatrix} 1.7 \\ 0 \end{pmatrix}.$$

Thus the line $y = 1.7t$ is the least squares line. The error E may be computed directly as $\|Ax_0 - y\|^2 = 0.3$.

The method above may also be applied if the experimenter wants to fit a parabola $y = ct^2 + dt + e$ to the data. In this case, he or she would use

$$\begin{pmatrix} t_1^2 & t_1 & 1 \\ \vdots & \vdots & \vdots \\ t_m^2 & t_m & 1 \end{pmatrix}$$

as the matrix A.

Finally, suppose in the linear case that the experimenter chose the times, t_i, to satisfy

$$\sum_{i=1}^{m} t_i = 0.$$

Then the two columns of A would be orthogonal, so A^*A would be a diagonal matrix (see Exercise 17). This, of course, would greatly simplify the computations.

Minimal Solutions

In the preceding discussion we showed that if $\text{rank}(A) = n$, then there exists a unique $x_0 \in \mathsf{F}^n$ such that Ax_0 is the point in W that is closest to y. Of course, if $\text{rank}(A) < n$, there will be infinitely many such vectors. It is often desirable to find such a vector of minimal norm. For what follows, we let $b = Ax_0$ as above. Then the system $AX = b$ has at least one solution. A solution s is called a *minimal solution* if $\|s\| \leqslant \|u\|$ for all other solutions u of $AX = b$.

Theorem 6.13. *Let $A \in \mathsf{M}_{m\times n}(F)$ and $b \in \mathsf{F}^m$. Suppose that $AX = b$ has at least one solution. Then*

(a) *There exists exactly one minimal solution s of $AX = b$, and $s \in \mathsf{R}(\mathsf{L}_{A^*})$.*

(b) *s is the only solution of $AX = b$ that lies in $\mathsf{R}(\mathsf{L}_{A^*})$; that is, if u is a solution of $(AA^*)X = b$, then $s = A^*u$.*

Proof. For simplicity of notation, we let $\mathsf{W} = \mathsf{R}(\mathsf{L}_{A^*})$ and $\mathsf{W}' = \mathsf{N}(\mathsf{L}_A)$. Let x be any solution of $AX = b$. By Proposition 6.6 we can write $x = s + y$ with $s \in \mathsf{W}$ and $y \in \mathsf{W}^\perp$, where $\mathsf{W}^\perp = \mathsf{W}'$ by Exercise 12. Note that $b = Ax = As + Ay = As$, so that s is a solution of $AX = b$ that lies in W. To prove (a) we need only show that s is the unique minimal solution. Let v be any solution of $AX = b$. By Theorem 3.9 we have that $v = s + u$, where $u \in \mathsf{W}'$. Since $s \in \mathsf{W}$, which equals $\mathsf{W}'^\perp$ by Exercise 12, we have by Exercise 10 of Section 6.1 that

$$\|v\|^2 = \|s + u\|^2 = \|s\|^2 + \|u\|^2 \geqslant \|s\|^2.$$

Thus s is a minimal solution. We can also see from the calculation above that if $\|v\| = \|s\|$, then $u = 0$ and $v = s$. Hence s is the unique minimal solution of $AX = b$, proving (a).

To prove (b), we assume that v is also a solution of $AX = b$ that lies in W. Then

$$v - s \in \mathsf{W} \cap \mathsf{W}' = \mathsf{W}'^{\perp} \cap \mathsf{W}' = \{0\},$$

so $v = s$. ∎

Example 3

Consider the system

$$\begin{cases} x + 2y + z = 4 \\ x - y + 2z = -11 \\ x + 5y = 19. \end{cases}$$

Let

$$A = \begin{pmatrix} 1 & 2 & 1 \\ 1 & -1 & 2 \\ 1 & 5 & 0 \end{pmatrix} \quad \text{and} \quad b = \begin{pmatrix} 4 \\ -11 \\ 19 \end{pmatrix}.$$

To find the minimal solution of this system, we must find a solution of $AA^*X = b$. Now

$$AA^* = \begin{pmatrix} 6 & 1 & 11 \\ 1 & 6 & -4 \\ 11 & -4 & 26 \end{pmatrix};$$

so we consider the system

$$\begin{cases} 6x + y + 11z = 4 \\ x + 6y - 4z = -11 \\ 11x - 4y + 26z = 19, \end{cases}$$

for which a solution is

$$u = \begin{pmatrix} 1 \\ -2 \\ 0 \end{pmatrix}.$$

(Any solution will suffice.) Hence

$$s = A^*u = \begin{pmatrix} -1 \\ 4 \\ -3 \end{pmatrix}$$

is the minimal solution of the given system. ∎

EXERCISES

1. Label the following statements as being true or false. Assume that the underlying inner product spaces are finite-dimensional.
   (a) Every linear operator has an adjoint.
   (b) Every linear operator on V has the form $x \to \langle x, y \rangle$ for some $y \in \mathsf{V}$.
   (c) For every linear operator T on V and every basis β of V, we have $[\mathsf{T}^*]_\beta = ([\mathsf{T}]_\beta)^*$.
   (d) The adjoint of a linear operator is always unique.
   (e) For any operators T and U and scalars a and b,
   $$(a\mathsf{T} + b\mathsf{U})^* = a\mathsf{T}^* + b\mathsf{U}^*.$$
   (f) For any $n \times n$ matrix A, $(\mathsf{L}_A)^* = \mathsf{L}_{A^*}$.
   (g) For any operator T, $(\mathsf{T}^*)^* = \mathsf{T}$.
2. For each of the following inner product spaces V (over F) and linear transformations $\mathsf{g}\colon \mathsf{V} \to F$, find a vector y such that $\mathsf{g}(x) = \langle x, y \rangle$ for all $x \in \mathsf{V}$.
   (a) $\mathsf{V} = \mathsf{R}^3$, $\mathsf{g}(a_1, a_2, a_3) = a_1 - 2a_2 + 4a_3$
   (b) $\mathsf{V} = \mathsf{C}^2$, $\mathsf{g}(z_1, z_2) = z_1 - 2z_2$
   (c) $\mathsf{V} = \mathsf{P}_2(R)$ with $\langle f, h \rangle = \int_0^1 f(t)h(t)\,dt$, $\mathsf{g}(f) = f(0) + f'(1)$
3. For each of the following inner product spaces V and linear operators T on V, evaluate T* at the given element of V.
   (a) $\mathsf{V} = \mathsf{R}^2$, $\mathsf{T}(a, b) = (2a + b, a - 3b)$, $x = (3, 5)$
   (b) $\mathsf{V} = \mathsf{C}^2$, $\mathsf{T}(z_1, z_2) = (2z_1 + iz_2, (1 - i)z_1)$, $x = (3 - i, 1 + 2i)$
   (c) $\mathsf{V} = \mathsf{P}_2(R)$ with
   $$\langle f, g \rangle = \int_0^1 f(t)g(t)\,dt,$$
   $$\mathsf{T}(f) = f' + 3f, \qquad f(x) = 4 - x + 3x^2$$
4. Complete the proof of Theorem 6.11.
5. Complete the proof of the corollary to Theorem 6.11 in two ways. First use Theorem 6.11 as in the proof of (c). Then use the matrix definition of A^*.
6. Let T be a linear operator on an inner product space V. Let $\mathsf{U}_1 = \mathsf{T} + \mathsf{T}^*$ and $\mathsf{U}_2 = \mathsf{T}\mathsf{T}^*$. Prove that $\mathsf{U}_1 = \mathsf{U}_1^*$ and $\mathsf{U}_2 = \mathsf{U}_2^*$.
7. Give an example of a linear operator T on an inner product space V such that $\mathsf{N}(\mathsf{T}) \neq \mathsf{N}(\mathsf{T}^*)$.
8. Let V be a finite-dimensional inner product space, and let T be a linear operator on V. Prove that if T is invertible, then T* is invertible and $(\mathsf{T}^*)^{-1} = (\mathsf{T}^{-1})^*$.
9. Prove that if $\mathsf{V} = \mathsf{W} \oplus \mathsf{W}^\perp$ and T is the projection on W with $\mathsf{N}(\mathsf{T}) = \mathsf{W}^\perp$, then $\mathsf{T} = \mathsf{T}^*$. (For definitions, see the exercises of Sections 1.3 and 2.1.)

10. Let T be a linear operator on an inner product space V. Prove that $(T(x)\| = \|x\|$ for all $x \in V$ if and only if $\langle T(x), T(y)\rangle = \langle x, y\rangle$ for all $x, y \in V$. *Hint:* Use Exercise 20 of Section 6.1.

11. For a linear operator T on an inner product space V, prove that $T^*T = T_0$ implies $T = T_0$. Is the same result true if we assume that $TT^* = T_0$?

12. Let V be a finite-dimensional inner product space, and let T be a linear operator on V. Prove that $R(T^*) = N(T)^{\perp}$. *Hint:* Prove that $R(T^*)^{\perp} = N(T)$, and then use Exercise 12(c) of Section 6.2.

13. Let T be a linear operator on a finite-dimensional inner product space V. Prove the following.

(a) $N(T^*T) = N(T)$. Deduce that $\text{rank}(T^*T) = \text{rank}(T)$.

(b) $\text{rank}(T) = \text{rank}(T^*)$. Deduce from (a) that $\text{rank}(TT^*) = \text{rank}(T)$.

(c) For any $n \times n$ matrix A, $\text{rank}(A^*A) = \text{rank}(AA^*) = \text{rank}(A)$.

14. Let V be an inner product space, and let $y, z \in V$. Define $T: V \to V$ by $T(x) = \langle x, y\rangle z$ for all $x \in V$. First prove that T is linear. Then show that T^* exists, and define it explicitly.

15. Let $T: V \to W$ be a linear transformation between finite-dimensional inner product spaces V and W.

(a) Prove that there exists a unique linear transformation $T^*: W \to V$ such that $\langle T(x), y\rangle = \langle x, T^*(y)\rangle$ for all $x \in V$ and $y \in W$.

(b) Let β and γ be orthonormal bases for V and W, respectively. Prove that $[T^*]_\gamma^\beta = ([T]_\beta^\gamma)^*$.

16.† Let A be an $n \times n$ matrix. Prove that $\det(A^*) = \overline{\det(A)}$.

17. Prove that if A is an $m \times n$ matrix whose columns are orthogonal, then A^*A is a diagonal matrix.

18. For the data $(-3, 9)$, $(-2, 6)$, $(0, 2)$, and $(1, 1)$ find the line and the parabola that provide the least squares fit. Compute E in both cases.

19. In physics, *Hooke's law* states that (within certain limits) there is a linear relation between the length x of a spring and the force y applied to (or exerted by) the spring. That is, $y = cx + d$, where c is called the *spring constant*. Use the following data to estimate the spring constant. (The length is given in inches and the force is given in pounds.)

Length, x	Force, y
3.5	1.0
4.0	2.2
4.5	2.8
5.0	4.3

20. Find the minimal solution of

$$\begin{cases} x + 2y - z = 1 \\ 2x + 3y + z = 2 \\ 4x + 7y - z = 4. \end{cases}$$

21. For the least squares line $y = ct + d$ corresponding to the m observations $(t_1, y_1), \ldots, (t_m, y_m)$, use Theorem 6.12 to derive the *normal equations:*

$$\left(\sum_{i=1}^{m} t_i^2\right)c + \left(\sum_{i=1}^{m} t_i\right)d = \sum_{i=1}^{m} t_i y_i$$

and

$$\left(\sum_{i=1}^{m} t_i\right)c + md = \sum_{i=1}^{m} y_i.$$

These equations may also be obtained by setting each of the partial derivatives of the error E to zero.

6.4 NORMAL AND SELF-ADJOINT OPERATORS

We have seen the importance of diagonalizable operators in Chapter 5. For these operators it is necessary and sufficient for the vector space V to possess a basis of eigenvectors. As V is an inner product space in this chapter, it is reasonable to seek conditions that will guarantee that V has an orthonormal basis of eigenvectors. A very important result that will help us achieve our goal is Schur's theorem (Theorem 6.14). The formulation below is in terms of linear operators. The next section will contain the more conventional matrix form. We begin with a lemma.

Lemma. *Let* T *be a linear operator on a finite-dimensional inner product space* V. *If* T *has an eigenvector, then so does* T*.

Proof. Let $A = [\mathsf{T}]_\beta$, where β is an orthonormal basis of V. Let λ be an eigenvalue of T, and hence of A. Then $\det(A - \lambda I) = 0$. So by Exercise 16 of Section 6.3 and the corollary to Theorem 6.11, we also have that $\det(A^* - \bar{\lambda} I) = 0$. So $\bar{\lambda}$ is an eigenvalue of A^* and hence of T*. In particular, T* has an eigenvector. ∎

Recall (Exercise 26 of Section 2.1) that a subspace W of V is T-invariant if T(W) is contained in W. If W is T-invariant, we may define the restriction $\mathsf{T}_\mathsf{W}: \mathsf{W} \to \mathsf{W}$ by $\mathsf{T}_\mathsf{W}(x) = \mathsf{T}(x)$ for all x in W. It is clear that T_W is a linear operator on W. Recall also from Section 5.2 that a polynomial splits if it factors into linear polynomials.

Theorem 6.14 (Schur). *Let T be a linear operator on a finite-dimensional inner product space V. Suppose that the characteristic polynomial of T splits. Then there exists an orthonormal basis β such that the matrix $[\mathsf{T}]_\beta$ is upper triangular.*

Proof. The proof will be by induction on the dimension n of V. The result is immediate if $n = 1$. So suppose that the result is true for linear operators on $(n-1)$-dimensional inner product spaces whose characteristic polynomials split. By the lemma we can assume that T^* has a unit eigenvector z. Suppose that $\mathsf{T}^*(z) = \lambda z$ and that $\mathsf{W} = \text{span}(\{z\})$. We now show that $\mathsf{W}^\perp$ is T-invariant. If $y \in \mathsf{W}^\perp$ and $x = cz \in \mathsf{W}$, then

$$\begin{aligned}\langle \mathsf{T}(y), x\rangle &= \langle \mathsf{T}(y), cz\rangle = \langle y, \mathsf{T}^*(cz)\rangle = \langle y, c\mathsf{T}^*(z)\rangle = \langle y, c\lambda z\rangle \\ &= \bar{c}\bar{\lambda}\langle y, z\rangle = \bar{c}\bar{\lambda}(0) = 0.\end{aligned}$$

So $\mathsf{T}(y) \in \mathsf{W}^\perp$. Now let T_1 be the restriction of T to $\mathsf{W}^\perp$. It is easy to show (see Theorem 5.26) that the characteristic polynomial of T_1 divides the characteristic polynomial of T and hence splits. By Theorem 6.7(c) $\dim(\mathsf{W}^\perp) = n - 1$, so we may apply the induction hypothesis to T_1 and obtain an orthonormal basis γ of $\mathsf{W}^\perp$ such $[\mathsf{T}_1]_\gamma$ is upper triangular. Clearly, $\beta = \gamma \cup \{z\}$ is an orthonormal basis of V such that $[\mathsf{T}]_\beta$ is upper triangular. ∎

We now return to our original goal of finding an orthonormal basis of eigenvectors of a linear operator T on a finite-dimensional inner product space V. Note that if such an orthonormal basis β exists, then $[\mathsf{T}]_\beta$ is a diagonal matrix. But by Theorem 6.10, we have that $[\mathsf{T}^*]_\beta = [\mathsf{T}]_\beta^*$ is also a diagonal matrix. Because diagonal matrices commute, we conclude that T and T^* commute. Thus if V possesses an orthonormal basis of eigenvectors of T, then $\mathsf{TT}^* = \mathsf{T}^*\mathsf{T}$.

Definitions. *Let V be an inner product space, and let T be a linear operator on V. We say that T is* normal *if $\mathsf{TT}^* = \mathsf{T}^*\mathsf{T}$. An $n \times n$ matrix A is* normal *if $AA^* = A^*A$.*

It follows immediately that T is normal if and only if $[\mathsf{T}]_\beta$ is normal, where β is an orthonormal basis.

Example 1

Let $\mathsf{T}: \mathsf{R}^2 \to \mathsf{R}^2$ be rotation by θ, where $0 < \theta < \pi$. The matrix representation of T in the standard ordered basis is given by

$$A = \begin{pmatrix} \cos\theta & -\sin\theta \\ \sin\theta & \cos\theta \end{pmatrix}.$$

Note that $AA^* = I = A^*A$, so A and hence T is normal. ∎

Clearly, the operator T in Example 1 does not even possess one eigenvector. So in the case of a real inner product space, we see that normality is not sufficient to guarantee an orthonormal basis of eigenvectors. All is not lost, however. We will show that normality suffices if V is a *complex* inner product space.

Before we prove the promised result for normal operators, we need some general properties of normal operators.

Theorem 6.15. *Let* V *be an inner product space, and let* T *be a normal operator on* V. *Then*

(a) $\|T(x)\| = \|T^*(x)\|$ *for all* $x \in V$.

(b) $T - cI$ *is normal for every* $c \in F$.

(c) *If* x *is an eigenvector of* T, *then* x *is also an eigenvector of* T^*. *In fact, if* $T(x) = \lambda x$, *then* $T^*(x) = \bar{\lambda} x$.

(d) *If* λ_1 *and* λ_2 *are distinct eigenvalues of* T *with corresponding eigenvectors* x_1 *and* x_2, *then* x_1 *and* x_2 *are orthogonal.*

Proof. (a) For any $x \in V$, we have

$$\|T(x)\|^2 = \langle T(x), T(x)\rangle = \langle T^*T(x), x\rangle = \langle TT^*(x), x\rangle$$
$$= \langle T^*(x), T^*(x)\rangle = \|T^*(x)\|^2.$$

The proof of (b) is left as an exercise.

(c) Suppose that $T(x) = \lambda x$ for some $x \in V$. Let $U = T - \lambda I$. Then $U(x) = 0$, and by (b) U is normal. Thus (a) implies that

$$0 = \|U(x)\| = \|U^*(x)\| = \|(T^* - \bar{\lambda} I)(x)\| = \|T^*(x) - \bar{\lambda} x\|.$$

Hence $T^*(x) = \bar{\lambda} x$. So x is an eigenvector of T^*.

(d) Let λ_1 and λ_2 be distinct eigenvalues of T with corresponding eigenvectors x_1 and x_2. Then, using (c), we have

$$\lambda_1\langle x_1, x_2\rangle = \langle \lambda_1 x_1, x_2\rangle = \langle T(x_1), x_2\rangle = \langle x_1, T^*(x_2)\rangle$$
$$= \langle x_1, \bar{\lambda}_2 x_2\rangle = \lambda_2\langle x_1, x_2\rangle.$$

Since $\lambda_1 \neq \lambda_2$, we conclude that $\langle x_1, x_2\rangle = 0$. ∎

Theorem 6.16. *Let* T *be a linear operator on a finite-dimensional complex inner product space* V. *Then* T *is normal if and only if there exists an orthonormal basis of eigenvectors of* T.

Proof. Suppose that T is normal. By the fundamental theorem of algebra (Theorem D.4) the characteristic polynomial of T splits. So we may apply Schur's theorem to obtain an orthonormal basis $\beta = \{x_1, \ldots, x_n\}$ of V such that $[T]_\beta = A$ is upper triangular. We know that x_1 is an eigenvector of T because A is upper triangular. Assume that $x_1, \ldots, x_{k-1}$ are eigenvectors of T. We will show that x_k is also an eigenvector of T. It will then follow by mathematical induction on k that all of the x_i's are eigenvectors of T, or equivalently, that A is a diagonal matrix.

We have

$$A = \begin{pmatrix} B & C \\ O & E \end{pmatrix} \quad \text{and} \quad A^* = \begin{pmatrix} B^* & O \\ C^* & E^* \end{pmatrix},$$

where B is a $(k-1) \times (k-1)$ diagonal matrix. Because A is upper triangular, $A_{jk} = 0$ for $j > k$. To show that x_k is an eigenvector of T, we need only show that $A_{jk} = 0$ for $j < k$. Note that by Theorem 6.15(c), $x_1, \ldots, x_{k-1}$ are also eigenvectors of T*. But $A^* = [\mathsf{T}^*]_\beta$, so $C^* = O$. Thus $(A^*)_{kj} = 0$ for $j < k$, and so $A_{jk} = 0$ for $j < k$. Therefore, the vector x_k is an eigenvector of T, so by induction, all the vectors of β are eigenvectors of T.

The converse was already proved on page 326. ∎

Interestingly, as the next example shows, Theorem 6.16 does not extend to infinite-dimensional complex inner product spaces.

Example 2

Consider the inner product space H defined earlier, and let $x_k = e^{ikx}$. Suppose that $\mathsf{V} = \text{span}(\{x_k\colon k \text{ is an integer}\})$. Clearly, $\beta = \{x_k\colon k \text{ is an integer}\}$ is an orthonormal basis of V. Now let T and U be linear operators on V such that

$$\mathsf{T}(x_k) = x_{k+1} \quad \text{and} \quad \mathsf{U}(x_k) = x_{k-1}$$

for all integers k. Then

$$\langle \mathsf{T}(x_i), x_j \rangle = \langle x_{i+1}, x_j \rangle = \delta_{(i+1)j} = \delta_{i(j-1)} = \langle x_i, x_{j-1} \rangle = \langle x_i, \mathsf{U}(x_j) \rangle.$$

It follows that $\mathsf{U} = \mathsf{T}^*$. Furthermore, $\mathsf{TT}^* = \mathsf{I} = \mathsf{T}^*\mathsf{T}$, so T is normal.

We will show that T has no eigenvectors. Suppose that x is an eigenvector of T. We may write

$$x = \sum_{i=n}^{m} a_i x_i \qquad \text{where } a_m \neq 0.$$

Because $\mathsf{T}(x) = \lambda x$ for some λ, we have

$$\sum_{i=n}^{m} a_i x_{i+1} = \sum_{i=n}^{m} \lambda a_i x_i.$$

Since $a_m \neq 0$, we can write x_{m+1} as a linear combination of $x_n, x_{n+1}, \ldots, x_m$. But this is a contradiction because β is linearly independent. ∎

Example 1 illustrates that normality is not sufficient to guarantee the existence of an orthonormal basis of eigenvectors for real inner product spaces. For real inner product spaces we must replace normality by the stronger condition that $\mathsf{T} = \mathsf{T}^*$.

Definitions. *Let* V *be an inner product space, and let* T *be a linear operator on* V. *We say that* T *is* self-adjoint *if* $\mathsf{T} = \mathsf{T}^*$. *An* $n \times n$ *matrix* A *is* self-adjoint *if* $A = A^*$.

It follows immediately that T is self-adjoint if and only if $[\mathsf{T}]_\beta$ is self-adjoint, where β is an orthonormal basis. For real matrices, this condition reduces to the requirement that A be symmetric.

Before we state our main result for self-adjoint operators, we need some preliminary work.

Lemma. *Let* T *be a self-adjoint operator on a finite-dimensional inner product space* V. *Then*

(a) *Every eigenvalue of* T *is real.*

(b) *The characteristic polynomial of* T *splits.*

Proof. (a) Suppose that $\mathsf{T}(x) = \lambda x$ for $x \neq 0$. Because a self-adjoint operator is also normal, we can apply Theorem 6.15(c) to obtain

$$\lambda x = \mathsf{T}(x) = \mathsf{T}^*(x) = \bar{\lambda} x.$$

So $\lambda = \bar{\lambda}$; that is, λ is real.

(b) Let β be an orthonormal basis of V and let $A = [\mathsf{T}]_\beta$. Then A is self-adjoint. Define $\mathsf{T}_A\colon \mathsf{C}^n \to \mathsf{C}^n$ by $\mathsf{T}_A(x) = Ax$. Now T_A is self-adjoint because $[\mathsf{T}_A]_\gamma = A$, where γ is the standard ordered (orthonormal) basis of C^n. So by (a) the eigenvalues of T_A are real. By the fundamental theorem of algebra the characteristic polynomial of T_A and hence of A splits into factors of the form $x - \lambda$, where λ is real. ∎

We are now ready to establish one of the major results of this chapter.

Theorem 6.17. *Let* T *be a linear operator on a finite-dimensional real inner product space* V. *Then* T *is self-adjoint if and only if there exists an orthonormal basis* β *of eigenvectors of* T.

Proof. Suppose that T is self-adjoint. By the lemma we may apply Schur's theorem to obtain an orthonormal basis β for V such that the matrix $A = [\mathsf{T}]_\beta$ is upper triangular. But

$$A^* = [\mathsf{T}]_\beta^* = [\mathsf{T}^*]_\beta = [\mathsf{T}]_\beta = A.$$

So A and A^* are both upper triangular, and therefore A is a diagonal matrix. Thus β must consist of eigenvectors of T.

The converse is left as an exercise. ∎

Theorem 6.17 is used extensively in many areas of mathematics and statistics. We will restate this theorem in matrix form in the next section.

Example 3

As we noted earlier, real self-adjoint matrices are symmetric, and self-adjoint matrices are normal. However, the matrix below is complex and symmetric, but not normal.

$$A = \begin{pmatrix} i & i \\ i & 1 \end{pmatrix} \quad \text{and} \quad A^* = \begin{pmatrix} -i & -i \\ -i & 1 \end{pmatrix}$$

A is not normal because $(AA^*)_{12} = 1 + i$, but $(A^*A)_{12} = 1 - i$. ∎

EXERCISES

1. Label the following statements as being true or false. Assume that the underlying inner product spaces are finite-dimensional.

(a) Every self-adjoint operator is normal.
(b) Operators and their adjoints have the same eigenvectors.
(c) If T is an operator on an inner product space V, then T is normal if and only if $[\mathsf{T}]_\beta$ is normal, where β is any ordered basis for V.
(d) A matrix A is normal if and only if L_A is normal.
(e) The eigenvalues of a self-adjoint operator must all be real.
(f) The identity and zero operators are self-adjoint.
(g) Every normal operator is diagonalizable.
(h) Every self-adjoint operator is diagonalizable.

2. For each of the linear operators below, determine whether it is normal, self-adjoint, or neither.

(a) $\mathsf{T}: \mathsf{R}^2 \to \mathsf{R}^2$ defined by $\mathsf{T}(a, b) = (2a - 2b, -2a + 5b)$
(b) $\mathsf{T}: \mathsf{C}^2 \to \mathsf{C}^2$ defined by $\mathsf{T}(a, b) = (2a + ib, a + 2b)$
(c) $\mathsf{T}: \mathsf{P}_2(R) \to \mathsf{P}_2(R)$ defined by $\mathsf{T}(f) = f'$ where $\langle f, g \rangle = \int_0^1 f(t)g(t)\,dt$

For (a), find an orthonormal basis for R^2 consisting of eigenvectors of T.

3. Let T and U be self-adjoint operators on an inner product space. Prove that TU is self-adjoint if and only if TU = UT.

4. Prove (b) of Theorem 6.15.

5. Let V be a complex inner product space, and let T be a linear operator on V. Define

$$\mathsf{T}_1 = \frac{1}{2}(\mathsf{T} + \mathsf{T}^*) \quad \text{and} \quad \mathsf{T}_2 = \frac{1}{2i}(\mathsf{T} - \mathsf{T}^*).$$

(a) Prove that T_1 and T_2 are self-adjoint and that $\mathsf{T} = \mathsf{T}_1 + i\mathsf{T}_2$.
(b) Suppose also that $\mathsf{T} = \mathsf{U}_1 + i\mathsf{U}_2$, where U_1 and U_2 are self-adjoint. Prove that $\mathsf{U}_1 = \mathsf{T}_1$ and $\mathsf{U}_2 = \mathsf{T}_2$.
(c) Prove that T is normal if and only if $\mathsf{T}_1\mathsf{T}_2 = \mathsf{T}_2\mathsf{T}_1$.

6. Let T be a linear operator on an inner product space V, and let W be a T-invariant subspace of V. Prove the following.

(a) If T is self-adjoint, then T_W is self-adjoint.

(b) $W^\perp$ is T*-invariant.

(c) If W is both T- and T*-invariant, then $(T_W)^* = (T^*)_W$.

(d) If W is both T- and T*-invariant and T is normal, then T_W is normal.

7. Let T be a normal operator on a finite-dimensional complex inner product space V, and let W be a subspace of V. Prove that if W is T-invariant, then W is also T*-invariant. *Hint:* Use Exercise 24 of Section 5.4.

8. Let T be a normal operator on a finite-dimensional inner product space V. Prove that $N(T) = N(T^*)$ and $R(T) = R(T^*)$. *Hint:* Use Theorem 6.15 and Exercise 12 of Section 6.3.

9. Let T be a self-adjoint operator on a finite-dimensional inner product space V. Prove that for all $x \in V$

$$\|T(x) \pm ix\|^2 = \|T(x)\|^2 + \|x\|^2.$$

Deduce that $(T - iI)$ is invertible and that $[(T - iI)^{-1}]^* = (T + iI)^{-1}$.

10. Assume that T is a linear operator on a complex (not necessarily finite-dimensional) inner product space V with an adjoint T*. Prove

(a) If T is self-adjoint, then $\langle T(x), x\rangle$ is real for all $x \in V$.

(b) If T satisfies $\langle T(x), x\rangle = 0$ for all $x \in V$, then $T = T_0$. *Hint:* Replace x by $x + y$ and then by $x + iy$ and expand the resulting inner products.

(c) If $\langle T(x), x\rangle$ is real for all $x \in V$, then $T = T^*$.

11. Let T be a normal operator on a finite-dimensional real inner product space V whose characteristic polynomial splits. Prove that V has an orthonormal basis of eigenvectors of T. Hence prove that T is self-adjoint.

12. Let A be an $n \times n$ real matrix. A is said to be a *Gramian* matrix if there exists a real (square) matrix B such that $A = B^tB$. Prove that A is a Gramian matrix if and only if A is symmetric and all of its eigenvalues are nonnegative. *Hint:* Apply Theorem 6.17 to L_A to obtain an orthonormal basis $\{x_1, \ldots, x_n\}$ of eigenvectors with the associated eigenvalues $\lambda_1, \ldots, \lambda_n$. Define the linear operator U by $U(x_i) = \sqrt{\lambda_i}x_i$ and complete the proof.

13. Let T be a self-adjoint operator on an n-dimensional inner product space V, and let $A = [T]_\beta$, where β is an orthonormal basis for V. T is said to be *positive definite* [*semidefinite*] if $\langle T(x), x\rangle > 0$ for all $x \neq 0$ [$\langle T(x), x\rangle \geqslant 0$ for all x]. Prove

(a) T is positive definite [semidefinite] if and only if all of its eigenvalues are positive [nonnegative].

(b) T is positive definite [semidefinite] if and only if L_A is also.

(c) T is positive definite if and only if

$$\sum_{i,j} A_{ij}a_i\overline{a_j} > 0 \qquad \text{for all nonzero } n\text{-tuples } (a_1, \ldots, a_n).$$

(d) T is positive semidefinite if and only if A is a Gramian matrix (as defined in Exercise 12).

Is the composition of two positive definite operators positive definite?

14. *Simultaneous Diagonalization.*
    (a) Let V be a finite-dimensional real inner product space, and let U and T be self-adjoint operators on V such that $\mathsf{UT} = \mathsf{TU}$. Prove that there exists an orthonormal basis for V consisting of vectors that are eigenvectors of both U and T. (The complex version of this result appears as Exercise 10 of Section 6.6). *Hint:* For any eigenspace $\mathsf{W} = \mathsf{E}_\lambda$ of T we have that W is both T- and U-invariant. By Exercise 6 we have that $\mathsf{W}^\perp$ is both T- and U-invariant. Apply Theorem 6.17 and Proposition 6.6.
    (b) State and prove the analogous result about commuting symmetric (real) matrices.

15. Prove the *Cayley–Hamilton theorem* for a complex $n \times n$ matrix A; that is, if $f(t)$ is the characteristic polynomial of A, prove that $f(A) = O$. *Hint:* By Schur's theorem show that you may assume that A is upper triangular, in which case

$$f(t) = \prod_{i=1}^{n} (A_{ii} - t).$$

Now if $\mathsf{T} = \mathsf{L}_A$, we have $(A_{jj}I - \mathsf{T})(x_j) \in \text{span}(\{x_1, \ldots, x_{j-1}\})$ for $j \geqslant 2$, where $\{x_1, \ldots, x_n\}$ is the standard ordered basis of C^n. (The general case is proved in Section 5.4.)

Exercises 16 through 20 use the definition found in Exercise 13.

16. Let T and U be positive definite operators on an inner product space V. Prove
    (a) $\mathsf{T} + \mathsf{U}$ is positive definite.
    (b) If $c > 0$, then $c\mathsf{T}$ is positive definite.
    (c) T^{-1} is positive definite.

17. Let V be an inner product space with inner product $\langle \cdot, \cdot \rangle$ and let T be a positive definite linear operator on V. Prove that $\langle x, y \rangle' = \langle \mathsf{T}(x), y \rangle$ defines another inner product on V.

18. Let V be a finite-dimensional inner product space and let T and U be self-adjoint operators on V such that U is positive definite. Prove that both TU and UT are diagonalizable linear operators that have only real eigenvalues. *Hint:* Show that TU is self-adjoint with respect to the inner product $\langle x, y \rangle' = \langle \mathsf{U}(x), y \rangle$. To show that UT is self-adjoint, repeat the argument with U^{-1} in place of U.

19. Prove the converse of Exercise 17: Let V be a finite-dimensional inner product space with inner product $\langle \cdot, \cdot \rangle$, and let $\langle \cdot, \cdot \rangle'$ be any other inner product on V.

(a) Prove that there exists a unique linear operator T on V such that $\langle x, y\rangle' = \langle T(x), y\rangle$ for all x and y in V. *Hint:* Let $\beta = \{x_1, \ldots, x_n\}$ be an orthonormal basis for V with respect to $\langle\cdot,\cdot\rangle$ and define the matrix A by $A_{ij} = \langle x_j, x_i\rangle$ for all i and j. Let T be the unique linear operator on V such that $[T]_\beta = A$.

(b) Prove that the operator T of part (a) is positive definite with respect to both inner products.

20. Let U be a diagonalizable linear operator on a finite-dimensional inner product space V such that all of the eigenvalues of U are real. Prove that there exist positive definite linear operators T_1 and T_1' and self-adjoint linear operators T_2 and T_2' such that $U = T_2T_1 = T_1'T_2'$. *Hint:* Let $\langle\cdot,\cdot\rangle$ be the inner product associated with V, β a basis of eigenvectors for U, $\langle\cdot,\cdot\rangle'$ the inner product on V with respect to which β is orthonormal [see Exercise 24(a) of Section 6.1], and T_1 the positive definite operator according to Exercise 19. Show that U is self-adjoint with respect to $\langle\cdot,\cdot\rangle'$, and $U = T_1^{-1}U^*T_1$ (the adjoint is with respect to $\langle\cdot,\cdot\rangle$). Let $T_2 = T_1^{-1}U^*$.

6.5 UNITARY AND ORTHOGONAL OPERATORS AND THEIR MATRICES

In this section we will continue our analogy between complex numbers and linear operators. Recall that the adjoint of a linear operator acts similarly to the conjugate of a complex number (see, for example, Theorem 6.11). A complex number z has length 1 if $z\bar{z} = 1$. In this section we will study those linear operators T on a vector space V such that $TT^* = T^*T = I$. We will see that these are precisely the linear operators that "preserve length" in the sense that $\|T(x)\| = \|x\|$ for all $x \in V$. As another characterization, we will prove that on a finite-dimensional complex inner product space these are the normal operators whose eigenvalues all have absolute value 1.

In past chapters we were interested in studying those functions that preserve the structure of the underlying space. In particular, linear operators preserve the operations of vector addition and scalar multiplication, and isomorphisms preserve all the vector space structure. It is now natural to consider those linear operators T on an inner product space that preserve length. We will see that this condition guarantees, in fact, that T preserves the inner product.

Definitions. *Let* V *be an inner product space* (*over* F), *and let* T *be a linear operator on* V. *If* $\|T(x)\| = \|x\|$ *for all* $x \in V$, *we call* T *a* unitary operator *if* F = C *and an* orthogonal operator *if* F = R.

Clearly, any rotation or reflection in R^2 preserves length and hence is an orthogonal operator. We will study these operators in much more detail in Section 6.10.

Example 1

Let $\mathsf{V} = \mathsf{H}$, and let $h \in \mathsf{V}$ with $|h(x)| = 1$ for all x. Define $\mathsf{T}: \mathsf{V} \to \mathsf{V}$ by $\mathsf{T}(f) = hf$. Then

$$\|\mathsf{T}(f)\|^2 = \|hf\|^2 = \frac{1}{2\pi}\int_0^{2\pi} h(t)f(t)\overline{h(t)f(t)}\,dt = \|f\|^2$$

since $|h(t)|^2 = 1$ for all t. So T is a unitary operator. ∎

Theorem 6.18. *Let* V *be a finite-dimensional inner product space, and let* T *be a linear operator on* V. *Then the following are equivalent:*

(a) $\mathsf{TT}^* = \mathsf{T}^*\mathsf{T} = \mathsf{I}$.

(b) $\langle \mathsf{T}(x), \mathsf{T}(y)\rangle = \langle x, y\rangle$ *for all* $x, y \in \mathsf{V}$.

(c) *If* β *is an orthonormal basis for* V, then $\mathsf{T}(\beta)$ *is an orthonormal basis for* V.

(d) *There exists an orthonormal basis* β *for* V *such that* $\mathsf{T}(\beta)$ *is an orthonormal basis for* V.

(e) $\|\mathsf{T}(x)\| = \|x\|$ *for all* $x \in \mathsf{V}$.

Thus all the conditions above are equivalent to the definition of a unitary or orthogonal operator. From (a) it follows that unitary or orthogonal operators are normal.

Before proving the theorem, we first prove the following lemma. Compare this lemma to Exercise 10(b) of Section 6.4.

Lemma. *Let* V *be a finite-dimensional inner product space, and let* U *be a self-adjoint operator on* V. *If* $\langle x, \mathsf{U}(x)\rangle = 0$ *for all* $x \in \mathsf{V}$, *then* $\mathsf{U} = \mathsf{T}_0$.

Proof. By either Theorem 6.16 or 6.17 we may choose an orthonormal basis β of eigenvectors of U. If $x \in \beta$, then $\mathsf{U}(x) = \lambda x$ for some λ. Thus

$$0 = \langle x, \mathsf{U}(x)\rangle = \langle x, \lambda x\rangle = \bar{\lambda}\langle x, x\rangle;$$

so $\bar{\lambda} = 0$. Hence $\mathsf{U}(x) = 0$ for all $x \in \beta$, and thus $\mathsf{U} = \mathsf{T}_0$. ∎

Proof of Theorem 6.18. We first prove that (a) implies (b). Let $x, y \in \mathsf{V}$. Then $\langle x, y\rangle = \langle (\mathsf{T}^*\mathsf{T})(x), y\rangle = \langle \mathsf{T}(x), \mathsf{T}(y)\rangle$.

Second, we prove that (b) implies (c). Let $\beta = \{x_1, \ldots, x_n\}$ be an orthonormal basis for V. Then $\mathsf{T}(\beta) = \{\mathsf{T}(x_1), \ldots, \mathsf{T}(x_n)\}$. Now $\langle \mathsf{T}(x_i), \mathsf{T}(x_j)\rangle = \langle x_i, x_j\rangle = \delta_{ij}$. So $\mathsf{T}(\beta)$ is an orthonormal basis of V.

That (c) implies (d) is obvious.

Next we prove that (d) implies (e). Let $x \in \mathsf{V}$, and let $\beta = \{x_1, \ldots, x_n\}$. Now

$$x = \sum_{i=1}^{n} a_i x_i$$

for some scalars a_i, and so

$$\|x\|^2 = \left\langle \sum_{i=1}^{n} a_i x_i, \sum_{j=1}^{n} a_j x_j \right\rangle = \sum_{i=1}^{n} \sum_{j=1}^{n} a_i \overline{a_j} \langle x_i, x_j \rangle$$

$$= \sum_{i=1}^{n} \sum_{j=1}^{n} a_i \overline{a_j} \delta_{ij} = \sum_{i=1}^{n} |a_i|^2$$

since β is orthonormal.

Applying the same manipulations to

$$\mathsf{T}(x) = \sum_{i=1}^{n} a_i \mathsf{T}(x_i)$$

and using the fact that $\mathsf{T}(\beta)$ is also orthonormal, we obtain

$$\|\mathsf{T}(x)\|^2 = \sum_{i=1}^{n} |a_i|^2.$$

Hence $\|\mathsf{T}(x)\| = \|x\|$.

Finally, we prove that (e) implies (a). For any $x \in \mathsf{V}$ we have

$$\langle x, x \rangle = \|x\|^2 = \|\mathsf{T}(x)\|^2 = \langle \mathsf{T}(x), \mathsf{T}(x) \rangle = \langle x, (\mathsf{T}^*\mathsf{T})(x) \rangle.$$

So $\langle x, (\mathsf{I} - \mathsf{T}^*\mathsf{T})(x) \rangle = 0$ for all $x \in \mathsf{V}$. Let $\mathsf{U} = \mathsf{I} - \mathsf{T}^*\mathsf{T}$; then U is self-adjoint, and $\langle x, \mathsf{U}(x) \rangle = 0$ for $x \in \mathsf{V}$. So by the lemma we have $\mathsf{T}_0 = \mathsf{U} = \mathsf{I} - \mathsf{T}^*\mathsf{T}$, and hence $\mathsf{T}^*\mathsf{T} = \mathsf{I}$. Thus since V is finite-dimensional, $\mathsf{T}^* = \mathsf{T}^{-1}$, and so $\mathsf{T}\mathsf{T}^* = \mathsf{I}$. ∎

It follows immediately from the definition that every eigenvalue of a unitary or orthogonal operator has absolute value 1. In fact, even more is true.

Corollary 1. *Let T be a linear operator on a finite-dimensional real inner product space V. V has an orthonormal basis of eigenvectors of T with corresponding eigenvalues of absolute value 1 if and only if T is both self-adjoint and orthogonal.*

Proof. Suppose that V has an orthonormal basis $\{x_1, \ldots, x_n\}$ such that $\mathsf{T}(x_i) = \lambda_i x_i$ and $|\lambda_i| = 1$ for all i. By Theorem 6.17 T is self-adjoint. Thus $(\mathsf{T}\mathsf{T}^*)(x_i) = \mathsf{T}(\lambda_i x_i) = \lambda_i \lambda_i x_i = \lambda_i^2 x_i = x_i$ for each i. So $\mathsf{T}\mathsf{T}^* = \mathsf{I}$, and by (a) of Theorem 6.18, T is orthogonal.

If T is self-adjoint, then by Theorem 6.17 we have that V possesses an orthonormal basis $\{x_1, \ldots, x_n\}$ such that $\mathsf{T}(x_i) = \lambda_i x_i$ for all i. If T is also orthogonal, we have

$$|\lambda_i| \cdot \|x_i\| = \|\lambda_i x_i\| = \|\mathsf{T}(x_i)\| = \|x_i\|,$$

so $|\lambda_i| = 1$ for every i. ∎

Corollary 2. *Let* T *be a linear operator on a finite-dimensional complex inner product space* V. *Then* V *has an orthonormal basis of eigenvectors of* T *with corresponding eigenvalues of absolute value* 1 *if and only if* T *is unitary.*

Proof. The proof is similar to the proof of Corollary 1. ∎

Example 2

Let $\mathsf{T}: \mathsf{R}^2 \to \mathsf{R}^2$ be a rotation by θ, where $0 < \theta < \pi$. It is clear geometrically that T "preserves length," i.e., that $\|\mathsf{T}(x)\| = \|x\|$ for all $x \in \mathsf{R}^2$. The fact that rotations by a fixed angle preserve perpendicularity not only can be seen geometrically but now follows from (b) of Theorem 6.18. Perhaps the fact that such a transformation preserves the inner product is not so obvious geometrically; however, we obtain this fact from (b) also. Finally, an inspection of the matrix

$$\begin{pmatrix} \cos\theta & -\sin\theta \\ \sin\theta & \cos\theta \end{pmatrix}$$

reveals that T is not self-adjoint for the given restriction on θ. As we mentioned earlier, this fact also follows from the geometric observation that T has no eigenvectors and from Theorem 6.15. It can be seen easily from the matrix above that T^* is a rotation by $-\theta$. ∎

We now examine the matrices that represent unitary and orthogonal transformations.

Definitions. *Let* A *be an* n × n *matrix that satisfies* $AA^* = A^*A = I$. *We call* A *a* unitary matrix *if* A *has complex entries, and we call* A *an* orthogonal matrix *if* A *has real entries.*

Note that the condition $AA^* = I$ is equivalent to the statement that the rows $A_{(1)}, \ldots, A_{(n)}$ of A form an orthonormal basis for F^n because

$$\delta_{ij} = I_{ij} = (AA^*)_{ij} = \sum_{k=1}^{n} A_{ik}(A^*)_{kj} = \sum_{k=1}^{n} A_{ik}\overline{A_{jk}} = \langle A_{(i)}, A_{(j)} \rangle.$$

A similar remark can be made about the columns of A and the condition $A^*A = I$.

It also follows from the definition above that if V is an inner product space and T is a linear operator on V, then T is unitary [orthogonal] if and only if $[\mathsf{T}]_\beta$ is unitary [orthogonal] for some orthonormal basis β of V.

Example 3

From Example 2 the matrix

$$\begin{pmatrix} \cos\theta & -\sin\theta \\ \sin\theta & \cos\theta \end{pmatrix}$$

is clearly orthogonal. One can easily see that the rows of the matrix form an orthonormal basis for R^2. ∎

We know that for a complex normal [real symmetric] matrix A there is an orthonormal basis β for F^n consisting of eigenvectors of A. Hence A is similar to a diagonal matrix D. By Theorem 5.1 the matrix Q whose columns are the vectors in β is such that $D = Q^{-1}AQ$. But since the columns of Q are an orthonormal basis for F^n, it follows that Q is unitary [orthogonal]. In this case we say that A is *unitarily equivalent* [*orthogonally equivalent*] to D. It is easily seen (see Exercise 17) that this relation is an equivalence relation on $\mathsf{M}_{n\times n}(C)$ [$\mathsf{M}_{n\times n}(R)$]. More generally, A and B are unitarily equivalent [orthogonally equivalent] if and only if there exists a unitary [orthogonal] matrix P such that $A = P^*BP$.

The preceding paragraph has proved half of each of the following two theorems.

Theorem 6.19. *Let* A *be a complex* n × n *matrix. Then* A *is normal if and only if* A *is unitarily equivalent to a diagonal matrix.*

Proof. By the remarks above we need only prove that if A is unitarily equivalent to a diagonal matrix, then A is normal.

Suppose that $A = P^*DP$, where P is a unitary matrix and D is a diagonal matrix. Then

$$AA^* = (P^*DP)(P^*DP)^* = (P^*DP)(P^*D^*P) = P^*DID^*P = P^*DD^*P.$$

Similarly, $A^*A = P^*D^*DP$. Since D is a diagonal matrix, however, we have $DD^* = D^*D$. Thus $AA^* = A^*A$. ∎

Theorem 6.20. *Let* A *be a real* n × n *matrix. Then* A *is symmetric if and only if* A *is orthogonally equivalent to a real diagonal matrix.*

Proof. The proof is similar to the proof of Theorem 6.19 and is left as an exercise. ∎

Example 4

Let

$$A = \begin{pmatrix} 4 & 2 & 2 \\ 2 & 4 & 2 \\ 2 & 2 & 4 \end{pmatrix}.$$

Since A is symmetric, Theorem 6.20 tells us that A is orthogonally equivalent to a diagonal matrix. We will find an orthogonal matrix P and a diagonal matrix D such that $P^tAP = D$.

To find P we must first obtain an orthonormal basis of eigenvectors. It is easy to show that the eigenvalues of A are 2 and 8. Eigenvectors corresponding to 2 are $(-1, 1, 0)$ and $(-1, 0, 1)$. Because these vectors are not orthogonal, we apply the Gram–Schmidt process to obtain the orthogonal eigenvectors $(-1, 1, 0)$ and $-\frac{1}{2}(1, 1, -2)$ corresponding to 2. An eigenvector corresponding to 8 is $(1, 1, 1)$. Notice $(1, 1, 1)$ is orthogonal to the preceding two vectors. This observation confirms Theorem 6.15(d). Therefore, an orthonormal basis of eigenvectors is

$$\left\{\frac{1}{\sqrt{2}}(-1, 1, 0), \frac{1}{\sqrt{6}}(1, 1, -2), \frac{1}{\sqrt{3}}(1, 1, 1)\right\}.$$

Thus we have that

$$P = \begin{pmatrix} \frac{-1}{\sqrt{2}} & \frac{1}{\sqrt{6}} & \frac{1}{\sqrt{3}} \\ \frac{1}{\sqrt{2}} & \frac{1}{\sqrt{6}} & \frac{1}{\sqrt{3}} \\ 0 & \frac{-2}{\sqrt{6}} & \frac{1}{\sqrt{3}} \end{pmatrix} \quad \text{and} \quad D = \begin{pmatrix} 2 & 0 & 0 \\ 0 & 2 & 0 \\ 0 & 0 & 8 \end{pmatrix}. \quad \blacksquare$$

Because of Schur's theorem (Theorem 6.14), the next result is immediate. As it is the matrix form of Schur's theorem, we refer to it also as Schur's theorem.

Theorem 6.21 (Schur). *Let* A *be an* $n \times n$ *matrix with entries from* F *and whose characteristic polynomial splits over* F.

(a) *If* $F = C$, *then* A *is a unitarily equivalent to a complex upper triangular matrix.*

(b) *If* $F = R$, *then* A *is an orthogonally equivalent to a real upper triangular matrix.*

Rigid Motions in the Plane

The purpose of this application is to characterize the so-called "rigid motions" of R^2. One may think intuitively of such a motion as a transformation that does not affect the shape of a figure under its action, hence the name "rigid." For example, reflections, rotations, and translations $(x \to x + x_0)$ are examples of rigid motions. In fact, we will prove that every rigid motion is a composition of these three transformations. The general situation in R^n will be handled in Section 6.10 and will use the results obtained here.

Definition. *Let* V *be a real inner product space. A function* $f: V \to V$ *is a* rigid motion *if*

$$\|f(x) - f(y)\| = \|x - y\|$$

for all $x, y \in V$.

Although we will prove a number of general results about rigid motions, our main result is in the setting of R^2.

Theorem 6.22. *Every rigid motion in R^2 is one of two types: a rotation (about the origin) followed by a translation, or a reflection (about the x-axis) followed by a rotation (about the origin) followed by a translation.*

Throughout we will assume that f is a rigid motion on a real inner product space V and that $\mathsf{T}: \mathsf{V} \to \mathsf{V}$ is defined by

$$\mathsf{T}(x) = f(x) - f(0)$$

for all $x \in \mathsf{V}$.

Lemma 1. *For all* $x, y \in \mathsf{V}$ *and* $a \in R$, *we have:*
(a) $\|\mathsf{T}(x)\| = \|x\|$.
(b) $\|\mathsf{T}(x) - \mathsf{T}(y)\| = \|x - y\|$.
(c) $\langle \mathsf{T}(x), \mathsf{T}(y)\rangle = \langle x, y\rangle$.
(d) $\|\mathsf{T}(x + ay) - \mathsf{T}(x) - a\mathsf{T}(y)\| = 0$.
Hence T *is an orthogonal linear operator.*

Proof. (a) Because f is a rigid motion, we have

$$\|\mathsf{T}(x)\| = \|f(x) - f(0)\| = \|x - 0\| = \|x\|$$

for all $x \in \mathsf{V}$.

(b) For all $x, y \in \mathsf{V}$ we have

$$\|\mathsf{T}(x) - \mathsf{T}(y)\| = \|[f(x) - f(0)] - [f(y) - f(0)]\|$$
$$= \|f(x) - f(y)\| = \|x - y\|.$$

(c) For all $x, y \in \mathsf{V}$ we have

$$\|\mathsf{T}(x) - \mathsf{T}(y)\|^2 = \|\mathsf{T}(x)\|^2 - 2\langle \mathsf{T}(x), \mathsf{T}(y)\rangle + \|\mathsf{T}(y)\|^2$$

and

$$\|x - y\|^2 = \|x\|^2 - 2\langle x, y\rangle + \|y\|^2.$$

Part (c) follows from the two equations above and parts (a) and (b).

(d) For all $x, y \in \mathsf{V}$ and $a \in R$, we have by parts (b), (a), and (c):

$$\|\mathsf{T}(x + ay) - \mathsf{T}(x) - a\mathsf{T}(y)\|^2 = \|[\mathsf{T}(x + ay) - \mathsf{T}(x)] - a\mathsf{T}(y)\|^2$$
$$= \|\mathsf{T}(x + ay) - \mathsf{T}(x)\|^2 + a^2\|\mathsf{T}(y)\|^2 - 2a\langle \mathsf{T}(x + ay) - \mathsf{T}(x), \mathsf{T}(y)\rangle$$
$$= \|(x + ay) - x\|^2 + a^2\|y\|^2 - 2a[\langle \mathsf{T}(x + ay), \mathsf{T}(y)\rangle - \langle \mathsf{T}(x), \mathsf{T}(y)\rangle]$$
$$= a^2\|y\|^2 + a^2\|y\|^2 - 2a[\langle x + ay, y\rangle - \langle x, y\rangle]$$
$$= 2a^2\|y\|^2 - 2a[\langle x, y\rangle + a\|y\|^2 - \langle x, y\rangle]$$
$$= 0. \quad \blacksquare$$

Lemma 2. *Every rigid motion is an orthogonal operator followed by a translation.*

Proof. By Lemma 1, T is an orthogonal operator. If we define $\mathsf{U}: \mathsf{V} \to \mathsf{V}$ by $\mathsf{U}(x) = x + f(0)$ for all $x \in \mathsf{V}$, then U is a translation. Also,

$$(\mathsf{UT})(x) = \mathsf{U}(\mathsf{T}(x)) = \mathsf{T}(x) + f(0) = f(x). \quad \blacksquare$$

Lemma 3. *If* V *is finite-dimensional, then* $det(\mathsf{T}) = \pm 1$.

Proof. Let β be an orthonormal basis for V. Then by Theorem 6.10 and Exercise 16 of Section 6.3, we have

$$\det(\mathsf{T}^*) = \det([\mathsf{T}^*]_\beta) = \det([\mathsf{T}]_\beta^*) = \det([\mathsf{T}]_\beta) = \det(\mathsf{T}).$$

Because T is orthogonal by Lemma 1(a), we have that $\mathsf{I} = \mathsf{T}^*\mathsf{T}$ by Theorem 6.18(a). So

$$1 = \det(\mathsf{I}) = \det(\mathsf{T}^*\mathsf{T}) = \det(\mathsf{T}^*)\cdot\det(\mathsf{T}) = \det(\mathsf{T})\cdot\det(\mathsf{T}) = \det(\mathsf{T})^2. \quad \blacksquare$$

Lemma 4. *Suppose that* $\mathsf{V} = \mathsf{R}^2$ *and that* β *is the standard ordered basis for* R^2. *Then there exists an angle* θ $(0 \leqslant \theta < 2\pi)$ *such that*

$$[\mathsf{T}]_\beta = \begin{pmatrix} \cos\theta & -\sin\theta \\ \sin\theta & \cos\theta \end{pmatrix} \qquad \text{if } det(\mathsf{T}) = 1$$

and

$$[\mathsf{T}]_\beta = \begin{pmatrix} \cos\theta & \sin\theta \\ \sin\theta & -\cos\theta \end{pmatrix} \qquad \text{if } det(\mathsf{T}) = -1.$$

Proof. Let $A = [\mathsf{T}]_\beta$. Because T is an orthogonal operator by Lemma 1, we conclude from Theorem 6.18(c) that $\mathsf{T}(\beta) = \{\mathsf{T}(e_1), \mathsf{T}(e_2)\}$ is an orthonormal basis of R^2. Because $\mathsf{T}(e_1)$ is a unit vector, there exists an angle θ $(0 \leqslant \theta < 2\pi)$ such that $\mathsf{T}(e_1) = (\cos\theta, \sin\theta)$. Since $\mathsf{T}(e_2)$ is orthogonal to $\mathsf{T}(e_1)$, there are only two possible choices for $\mathsf{T}(e_2)$. Either

$$\mathsf{T}(e_2) = (-\sin\theta, \cos\theta) \quad \text{or} \quad \mathsf{T}(e_2) = (\sin\theta, -\cos\theta).$$

If $\det(\mathsf{T}) = 1$, we must have the first case; and if $\det(\mathsf{T}) = -1$, we must have the second case. $\blacksquare$

Proof of Theorem 6.22. By Lemma 2 we need only analyze the orthogonal operator T. By Lemma 3 $\det(\mathsf{T}) = \pm 1$. By Lemma 4, if $\det(\mathsf{T}) = 1$, we see that T is a rotation by θ. If $\det(\mathsf{T}) = -1$, then using

$$[\mathsf{T}]_\beta = \begin{pmatrix} \cos\theta & \sin\theta \\ \sin\theta & -\cos\theta \end{pmatrix} = \begin{pmatrix} \cos\theta & -\sin\theta \\ \sin\theta & \cos\theta \end{pmatrix}\begin{pmatrix} 1 & 0 \\ 0 & -1 \end{pmatrix},$$

we have that T is a reflection about the x-axis followed by a rotation. $\blacksquare$

Conic Sections

As an application of Theorem 6.20, we consider the quadratic equation

$$ax^2 + 2bxy + cy^2 + dx + ey + f = 0. \tag{3}$$

For special choices of the coefficients in (3), we obtain the various conic sections. For example, if $a = c = 1$, $b = d = e = 0$, and $f = -1$, we obtain the circle $x^2 + y^2 = 1$ with center at the origin. The remaining conic sections, namely, the ellipse, parabola, and hyperbola, are obtained by other choices of the coefficients. The absence of the xy-term allows easy graphing of these conics by the method of completing the square. For example, the equation $x^2 + 2x + y^2 + 4y + 2 = 0$ may be rewritten as $(x + 1)^2 + (y + 2)^2 = 3$, a circle with center at $(-1, -2)$ in the x, y-coordinate system and radius $\sqrt{3}$. If we consider the transformation of coordinates $(x, y) \to (x', y')$, where $x' = x + 1$ and $y' = y + 2$, then our equation simplifies to $(x')^2 + (y')^2 = 3$. This change of variable allows us to eliminate the x- and y-terms.

We now concentrate solely on the elimination of the xy-term. To accomplish this, we consider the expression

$$ax^2 + 2bxy + cy^2, \tag{4}$$

which is called the *associated quadratic form* of (3). Quadratic forms will be studied in more generality in Section 6.7.

If we let

$$A = \begin{pmatrix} a & b \\ b & c \end{pmatrix} \quad \text{and} \quad X = \begin{pmatrix} x \\ y \end{pmatrix},$$

then (4) may be rewritten as $X^tAX = \langle AX, X \rangle$. For example, the quadratic form $3x^2 + 4xy + 6y^2$ may be written as

$$X^t \begin{pmatrix} 3 & 2 \\ 2 & 6 \end{pmatrix} X.$$

The fact that A is symmetric is crucial in our discussion. For, by Theorem 6.20, we may choose an orthogonal matrix P and a diagonal matrix D with real diagonal entries λ_1 and λ_2 such that $P^tAP = D$. Now define

$$X' = \begin{pmatrix} x' \\ y' \end{pmatrix}$$

by $X' = P^tX$ or, equivalently, by $PX' = PP^tX = X$. Then

$$X^tAX = (PX')^tA(PX') = X'^t(P^tAP)X' = X'^tDX' = \lambda_1(x')^2 + \lambda_2(y')^2.$$

Thus the transformation $(x, y) \to (x', y')$ allows us to eliminate the xy-term in (4) and hence in (3).

Furthermore, since P is orthogonal, we have by Lemma 3 to Theorem 6.22 that $\det(P) = \pm 1$. If $\det(P) = -1$, we may interchange the columns of P to obtain a matrix Q. Because the columns of P form an orthonormal basis of eigenvectors of A, the same is true of the columns of Q. Therefore,

$$Q^tAQ = \begin{pmatrix} \lambda_2 & 0 \\ 0 & \lambda_1 \end{pmatrix}.$$

Notice that $\det(Q) = -\det(P) = 1$. Hence we may as well assume that $\det(P) = 1$. By Lemma 4 to Theorem 6.22, it follows that matrix P represents a rotation.

In summary, the xy-term in (3) may be eliminated by a rotation of the x-axis and y-axis to new axes x' and y' given by $X = PX'$, where P is an orthogonal matrix and $\det(P) = 1$. Furthermore, the coefficients of $(x')^2$ and $(y')^2$ are the eigenvalues of

$$A = \begin{pmatrix} a & b \\ b & c \end{pmatrix}.$$

This result is a restatement of the *principal axis theorem* for R^2. The arguments above, of course, are easily extended to quadratic equations in n variables. For example, in the case $n = 3$, by special choices of the coefficients we obtain the quadric surfaces—the elliptic cone, the ellipsoid, the hyperbolic paraboloid, etc.

As an example, consider the quadratic equation

$$2x^2 - 4xy + 5y^2 - 36 = 0,$$

for which the associated quadratic form is $2x^2 - 4xy + 5y^2$. In the notation above

$$A = \begin{pmatrix} 2 & -2 \\ -2 & 5 \end{pmatrix},$$

so that the eigenvalues of A are 1 and 6 with associated eigenvectors

$$\begin{pmatrix} 2 \\ 1 \end{pmatrix} \quad \text{and} \quad \begin{pmatrix} -1 \\ 2 \end{pmatrix}.$$

As expected [from Theorem 6.15(d)], these vectors are orthogonal. The corresponding basis of eigenvectors

$$\left\{ \begin{pmatrix} \frac{2}{\sqrt{5}} \\ \frac{1}{\sqrt{5}} \end{pmatrix}, \begin{pmatrix} \frac{-1}{\sqrt{5}} \\ \frac{2}{\sqrt{5}} \end{pmatrix} \right\}$$

determines new axes x' and y' as in Figure 6.3. Hence if

$$P = \begin{pmatrix} \frac{2}{\sqrt{5}} & \frac{-1}{\sqrt{5}} \\ \frac{1}{\sqrt{5}} & \frac{2}{\sqrt{5}} \end{pmatrix} = \frac{1}{\sqrt{5}}\begin{pmatrix} 2 & -1 \\ 1 & 2 \end{pmatrix},$$

then

$$P^t AP = \begin{pmatrix} 1 & 0 \\ 0 & 6 \end{pmatrix}.$$

Under the transformation $X = PX'$ or

$$x = \frac{2}{\sqrt{5}}x' - \frac{1}{\sqrt{5}}y'$$

$$y = \frac{1}{\sqrt{5}}x' + \frac{2}{\sqrt{5}}y',$$

we have the new quadratic form $(x')^2 + 6(y')^2$. Thus the original equation $2x^2 - 4xy + 5y^2 = 36$ may be written in the form $(x')^2 + 6(y')^2 = 36$ relative to

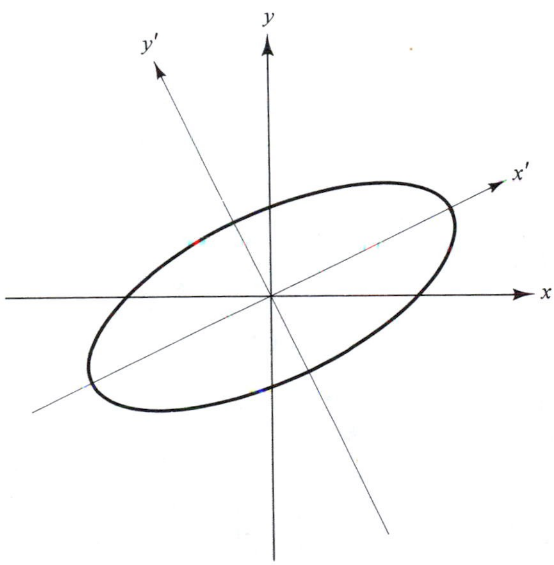

Figure 6.3

a new coordinate system with the x'- and y'- axes in the directions of

$$\begin{pmatrix} \dfrac{2}{\sqrt{5}} \\ \dfrac{1}{\sqrt{5}} \end{pmatrix} \quad \text{and} \quad \begin{pmatrix} \dfrac{-1}{\sqrt{5}} \\ \dfrac{2}{\sqrt{5}} \end{pmatrix},$$

respectively. It is clear that this equation represents an ellipse (see Figure 6.3).

EXERCISES

1. Label the following statements as being true or false. Assume that the underlying inner product spaces are finite-dimensional.

(a) Every unitary operator is normal.

(b) Every orthogonal operator is diagonalizable.

(c) A matrix is unitary if and only if it is invertible.

(d) If two matrices are unitarily equivalent, then they are also similar.

(e) The sum of unitary matrices is unitary.

(f) The adjoint of a unitary operator is unitary.

(g) If T is an orthogonal operator on V, then $[\mathsf{T}]_\beta$ is an orthogonal matrix for any ordered basis β for V.

(h) If all the eigenvalues of an operator are 1, then the operator must be unitary or orthogonal.

(i) An operator may preserve the norm but not the inner product.

2. For each of the following matrices A, find an orthogonal or unitary matrix P and a diagonal matrix D such that $P^*AP = D$.

(a) $A = \begin{pmatrix} 1 & 2 \\ 2 & 1 \end{pmatrix}$ **(b)** $A = \begin{pmatrix} 0 & -1 \\ 1 & 0 \end{pmatrix}$ **(c)** $A = \begin{pmatrix} 2 & 3-3i \\ 3+3i & 5 \end{pmatrix}$

(d) $A = \begin{pmatrix} 0 & 2 & 2 \\ 2 & 0 & 2 \\ 2 & 2 & 0 \end{pmatrix}$ **(e)** $A = \begin{pmatrix} 2 & 1 & 1 \\ 1 & 2 & 1 \\ 1 & 1 & 2 \end{pmatrix}$

3. Prove that the composition of unitary [orthogonal] operators is unitary [orthogonal].

4. For $z \in C$ define $\mathsf{T}_z: C \to C$ by $\mathsf{T}_z(u) = zu$. Characterize those z for which T_z is normal, self-adjoint, or unitary.

5. Which of the following pairs of matrices are unitarily equivalent?

(a) $\begin{pmatrix} 1 & 0 \\ 0 & 1 \end{pmatrix}$ and $\begin{pmatrix} 0 & 1 \\ 1 & 0 \end{pmatrix}$ **(b)** $\begin{pmatrix} 0 & 1 \\ 1 & 0 \end{pmatrix}$ and $\begin{pmatrix} 0 & \frac{1}{2} \\ \frac{1}{2} & 0 \end{pmatrix}$

(c) $\begin{pmatrix} 0 & 1 & 0 \\ -1 & 0 & 0 \\ 0 & 0 & 1 \end{pmatrix}$ and $\begin{pmatrix} 2 & 0 & 0 \\ 0 & -1 & 0 \\ 0 & 0 & 0 \end{pmatrix}$

(d) $\begin{pmatrix} 0 & 1 & 0 \\ -1 & 0 & 0 \\ 0 & 0 & 1 \end{pmatrix}$ and $\begin{pmatrix} 1 & 0 & 0 \\ 0 & i & 0 \\ 0 & 0 & -i \end{pmatrix}$

(e) $\begin{pmatrix} 1 & 1 & 0 \\ 0 & 2 & 2 \\ 0 & 0 & 3 \end{pmatrix}$ and $\begin{pmatrix} 1 & 0 & 0 \\ 0 & 2 & 0 \\ 0 & 0 & 3 \end{pmatrix}$

6. Let V be the inner product space of complex-valued continuous functions on $[0, 1]$ with the inner product

$$\langle f, g \rangle = \int_0^1 f(t)\overline{g(t)}\, dt.$$

Let $h \in \mathsf{V}$, and define $\mathsf{T}\colon \mathsf{V} \to \mathsf{V}$ by $\mathsf{T}(f) = hf$. Prove that T is a unitary operator if and only if $|h(t)| = 1$ for $0 \leqslant t \leqslant 1$.

7. Prove that if T is a unitary operator on a finite-dimensional inner product space, then T has a "square root"; that is, there exists a unitary operator U such that $\mathsf{T} = \mathsf{U}^2$.

8. Let V be an inner product space, and let $\mathsf{T}\colon \mathsf{V} \to \mathsf{V}$ be self-adjoint. If $\mathsf{U} = (\mathsf{T} + i\mathsf{I})(\mathsf{T} - i\mathsf{I})^{-1}$, prove, using Exercise 9 of Section 6.4, that U is unitary.

9. Let U be a linear operator on a finite-dimensional inner product space V. If $\|\mathsf{U}(x)\| = \|x\|$ for all x in some orthonormal basis for V, must U be unitary? Prove or give a counterexample.

10. Let A be a complex normal or real symmetric $n \times n$ matrix with eigenvalues $\lambda_1, \ldots, \lambda_n$ (not necessarily distinct). Prove that

$$\operatorname{tr}(A) = \sum_{i=1}^{n} \lambda_i \quad \text{and} \quad \operatorname{tr}(A^*A) = \sum_{i=1}^{n} |\lambda_i|^2.$$

11. Find an orthogonal matrix whose first row is $(\frac{1}{3}, \frac{2}{3}, \frac{2}{3})$.

12. Let A be an $n \times n$ real symmetric or complex normal matrix. Prove that

$$\det(A) = \prod_{i=1}^{n} \lambda_i,$$

where the λ_i's are the (not necessarily distinct) eigenvalues of A.

13. Suppose that A and B are diagonalizable matrices. Prove or disprove that A is similar to B if and only if A and B are unitarily equivalent.

14. Let U be a unitary operator on an inner product space V, and let W be a finite-dimensional U-invariant subspace of V. Prove
(a) $U(W) = W$.
(b) $W^{\perp}$ is U-invariant.
Contrast part (b) with Exercise 15.

15. Find an example of a unitary operator U on an inner product space and a U-invariant subspace W such that $W^{\perp}$ is not U-invariant.

16. Prove that a matrix that is both unitary and upper triangular must be a diagonal matrix.

17. Show that "is unitarily equivalent to" is an equivalence relation on $M_{n \times n}(C)$.

18. Let W be a finite-dimensional subspace of an inner product space V. By Theorem 6.7 and the exercises of Section 1.3, $V = W \oplus W^{\perp}$. Define $U: V \to V$ by $U(x_1 + x_2) = x_1 - x_2$, where $x_1 \in W$ and $x_2 \in W^{\perp}$. Prove that U is a self-adjoint unitary operator.

19. Let V be a finite-dimensional inner product space. A linear operator U on V is called a *partial isometry* if there exists a subspace W of V such that $\|U(x)\| = \|x\|$ for all $x \in W$ and $U(x) = 0$ for all $x \in W^{\perp}$. Observe that W need *not* be U-invariant. Suppose that U is such an operator and $\{x_1, \ldots, x_k\}$ is an orthonormal basis of W. Prove the following.
(a) $\langle U(x), U(y) \rangle = \langle x, y \rangle$ for all $x, y \in W$. *Hint:* Use Exercise 20 of Section 6.1.
(b) $\{U(x_1), \ldots, U(x_k)\}$ is an orthonormal basis for $R(U)$.
(c) There exists an orthonormal basis γ for V such that the first k columns of $[U]_{\gamma}$ form an orthonormal set and the remaining columns are zero.
(d) Let $\{y_1, \ldots, y_j\}$ be an orthonormal basis for $R(U)^{\perp}$. Let $\beta = \{U(x_1), \ldots, U(x_k), y_1, \ldots, y_j\}$. Then β is an orthonormal basis for V.
(e) Define T to be the linear operator on V that satisfies $T(U(x_i)) = x_i$ $(1 \leqslant i \leqslant k)$ and $T(y_i) = 0$ $(1 \leqslant i \leqslant j)$. Prove that T is well-defined and that $T = U^*$. *Hint:* Show that $\langle U(x), y \rangle = \langle x, T(y) \rangle$ for all $x, y \in \beta$. There are four cases.
(f) Prove that U^* is a partial isometry.
This exercise is continued in Exercise 9 of Section 6.6.

20. Let A and B be $n \times n$ matrices that are unitarily equivalent.
(a) Prove that $\mathrm{tr}(A^*A) = \mathrm{tr}(B^*B)$.
(b) Use part (a) to prove that

$$\sum_{i,j=1}^{n} |A_{ij}|^2 = \sum_{i,j=1}^{n} |B_{ij}|^2.$$

21. Find new coordinates x', y' so that the following quadratic forms can be written as $\lambda_1(x')^2 + \lambda_2(y')^2$.
(a) $x^2 + 4xy + y^2$
(b) $2x^2 + 2xy + 2y^2$

(c) $x^2 - 12xy - 4y^2$
(d) $3x^2 + 2xy + 3y^2$
(e) $x^2 - 2xy + y^2$

22. Consider the expression X^tAX, where $X^t = (x, y, z)$ and A is as defined in Exercise 2(e). Find a change of coordinates x', y', z' so that the expression above can be written in the form $\lambda_1(x')^2 + \lambda_2(y')^2 + \lambda_3(z')^2$.

23. Let $y_1, \ldots, y_n$ be linearly independent vectors in F^n, and let $x_1, \ldots, x_n$ be the orthogonal vectors obtained from $y_1, \ldots, y_n$ by the Gram–Schmidt orthogonalization process. Let $z_1, \ldots, z_n$ be the orthonormal basis obtained by defining

$$z_k = \frac{1}{\|x_k\|} x_k.$$

(a) By solving (1) in Section 6.2 for y_k in terms of z_k, show that

$$y_k = \|x_k\| z_k + \sum_{j=1}^{k-1} \langle y_k, z_j \rangle z_j \qquad (1 \leqslant k \leqslant n).$$

(b) Let A and Q denote the $n \times n$ matrices in which the kth columns are y_k and z_k, respectively. Define $R \in \mathsf{M}_{n \times n}(F)$ by

$$R_{jk} = \begin{cases} \|x_j\| & \text{if } j = k \\ \langle y_k, z_j \rangle & \text{if } j < k \\ 0 & \text{if } j > k. \end{cases}$$

Prove that $A = QR$.

(c) Compute Q and R as in part (b) for the 3×3 matrix whose columns are the vectors y_1, y_2, and y_3, respectively, in Example 4 of Section 6.2.

(d) Since Q is unitary [orthogonal] and R is upper triangular in part (b), we have shown that every invertible matrix is the product of a unitary [orthogonal] matrix and an upper triangular matrix. Suppose that $A \in \mathsf{M}_{n \times n}(F)$ is invertible and $A = Q_1R_1 = Q_2R_2$, where $Q_1, Q_2 \in \mathsf{M}_{n \times n}(F)$ are unitary and $R_1, R_2 \in \mathsf{M}_{n \times n}(F)$ are upper triangular. Prove that $D = R_2R_1^{-1}$ is a unitary diagonal matrix. *Hint:* Use Exercise 16.

(e) The QR factorization described in (b) provides an orthogonalization method for solving a linear system $AX = B$ when A is invertible: Decompose A to QR, by the Gram–Schmidt process or other means, where Q is unitary and R is upper triangular. Then $QRX = B$, and hence $RX = Q^*B$. This last system can be easily solved since R is upper triangular.

At one time, because of its great stability, this method for solving large systems of linear equations with a computer was being advocated as a better method than Gaussian elimination even though it requires about three times as much work. (Later, however, J. H. Wilkinson showed that if Gaussian elimination is done properly, then it is nearly as stable as the orthogonalization method.)

Use the orthogonalization method and part (c) to solve the system

$$\begin{cases} x_1 + 2x_2 + 2x_3 = 1 \\ x_1 \qquad\;\; + 2x_3 = 11 \\ \qquad x_2 + \;\; x_3 = -1. \end{cases}$$

24. Suppose that β and γ are ordered bases for an n-dimensional real (complex) inner product space V. Prove that if Q is an orthogonal (unitary) $n \times n$ matrix that changes γ-coordinates into β-coordinates, then β is orthonormal if and only if γ is orthonormal.

6.6 ORTHOGONAL PROJECTIONS AND THE SPECTRAL THEOREM

In this section we rely heavily on Theorems 6.16 and 6.17 to develop an elegant representation of a normal or self-adjoint operator T on a finite-dimensional inner product space. We will prove that such an operator can be written in the form $\lambda_1 T_1 + \cdots + \lambda_k T_k$, where $\lambda_1, \ldots, \lambda_k$ are the distinct eigenvalues of T and $T_1, \ldots, T_k$ are "orthogonal projections." We first must develop some results about these special projections.

We assume that the reader is familiar with the results about direct sums developed at the end of Section 5.2. The special case where V is the direct sum of two subspaces is considered in the exercises of Section 1.3.

Recall from the exercises of Section 2.1 that a linear operator T on V is a *projection* on a subspace W_1 of V if there exists a subspace W_2 of V such that $V = W_1 \oplus W_2$, and for $x = x_1 + x_2$, where $x_1 \in W_1$ and $x_2 \in W_2$, we have $T(x) = x_1$. By Exercises 22 and 23 of Section 2.1 we have

$$R(T) = W_1 = \{x \colon T(x) = x\} \quad \text{and} \quad N(T) = W_2.$$

So $V = R(T) \oplus N(T)$. Thus every projection is a projection on its range, and so we may simply refer to T as a projection. In fact, it can be shown (see Exercise 14 of Section 2.3) that T is a projection if and only if $T = T^2$. Because $V = W_1 \oplus W_2 = W_1 \oplus W_3$ does *not* imply that $W_2 = W_3$, we see that W_1 does not uniquely determine T. For an *orthogonal* projection T, however, T is uniquely determined by its range.

Definition. *Let V be an inner product space, and let $T\colon V \to V$ be a projection. We say that T is an* orthogonal projection *if $R(T)^\perp = N(T)$ and $N(T)^\perp = R(T)$.*

Note that by Exercise 12(c) of Section 6.2 if V is finite-dimensional, we need only assume that one of the conditions above holds. For example, if $\mathsf{R}(\mathsf{T})^\perp = \mathsf{N}(\mathsf{T})$, then $\mathsf{R}(\mathsf{T}) = \mathsf{R}(\mathsf{T})^{\perp\perp} = \mathsf{N}(\mathsf{T})^\perp$.

Now assume that W is a finite-dimensional subspace of an inner product space V. Proposition 6.6 guarantees that there exists an orthogonal projection on W. We can say even more—there exists exactly one orthogonal projection on W. For if T and U are orthogonal projections on W, then $\mathsf{R}(\mathsf{T}) = \mathsf{W} = \mathsf{R}(\mathsf{U})$. Hence $\mathsf{N}(\mathsf{T}) = \mathsf{R}(\mathsf{T})^\perp = \mathsf{R}(\mathsf{U})^\perp = \mathsf{N}(\mathsf{U})$, and since all projections are uniquely determined by their range and null space, we have that $\mathsf{T} = \mathsf{U}$. We call T the *orthogonal projection on* W. To understand the geometric difference between an arbitrary projection on W and the orthogonal projection on W, let $\mathsf{V} = \mathsf{R}^2$ and $\mathsf{W} = \operatorname{span}\{(1, 1)\}$. Define U and T as in Figure 6.4, where $\mathsf{T}(v)$ is the foot of a perpendicular from v on the line $y = x$ and $\mathsf{U}(a_1, a_2) = (a_1, a_1)$. Then T is the orthogonal projection on W, and U is a projection on W that is not orthogonal. Note that $v - \mathsf{T}(v) \in \mathsf{W}^\perp$, whereas $v - \mathsf{U}(v) \notin \mathsf{W}^\perp$.

From Figure 6.4 we see that $\mathsf{T}(v)$ is the "best approximation in W to v"; that is, if $w \in \mathsf{W}$, then $\|w - v\| \geqslant \|\mathsf{T}(v) - v\|$. In fact, this approximation property characterizes T. These results follow immediately from the corollary to Proposition 6.6.

As an application of this result to Fourier analysis, recall the inner product space H of continuous (complex-valued) functions on the interval $[0, 2\pi]$ introduced in Section 6.1. Define a *trigonometric polynomial of degree n* to be a function $g \in \mathsf{H}$ of the form

$$g(x) = \sum_{j=-n}^{n} a_j e^{ijx},$$

where a_n or a_{-n} is nonzero.

Let $f \in \mathsf{H}$. We will show that the best approximation to f by a trigonometric polynomial of degree less than or equal to n is the polynomial whose coefficients are the Fourier coefficients of f relative to the orthonormal set $\{e^{ijx} : j \text{ is an integer}\}$.

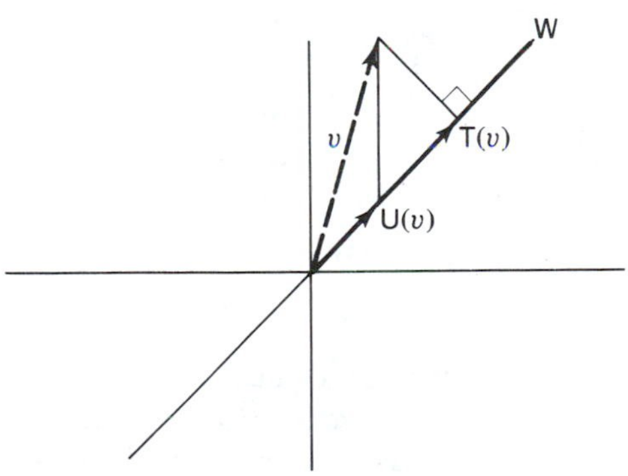

Figure 6.4

For this result, let $W = \text{span}(\{e^{ijx}: |j| \leqslant n\})$, and let T be the orthogonal projection on W. The corollary to Proposition 6.6 tells us that

$$T(f) = \sum_{j=-n}^{n} \langle f, e^{ijx} \rangle e^{ijx}$$

is the best approximation to f in H.

An algebraic characterization of orthogonal projections follows in the next theorem.

Theorem 6.23. *Let V be an inner product space, and let T be a linear operator on V. Then T is an orthogonal projection if and only if* $T^2 = T = T^*$.

Proof. Suppose that T is an orthogonal projection. Since $T = T^2$ because T is a projection, we need only show that $T = T^*$. Now $V = R(T) \oplus N(T)$ and $R(T)^\perp = N(T)$. If $x, y \in V$, then $x = x_1 + x_2$ and $y = y_1 + y_2$, where $x_1, y_1 \in R(T)$ and $x_2, y_2 \in N(T)$. Hence

$$\langle x, T(y) \rangle = \langle x_1 + x_2, y_1 \rangle = \langle x_1, y_1 \rangle + \langle x_2, y_1 \rangle = \langle x_1, y_1 \rangle$$

and

$$\langle x, T^*(y) \rangle = \langle T(x), y \rangle = \langle x_1, y_1 + y_2 \rangle = \langle x_1, y_1 \rangle + \langle x_1, y_2 \rangle = \langle x_1, y_1 \rangle.$$

So $\langle x, T(y) \rangle = \langle x, T^*(y) \rangle$ for all $x, y \in V$, and thus $T = T^*$.

Now suppose that $T = T^2 = T^*$. Then T is a projection by Exercise 14 of Section 2.3, and hence we must show that $R(T) = N(T)^\perp$ and $R(T)^\perp = N(T)$. Let $x \in R(T)$ and $y \in N(T)$. Then $x = T(x) = T^*(x)$, and so

$$\langle x, y \rangle = \langle T^*(x), y \rangle = \langle x, T(y) \rangle = \langle x, 0 \rangle = 0.$$

Therefore, $x \in N(T)^\perp$, from which it follows that $R(T) \subseteq N(T)^\perp$.

Let $y \in N(T)^\perp$. We must show that $y \in R(T)$, that is, that $T(y) = y$. Now

$$\begin{aligned} \|y - T(y)\|^2 &= \langle y - T(y), y - T(y) \rangle \\ &= \langle y, y - T(y) \rangle - \langle T(y), y - T(y) \rangle. \end{aligned}$$

Since $y - T(y) \in N(T)$, the first term is zero. But also

$$\langle T(y), y - T(y) \rangle = \langle y, T^*(y - T(y)) \rangle = \langle y, T(y - T(y)) \rangle = \langle y, 0 \rangle = 0.$$

Thus $y - T(y) = 0$; that is, $y = T(y) \in R(T)$. Hence $R(T) = N(T)^\perp$.

Using the results above, we have that $R(T)^\perp = N(T)^{\perp\perp} \supseteq N(T)$ (by Exercise 12(b) of Section 6.2). We need only show that if $x \in R(T)^\perp$, then $x \in N(T)$. For $y \in V$, we have $\langle T(x), y \rangle = \langle x, T^*(y) \rangle = \langle x, T(y) \rangle = 0$. So $T(x) = 0$, and thus $x \in N(T)$. ∎

Let V be a finite-dimensional inner product space, W be a subspace of V, and T be the orthogonal projection on W. We may choose an orthonormal basis $\beta = \{x_1, \ldots, x_n\}$ for V such that $\{x_1, \ldots, x_k\}$ is a basis for W. Then $[T]_\beta$ is a diagonal matrix with ones along the first k diagonal entries and zeros elsewhere. In fact, $[\mathsf{T}]_\beta$ has the form

$$\begin{pmatrix} I_k & O_1 \\ O_2 & O_3 \end{pmatrix}.$$

If U is any projection on W, we may choose a basis γ for V such that $[\mathsf{U}]_\gamma$ has the form above; however, γ will not necessarily be orthonormal.

We are now ready for the principal theorem of this section.

Theorem 6.24 (The Spectral Theorem). *Suppose that* T *is a linear operator on a finite-dimensional inner product space* V *over* F. *Assume that* T *is normal if* $\mathsf{F} = \mathsf{C}$ *and that* T *is self-adjoint if* $\mathsf{F} = \mathsf{R}$. *If* $\lambda_1, \ldots, \lambda_k$ *are the distinct eigenvalues of* T, *let* W_i *be the eigenspace of* T *corresponding to the eigenvalue* λ_i $(1 \leqslant i \leqslant k)$, *and let* T_i *be the orthogonal projection on* W_i $(1 \leqslant i \leqslant k)$. *Then*

(a) $\mathsf{V} = \mathsf{W}_1 \oplus \cdots \oplus \mathsf{W}_k$.

(b) *If* W_i' *denotes the direct sum of the subspaces of* W_j, $j \neq i$, *then* $\mathsf{W}_i^\perp = \mathsf{W}_i'$.

(c) $\mathsf{T}_i\mathsf{T}_j = \delta_{ij}\mathsf{T}_i$ *for* $1 \leqslant i, j \leqslant k$.

(d) $\mathsf{I} = \mathsf{T}_1 + \cdots + \mathsf{T}_k$.

(e) $\mathsf{T} = \lambda_1\mathsf{T}_1 + \cdots + \lambda_k\mathsf{T}_k$.

Proof. (a) By Theorems 6.16 and 6.17, T is diagonalizable, so

$$\mathsf{V} = \mathsf{W}_1 \oplus \cdots \oplus \mathsf{W}_k$$

by Theorem 5.16.

(b) If $x \in \mathsf{W}_i$ and $y \in \mathsf{W}_j$ for some i and j, then $\langle x, y \rangle = 0$ by Theorem 6.15(d). It follows easily from this that $\mathsf{W}_i' \subseteq \mathsf{W}_i^\perp$. Now from (a) we have that

$$\dim(\mathsf{W}_i') = \sum_{j \neq i} \dim(\mathsf{W}_j) = \dim(\mathsf{V}) - \dim(\mathsf{W}_i).$$

On the other hand, we have that $\dim(\mathsf{W}_i^\perp) = \dim(\mathsf{V}) - \dim(\mathsf{W}_i)$ by Theorem 6.7(c). Hence $\mathsf{W}_i' = \mathsf{W}_i^\perp$, proving (b).

The proof of part (c) is left as an exercise.

Since T_i is the orthogonal projection on W_i, we have from (b) that $\mathsf{N}(\mathsf{T}_i) = \mathsf{R}(\mathsf{T}_i)^\perp = \mathsf{W}_i^\perp = \mathsf{W}_i'$. Hence for $x \in \mathsf{V}$ we have that $x = x_1 + \cdots + x_k$, where $x_j \in \mathsf{W}_j$ and $\mathsf{T}_i(x) = x_i$, proving (d).

(e) For $x \in \mathsf{V}$, write $x = x_1 + \cdots + x_k$, where $x_j \in \mathsf{W}_j$ $(1 \leqslant j \leqslant k)$. Then

$$\begin{aligned} \mathsf{T}(x) &= \mathsf{T}(x_1) + \cdots + \mathsf{T}(x_k) = \lambda_1 x_1 + \cdots + \lambda_k x_k \\ &= \lambda_1\mathsf{T}_1(x) + \cdots + \lambda_k\mathsf{T}_k(x) = (\lambda_1\mathsf{T}_1 + \cdots + \lambda_k\mathsf{T}_k)(x). \quad \blacksquare \end{aligned}$$

The set $\{\lambda_1, \ldots, \lambda_k\}$ of eigenvalues of T is called the *spectrum of* T, the sum $\mathsf{I} = \mathsf{T}_1 + \cdots + \mathsf{T}_k$ in (d) is called the *resolution of the identity operator induced by*

T, and the sum $T = \lambda_1 T_1 + \cdots + \lambda_k T_k$ in (e) is called the *spectral decomposition of* T. Since the distinct eigenvalues of T are uniquely determined (up to order) by the subspaces W_i (and hence by the orthogonal projections T_i), the spectral decomposition of T is unique.

With the notation above, let β be the union of orthonormal bases of the W_i's and let $m_i = \dim(W_i)$. (Thus m_i is the multiplicity of λ_i.) Then $[T]_\beta$ has the form

$$\begin{pmatrix} \lambda_1 I_{m_1} & O & \cdots & O \\ O & \lambda_2 I_{m_2} & \cdots & O \\ \vdots & \vdots & & \vdots \\ O & O & \cdots & \lambda_k I_{m_k} \end{pmatrix};$$

that is, $[T]_\beta$ is a diagonal matrix in which the diagonal entries are the eigenvalues λ_i of T, and each λ_i is repeated m_i times. If $T = \lambda_1 T_1 + \cdots + \lambda_k T_k$ as in (e) of the spectral theorem, then it follows (from Exercise 7) that $g(T) = g(\lambda_1)T_1 + \cdots + g(\lambda_k)T_k$ for any polynomial g. This fact will be used below.

We now list several interesting corollaries of the spectral theorem; many more results are found in the exercises. For what follows we assume that V is a finite-dimensional inner product space over F and that T is a linear operator on V.

Corollary 1. *If* $F = C$, *then* T *is normal if and only if* $T^* = g(T)$ *for some polynomial* g.

Proof. Suppose first that T is normal. Let $T = \lambda_1 T_1 + \cdots + \lambda_k T_k$ be the spectral decomposition of T. Taking the adjoint of both sides of the equation above, we have $T^* = \bar{\lambda}_1 T_1 + \cdots + \bar{\lambda}_k T_k$ since each T_i is self-adjoint. Using the Lagrange interpolation formula (see Section 1.6), we may choose a polynomial g such that $g(\lambda_i) = \bar{\lambda}_i$ for $1 \leqslant i \leqslant k$. Then

$$\begin{aligned} g(T) &= g(\lambda_1)T_1 + \cdots + g(\lambda_k)T_k \\ &= \bar{\lambda}_1 T_1 + \cdots + \bar{\lambda}_k T_k \\ &= T^*. \end{aligned}$$

Conversely, if $T^* = g(T)$ for some polynomial g, then T commutes with T^* since T commutes with every polynomial in T. ∎

Corollary 2. *If* $F = C$, *then* T *is unitary if and only if* T *is normal and* $|\lambda| = 1$ *for every eigenvalue* λ *of* T.

Proof. Suppose first that T is unitary and hence normal. Then if $T(x) = \lambda x$, we have $|\lambda| \cdot \|x\| = \|\lambda x\| = \|T(x)\| = \|x\|$, and hence $|\lambda| = 1$ if $x \neq 0$.

Now suppose that $|\lambda| = 1$ for every eigenvalue λ of T, and let $\mathsf{T} = \lambda_1\mathsf{T}_1 + \cdots + \lambda_k\mathsf{T}_k$ be the spectral decomposition of T. Then by (c) of the spectral theorem

$$\begin{aligned}\mathsf{TT}^* &= (\lambda_1\mathsf{T}_1 + \cdots + \lambda_k\mathsf{T}_k)(\bar{\lambda}_1\mathsf{T}_1 + \cdots + \bar{\lambda}_k\mathsf{T}_k)\\ &= |\lambda_1|^2\mathsf{T}_1 + \cdots + |\lambda_k|^2\mathsf{T}_k\\ &= \mathsf{T}_1 + \cdots + \mathsf{T}_k\\ &= \mathsf{I}.\end{aligned}$$

Hence T is unitary. ∎

Corollary 3. *If* $\mathsf{F} = \mathsf{C}$ *and* T *is normal, then* T *is self-adjoint if and only if every eigenvalue of* T *is real.*

Proof. Let $\mathsf{T} = \lambda_1\mathsf{T}_1 + \cdots + \lambda_k\mathsf{T}_k$ be the spectral decomposition of T. Suppose that every eigenvalue of T is real. Then $\mathsf{T}^* = \bar{\lambda}_1\mathsf{T}_1 + \cdots + \bar{\lambda}_k\mathsf{T}_k = \lambda_1\mathsf{T}_1 + \cdots + \lambda_k\mathsf{T}_k = \mathsf{T}$.

The converse has been in proved in the lemma to Theorem 6.17. ∎

Corollary 4. *Let* T *be as in the spectral theorem with spectral decomposition* $\mathsf{T} = \lambda_1\mathsf{T}_1 + \cdots + \lambda_k\mathsf{T}_k$. *Then each* T_j *is a polynomial in* T.

Proof. Choose a polynomial g_j $(1 \leqslant j \leqslant k)$ such that $g_j(\lambda_i) = \delta_{ij}$. Then $g_j(\mathsf{T}) = g_j(\lambda_1)\mathsf{T}_1 + \cdots + g_j(\lambda_k)\mathsf{T}_k = \delta_{1j}\mathsf{T}_1 + \cdots + \delta_{kj}\mathsf{T}_k = \mathsf{T}_j$. ∎

EXERCISES

1. Label the following statements as being true or false. Assume that the underlying inner product spaces are finite-dimensional.
   (a) All projections are self-adjoint.
   (b) An orthogonal projection is uniquely determined by its range.
   (c) Every self-adjoint operator is a linear combination of orthogonal projections.
   (d) If an operator possesses a spectral decomposition, then so does its adjoint.
   (e) If T is a projection on W, then $\mathsf{T}(x)$ is the vector in W that is closest to x.
   (f) Every orthogonal projection is a unitary operator.
2. Let $\mathsf{V} = \mathsf{R}^2$, $\mathsf{W} = \text{span}(\{(1, 2)\})$, and β be the standard ordered basis for V. Compute $[\mathsf{T}]_\beta$, where T is the orthogonal projection on W. Do the same for $\mathsf{V} = \mathsf{R}^3$ and $\mathsf{W} = \text{span}(\{(1, 0, 1)\})$.
3. For each of the matrices A in Exercise 2 of Section 6.5:
   (1) Prove that L_A possesses a spectral decomposition.

(2) Explicitly define each of the orthogonal projections on the eigenspaces of L_A.

(3) Verify your results using the spectral theorem.

4. Let W be a finite-dimensional subspace of an inner product space V. Show that if T is the orthogonal projection on W, then I − T is the orthogonal projection on $W^\perp$.

5. Let V be a finite-dimensional inner product space, and let T: V → V be a projection.

(a) If T is an orthogonal projection, prove that $\|T(x)\| \leqslant \|x\|$ for all $x \in V$. Give an example of a projection T for which this inequality does not hold. If equality holds, what can be concluded about T?

(b) If T is also normal and V is complex, prove that T must be an orthogonal projection.

6. Let T be a projection on a finite-dimensional inner product space V. Prove that if $\|T(x)\| \leqslant \|x\|$ for all $x \in V$, then T is an orthogonal projection.

7. Let T be a normal operator and U a linear operator on a finite-dimensional complex inner product space V. Use the spectral decomposition $\lambda_1 T_1 + \cdots + \lambda_k T_k$ of T to prove the following.

(a) If g is a polynomial, then

$$g(T) = \sum_{i=1}^{k} g(\lambda_i)T_i.$$

(b) If $T^n = T_0$ for some n, then $T = T_0$.

(c) U commutes with T if and only if U commutes with each T_i.

(d) If U is normal and commutes with T, then $U = \mu_1 T_1 + \cdots + \mu_r T_r$, where $\mu_1, \ldots, \mu_r$ are the (not necessarily distinct) eigenvalues of U. *Hint:* Show that the eigenspaces of T are invariant under U.

(e) There exists a normal operator U on V such that $U^2 = T$.

(f) T is invertible if $\lambda_i \neq 0$ for $1 \leqslant i \leqslant k$.

(g) T is a projection if and only if every eigenvalue of T is 1 or 0.

(h) $T = -T^*$ (such a T is called *skew-symmetric*) if and only if every λ_i is an imaginary number.

8. Use Corollary 1 of the spectral theorem to show that if T is a normal operator on a complex finite-dimensional inner product space and U is a linear operator that commutes with T, then U commutes with T^*.

9. Referring to Exercise 19 of Section 6.5, prove the following facts about U.

(a) U^*U is an orthogonal projection on W.

(b) $UU^*U = U$.

10. *Simultaneous Diagonalization.* Let V be a finite-dimensional complex inner product space, and let U, T: V → V be normal operators such that TU = UT. Prove that there exists an orthonormal basis for V consisting of vectors that are eigenvectors of both T and U. *Hint:* Use the hint of Exercise 14 of Section 6.4 along with Exercise 8.

11. Prove part (c) of the spectral theorem.

6.7 BILINEAR AND QUADRATIC FORMS

There is a certain class of scalar-valued functions of two variables defined on a vector space that is often considered in the study of such diverse subjects as geometry and multivariable calculus. This is the class of "bilinear forms." We will now study the basic properties of this class with a special emphasis on symmetric bilinear forms and will consider some of its applications to quadratic surfaces and multivariable calculus.

Throughout this section all bases should be regarded as ordered bases.

Bilinear Forms

Definition. *Let* V *be a vector space over a field* F. *A function* H *from the set* $V \times V$ *of ordered pairs of vectors in* V *to* F *is called a* bilinear form *on* V *if* H *is linear in each variable when the other variable is held fixed, that is, if*

(a) $H(ax_1 + x_2, y) = aH(x_1, y) + H(x_2, y)$ *for all* $x_1, x_2, y \in V$ *and* $a \in F$.

(b) $H(x, ay_1 + y_2) = aH(x, y_1) + H(x, y_2)$ *for all* $x, y_1, y_2 \in V$ *and* $a \in F$.

We denote the set of all bilinear forms on V by $\mathscr{B}(V)$. Observe that an inner product on a real vector space V is a bilinear form.

Example 1

Define a function $H: R^2 \times R^2 \to R$ by

$$H\left(\begin{pmatrix} a_1 \\ a_2 \end{pmatrix}, \begin{pmatrix} b_1 \\ b_2 \end{pmatrix}\right) = 2a_1b_1 + 3a_1b_2 + 4a_2b_1 - a_2b_2 \qquad \text{for } \begin{pmatrix} a_1 \\ a_2 \end{pmatrix}, \begin{pmatrix} b_1 \\ b_2 \end{pmatrix} \in R^2.$$

We may verify directly that H is a bilinear form on R^2. It will prove more enlightening and less tedious, however, to observe that if

$$A = \begin{pmatrix} 2 & 3 \\ 4 & -1 \end{pmatrix}, \qquad x = \begin{pmatrix} a_1 \\ a_2 \end{pmatrix}, \quad \text{and} \quad y = \begin{pmatrix} b_1 \\ b_2 \end{pmatrix},$$

then

$$H(x, y) = x^t A y.$$

The bilinearity of H now follows directly from the distributive property of matrix multiplication over matrix addition. ∎

The bilinear form above is a special case of the following more general situation.

Example 2

Let $V = F^n$, where the elements are considered as column vectors. For any $n \times n$ matrix A with entries from F define $H: V \times V \to F$ by

$$H(x, y) = x^t A y \qquad \text{for } x, y \in V.$$

Notice that since x and y are $n \times 1$ matrices and A is an $n \times n$ matrix, $H(x, y)$ is a 1×1 matrix for any $x, y \in \mathsf{V}$. We identify this matrix with its single entry. As in Example 1 the bilinearity of H follows from the distributive property of matrix multiplication over matrix addition. For example, if $a \in F$ and $x_1, x_2, y \in \mathsf{V}$, then

$$\begin{aligned} H(ax_1 + x_2, y) &= (ax_1 + x_2)^t A y = (ax_1^t + x_2^t)Ay \\ &= ax_1^t A y + x_2^t A y \\ &= aH(x_1, y) + H(x_2, y). \quad \blacksquare \end{aligned}$$

We now list several properties possessed by all bilinear forms. Their proofs are left to the reader (see Exercise 2).

For any bilinear form H on a vector space V over a field F:

1. If, for any $x \in \mathsf{V}$, functions $\mathsf{L}_x, \mathsf{R}_x \colon \mathsf{V} \to F$ are defined by

$$\mathsf{L}_x(y) = H(x, y) \quad \text{and} \quad \mathsf{R}_x(y) = H(y, x) \qquad \text{for all } y \in \mathsf{V},$$

   then L_x and R_x are linear.
2. $H(0, x) = H(x, 0) = 0$ for all $x \in \mathsf{V}$.
3. If $x, y, z, w \in \mathsf{V}$, then

$$H(x + y, z + w) = H(x, z) + H(x, w) + H(y, z) + H(y, w).$$

4. If $J \colon \mathsf{V} \times \mathsf{V} \to F$ is defined by $J(x, y) = H(y, x)$, then J is a bilinear form.

For a vector space V, bilinear forms $H_1, H_2 \in \mathscr{B}(\mathsf{V})$, and any scalar a, we define the *sum* $H_1 + H_2$ and the *product* aH_1 by the equations

$$(H_1 + H_2)(x, y) = H_1(x, y) + H_2(x, y)$$

and

$$(aH_1)(x, y) = a(H_1(x, y)) \qquad \text{for all } x, y \in \mathsf{V}.$$

It is a simple exercise to verify that $H_1 + H_2$ and aH_1 are again bilinear forms. It is not surprising that with respect to these operations $\mathscr{B}(\mathsf{V})$ is a vector space.

Theorem 6.25. *For any vector space* V, $\mathscr{B}(\mathsf{V})$ *is a vector space with respect to the definitions of sum and product above.*

Proof. Exercise. ∎

Let V be an n-dimensional vector space with basis $\beta = \{x_1, x_2, \ldots, x_n\}$. For any bilinear form $H \in \mathscr{B}(\mathsf{V})$ we can associate with H an $n \times n$ matrix A whose entry in row i column j is defined by

$$A_{ij} = H(x_i, x_j) \qquad \text{for all } i, j = 1, 2, \ldots, n.$$

Definition. *The matrix* A *above is called the* matrix representation of H with respect to the basis β.

We can therefore define a mapping ψ_β from $\mathscr{B}(\mathsf{V})$ to $\mathsf{M}_{n\times n}(F)$, where F is the field of scalars for V, such that for any $H \in \mathscr{B}(\mathsf{V})$, $\psi_\beta(H) = A$, where A is the matrix representation of H with respect to β.

Example 3

Consider the bilinear form H of Example 1. Let

$$\beta = \left\{\begin{pmatrix}1\\1\end{pmatrix}, \begin{pmatrix}1\\-1\end{pmatrix}\right\} \quad \text{and} \quad B = \psi_\beta(H).$$

Then

$$B_{11} = H\left(\begin{pmatrix}1\\1\end{pmatrix}, \begin{pmatrix}1\\1\end{pmatrix}\right) = 2 + 3 + 4 - 1 = 8,$$

$$B_{12} = H\left(\begin{pmatrix}1\\1\end{pmatrix}, \begin{pmatrix}1\\-1\end{pmatrix}\right) = 2 - 3 + 4 + 1 = 4,$$

$$B_{21} = H\left(\begin{pmatrix}1\\-1\end{pmatrix}, \begin{pmatrix}1\\1\end{pmatrix}\right) = 2 + 3 - 4 + 1 = 2,$$

$$B_{22} = H\left(\begin{pmatrix}1\\-1\end{pmatrix}, \begin{pmatrix}1\\-1\end{pmatrix}\right) = 2 - 3 - 4 - 1 = -6.$$

So

$$\psi_\beta(H) = \begin{pmatrix}8 & 4\\2 & -6\end{pmatrix}.$$

If γ is the standard basis for R^2, the reader can verify that

$$\psi_\gamma(H) = \begin{pmatrix}2 & 3\\4 & -1\end{pmatrix}. \quad \blacksquare$$

Theorem 6.26. *For any* n*-dimensional vector space* V *over a field* F *and any basis* β *for* V, ψ_β *is a vector space isomorphism from* $\mathscr{B}(\mathsf{V})$ *onto* $\mathsf{M}_{n\times n}(\mathrm{F})$.

Proof. We leave it to the reader to verify that ψ_β is a linear transformation.

To show that ψ_β is one-to-one, suppose that $H \in \mathscr{B}(\mathsf{V})$ and $\psi_\beta(H) = O$, the zero matrix. We wish to show that H is trivial, i.e., $H(x, y) = 0$ for all $x, y \in \mathsf{V}$. Fix an $x_i \in \beta$, and recall the function $\mathsf{L}_{x_i}\colon \mathsf{V} \to F$ defined by $\mathsf{L}_{x_i}(x) = H(x_i, x)$ for all $x \in \mathsf{V}$. By property 1 on p. 356, L_{x_i} is linear; by hypothesis, $\mathsf{L}_{x_i}(x_j) = H(x_i, x_j) = 0$ for all $x_j \in \beta$. Hence L_{x_i} is the zero transformation from V to F. So

$$H(x_i, x) = \mathsf{L}_{x_i}(x) = 0 \qquad \text{for all } x \in \mathsf{V} \text{ and } x_i \in \beta. \tag{5}$$

Next fix an arbitrary $y \in V$, and recall the mapping $R_y: V \to F$ defined by $R_y(x) = H(x, y)$ for all $x \in V$. Again R_y is linear. But by (5) $R_y(x_i) = H(x_i, y) = 0$ for any $x_i \in \beta$. Thus R_y is trivial, and we conclude that $H(x, y) = R_y(x) = 0$ for all $x, y \in V$. So H is trivial, and ψ_β is therefore one-to-one.

To show that ψ_β is onto, let $A \in M_{n \times n}(F)$. Recall the isomorphism $\phi_\beta: V \to F^n$ as defined in Section 2.4. For $x \in V$ we view $\phi_\beta(x) \in F^n$ as a column vector. Define a mapping $H: V \times V \to F$ by

$$H(x, y) = [\phi_\beta(x)]^t A[\phi_\beta(y)] \qquad \text{for all } x, y \in V.$$

By Example 2, $H \in \mathscr{B}(V)$. We will show that $\psi_\beta(H) = A$. If $x_i, x_j \in \beta$, then $\phi_\beta(x_i) = e_i$ and $\phi_\beta(x_j) = e_j$. Consequently, for any i and j,

$$H(x_i, x_j) = [\phi_\beta(x_i)]^t A[\phi_\beta(x_j)] = e_i^t A e_j = A_{ij}.$$

We conclude that $\psi_\beta(H) = A$, and thus ψ_β is onto. ∎

Corollary 1. *For any* n*-dimensional vector space* V, $\mathscr{B}(V)$ *is of dimension* n^2.

Proof. Exercise. ∎

The following corollary is easily established by reviewing the proof of Theorem 6.26.

Corollary 2. *Let* V *be an* n*-dimensional vector space over the field* F *with basis* β. *If* $H \in \mathscr{B}(V)$ *and* $A \in M_{n \times n}(F)$, *then* $\psi_\beta(H) = A$ *if and only if* $H(x, y) = [\phi_\beta(x)]^t A[\phi_\beta(y)]$ *for all* $x, y \in V$.

The following is now an immediate consequence of Corollary 2.

Corollary 3. *For any field* F, *positive integer* n, *and* $H \in \mathscr{B}(F^n)$, *there exists a unique matrix* $A \in M_{n \times n}(F)$, *namely* $A = \psi_\beta(H)$, *such that*

$$H(x, y) = x^t A y \qquad \textit{for all } x, y \in F^n,$$

where β *is the standard basis for* F^n.

There appears to be an analogy between bilinear forms and linear operators in that each is associated with a unique square matrix and also this correspondence depends on the choice of a basis for the vector space. As in the case of operators one can pose the question: How is the matrix corresponding to a fixed bilinear form modified when the basis is changed? As we have seen, when this question was posed for linear operators, it led to the study of the similarity relation on square matrices. In the case of bilinear forms we are led to the study of another relation on square matrices, the “congruence” relation.

Definition. *Two matrices* A, B $\in \mathsf{M}_{n \times n}(F)$ *are said to be* congruent *if there exists an invertible matrix* Q $\in \mathsf{M}_{n \times n}(F)$ *such that*

$$Q^t A Q = B.$$

It is easily seen that congruence is an equivalence relation (see Exercise 11).

The following theorem relates congruence to the matrix representation of a bilinear form.

Theorem 6.27. *Let* V *be a finite-dimensional vector space with bases* $\beta = \{x_1, x_2, \ldots, x_n\}$ *and* $\gamma = \{y_1, y_2, \ldots, y_n\}$, *and let* Q *be the change of coordinate matrix changing* γ*-coordinates to* β*-coordinates. Then, for any* $H \in \mathscr{B}(V)$, $\psi_\gamma(H) = Q^t \psi_\beta(H) Q$. *In particular,* $\psi_\gamma(H)$ *and* $\psi_\beta(H)$ *are congruent.*

Proof. There are essentially two proofs of this theorem. One involves a direct computation, while the other follows immediately from a certain clever observation. We present the former proof here and leave the latter proof as an exercise (see Exercise 12).

Suppose that $A = \psi_\beta(H)$ and $B = \psi_\gamma(H)$. Then for any i and j such that $1 \leqslant i, j \leqslant n$

$$y_i = \sum_{k=1}^{n} Q_{ki} x_k \quad \text{and} \quad y_j = \sum_{r=1}^{n} Q_{rj} x_r.$$

Thus

$$\begin{aligned}
B_{ij} = H(y_i, y_j) &= H\left(\sum_{k=1}^{n} Q_{ki} x_k, y_j\right) \\
&= \sum_{k=1}^{n} Q_{ki} H(x_k, y_j) \\
&= \sum_{k=1}^{n} Q_{ki} H\left(x_k, \sum_{r=1}^{n} Q_{rj} x_r\right) \\
&= \sum_{k=1}^{n} Q_{ki} \sum_{r=1}^{n} Q_{rj} H(x_k, x_r) \\
&= \sum_{k=1}^{n} Q_{ki} \sum_{r=1}^{n} Q_{rj} A_{kr} \\
&= \sum_{k=1}^{n} Q_{ki} \sum_{r=1}^{n} A_{kr} Q_{rj} \\
&= \sum_{k=1}^{n} Q_{ki} (AQ)_{kj} \\
&= \sum_{k=1}^{n} Q^t_{ik} (AQ)_{kj} = (Q^t A Q)_{ij}.
\end{aligned}$$

Thus $B = Q^t A Q$. ∎

The following is a converse to Theorem 6.27.

Corollary. *Let* V *be an* n*-dimensional vector space with basis* $\beta = \{x_1, \ldots, x_n\}$. *Suppose that* H *be a bilinear form on* V *and* B *is an* n × n *matrix. If* B *is congruent to* $\psi_\beta(H)$, *then there exists a basis* γ *for* V *such that* $\psi_\gamma(H) = B$. *In fact, if* $B = Q^t\psi_\beta(H)Q$ *for an invertible matrix* Q, *then* Q *changes* γ*-coordinates into* β*-coordinates.*

Proof. Suppose that $B = Q^t\psi_\beta(H)Q$ for some invertible matrix Q. Let $\gamma = \{y_1, \ldots, y_n\}$, where

$$y_j = \sum_{i=1}^{n} Q_{ij}x_i \qquad \text{for } 1 \leqslant j \leqslant n.$$

Since Q is invertible, γ is an ordered basis for V and Q is the change of coordinate matrix that changes γ-coordinates to β-coordinates. Therefore, by Theorem 6.26

$$B = Q^t\psi_\beta(H)Q = \psi_\gamma(H). \quad \blacksquare$$

Symmetric Bilinear Forms

Like the diagonalization problem for linear operators, there is an analogous "diagonalization" problem for bilinear forms, namely the problem of determining those bilinear forms for which there are diagonal matrix representations. As we will see, the "diagonalizable" bilinear forms are those that are "symmetric."

Definition. *A bilinear form* H *over a vector space* V *is called* symmetric *if* $H(x, y) = H(y, x)$ *for all* $x, y \in V$.

As the name suggests, symmetric bilinear forms correspond to symmetric matrices.

Theorem 6.28. *Let* V *be a finite-dimensional vector space. For* $H \in \mathscr{B}(V)$ *the following are equivalent:*

(a) H *is symmetric.*

(b) *For any basis* γ *for* V, $\psi_\gamma(H)$ *is a symmetric matrix.*

(c) *There exists a basis* β *for* V *such that* $\psi_\beta(H)$ *is a symmetric matrix.*

Proof. First we prove that (a) implies (b). Suppose H is symmetric. Let $\gamma = \{y_1, y_2, \ldots, y_n\}$ be a basis for V, and let $B = \psi_\gamma(H)$. Then, for any i and j, $B_{ij} = H(y_i, y_j) = H(y_j, y_i) = B_{ji}$. Thus B is a symmetric matrix, proving (b).

Clearly (b) implies (c).

Finally, we prove that (c) implies (a). Suppose that, for some basis $\beta = \{x_1, x_2, \ldots, x_n\}$, $\psi_\beta(H) = A$ is a symmetric matrix. Define $J: V \times V \to F$, where F is the field of scalars for V, by $J(x, y) = H(y, x)$

for all $x, y \in V$. By property 4 on p. 356, $J \in \mathscr{B}(V)$. Let $C = \psi_\beta(J)$. Then for any i and j

$$C_{ij} = J(x_i, x_j) = H(x_j, x_i) = A_{ji} = A_{ij}.$$

Thus $C = A$. Since ψ_β is one-to-one, we conclude that $J = H$. Hence $H(y, x) = J(x, y) = H(x, y)$ for all $x, y \in V$, so H is symmetric, proving (a). ∎

Definition. *A bilinear form* H *on a finite-dimensional vector space* V *is called* diagonalizable *if there exists a basis* β *for* V *such that* $\psi_\beta(H)$ *is a diagonal matrix.*

Corollary. *Let* V *be a finite-dimensional vector space. For any* $H \in \mathscr{B}(V)$, *if* H *is diagonalizable, then* H *is symmetric.*

Proof. Suppose that H is diagonalizable. Then there exists a basis β for V such that $\psi_\beta(H) = D$, a diagonal matrix. Trivially, D is a symmetric matrix. So, by Theorem 6.28, H is symmetric. ∎

Unfortunately, the converse is not true, as illustrated by the following example.

Example 4

Let $F = Z_2$ (see Appendix C), and let $V = F^2$. Define $H: V \times V \to F$ by

$$H\left(\begin{pmatrix} a_1 \\ a_2 \end{pmatrix}, \begin{pmatrix} b_1 \\ b_2 \end{pmatrix}\right) = a_1b_2 + a_2b_1.$$

Clearly, H is symmetric. In fact, if β is the standard basis for F^2, then

$$A = \psi_\beta(H) = \begin{pmatrix} 0 & 1 \\ 1 & 0 \end{pmatrix},$$

a symmetric matrix. We will assume that H is diagonalizable and obtain a contradiction.

Suppose that H is diagonalizable. Then there exists a basis γ for F^2 such that $B = \psi_\gamma(H)$ is a diagonal matrix. Thus by Theorem 6.27 there exists an invertible matrix Q such that $B = Q^tAQ$. Since Q is invertible, $\text{rank}(B) = \text{rank}(A) = 2$. So B is a diagonal matrix whose diagonal entries are nonzero. Since the only nonzero element of F is 1,

$$B = \begin{pmatrix} 1 & 0 \\ 0 & 1 \end{pmatrix}.$$

Suppose that

$$Q = \begin{pmatrix} a & b \\ c & d \end{pmatrix}.$$

Then

$$\begin{pmatrix} 1 & 0 \\ 0 & 1 \end{pmatrix} = B = Q^t A Q$$

$$= \begin{pmatrix} a & c \\ b & d \end{pmatrix}\begin{pmatrix} 0 & 1 \\ 1 & 0 \end{pmatrix}\begin{pmatrix} a & b \\ c & d \end{pmatrix} = \begin{pmatrix} ac + ac & bc + ad \\ bc + ad & bd + bd \end{pmatrix}.$$

But $p + p = 0$ for all $p \in F$, and so $ac + ac = 0$. Thus, comparing the row 1, column 1 entries of the matrices in the equation above, we conclude that $1 = 0$, a contradiction. Consequently, H is not diagonalizable. ∎

The bilinear form of Example 4 is an anomaly. Its failure to be diagonalizable stems from the fact that the scalar field Z_2 is of characteristic two. If F is not of characteristic two, then $1 + 1$ is invertible. Under these circumstances we denote "1 + 1" by "2" and its multiplicative inverse by $\frac{1}{2}$.

Prior to proving the converse of the corollary to Theorem 6.28 for scalar fields other than those of characteristic two, we must establish the following lemma.

Lemma. *Let* H *be a nontrivial symmetric bilinear form on a vector space* V *over a field* F *not of characteristic two. Then there exists an element* $x \in V$ *such that* $H(x, x) \neq 0$.

Proof. Suppose that for some $v, w \in V$, $H(v, w) \neq 0$. If $H(v, v) \neq 0$ or $H(w, w) \neq 0$, there is nothing to prove. Otherwise, suppose that $H(v, v) = H(w, w) = 0$. Setting $x = v + w$, we have

$$H(x, x) = H(v, v) + H(v, w) + H(w, v) + H(w, w)$$
$$= 2H(v, w) \neq 0$$

since $2 \neq 0$ and $H(v, w) \neq 0$. ∎

Theorem 6.29. *Let* V *be a finite-dimensional vector space over a field* F *not of characteristic two. Then every symmetric bilinear form on* V *is diagonalizable.*

Proof. We use mathematical induction on $n = \dim(V)$. If $n = 1$, then every member of $\mathscr{B}(V)$ is diagonalizable. Suppose that the theorem is valid for vector spaces of dimension less than n for some fixed integer $n > 1$. If H is the trivial bilinear form, then certainly H is diagonalizable. Suppose then that H is nontrivial and symmetric. Then by the lemma there exists an element $x \in V$ (necessarily nonzero) such that $H(x, x) \neq 0$. Define

$$L: V \to F \text{ by } L(z) = H(x, z) \qquad \text{for all } z \in V.$$

Then L is linear, and since $L(x) = H(x, x) \neq 0$, L is nontrivial. Consequently $\text{rank}(L) = 1$, and hence $\dim(N(L)) = n - 1$. The restriction of H to $N(L)$ is obviously a symmetric bilinear form on a vector space of dimension $n-1$. Thus by the induction hypothesis there exists a basis $\{x_1, x_2, \ldots, x_{n-1}\}$ for $\mathsf{N}(\mathsf{L})$ such that $H(x_i, x_j) = 0$ for $i \neq j$ $(1 \leqslant i, j \leqslant n-1)$. Set $x_n = x$. Then $x_n \notin \mathsf{N}(\mathsf{L})$, and hence $\beta = \{x_1, x_2, \ldots, x_n\}$ is a basis for V. In addition, $H(x_i, x_n) = H(x_n, x_i) = 0$ for $i = 1, 2, \ldots, n-1$. We conclude that $\psi_\beta(H)$ is a diagonal matrix, so H is diagonalizable. ∎

Corollary. *Let* F *be a field that is not of characteristic two. If* $A \in \mathsf{M}_{n \times n}(F)$ *is a symmetric matrix, then* A *is congruent to a diagonal matrix.*

Proof. Exercise. ∎

Let A be a symmetric $n \times n$ matrix with entries from a field not of characteristic two. By the corollary to Theorem 6.29 A is congruent to a diagonal matrix. We will show how to find a diagonal matrix D and an invertible matrix Q such that $Q^tAQ = D$. The reader may wish to review Section 3.1 to recall the relationship between elementary matrices and the elementary matrix operations.

If E is an elementary $n \times n$ matrix, then AE is obtained from A by means of a certain elementary column operation on A. By Exercise 20, E^tA is obtained from A by means of the same operation performed on the rows rather than on the columns of A. Thus E^tAE is obtained from A by perfoming an elementary operation on the columns of A and then performing the same operation on the rows of the matrix AE. (Note that the order of the operations can be reversed.) Now suppose that Q is an invertible matrix and D is a diagonal matrix such that $Q^tAQ = D$. By Corollary 3 to Theorem 3.6, Q is a product of elementary matrices, say $Q = E_1E_2 \cdots E_k$. Thus $D = Q^tAQ = E_k^tE_{k-1}^t \cdots E_1^tAE_1E_2 \cdots E_k$.

On the basis of the equation above, we conclude that by means of several elementary column operations and the corresponding row operations, A can be transformed into a diagonal matrix D. Furthermore, if $E_1, E_2, \ldots, E_k$ are the elementary matrices corresponding to the elementary column operations (indexed in the order performed) and if $Q = E_1E_2 \cdots E_k$, then $Q^tAQ = D$.

The statement above provides the key for finding D and Q for a given A.

Example 5

Suppose that

$$A = \begin{pmatrix} 1 & 2 \\ 2 & 3 \end{pmatrix}.$$

We begin by using elementary column operations to insert a zero in the first row, second column; in this case we must subtract twice the first column of A from the second column of A. The corresponding row operation would involve subtracting twice the first row from the second row. Let E_1 be the elementary matrix corresponding to the elementary column operation above. Then

$$E_1 = \begin{pmatrix} 1 & -2 \\ 0 & 1 \end{pmatrix};$$

consequently,

$$AE_1 = \begin{pmatrix} 1 & 0 \\ 2 & -1 \end{pmatrix} \quad \text{and} \quad E_1^t AE = \begin{pmatrix} 1 & 0 \\ 0 & -1 \end{pmatrix}.$$

Observe that since E_1 produced a zero in row 1, column 2, E_1^t produced a zero in row 2, column 1. Thus for

$$Q = \begin{pmatrix} 1 & -2 \\ 0 & 1 \end{pmatrix} \quad \text{and} \quad D = \begin{pmatrix} 1 & 0 \\ 0 & -1 \end{pmatrix}, \qquad Q^t AQ = D.$$

Next consider

$$A = \begin{pmatrix} 0 & 1 & 2 \\ 1 & -1 & 3 \\ 2 & 3 & 4 \end{pmatrix}.$$

It is desirable to have ± 1 in the row 1, column 1 position and use it to eliminate all other entries in the first row and first column of A. Thus we begin by interchanging the first and second columns of A. The elementary matrix corresponding to this column operation is

$$E_1 = \begin{pmatrix} 0 & 1 & 0 \\ 1 & 0 & 0 \\ 0 & 0 & 1 \end{pmatrix}.$$

Clearly

$$E_1^t AE_1 = \begin{pmatrix} -1 & 1 & 3 \\ 1 & 0 & 2 \\ 3 & 2 & 4 \end{pmatrix}.$$

This matrix is obtained by interchanging the first two columns of A to obtain AE and then interchanging the first two rows of AE. Next we produce a zero in the first row, second column and in the second row, first column of $E_1^t AE_1$ by adding the first column of $E_1^t AE_1$ to the second column of $E_1^t AE_1$ and following this operation with the corresponding row operation. Finally we add three times the first column to the third column and follow with the corresponding row operation. Note that the column operations can be performed in succession prior to performing the row operations. Thus if

$$E_2 = \begin{pmatrix} 1 & 1 & 0 \\ 0 & 1 & 0 \\ 0 & 0 & 1 \end{pmatrix} \quad \text{and} \quad E_3 = \begin{pmatrix} 1 & 0 & 3 \\ 0 & 1 & 0 \\ 0 & 0 & 1 \end{pmatrix},$$

$$E_3^tE_2^t(E_1^tAE_1)E_2E_3 = \begin{pmatrix} -1 & 0 & 0 \\ 0 & 1 & 5 \\ 0 & 5 & 13 \end{pmatrix}.$$

The reader can now easily see that by setting

$$E_4 = \begin{pmatrix} 1 & 0 & 0 \\ 0 & 1 & -5 \\ 0 & 0 & 1 \end{pmatrix},$$

we have

$$E_4^t(E_3^tE_2^tE_1^tAE_1E_2E_3)E_4 = \begin{pmatrix} -1 & 0 & 0 \\ 0 & 1 & 0 \\ 0 & 0 & -12 \end{pmatrix}.$$

Thus with $Q = E_1E_2E_3E_4$ and

$$D = \begin{pmatrix} -1 & 0 & 0 \\ 0 & 1 & 0 \\ 0 & 0 & -12 \end{pmatrix},$$

$$Q^tAQ = D. \quad \blacksquare$$

The reader should justify the following method (similar to that introduced in Section 3.2 to compute the inverse of a matrix) for computing Q^t (and hence Q) without recording each elementary matrix separately: Use a sequence of elementary column operations followed by the corresponding elementary row operations to change the augmented matrix $(A \mid I)$ into the form $(D \mid B)$, where D is a diagonal matrix. Then $B = Q^t$.

In the preceding example this method produces the following sequence of matrices:

$$(A \mid I) = \left(\begin{array}{ccc|ccc} 0 & 1 & 2 & 1 & 0 & 0 \\ 1 & -1 & 3 & 0 & 1 & 0 \\ 2 & 3 & 4 & 0 & 0 & 1 \end{array}\right), \quad \left(\begin{array}{ccc|ccc} 1 & 0 & 2 & 1 & 0 & 0 \\ -1 & 1 & 3 & 0 & 1 & 0 \\ 3 & 2 & 4 & 0 & 0 & 1 \end{array}\right),$$

$$\left(\begin{array}{ccc|ccc} -1 & 1 & 3 & 0 & 1 & 0 \\ 1 & 0 & 2 & 1 & 0 & 0 \\ 3 & 2 & 4 & 0 & 0 & 1 \end{array}\right), \quad \left(\begin{array}{ccc|ccc} -1 & 0 & 3 & 0 & 1 & 0 \\ 1 & 1 & 2 & 1 & 0 & 0 \\ 3 & 5 & 4 & 0 & 0 & 1 \end{array}\right),$$

$$\left(\begin{array}{ccc|ccc} -1 & 0 & 3 & 0 & 1 & 0 \\ 0 & 1 & 5 & 1 & 1 & 0 \\ 3 & 5 & 4 & 0 & 0 & 1 \end{array}\right), \quad \left(\begin{array}{ccc|ccc} -1 & 0 & 0 & 0 & 1 & 0 \\ 0 & 1 & 5 & 1 & 1 & 0 \\ 3 & 5 & 13 & 0 & 0 & 1 \end{array}\right),$$

$$\left(\begin{array}{ccc|ccc} -1 & 0 & 0 & 0 & 1 & 0 \\ 0 & 1 & 5 & 1 & 1 & 0 \\ 0 & 5 & 13 & 0 & 3 & 1 \end{array}\right), \quad \left(\begin{array}{ccc|ccc} -1 & 0 & 0 & 0 & 1 & 0 \\ 0 & 1 & 0 & 1 & 1 & 0 \\ 0 & 5 & -12 & 0 & 3 & 1 \end{array}\right),$$

and

$$\left(\begin{array}{ccc|ccc} -1 & 0 & 0 & 0 & 1 & 0 \\ 0 & 1 & 0 & 1 & 1 & 0 \\ 0 & 0 & -12 & -5 & -2 & 1 \end{array}\right) = (D \mid Q^t).$$

Hence

$$D = \begin{pmatrix} -1 & 0 & 0 \\ 0 & 1 & 0 \\ 0 & 0 & -12 \end{pmatrix}, \quad Q^t = \begin{pmatrix} 0 & 1 & 0 \\ 1 & 1 & 0 \\ -5 & -2 & 1 \end{pmatrix}, \quad \text{and} \quad Q = \begin{pmatrix} 0 & 1 & -5 \\ 1 & 1 & -2 \\ 0 & 0 & 1 \end{pmatrix}.$$

Quadratic Forms

Associated with symmetric bilinear forms are functions called "quadratic forms."

Definition. *Let* V *be a vector space over a field* F. *A function* $\mathsf{K}: \mathsf{V} \to \mathsf{F}$ *is called a* quadratic form *if there exists a symmetric bilinear form* $\mathsf{H} \in \mathscr{B}(\mathsf{V})$ *such that*

$$\mathsf{K}(\mathsf{x}) = \mathsf{H}(\mathsf{x}, \mathsf{x}) \qquad \textit{for all } \mathsf{x} \in \mathsf{V}. \tag{6}$$

If the field F is not of characteristic two, there is a one-to-one correspondence between symmetric bilinear forms and quadratic forms given by (6). In fact, if K is a quadratic form on a vector space V over a field F not of characteristic two, and if $K(x) = H(x, x)$ for some symmetric bilinear form on V, then

$$H(x, y) = \tfrac{1}{2}[K(x + y) - K(x) - K(y)] \tag{7}$$

(see Exercise 15 for details).

Example 6

Certainly, the classical example of a quadratic form is the homogeneous second-degree polynomial of several variables. Given variables $t_1, t_2, \ldots, t_n$ that take values in a field F not of characteristic two and (not necessarily distinct) scalars a_{ij} $(1 \leqslant i \leqslant j \leqslant n)$, define the polynomial

$$f(t_1, t_2, \ldots, t_n) = \sum_{i \leqslant j} a_{ij} t_i t_j.$$

Let $K: \mathsf{F}^n \to F$ be the quadratic form defined by $K(c_1, c_2, \ldots, c_n) = f(c_1, c_2, \ldots, c_n)$.

Any polynomial of the form above is called a *homogeneous polynomial of the second degree in* n *variables.* In fact, if β is the standard basis for F^n, then the symmetric bilinear form corresponding to the quadratic form above is H, where $\psi_\beta(H) = A$ and

$$A_{ij} = A_{ji} = \begin{cases} a_{ii} & \text{if } i = j \\ \frac{1}{2}a_{ij} & \text{if } i \neq j. \end{cases}$$

To see this, simply apply (7) to obtain $H(e_i, e_j) = A_{ij}$ from the quadratic form K, and verify that $f(t_1, \ldots, t_n)$ is computable from H by means of (6).

In particular, given the polynomial

$$f(t_1, t_2, t_3) = 2t_1^2 - t_2^2 + 6t_1t_2 - 4t_2t_3$$

with real coefficients, let

$$A = \begin{pmatrix} 2 & 3 & 0 \\ 3 & -1 & -2 \\ 0 & -2 & 0 \end{pmatrix}.$$

Setting $H(x, y) = x^tAy$ for all $x, y \in \mathsf{R}^3$, we see that

$$f(t_1, t_2, t_3) = (t_1, t_2, t_3)A\begin{pmatrix} t_1 \\ t_2 \\ t_3 \end{pmatrix} \qquad \text{for} \begin{pmatrix} t_1 \\ t_2 \\ t_2 \end{pmatrix} \in \mathsf{R}^3. \quad \blacksquare$$

Quadratic Forms over the Field *R*

Since symmetric matrices over R are "orthogonally diagonalizable" (see Theorem 6.20), the theory of symmetric bilinear forms and quadratic forms on finite-dimensional vector spaces over R is especially nice. The following theorem and its corollary are certainly among the most useful results in the theory of bilinear and quadratic forms.

Theorem 6.30. *Let* V *be a finite-dimensional real inner product space, and let* H *be a symmetric bilinear form on* V. *Then there exists an orthonormal basis* β *for* V *such that* $\psi_\beta(\mathrm{H})$ *is a diagonal matrix.*

Proof. Choose any orthonormal basis $\gamma = \{x_1, \ldots, x_n\}$ for V, and let $A = \psi_\gamma(\mathrm{H})$. Since A is symmetric, there exists an orthogonal matrix Q and a diagonal matrix D such that $D = Q^tAQ$ by Theorem 6.20. Let $\beta = \{y_1, \ldots, y_n\}$ be defined by

$$y_j = \sum_{i=1}^{n} Q_{ij}x_i \qquad \text{for } 1 \leqslant j \leqslant n.$$

By Theorem 6.27, $\psi_\beta(H) = D$. Furthermore, since Q is orthogonal and γ is orthonormal, β is orthonormal by Exercise 24 of Section 6.5. $\blacksquare$

Corollary. *Let* K *be a quadratic form on a finite-dimensional real inner product space* V*. There exists an orthonormal basis* $\beta = \{x_1, x_2, \ldots, x_n\}$ *for* V *and scalars* $\lambda_1, \lambda_2, \ldots, \lambda_n$ *(not necessarily distinct) such that if* $x \in V$ *and*

$$x = \sum_{i=1}^{n} s_i x_i, \qquad s_i \in R,$$

then

$$K(x) = \sum_{i=1}^{n} \lambda_i s_i^2.$$

In fact, if H *is the symmetric bilinear form determined by* K*, then* β *can be chosen to be any orthonormal basis for* V *for which* $\psi_\beta(H)$ *is a diagonal matrix.*

Proof. Let H be the symmetric bilinear form for which $K(x) = H(x, x)$ for all $x \in V$. By Theorem 6.30 there exists an orthonormal basis $\beta = \{x_1, x_2, \ldots, x_n\}$ for V for which

$$\psi_\beta(H) = D = \begin{pmatrix} \lambda_1 & 0 & \cdots & 0 \\ 0 & \lambda_2 & \cdots & 0 \\ \vdots & \vdots & & \vdots \\ 0 & 0 & \cdots & \lambda_n \end{pmatrix}.$$

Let $x \in V$ and suppose that

$$x = \sum_{i=1}^{n} s_i x_i.$$

Then

$$K(x) = H(x, x) = [\phi_\beta(x)]^t D[\phi_B(x)] = (s_1, \ldots, s_n)D \begin{pmatrix} s_1 \\ \vdots \\ s_n \end{pmatrix} = \sum_{i=1}^{n} \lambda_i s_i^2. \quad \blacksquare$$

Example 7

For the homogeneous real polynomial of degree 2

$$f(t_1, t_2) = 5t_1^2 + 2t_2^2 + 4t_1 t_2, \tag{8}$$

we will find an orthonormal basis $\beta = \{x_1, x_2\}$ for R^2 and scalars λ_1 and λ_2 such that if

$$\begin{pmatrix} t_1 \\ t_2 \end{pmatrix} \in R^2 \quad \text{and} \quad \begin{pmatrix} t_1 \\ t_2 \end{pmatrix} = s_1 x_1 + s_2 x_2,$$

then $f(t_1, t_2) = \lambda_1 s_1^2 + \lambda_2 s_2^2$. We may think of s_1 and s_2 as the coordinates of

$$\begin{pmatrix} t_1 \\ t_2 \end{pmatrix}$$

relative to β. Thus the polynomial $f(t_1, t_2)$, as an expression involving the coordinates of a point with respect to the standard basis for R^2, is transformed into a new polynomial $g(s_1, s_2) = \lambda_1 s_1^2 + \lambda_2 s_2^2$ interpreted as an expression involving the coordinates of a point relative to the new basis β.

Let H denote the symmetric bilinear form corresponding to the quadratic form defined by (8). If γ is the standard basis for R^2, then

$$\psi_\gamma(H) = \begin{pmatrix} 5 & 2 \\ 2 & 2 \end{pmatrix}.$$

Next we find an orthogonal matrix Q for which $Q^t A Q$ is a diagonal matrix. We begin by computing an orthonormal basis of eigenvectors of L_A. The characteristic polynomial $h(t)$ of A is

$$h(t) = \det\begin{pmatrix} 5-t & 2 \\ 2 & 2-t \end{pmatrix} = (t-6)(t-1).$$

Thus $\lambda_1 = 6$ and $\lambda_2 = 1$ are the eigenvalues of A, and each has multiplicity 1. A simple computation yields corresponding eigenvectors of norm one,

$$x_1 = \frac{1}{\sqrt{5}}\begin{pmatrix} 2 \\ 1 \end{pmatrix} \quad \text{and} \quad x_2 = \frac{1}{\sqrt{5}}\begin{pmatrix} 1 \\ -2 \end{pmatrix}.$$

Since x_1 and x_2 are orthogonal, $\beta = \{x_1, x_2\}$ is an orthonormal basis for R^2. Setting

$$Q = \frac{1}{\sqrt{5}}\begin{pmatrix} 2 & 1 \\ 1 & -2 \end{pmatrix},$$

we see that Q is an orthogonal matrix and

$$Q^t A Q = \begin{pmatrix} 6 & 0 \\ 0 & 1 \end{pmatrix}.$$

Clearly Q is also a change of coordinate matrix. Consequently,

$$\psi_\beta(H) = Q^t \psi_\gamma(H) Q = Q^t A Q = \begin{pmatrix} 6 & 0 \\ 0 & 1 \end{pmatrix}.$$

Thus by the corollary to Theorem 6.30, for any $x = s_1 x_1 + s_2 x_2 \in \mathsf{R}^2$,

$$K(x) = 6s_1^2 + s_2^2.$$

So $g(s_1, s_2) = 6s_1^2 + s_2^2$. ∎

The following example illustrates how the theory of quadratic forms can be applied to the problem of describing quadratic surfaces in R^3.

Example 8

Consider the surface $\mathscr{S}$ in R^3 defined by the equation

$$2t_1^2 + 6t_1t_2 + 5t_2^2 - 2t_2t_3 + 2t_3^2 + 3t_1 - 2t_2 - t_3 + 14 = 0; \tag{9}$$

i.e., $\mathscr{S}$ is the set of all elements

$$\begin{pmatrix} t_1 \\ t_2 \\ t_3 \end{pmatrix} \in \mathsf{R}^3$$

that satisfy (9). If γ is the standard basis for R^3, then (9) is an equation involving the coordinates of points in $\mathscr{S}$ relative to γ. We would like to select a new orthonormal basis β for R^3 such that the equation describing the coordinates of any point of $\mathscr{S}$ relative to β is considerably simpler than (9).

We begin with the observation that the terms of second degree on the left-hand side of (9) add to form a quadratic form on R^3:

$$K\begin{pmatrix} t_1 \\ t_2 \\ t_3 \end{pmatrix} = 2t_1^2 + 6t_1t_2 + 5t_2^2 - 2t_2t_3 + 2t_3^2.$$

Next we diagonalize K. If H is the symmetric bilinear form corresponding to K and $A = \psi_\gamma(H)$, then

$$A = \begin{pmatrix} 2 & 3 & 0 \\ 3 & 5 & -1 \\ 0 & -1 & 2 \end{pmatrix}.$$

Now the characteristic polynomial of A is

$$h(t) = \det\begin{pmatrix} 2-t & 3 & 0 \\ 3 & 5-t & -1 \\ 0 & -1 & 2-t \end{pmatrix} = -1(t-2)(t-7)t,$$

and consequently, A has eigenvalues $\lambda_1 = 2$, $\lambda_2 = 7$, and $\lambda_3 = 0$. A simple calculation yields eigenvectors

$$x_1 = \frac{1}{\sqrt{10}}\begin{pmatrix} 1 \\ 0 \\ 3 \end{pmatrix}, \quad x_2 = \frac{1}{\sqrt{35}}\begin{pmatrix} 3 \\ 5 \\ -1 \end{pmatrix}, \quad \text{and} \quad x_3 = \frac{1}{\sqrt{14}}\begin{pmatrix} -3 \\ 2 \\ 1 \end{pmatrix}$$

of norm 1 corresponding to the respective eigenvalues.

Now set $\beta = \{x_1, x_2, x_3\}$ and

$$Q = \begin{pmatrix} \frac{1}{\sqrt{10}} & \frac{3}{\sqrt{35}} & -\frac{3}{\sqrt{14}} \\ 0 & \frac{5}{\sqrt{35}} & \frac{2}{\sqrt{14}} \\ \frac{3}{\sqrt{10}} & -\frac{1}{\sqrt{35}} & \frac{1}{\sqrt{14}} \end{pmatrix}.$$

As in Example 7, Q is the change of coordinate matrix changing β-coordinates to γ-coordinates and

$$\psi_\beta(H) = Q^t\psi_\gamma(H)Q = Q^tAQ = \begin{pmatrix} 2 & 0 & 0 \\ 0 & 7 & 0 \\ 0 & 0 & 0 \end{pmatrix}.$$

By the corollary to Theorem 6.30 if $x = s_1x_1 + s_2x_2 + s_3x_3$, then

$$K(x) = 2s_1^2 + 7s_2^2. \tag{10}$$

We are now ready to transform (10) into an equation involving coordinates relative to β.

Let

$$x = \begin{pmatrix} t_1 \\ t_2 \\ t_3 \end{pmatrix} \in \mathsf{R}^3.$$

If $x = s_1x_1 + s_2x_2 + s_3x_3$, we have

$$\begin{pmatrix} t_1 \\ t_2 \\ t_3 \end{pmatrix} = s_1\begin{pmatrix} \frac{1}{\sqrt{10}} \\ 0 \\ \frac{3}{\sqrt{10}} \end{pmatrix} + s_2\begin{pmatrix} \frac{3}{\sqrt{35}} \\ \frac{5}{\sqrt{35}} \\ \frac{-1}{\sqrt{35}} \end{pmatrix} + s_3\begin{pmatrix} \frac{-3}{\sqrt{14}} \\ \frac{2}{\sqrt{14}} \\ \frac{1}{\sqrt{14}} \end{pmatrix}.$$

Thus

$$t_1 = \frac{s_1}{\sqrt{10}} + \frac{3s_2}{\sqrt{35}} - \frac{3s_3}{\sqrt{14}},$$

$$t_2 = \frac{5s_2}{\sqrt{35}} + \frac{2s_3}{\sqrt{14}},$$

$$t_3 = \frac{3s_1}{\sqrt{10}} - \frac{s_2}{\sqrt{35}} + \frac{s_3}{\sqrt{14}}.$$

Therefore,

$$3t_1 - 2t_2 - t_3 = -\frac{14s_3}{\sqrt{14}} = -\sqrt{14}s_3.$$

Combining (9), (10), and the equation above, we conclude that if $x \in \mathsf{R}^3$ and $x = s_1x_1 + s_2x_2 + s_3x_3$, then $x \in \mathscr{S}$ if and only if

$$2s_1^2 + 7s_2^2 - \sqrt{14}s_3 + 14 = 0 \quad \text{or} \quad s_3 = \frac{\sqrt{14}}{7}s_1^2 + \frac{\sqrt{14}}{2}s_2^2 + \sqrt{14}.$$

Consequently, if we draw new axes x', y', and z' in the directions of x_1, x_2, and x_3, respectively, the graph of the equation above rewritten as

$$z' = \frac{\sqrt{14}}{7}(x')^2 + \frac{\sqrt{14}}{2}(y')^2 + \sqrt{14}$$

will coincide with the surface $\mathscr{S}$. Thus we recognize $\mathscr{S}$ to be an elliptic paraboloid (see Figure 6.5). ∎

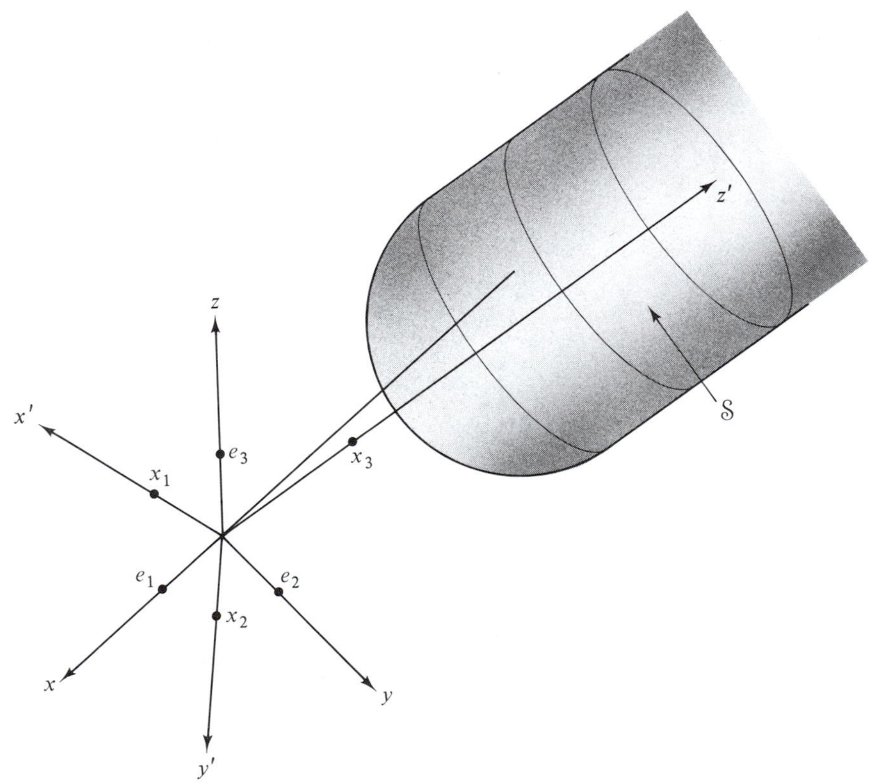

Figure 6.5

The Second Derivative Test for Functions of Several Variables

We now consider an application of the theory of quadratic forms to multivariable calculus—the derivation of the second derivative test for local extrema of a function of several variables. We assume an acquaintance with the calculus of functions of several variables to the extent of Taylor's theorem. The reader is undoubtedly familiar with the one-variable version of Taylor's theorem. For a statement and proof of the multivariable version, consult, for example, *Introduction to Analysis*, by Maxwell Rosenlicht (Dover, New York, 1986).

Let $z = f(t_1, t_2, \ldots, t_n)$ be a real-valued function of n real variables for which all third-order partial derivatives exist and are continuous. The function f is said to have a *local maximum* at a point $p \in \mathsf{R}^n$ if there exists a positive number δ for which $f(p) \geqslant f(x)$ whenever $\|x - p\| < \delta$. Likewise f is said to have a *local minimum* at $p \in \mathsf{R}^n$ if, for some $\delta > 0$, $f(p) \leqslant f(x)$ whenever $\|x - p\| < \delta$. If f has either a local maximum or a local minimum at p, we say that f has a *local extremum* at p. A point $p \in \mathsf{R}^n$ is called a *critical point* of f if $\partial f(p)/\partial t_i = 0$ for $i = 1, 2, \ldots, n$. It is a well-known fact that if f has a local extremum at a point $p \in \mathsf{R}^n$, then p is a critical point of f. For, if f has a local extremum at p, then for any $i = 1, 2, \ldots, n$ we may define a real-valued function ϕ_i of one variable by $\phi_i(t) = f(p_1, p_2, \ldots, p_{i-1}, t, p_{i+1}, \ldots, p_n)$, where p_j is the jth coordinate of p for each j. Obviously, ϕ_i has a local extremum at $t = p_i$. So by ordinary one-variable calculus arguments

$$\frac{\partial f(p)}{\partial t_i} = \frac{d\phi_i(p_i)}{dt} = 0.$$

So p is a critical point of f. Unfortunately critical points are not necessarily local extrema. The second derivative test gives us additional conditions under which critical points are local extrema.

Theorem 6.31 (The Second Derivative Test). *Let* $\mathrm{f}(\mathrm{t}_1, \mathrm{t}_2, \ldots, \mathrm{t}_n)$ *be a real-valued function of* n *real variables for which all the third-order partial derivatives exist and are continuous. Let* p *be a critical point of* f, *and let* A *denote the* $\mathrm{n} \times \mathrm{n}$ *matrix whose entries are given by*

$$\mathrm{A}_{\mathrm{ij}} = \frac{\partial^2 \mathrm{f}(\mathrm{p})}{(\partial \mathrm{t}_\mathrm{i})(\partial \mathrm{t}_\mathrm{j})}.$$

(*Note that* A *is a symmetric matrix and therefore has real eigenvalues.*)

(a) *If all the eigenvalues of* A *are positive, then* f *has a local minimum at* p.

(b) *If all the eigenvalues of* A *are negative, then* f *has a local maximum at* p.

(c) *If* A *has at least one positive and at least one negative eigenvalue, then* f *has no local extremum at* p (*i.e.,* p *is a saddle-point of* f).

(d) *If* rank(A) < n *and* A *does not have both positive and negative eigenvalues, then the second derivative test is inconclusive.*

Proof. If $p \neq 0$, we may define a function $g\colon \mathsf{R}^n \to R$ by

$$g(t_1, t_2, \ldots, t_n) = f(t_1 + p_1, t_2 + p_2, \ldots, t_n + p_n) - f(p_1, p_2, \ldots, p_n).$$

The following observations are easily verified:

1. The function f has a local maximum [minimum] at p if and only if g has a local maximum [minimum] at $0 = (0, 0, \ldots, 0)$.
2. The partial derivatives of g at 0 coincide with the corresponding partial derivatives of f at p.
3. 0 is a critical point of g.
4. $A_{ij} = \dfrac{\partial^2 g(0)}{(\partial t_i)(\partial t_j)}$ for all i and j.

In view of the observations above we may suppose without loss of generality that $p = 0$ and $f(p) = 0$.

We next apply Taylor's theorem to f at 0 and conclude that there exists a real-valued function S on R^n such that

$$\lim_{x \to 0} \frac{S(x)}{\|x\|^2} = \lim_{(t_1,\ldots,t_n)\to 0} \frac{S(t_1, \ldots, t_n)}{t_1^2 + \cdots + t_n^2} = 0, \tag{11}$$

and

$$f(t_1, \ldots, t_n) = f(0) + \sum_{i=1}^{n} \frac{\partial f(0)}{\partial t_i} t_i + \frac{1}{2}\left[\sum_{i,j=1}^{n} \frac{\partial^2 f(0)}{(\partial t_i)(\partial t_j)} t_i t_j\right] + S(t_1, \ldots, t_n). \tag{12}$$

Under the hypotheses that 0 is a critical point and $f(0) = 0$, (12) reduces to

$$f(t_1, \ldots, t_n) = \frac{1}{2}\left[\sum_{i,j=1}^{n} \frac{\partial^2 f(0)}{(\partial t_i)(\partial t_j)} t_i t_j\right] + S(t_1, \ldots, t_n). \tag{13}$$

Let us define a quadratic form $K\colon \mathsf{R}^n \to R$ by

$$K\begin{pmatrix} t_1 \\ \vdots \\ t_n \end{pmatrix} = \frac{1}{2} \sum_{i,j=1}^{n} \frac{\partial^2 f(0)}{(\partial t_i)(\partial t_j)} t_i t_j. \tag{14}$$

Let H be the symmetric bilinear form corresponding to K, and let γ be the standard basis for R^n. It is a simple matter to verify that $\psi_\gamma(H) = \frac{1}{2}A$. Since A is self-adjoint, Theorem 6.20 shows that there exists an orthogonal matrix Q such that

$$Q^tAQ = \begin{pmatrix} \lambda_1 & 0 & \cdots & 0 \\ 0 & \lambda_2 & \cdots & 0 \\ \vdots & \vdots & & \vdots \\ 0 & 0 & \cdots & \lambda_n \end{pmatrix}$$

is a diagonal matrix whose diagonal entries are the eigenvalues of A. Let $\beta = \{x_1, x_2, \ldots, x_n\}$ be the orthonormal basis for R^n whose ith member is $Q^{(i)}$, the ith column of Q. Then Q is the change of coordinate matrix changing β-coordinates to γ-coordinates, and by Theorem 6.27

$$\psi_\beta(H) = Q^t\psi_\gamma(H)Q = \frac{1}{2}Q^tAQ = \begin{pmatrix} \frac{\lambda_1}{2} & 0 & \cdots & 0 \\ 0 & \frac{\lambda_2}{2} & \cdots & 0 \\ \vdots & \vdots & & \vdots \\ 0 & 0 & \cdots & \frac{\lambda_n}{2} \end{pmatrix}.$$

Suppose that A is not the zero matrix. Then A has nonzero eigenvalues. Pick a positive number ϵ such that $\epsilon < |\lambda_i|/2$ for all $\lambda_i \neq 0$. By (11) there exists a positive number δ such that if $x \in \mathsf{R}^n$ and $0 < \|x\| < \delta$, then $|S(x)| < \epsilon\,\|x\|^2$. Now pick any $x \in \mathsf{R}^n$ for which $\|x\| < \delta$. Then by (13) and (14)

$$|f(x) - K(x)| = |S(x)| < \epsilon\|x\|^2$$

or

$$K(x) - \epsilon\|x\|^2 < f(x) < K(x) + \epsilon\|x\|^2. \tag{15}$$

If

$$x = \sum_{i=1}^{n} s_i x_i,$$

then

$$\|x\|^2 = \sum_{i=1}^{n} s_i^2 \quad \text{and} \quad K(x) = \frac{1}{2}\sum_{i=1}^{n} \lambda_i s_i^2. \tag{16}$$

Thus by (15) and (16)

$$\sum_{i=1}^{n} \left(\frac{1}{2}\lambda_i - \epsilon\right) s_i^2 < f(x) < \sum_{i=1}^{n} \left(\frac{1}{2}\lambda_i + \epsilon\right) s_i^2. \tag{17}$$

Now suppose that all the eigenvalues of A are positive. Then $\frac{1}{2}\lambda_i - \epsilon > 0$ for all i, and hence by the left inequality in (17)

$$f(0) = 0 \leqslant \sum_{i=1}^{n} \left(\frac{1}{2}\lambda_i - \epsilon\right) s_i^2 < f(x).$$

Thus for $\|x\| < \delta$, $f(0) \leqslant f(x)$. We conclude that f has a local minimum at 0. Similarly, by an argument involving the right inequality in (17) we conclude that if all the eigenvalues of A are negative, then f has a local maximum at 0. This establishes parts (a) and (b) of the theorem.

Next suppose that A has both a positive and a negative eigenvalue, say $\lambda_i > 0$ and $\lambda_j < 0$ for some i and j. Then $\frac{1}{2}\lambda_i - \epsilon > 0$ and $\frac{1}{2}\lambda_j + \epsilon < 0$. Let s be any real number such that $|s| < \delta$. Then by (17)

$$f(0) = 0 \leqslant (\tfrac{1}{2}\lambda_i - \epsilon)s^2 < f(sx_i) \quad \text{and} \quad f(sx_j) < (\tfrac{1}{2}\lambda_j + \epsilon)s^2 \leqslant 0 = f(0).$$

Since $\|sx_i\| = \|sx_j\| = |s|$, we conclude that f attains positive and negative values arbitrarily close to 0. Consequently, f has neither a local maximum nor a local minimum at 0. This establishes (c).

To illustrate that the second derivative test is inconclusive under the conditions stated in (d) of the theorem, consider the functions

$$f(t_1, t_2) = t_1^2 - t_2^4 \quad \text{and} \quad f(t_1, t_2) = t_1^2 + t_2^4$$

at $p = 0$. In either case

$$A = \begin{pmatrix} 2 & 0 \\ 0 & 0 \end{pmatrix},$$

but in the former case f does not have a local extremum at 0, while in the latter case f has a local minimum at 0. ∎

Sylvester's Law of Inertia

Any two matrix representations of a bilinear form have the same rank because rank is preserved under congruence. We can therefore define the *rank* of a bilinear form to be the rank of any one of its matrix representations. If a matrix representation is a diagonal matrix, then the rank is also equal to the number of nonzero diagonal entries of the matrix.

We confine our analysis to symmetric bilinear forms on finite-dimensional real vector spaces. Each such form has a diagonal representation, which may have positive and negative as well as zero diagonal entries. Although these entries are not unique, we will show that the number of entries that are positive and the number of entries that are negative are unique. That is, they are independent of the choice of diagonal representation. This result is called *Sylvester's law of inertia*. We will prove this law and apply it to describe the equivalence classes of congruent symmetric real matrices.

Sylvester's Law of Inertia. *Let* H *be a symmetric bilinear form on a finite-dimensional real vector space* V. *Then the number of positive diagonal entries and the number of negative diagonal entries of any diagonal representation of* H *are both independent of the diagonal representation.*

Proof. Suppose β and γ are ordered bases for V that determine diagonal representations of H. Without loss of generality we may assume that both β and γ are ordered so that the positive entries precede the negative entries, which in turn precede the zero entries on the diagonals of both representations. It suffices to show that both representations have the same number of positive entries because the number of negative entries is equal to the difference between the rank and the number of positive entries. Let p and q be the number of positive diagonal entries of the diagonal representations of H with respect to β and γ, respectively. We suppose that $p \neq q$ and arrive at a contradiction. Without loss of generality assume that $p < q$. Suppose that

$$\beta = \{x_1, x_2, \ldots, x_p, \ldots, x_r, \ldots, x_n\}$$

and

$$\gamma = \{y_1, y_2, \ldots, y_q, \ldots, y_r, \ldots, y_n\}$$

where r is the rank of H and n is the dimension of V. Let $\mathsf{L}\colon \mathsf{V} \to \mathsf{R}^{r+p-q}$ be the mapping defined by

$$\mathsf{L}(x) = (H(x, x_1), \ldots, H(x, x_p), H(x, y_{q+1}), \ldots, H(x, y_r)).$$

It is easy to verify that L is linear and $\operatorname{rank}(\mathsf{L}) \leqslant p + r - q$. Hence

$$\operatorname{nullity}(\mathsf{L}) \geqslant n - p - r + q > n - r.$$

Therefore, we can choose a nonzero vector $x_0 \in \mathsf{N}(\mathsf{L})$ so that x_0 is not in the span of $\{x_{r+1}, \ldots, x_n\}$. Since $x_0 \in \mathsf{N}(\mathsf{L})$, it follows that $H(x_0, x_i) = 0$ for $i \leqslant p$ and $H(x_0, y_i) = 0$ for $q < i \leqslant r$. Let

$$x_0 = \sum_{j=1}^{n} a_j x_j \quad \text{and} \quad x_0 = \sum_{j=1}^{n} b_j y_j.$$

For any $i \leqslant p$,

$$H(x_0, x_i) = H\left(\sum_{j=1}^{n} a_j x_j, x_i\right) = \sum_{j=1}^{n} a_j H(x_j, x_i) = a_i H(x_i, x_i).$$

But for $i \leqslant p$, we have $H(x_i, x_i) > 0$ and $H(x_0, x_i) = 0$. Therefore, $a_i = 0$ for every $i \leqslant p$. Similarly, $b_i = 0$ whenever $q + 1 \leqslant i \leqslant r$. Since x_0 is not in the span of $\{x_{r+1}, \ldots, x_n\}$, it follows that $a_i \neq 0$ for some $p < i \leqslant r$. Thus

$$H(x_0, x_0) = H\left(\sum_{j=1}^{n} a_j x_j, \sum_{i=1}^{n} a_i x_i\right)$$

$$= \sum_{j=1}^{n} (a_j)^2 H(x_j, x_j)$$

$$= \sum_{j=p+1}^{r} (a_j)^2 H(x_j, x_j)$$
$$< 0.$$

Furthermore,

$$H(x_0, x_0) = H\left(\sum_{j=1}^{n} b_j y_j, \sum_{i=1}^{n} b_i y_i \right)$$
$$= \sum_{j=1}^{n} (b_j)^2 H(y_j, y_j)$$
$$= \sum_{j=1}^{q} (b_j)^2 H(y_j, y_j)$$
$$\geqslant 0.$$

So $H(x_0, x_0) < 0$ and $H(x_0, x_0) \geqslant 0$, which is a contradiction. We conclude that $p = q$. ∎

Definitions. *The number of positive diagonal entries of a diagonal representation of a symmetric bilinear form on a real vector space is called the* index *of the form. The difference between the number of positive and the number of negative diagonal entries of a diagonal representation of a symmetric bilinear form is called the* signature *of the form. The three terms "rank," "index," and "signature" are called* invariants *of the bilinear form because they are invariant with respect to matrix representations. These same terms apply to the associated quadratic form. Notice that the values of any two of these invariants determine the value of the third.*

Example 9

The bilinear form corresponding to the quadratic form K of Example 8 has a 3×3 diagonal matrix representation with the diagonal entries 2, 7, and 0. Therefore, the rank, index, and signature of K are each 2. ∎

Example 10

The bilinear form corresponding to the quadratic form $K(x, y) = x^2 - y^2$ on R^2 has as its matrix representation with respect to the standard ordered basis the diagonal matrix with diagonal entries 1 and -1. Therefore, the rank of K is 2, the index of K is 1, and the signature of K is 0. ∎

Since the congruence relation is intimately associated with bilinear forms, we can apply Sylvester's law of inertia to study this relation on the set of real symmetric matrices.

Let A be an $n \times n$ real symmetric matrix, and suppose that D and E are each diagonal matrices congruent to A. By Corollary 3 to Theorem 6.26, A is the matrix representation of the bilinear form H on R^n defined by $H(x, y) = x^t Ay$ with respect to the standard ordered basis for R^n. By the corollary to Theorem 6.27, D and E are also matrix representations of H. Therefore, Sylvester's law of inertia tells us that D and E have the same number of positive as well as negative entries. We can formulate this fact as the matrix version of Sylvester's law.

Corollary 1 (Sylvester's Law of Inertia for Matrices). *Let* A *be a real symmetric matrix. Then the number of positive diagonal entries and the number of negative diagonal entries of any diagonal matrix congruent to* A *is independent of the choice of the diagonal matrix.*

Definitions. *Let* A *be a real symmetric matrix, and let* D *be a diagonal matrix that is congruent to* A. *The number of positive diagonal entries of* D *is called the* index *of* A. *The difference between the number of positive diagonal entries and the number of negative diagonal entries of* D *is called the* signature *of* A. *As before, the "rank," "index," and "signature" of a matrix are called* invariants *of the matrix, and the values of any two of these invariants determine the value of the third.*

Any two of these invariants can be used to determine an equivalence class of congruent real symmetric matrices.

Corollary 2. *Two real symmetric* $n \times n$ *matrices are congruent if and only if they have the same invariants.*

Proof. If A and B are congruent symmetric matrices, then they are both congruent to the same diagonal matrix, and it follows that they have the same invariants.

Conversely, suppose that A and B have the same invariants. Let D and E be diagonal matrices congruent to A and B, respectively. By Exercise 22 we may choose D and E so that the diagonal entries are in the order of positive, negative, and zero. Since A and B have the same invariants, so do D and E. Let p and r denote the index and the rank, respectively, of both D and E. Let d_i denote the ith diagonal entry of D, and let Q be the $n \times n$ diagonal matrix whose ith diagonal entry q_i is given by

$$q_i = \begin{cases} \dfrac{1}{\sqrt{d_i}} & \text{if } 1 \leqslant i \leqslant p \\[2ex] \dfrac{-1}{\sqrt{-d_i}} & \text{if } p < i \leqslant r \\[2ex] 0 & \text{if } r < i. \end{cases}$$

Then $Q^tDQ = J_{pr}$, where

$$J_{pr} = \begin{pmatrix} I_p & 0 & 0 \\ 0 & -I_{r-p} & 0 \\ 0 & 0 & 0 \end{pmatrix}.$$

It follows that A is congruent to J_{pr}. Similarly, B is congruent to J_{pr}, and hence A is congruent to B. ∎

The matrix J_{pr} acts as a *canonical form* for the theory of real symmetric matrices. The next corollary, whose proof is contained in the proof of Corollary 2, describes the role of J_{pr}.

Corollary 3. *A real symmetric* $n \times n$ *matrix* A *has index* p *and rank* r *if and only if* A *is congruent to* J_{pr} *(as defined above).*

Example 11

Let

$$A = \begin{pmatrix} 0 & 1 & 2 \\ 1 & -1 & 3 \\ 2 & 3 & 4 \end{pmatrix}, \qquad B = \begin{pmatrix} 1 & 2 & 3 \\ 2 & 4 & 5 \\ 3 & 5 & 6 \end{pmatrix}, \quad \text{and} \quad C = \begin{pmatrix} 1 & 0 & 1 \\ 0 & 1 & 2 \\ 1 & 2 & 1 \end{pmatrix}.$$

We apply Corollary 2 to determine the congruences among A, B, and C. The matrix A is the 3×3 matrix of Example 5, in which it is shown that A is congruent to the diagonal matrix with diagonal entries -1, 1, and -12. Therefore, A has rank 3 and index 1. Using the methods of that example (it is not necessary to compute Q), it can be shown that B and C are congruent, respectively, to the diagonal matrices D_B and D_C, where

$$D_B = \begin{pmatrix} 1 & 0 & 0 \\ 0 & \frac{1}{3} & 0 \\ 0 & 0 & -3 \end{pmatrix} \quad \text{and} \quad D_C = \begin{pmatrix} 1 & 0 & 0 \\ 0 & 1 & 0 \\ 0 & 0 & -4 \end{pmatrix}.$$

It follows that both B and C have rank 3 and index 2. We conclude that B and C are congruent, and that A is congruent to neither B nor C. ∎

EXERCISES

1. Label the following statements as being true or false.

(a) Every quadratic form is a bilinear form.

(b) If two matrices are congruent, they have the same eigenvalues.

(c) Symmetric bilinear forms have symmetric matrix representations.

(d) Any symmetric matrix is congruent to a diagonal matrix.

(e) The sum of two symmetric bilinear forms is a symmetric bilinear form.

(f) Two symmetric matrices with the same characteristic polynomial are matrix representations of the same bilinear form.

(g) There exists a bilinear form H such that $H(x, y) \neq 0$ for all x and y.

(h) If V is a vector space of dimension n, then $\dim(\mathscr{B}(\mathsf{V})) = 2n$.

(i) Let H be a bilinear form on a finite-dimensional vector space V. For any $x \in \mathsf{V}$ there exists a $y \in \mathsf{V}$ such that $y \neq 0$ but $H(x, y) = 0$.

(j) If H is any bilinear form on a finite-dimensional real inner product space V, there exists a basis β for V such that $\psi_\beta(H)$ is a diagonal matrix.

2. Prove properties 1, 2, 3, and 4 on page 356.

3. **(a)** Verify that the sum of two bilinear forms is a bilinear form.

(b) Verify that the product of a scalar and a bilinear form is a bilinear form.

(c) Prove Theorem 6.25.

4. Determine which of the following are bilinear forms.

(a) Let $\mathsf{V} = \mathsf{C}[0, 1]$ be the space of continuous real-valued functions on the closed interval $[0, 1]$. For $f, g \in \mathsf{V}$, define

$$H(f, g) = \int_0^1 f(t)g(t)\,dt.$$

(b) Let V be a vector space over a field F, and let $J \in \mathscr{B}(\mathsf{V})$ be nontrivial. Define $H\colon \mathsf{V} \times \mathsf{V} \to F$ by

$$H(x, y) = [J(x, y)]^2 \qquad \text{for all } x, y \in \mathsf{V}.$$

(c) Define $H\colon R \times R \to R$ by $H(t_1, t_2) = t_1 + 2t_2$.

(d) Consider the members of R^2 as column vectors. Define $H\colon \mathsf{R}^2 \to R$ for $x, y \in \mathsf{R}^2$ by $H(x, y) = \det(x, y)$, where $\det(x, y)$ denotes the determinant of the 2×2 matrix with x as its first column and y as its second column.

(e) Let V be a real inner product space. Define $H\colon \mathsf{V} \times \mathsf{V} \to R$ by $H(x, y) = \langle x, y\rangle$ for $x, y \in \mathsf{V}$.

(f) Let V be a complex inner product space. Define $H\colon \mathsf{V} \times \mathsf{V} \to C$ by $H(x, y) = \langle x, y\rangle$ for $x, y \in \mathsf{V}$.

5. Verify that each of the given mappings is a bilinear form. Then compute the matrix representation of H with respect to the given basis.

(a) $H\colon \mathsf{R}^3 \times \mathsf{R}^3 \to R$, where

$$H\left(\begin{pmatrix} a_1 \\ a_2 \\ a_3 \end{pmatrix}, \begin{pmatrix} b_1 \\ b_2 \\ b_3 \end{pmatrix}\right) = a_1b_1 - 2a_1b_2 + a_2b_1 - a_3b_3$$

with

$$\beta = \left\{\begin{pmatrix} 1 \\ 0 \\ 1 \end{pmatrix}, \begin{pmatrix} 1 \\ 0 \\ -1 \end{pmatrix}, \begin{pmatrix} 0 \\ 1 \\ 0 \end{pmatrix}\right\}.$$

(b) Let $V = M_{2\times 2}(R)$ with basis

$$\beta = \left\{ \begin{pmatrix} 1 & 0 \\ 0 & 0 \end{pmatrix}, \begin{pmatrix} 0 & 1 \\ 0 & 0 \end{pmatrix}, \begin{pmatrix} 0 & 0 \\ 1 & 0 \end{pmatrix}, \begin{pmatrix} 0 & 0 \\ 0 & 1 \end{pmatrix} \right\}.$$

Define $H: V \times V \to R$ by $H(A, B) = \text{tr}(A) \cdot \text{tr}(B)$.

(c) Let $\beta = \{\cos t, \sin t, \cos 2t, \sin 2t\}$ and $V = \text{span}(\beta)$. In the space of all continuous functions on R, V is a four-dimensional subspace with basis β. Define $H: V \times V \to R$ by $H(f, g) = f'(0) \cdot g''(0)$.

6. Let V and W be vector spaces over the same field, and let $T: V \to W$ be a linear transformation. For any $H \in \mathscr{B}(W)$, define $\hat{T}(H): V \times V \to F$ by $\hat{T}(H)(x, y) = H(T(x), T(y))$ for all $x, y \in V$. Prove that

(a) For $H \in \mathscr{B}(W)$, $\hat{T}(H) \in \mathscr{B}(V)$.

(b) $\hat{T}: \mathscr{B}(W) \to \mathscr{B}(V)$ is a linear transformation.

(c) If T is an isomorphism, then so is $\hat{T}$.

7. In the notation of Theorem 6.26:

(a) Prove that for any basis β, ψ_β is linear.

(b) Let V be an n-dimensional vector space over a field F with basis β, and let $\phi_\beta: V \to F^n$ be the standard representation of V with respect to β. Let $A \in M_{n\times n}(F)$. Define $H: V \times V \to F$ by $H(x, y) = [\phi_\beta(x)]^t A[\phi_\beta(y)]$. Prove that $H \in \mathscr{B}(V)$. Can you establish this as a corollary to Exercise 6?

(c) Prove the converse of part (b): Let H be a bilinear form on V. If $A = \psi_\beta(H)$, then $H(x, y) = [\phi_\beta(x)]^t A[\phi_\beta(y)]$ for all x and y in V.

8. (a) Prove Corollary 1 to Theorem 6.26.

(b) For a finite-dimensional vector space V, describe a method for finding a basis for $\mathscr{B}(V)$.

9. Prove Corollary 2 to Theorem 6.26.

10. Prove Corollary 3 to Theorem 6.26.

11. Prove that the relation of congruence is an equivalence relation.

12. The following outline provides an alternate proof to Theorem 6.27.

(a) If β and γ are bases for a finite-dimensional vector space V, and if Q is the change of coordinate matrix changing γ-coordinates into β-coordinates, prove that $\phi_\beta = L_Q\phi_\gamma$, where ϕ_β and ϕ_γ are the standard representations of V with respect to β and γ, respectively.

(b) Apply Corollary 2 to Theorem 6.26 to part (a) to obtain an alternate proof of Theorem 6.27.

13. Let V be a finite-dimensional vector space and $H \in \mathscr{B}(V)$. Prove that for any bases β and γ of V, $\text{rank}(\psi_\beta(H)) = \text{rank}(\psi_\gamma(H))$.

14. Prove the following.

(a) Any square diagonal matrix is symmetric.

(b) Any matrix congruent to a diagonal matrix is symmetric.

(c) Prove the corollary to Theorem 6.29.

15. Let V be a vector space over a field F not of characteristic two, and let H be a symmetric bilinear form on V. Prove that if $K(x) = H(x, x)$ is the quadratic form associated with H, then

$$H(x, y) = \tfrac{1}{2}[K(x + y) - K(x) - K(y)]$$

for all $x, y \in \mathsf{V}$.

16. For the following quadratic forms K over a real inner product space V, find a symmetric bilinear form H such that $K(x) = H(x, x)$ for all $x \in \mathsf{V}$. Then find an orthonormal basis β for V such that $\psi_\beta(H)$ is a diagonal matrix.

(a) $K\colon \mathsf{R}^2 \to R$ defined by

$$K\begin{pmatrix} t_1 \\ t_2 \end{pmatrix} = -2t_1^2 + 4t_1t_2 + t_2^2$$

(b) $K\colon \mathsf{R}^2 \to R$ defined by

$$K\begin{pmatrix} t_1 \\ t_2 \end{pmatrix} = 7t_1^2 - 8t_1t_2 + t_2^2$$

(c) $K\colon \mathsf{R}^3 \to R$ defined by

$$K\begin{pmatrix} t_1 \\ t_2 \\ t_3 \end{pmatrix} = 3t_1^2 + 3t_2^2 + 3t_3^2 - 2t_1t_3$$

17. Let $\mathcal{S}$ be the set of all

$$\begin{pmatrix} t_1 \\ t_2 \\ t_3 \end{pmatrix} \in \mathsf{R}^3$$

such that

$$3t_1^2 + 3t_2^2 + 3t_3^2 - 2t_1t_3 + 2\sqrt{2}(t_1 + t_3) + 1 = 0.$$

Find an orthonormal basis β for R^3 such that the equation relating the coordinates of points of $\mathcal{S}$ relative to β is simplified. Describe $\mathcal{S}$ geometrically.

18. Prove the following refinement of part (d) of Theorem 6.31.

(a) If $0 < \operatorname{rank}(A) < n$ and A has no negative eigenvalues, then f has no local maximum at p.

(b) If $0 < \operatorname{rank}(A) < n$ and A has no positive eigenvalues, then f has no local minimum at p.

19. Prove the following variation of the second derivative test for the case $n = 2$. Define

$$D = \left[\frac{\partial^2 f(p)}{(\partial t_1)^2}\right]\left[\frac{\partial^2 f(p)}{(\partial t_2)^2}\right] - \left[\frac{\partial^2 f(p)}{(\partial t_1)(\partial t_2)}\right]^2.$$

(a) If $D > 0$ and $\partial^2 f(p)/(\partial t_1)^2 > 0$, then f has a local minimum at p.
(b) If $D > 0$ and $\partial^2 f(p)/(\partial t_2)^2 < 0$, then f has a local maximum at p.
(c) If $D < 0$, then f has no local extremum at p.
(d) If $D = 0$, then the test is inconclusive.
Hint: Observe that $D = \det(A) = \lambda_1\lambda_2$, where λ_1 and λ_2 are the eigenvalues of A, and A is as in Theorem 6.31.

20. Let A be an $n \times n$ matrix over a field F, and let E be an elementary $n \times n$ matrix over F. In Section 3.1 it was shown that AE can be obtained from A by means of an elementary column operation. Prove that E^tA can be obtained from A by means of the same elementary operation but performed on the rows rather than on the columns of A. *Hint:* Note that $E^tA = (A^tE)^t$.

21. For each of the following matrices A with entries from the field of rational numbers, find a diagonal matrix D and an invertible matrix Q such that $Q^tAQ = D$.

(a)
$$A = \begin{pmatrix} 1 & 3 \\ 3 & 2 \end{pmatrix}$$

(b)
$$A = \begin{pmatrix} 0 & 1 \\ 1 & 0 \end{pmatrix}$$

Hint: Use an elementary operation other than that of interchanging columns.

(c)
$$A = \begin{pmatrix} 3 & 1 & 2 \\ 1 & 4 & 0 \\ 2 & 0 & -1 \end{pmatrix}$$

22. Prove that if the diagonal entries of a diagonal matrix are permuted, then the resulting diagonal matrix is congruent to the original one.

23. Let T be a linear operator on a real inner product space V, and define $H(x, y) = \langle x, \mathsf{T}(y)\rangle$ for all x, y in V.
(a) Prove that H is a bilinear form on V.
(b) Prove that H is symmetric if and only if T is self-adjoint.
(c) What properties must T have to require that H is an inner product on V?
(d) Explain why H may fail to be a bilinear form if V is a complex inner product space.

24. Prove the converse to Exercise 23: Let V be a finite-dimensional real inner product space, and let H be a bilinear form on V. Then there exists a unique linear operator T on V such that $H(x, y) = \langle x, \mathsf{T}(y)\rangle$ for all x and y in V. *Hint:* Choose an orthonormal basis β for V, let $A = \psi_\beta(H)$, and let T be the linear operator on V such that $[\mathsf{T}]_\beta = A$. Apply Exercise 7(c) of this section and Exercise 13 of Section 6.2.

25. Prove that the total number of distinct equivalence classes of congruent

$n \times n$ real symmetric matrices is given by

$$\frac{(n+1)(n+2)}{2}.$$

6.8* EINSTEIN'S SPECIAL THEORY OF RELATIVITY

As a result of physical experiments performed in the latter half of the nineteenth century (most notably the Michelson–Morley experiment of 1887) physicists concluded that *the results obtained in measuring the speed of light are independent of the velocity of the instrument used to measure the speed of light.* For example, suppose that while on Earth an experimenter measures the speed of light emitted from the sun and finds it to be 186,000 miles per second. Now suppose that the experimenter places the measuring equipment in a spaceship that leaves Earth traveling at 100,000 miles per second in a direction away from the sun. A repetition of the same experiment from the spaceship would yield the same result: Light is traveling at 186,000 miles per second relative to the spaceship rather than 86,000 miles per second as one might expect!

This revelation led to a new way of relating coordinate systems used to locate events in space-time. The result was Albert Einstein's *special theory of relativity*. We will develop via a linear algebra viewpoint the essence of Einstein's theory.

The basic problem is to compare two different inertial (nonaccelerating) coordinate systems that are in motion relative to each other under the assumption that the speed of light is the same when measured in either system. Suppose we are given two inertial coordinate systems S and S' in three-space (R^3) such that S' moves at a constant velocity in relation to S as measured from S (see Figure 6.6). To simplify matters, let us suppose that

1. The corresponding axes of S and S' (x and x', y and y', z and z') are parallel, and the origin of S' moves in the positive direction of the x-axis of S at a constant velocity $v > 0$ relative to S.
2. Two clocks C and C' are placed in space—the first stationary relative to coordinate system S, and the second stationary relative to S'. These clocks are designed to give as readings real numbers in units of seconds. The clocks are calibrated so that at the instant the origins of S and S' coincide both clocks give the reading zero.
3. Our unit of length is the *light second* (the distance light travels in 1 second) and our unit of time is the second. Note that with respect to these units the speed of light is 1 light second per second.

Given any event (something whose position and time of occurrence can be described), we may assign a set of "space-time coordinates" to it. For example, if

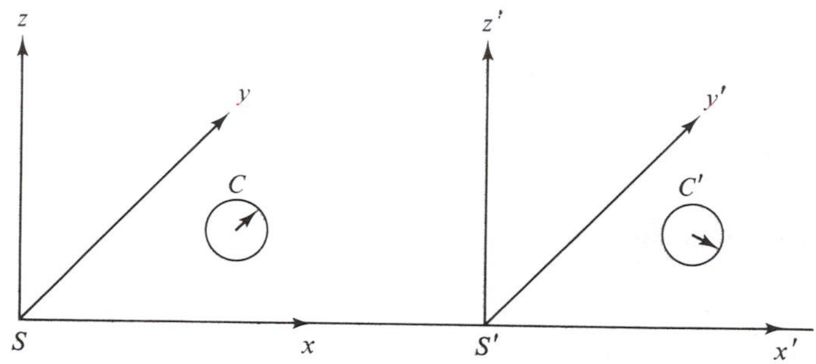

Figure 6.6

p is an event that occurs at position

$$\begin{pmatrix} x \\ y \\ z \end{pmatrix}$$

relative to S and at time t as read on clock C, we can assign to p the set of coordinates

$$\begin{pmatrix} x \\ y \\ z \\ t \end{pmatrix}.$$

This ordered 4-tuple is called *the space-time coordinates* of p relative to S and C. Likewise p has a set of space-time coordinates

$$\begin{pmatrix} x' \\ y' \\ z' \\ t' \end{pmatrix}$$

relative to S' and C'.

We can define a mapping $\mathsf{T}_v\colon \mathsf{R}^4 \to \mathsf{R}^4$ (which depends on the velocity v) as a consequence of the above such that, for any set of space-time coordinates

$$\begin{pmatrix} x \\ y \\ z \\ t \end{pmatrix}$$

measuring an event with respect to S and C,

$$\mathsf{T}_v \begin{pmatrix} x \\ y \\ z \\ t \end{pmatrix} = \begin{pmatrix} x' \\ y' \\ z' \\ t' \end{pmatrix}$$

is the set of space-time coordinates of this event with respect to S' and C'. Intuitively T_v is one-to-one and onto.

Einstein made certain assumptions about T_v that led to his special theory of relativity. We will formulate an equivalent set of assumptions.

Axioms of the Special Theory of Relativity

(R1) The speed of any light beam, when measured in either coordinate system using a clock stationary relative to that coordinate system, is 1.

(R2) The mapping $\mathsf{T}_v\colon \mathsf{R}^4 \to \mathsf{R}^4$ is an isomorphism.

(R3) For any

$$\begin{pmatrix} x \\ y \\ z \\ t \end{pmatrix} \in \mathsf{R}^4,$$

if

$$\mathsf{T}_v \begin{pmatrix} x \\ y \\ z \\ t \end{pmatrix} = \begin{pmatrix} x' \\ y' \\ z' \\ t' \end{pmatrix},$$

then $y' = y$ and $z' = z$.

(R4) For

$$\mathsf{T}_v \begin{pmatrix} x \\ y \\ z \\ t \end{pmatrix} = \begin{pmatrix} x' \\ y' \\ z' \\ t' \end{pmatrix},$$

x' and t' are independent of y and z; that is, if

$$\mathsf{T}_v \begin{pmatrix} x \\ y_1 \\ z_1 \\ t \end{pmatrix} = \begin{pmatrix} x' \\ y' \\ z' \\ t' \end{pmatrix} \quad \text{and} \quad \mathsf{T}_v \begin{pmatrix} x \\ y_2 \\ z_2 \\ t \end{pmatrix} = \begin{pmatrix} x'' \\ y'' \\ z'' \\ t'' \end{pmatrix},$$

then $x'' = x'$ and $t'' = t'$.

(R5) The origin of S moves in the negative direction of the x'-axis of S' at the constant velocity $-v < 0$ as measured from S'.

As we will see, these five axioms completely characterize T_v. The operator T_v is called the *Lorentz transformation in direction* x. We intend to compute T_v and use it to study the curious phenomenon of time contraction.

Theorem 6.32. *On* R^4

(a) $\mathsf{T}_v(e_i) = e_i$ *for* $i = 2, 3$.

(b) $\operatorname{span}(\{e_2, e_3\})$ *is* T_v*-invariant.*

(c) $\operatorname{span}(\{e_1, e_4\})$ *is* T_v*-invariant.*

(d) *Both* $\operatorname{span}(\{e_2, e_3\})$ *and* $\operatorname{span}(\{e_1, e_4\})$ *are* T_v^**-invariant.*

(e) $\mathsf{T}_v^*(e_i) = e_i$ *for* $i = 2, 3$.

Proof. (a) By axiom (R2)

$$\mathsf{T}_v\begin{pmatrix}0\\0\\0\\0\end{pmatrix} = \begin{pmatrix}0\\0\\0\\0\end{pmatrix},$$

and hence by axiom (R4) the first and fourth coordinates of

$$\mathsf{T}_v\begin{pmatrix}0\\a\\b\\0\end{pmatrix}$$

are both zero for any $a, b \in R$. Thus by axiom (R3)

$$\mathsf{T}_v\begin{pmatrix}0\\1\\0\\0\end{pmatrix} = \begin{pmatrix}0\\1\\0\\0\end{pmatrix} \quad \text{and} \quad \mathsf{T}_v\begin{pmatrix}0\\0\\1\\0\end{pmatrix} = \begin{pmatrix}0\\0\\1\\0\end{pmatrix}.$$

The proofs of (b), (c), and (d) are left as exercises.

(e) For any $j \neq 2$, $\langle \mathsf{T}_v^*(e_2), e_j\rangle = \langle e_2, \mathsf{T}_v(e_j)\rangle = 0$ by (a) and (c); for $j = 2$, $\langle \mathsf{T}_v^*(e_2), e_j\rangle = \langle e_2, \mathsf{T}_v(e_2)\rangle = \langle e_2, e_2\rangle = 1$ by (a). We conclude that $\mathsf{T}_v^*(e_2)$ is a multiple of e_2, i.e., that $\mathsf{T}_v^*(e_2) = \lambda e_2$ for some $\lambda \in R$. Thus $1 = \langle e_2, e_2\rangle = \langle e_2, \mathsf{T}_v(e_2)\rangle = \langle \mathsf{T}_v^*(e_2), e_2\rangle = \langle \lambda e_2, e_2\rangle = \lambda$, and hence $\mathsf{T}_v^*(e_2) = e_2$. Similarly $\mathsf{T}_v^*(e_3) = e_3$. ∎

Suppose that at the instant the origins of S and S' coincide a light flash is emitted from their common origin. The event of the light flash when measured either relative to S and C or relative to S' and C' has space-time coordinates

$$\begin{pmatrix} 0 \\ 0 \\ 0 \\ 0 \end{pmatrix}.$$

Let P be the set of all events whose space-time coordinates

$$\begin{pmatrix} x \\ y \\ z \\ t \end{pmatrix}$$

relative to S and C are such that the flash is observable from the point with coordinates

$$\begin{pmatrix} x \\ y \\ z \end{pmatrix}$$

(as measured relative to S) at the time t (as measured on C). Let us characterize P in terms of x, y, z, and t. Since the speed of light is 1, at any time $t \geqslant 0$ the light flash is observable from any point whose distance to the origin of S (as measured on S) is $t \cdot 1 = t$. These are precisely the points that lie on the sphere of radius t with center at the origin. The coordinates (relative to S) of such points satisfy the equation $x^2 + y^2 + z^2 = t^2$. Hence an event lies in P if and only if relative to S and C its space-time coordinates

$$\begin{pmatrix} x \\ y \\ z \\ t \end{pmatrix} \qquad (t \geqslant 0)$$

satisfy the equation $x^2 + y^2 + z^2 - t^2 = 0$. By virtue of axiom (R1) we can characterize P in terms of the space-time coordinates relative to S' and C' similarly: An event lies in P if and only if relative to S' and C' its space-time coordinates

$$\begin{pmatrix} x' \\ y' \\ z' \\ t' \end{pmatrix} \qquad (t' \geqslant 0)$$

satisfy the equation $(x')^2 + (y')^2 + (z')^2 - (t')^2 = 0.$

Let

$$A = \begin{pmatrix} 1 & 0 & 0 & 0 \\ 0 & 1 & 0 & 0 \\ 0 & 0 & 1 & 0 \\ 0 & 0 & 0 & -1 \end{pmatrix}.$$

Theorem 6.33. *For any* $w \in \mathsf{R}^4$ *if* $\langle \mathsf{L}_A(w), w\rangle = 0$, *then*

$$\langle \mathsf{T}_v^* \mathsf{L}_A \mathsf{T}_v(w), w\rangle = 0.$$

Proof. Let

$$w = \begin{pmatrix} x \\ y \\ z \\ t \end{pmatrix} \in \mathsf{R}^4,$$

and suppose that $\langle \mathsf{L}_A(w), w\rangle = 0$.

CASE 1. $t \geqslant 0$. Since $\langle \mathsf{L}_A(w), w\rangle = x^2 + y^2 + z^2 - t^2$, w is the set of coordinates of an event in P relative to S and C. Because

$$\begin{pmatrix} x \\ y \\ z \\ t \end{pmatrix} \quad \text{and} \quad \begin{pmatrix} x' \\ y' \\ z' \\ t' \end{pmatrix}$$

are the space-time coordinates of the same event relative to S' and C', the discussion preceding Theorem 6.33 yields

$$(x')^2 + (y')^2 + (z')^2 - (t')^2 = 0.$$

Thus $\langle \mathsf{T}_v^* \mathsf{L}_A \mathsf{T}_v(w), w\rangle = \langle \mathsf{L}_A \mathsf{T}_v(w), \mathsf{T}_v(w)\rangle = (x')^2 + (y')^2 + (z')^2 - (t')^2 = 0$, and the conclusion follows.

CASE 2. $t < 0$. The proof follows by applying case 1 to $-w$. ∎

We now proceed to deduce information about T_v. Let

$$w_1 = \begin{pmatrix} 1 \\ 0 \\ 0 \\ 1 \end{pmatrix} \quad \text{and} \quad w_2 = \begin{pmatrix} 1 \\ 0 \\ 0 \\ -1 \end{pmatrix}.$$

By Exercise 3, $\{w_1, w_2\}$ is an orthogonal basis for span($\{e_1, e_4\}$), and span($\{e_1, e_4\}$) is $\mathsf{T}_v^* \mathsf{L}_A \mathsf{T}_v$-invariant. The next result tells us even more.

Theorem 6.34. *There exist nonzero scalars* a *and* b *such that*

(a) $T_v^* L_A T_v(w_1) = aw_2$.

(b) $T_v^* L_A T_v(w_2) = bw_1$.

Proof. (a) Because $\langle L_A(w_1), w_1 \rangle = 0$, $\langle T_v^* L_A T_v(w_1), w_1 \rangle = 0$ by Theorem 6.33. Thus $T_v^* L_A T_v(w_1)$ is orthogonal to w_1. Since $\text{span}(\{e_1, e_4\}) = \text{span}(\{w_1, w_2\})$ is $T_v^* L_A T_v$-invariant, $T_v^* L_A T_v(w_1)$ must lie in this set. But $\{w_1, w_2\}$ is an orthogonal basis for this subspace, and so $T_v^* L_A T_v(w_1)$ must be a multiple of w_2. Thus $T_v^* L_A T_v(w_1) = aw_2$ for some scalar a. Since T_v and A are invertible, so is $T_v^* L_A T_v$. Thus $a \neq 0$, proving (a). The proof of (b) is similar. ∎

Corollary. *Let* $B_v = [T_v]_\beta$, *where* β *is the standard ordered basis for* R^4. *Then*

(a) $B_v^* A B_v = A$.

(b) $T_v^* L_A T_v = L_A$.

We leave the proof of the corollary as an exercise. For hints, see Exercise 4.

Now consider the situation 1 second after the origins of S and S' have coincided as measured by the clock C. Since the origin of S' is moving along the x-axis at a velocity v as measured in S, its space-time coordinates relative to S and C are

$$\begin{pmatrix} v \\ 0 \\ 0 \\ 1 \end{pmatrix}.$$

Similarly, the space-time coordinates for the origin of S' relative to S' and C' must be

$$\begin{pmatrix} 0 \\ 0 \\ 0 \\ t' \end{pmatrix}$$

for some $t' > 0$. Thus we have

$$T_v \begin{pmatrix} v \\ 0 \\ 0 \\ 1 \end{pmatrix} = \begin{pmatrix} 0 \\ 0 \\ 0 \\ t' \end{pmatrix} \qquad \text{for some } t' > 0. \tag{18}$$

By the corollary to Theorem 6.34,

$$\left\langle \mathsf{T}_v^*\mathsf{L}_A\mathsf{T}_v\begin{pmatrix} v\\0\\0\\1\end{pmatrix}, \begin{pmatrix} v\\0\\0\\1\end{pmatrix}\right\rangle = \left\langle \mathsf{L}_A\begin{pmatrix} v\\0\\0\\1\end{pmatrix}, \begin{pmatrix} v\\0\\0\\1\end{pmatrix}\right\rangle = v^2 - 1. \tag{19}$$

But also,

$$\left\langle \mathsf{T}_v^*\mathsf{L}_A\mathsf{T}_v\begin{pmatrix} v\\0\\0\\1\end{pmatrix}, \begin{pmatrix} v\\0\\0\\1\end{pmatrix}\right\rangle = \left\langle \mathsf{L}_A\mathsf{T}_v\begin{pmatrix} v\\0\\0\\1\end{pmatrix}, \mathsf{T}_v\begin{pmatrix} v\\0\\0\\1\end{pmatrix}\right\rangle$$

$$= \left\langle \mathsf{L}_A\begin{pmatrix} 0\\0\\0\\t'\end{pmatrix}, \begin{pmatrix} 0\\0\\0\\t'\end{pmatrix}\right\rangle = -(t')^2. \tag{20}$$

Combining (19) and (20), we conclude that

$$v^2 - 1 = -(t')^2, \quad \text{or} \quad t' = \sqrt{1 - v^2}. \tag{21}$$

Thus, from (18) and (21), we obtain

$$\mathsf{T}_v\begin{pmatrix} v\\0\\0\\1\end{pmatrix} = \begin{pmatrix} 0\\0\\0\\\sqrt{1 - v^2}\end{pmatrix}. \tag{22}$$

Next recall that the origin of S moves in the negative direction of the x'-axis of S' at the constant velocity $-v < 0$ as measured from S'. [This fact is axiom (R5).] Consequently, 1 second after the origins of S and S' have coincided as measured on clock C, there exists a time $t'' > 0$ as measured on clock C' such that

$$\mathsf{T}_v\begin{pmatrix} 0\\0\\0\\1\end{pmatrix} = \begin{pmatrix} -vt''\\0\\0\\t''\end{pmatrix}. \tag{23}$$

From (23) it follows in a manner similar to the derivation of (22) that

$$t'' = \frac{1}{\sqrt{1 - v^2}}, \tag{24}$$

and hence from (23) and (24)

$$\mathsf{T}_v \begin{pmatrix} 0 \\ 0 \\ 0 \\ 1 \end{pmatrix} = \begin{pmatrix} \dfrac{-v}{\sqrt{1-v^2}} \\ 0 \\ 0 \\ \dfrac{1}{\sqrt{1-v^2}} \end{pmatrix}. \tag{25}$$

The following result is now easily proved using (22) and (25) and Theorem 6.32.

Theorem 6.35. *Let β be the standard ordered basis for R^4. Then*

$$[\mathsf{T}_v]_\beta = B_v = \begin{pmatrix} \dfrac{1}{\sqrt{1-v^2}} & 0 & 0 & \dfrac{-v}{\sqrt{1-v^2}} \\ 0 & 1 & 0 & 0 \\ 0 & 0 & 1 & 0 \\ \dfrac{-v}{\sqrt{1-v^2}} & 0 & 0 & \dfrac{1}{\sqrt{1-v^2}} \end{pmatrix}.$$

Time Contraction

A most curious and paradoxical conclusion follows if we accept Einstein's theory of time contraction. Suppose that an astronaut leaves our solar system in a space vehicle traveling at a fixed velocity v as measured relative to our solar system. It follows from Einstein's theory that at the end of time t as measured on Earth the time that will have passed on the space vehicle is only $t\sqrt{1-v^2}$. To establish this result, consider the coordinate systems S and S' and clocks C and C' that we studied above. Suppose that the origin of S' coincides with the space vehicle and the origin of S coincides with a point in the solar system (stationary relative to the sun) so that the origin of S and S' coincide and clocks C and C' read zero at the moment the astronaut embarks on the trip.

As viewed from S, the space-time coordinates of the vehicle at any time $t > 0$ as measured by C are

$$\begin{pmatrix} vt \\ 0 \\ 0 \\ t \end{pmatrix},$$

whereas as viewed from S' the space-time coordinates of the vehicle at any time

$t' > 0$ as measured by C' are

$$\begin{pmatrix} 0 \\ 0 \\ 0 \\ t' \end{pmatrix}.$$

But if two sets of space-time coordinates

$$\begin{pmatrix} vt \\ 0 \\ 0 \\ t \end{pmatrix} \quad \text{and} \quad \begin{pmatrix} 0 \\ 0 \\ 0 \\ t' \end{pmatrix}$$

are to describe the same event, it must follow that

$$\mathsf{T}_v \begin{pmatrix} vt \\ 0 \\ 0 \\ t \end{pmatrix} = \begin{pmatrix} 0 \\ 0 \\ 0 \\ t' \end{pmatrix}.$$

Thus

$$\begin{pmatrix} \dfrac{1}{\sqrt{1-v^2}} & 0 & 0 & \dfrac{-v}{\sqrt{1-v^2}} \\ 0 & 1 & 0 & 0 \\ 0 & 0 & 1 & 0 \\ \dfrac{-v}{\sqrt{1-v^2}} & 0 & 0 & \dfrac{1}{\sqrt{1-v^2}} \end{pmatrix} \begin{pmatrix} vt \\ 0 \\ 0 \\ t \end{pmatrix} = \begin{pmatrix} 0 \\ 0 \\ 0 \\ t' \end{pmatrix}.$$

From the equation above it follows that

$$\frac{-v^2 t}{\sqrt{1-v^2}} + \frac{t}{\sqrt{1-v^2}} = t' \quad \text{or} \quad t' = t\sqrt{1-v^2}. \tag{26}$$

This is the desired result.

A dramatic consequence of time contraction is that distances are contracted along the line of motion (see Exercise 9).

Let us make one additional point. Suppose that we consider units of distance and time more commonly used than the light second and second, such as the mile and the hour or the kilometer and the second. Let c denote the speed of light relative to our chosen units of distance and time. It is easily seen that if an object travels at a velocity v relative to a set of units, then it is traveling at a velocity v/c in units of light seconds per second. Thus for an arbitrary set of units of distance and time, (26) becomes

$$t' = t\sqrt{1 - \frac{v^2}{c^2}}.$$

EXERCISES

1. Prove (b), (c), and (d) of Theorem 6.32.
2. Complete the proof of Theorem 6.33 for the case $t < 0$.
3. For

$$w_1 = \begin{pmatrix} 1 \\ 0 \\ 0 \\ 1 \end{pmatrix} \quad \text{and} \quad w_2 = \begin{pmatrix} 1 \\ 0 \\ 0 \\ -1 \end{pmatrix}$$

show that

(a) $\{w_1, w_2\}$ is an orthogonal basis for span($\{e_1, e_4\}$).

(b) span($\{e_1, e_4\}$) is $\mathsf{T}_v^*\mathsf{L}_A\mathsf{T}_v$-invariant.

4. Prove the corollary to Theorem 6.34.

Hints:

(a) Prove that

$$B_v^*AB_v = \begin{pmatrix} p & 0 & 0 & q \\ 0 & 1 & 0 & 0 \\ 0 & 0 & 1 & 0 \\ -q & 0 & 0 & -p \end{pmatrix},$$

where

$$p = \frac{a+b}{2} \quad \text{and} \quad q = \frac{a-b}{2}.$$

(b) Show that $q = 0$ by using the fact that $B_v^*AB_v$ is self-adjoint.

(c) Apply Theorem 6.33 to

$$w = \begin{pmatrix} 0 \\ 1 \\ 0 \\ 1 \end{pmatrix}$$

to show that $p = 1$.

5. Derive (24), and prove that

$$\mathsf{T}_v\begin{pmatrix}0\\0\\0\\1\end{pmatrix}=\begin{pmatrix}\dfrac{-v}{\sqrt{1-v^2}}\\0\\0\\\dfrac{1}{\sqrt{1-v^2}}\end{pmatrix}. \tag{25}$$

Hint: Use a technique similar to the derivation of (22).

6. Consider three coordinate systems, S, S', and S'' with corresponding axes (x, x', x''; y, y', y''; and z, z', z'') parallel and such that the x-, x'-, and x''-axes coincide. Suppose S' is moving past S at a velocity $v_1 > 0$ (as measured on S), S'' is moving past S' at a velocity $v_2 > 0$ (as measured on S'), and S'' is moving past S at a velocity $v_3 > 0$ (as measured on S), and that there are three clocks C, C', and C'' such that C is stationary relative to S, C' is stationary relative to S', and C'' is stationary relative to S''. Suppose that when measured on any of the three clocks all the origins of S, S', and S'' coincide at time 0. Assuming that $\mathsf{T}_{v_3} = \mathsf{T}_{v_2}\mathsf{T}_{v_1}$ (i.e., $B_{v_3} = B_{v_2}B_{v_1}$), prove that

$$v_3 = \frac{v_1 + v_2}{1 + v_1 v_2}.$$

Note that substituting $v_2 = 1$ in the equation above yields $v_3 = 1$. This tells us that the speed of light as measured in either S or S' is the same. Why would we be surprised if this were not the case?

7. Compute $(B_v)^{-1}$. Show $(B_v)^{-1} = B_{(-v)}$. Conclude that if S' moves at a negative velocity v relative to S, then $[\mathsf{T}_v]_\beta = B_v$, where B_v is of the form given in Theorem 6.35.

8. Suppose that an astronaut left Earth in the year 1776 and traveled to a star 99 light years away from Earth at 99% of the speed of light and that upon reaching the star immediately turned around and returned to Earth at the same speed. Assuming Einstein's special theory of relativity, show that if the astronaut was 20 years old at the time of departure, then he or she would return to Earth at age 48.2 in the year 1976. Explain the use of Exercise 7 in solving this problem.

9. Recall the moving space vehicle considered in the study of time contraction. Suppose that the vehicle is moving toward a fixed star located on the x-axis of S at a distance b units from the origin of S. If the space vehicle moves toward the star at velocity v, earthlings (who remain "almost" stationary relative to S) will compute the time it takes for the vehicle to reach the star as $t = b/v$. Due to the phenomenom of time contraction the astronaut will perceive a time span of $t' = t\sqrt{1-v^2} = (b/v)\sqrt{1-v^2}$. A paradox appears in that the

astronaut perceives a time span inconsistent with a distance of b and a velocity of v. The paradox is resolved by observing that the distance from the solar system to the star as measured by the astronaut is less than b.

Assuming that the coordinate systems S and S' and clocks C and C' are as in the discussion of time contraction,

(a) Argue that at time t (as measured on C) the space-time coordinates of the star relative to S and C are

$$\begin{pmatrix} b \\ 0 \\ 0 \\ t \end{pmatrix}.$$

(b) Show that at time t (as measured on C) the space-time coordinates of the star relative to S' and C' are

$$\begin{pmatrix} \dfrac{b - vt}{\sqrt{1 - v^2}} \\ 0 \\ 0 \\ \dfrac{t - bv}{\sqrt{1 - v^2}} \end{pmatrix}.$$

(c) Setting

$$x' = \frac{b - tv}{\sqrt{1 - v^2}} \quad \text{and} \quad t' = \frac{t - bv}{\sqrt{1 - v^2}},$$

show that $x' = b\sqrt{1 - v^2} - t'v$.

This result may be interpreted to mean that at time t' as measured by the astronaut, the distance from the astronaut to the star as measured by the astronaut (see Figure 6.7 on the next page) is

$$b\sqrt{1 - v^2} - t'v.$$

(d) Conclude from this that

(1) The speed of the space vehicle relative to the star as measured by the astronaut is v.

(2) The distance from Earth to the star as measured by the astronaut is $b\sqrt{1 - v^2}$.

Thus distances along the line of motion of the space vehicle appear to be contracted by a factor of $\sqrt{1 - v^2}$.

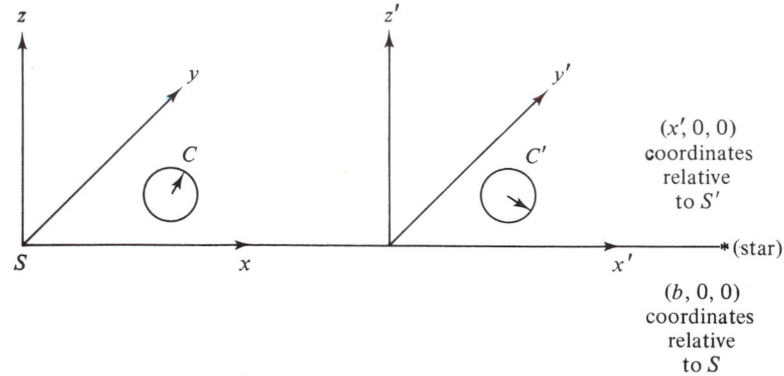

Figure 6.7

6.9* CONDITIONING AND THE RAYLEIGH QUOTIENT

In Section 3.4 we studied specific techniques that allow us to solve systems of linear equations in the form $AX = b$, where A is an $m \times n$ matrix and b is an $m \times 1$ vector. Such systems often arise through applications to the real world. The coefficients in the system are frequently obtained from experimental data, and in many cases both m and n are so large that a computer must be used in the calculation of the solution. Thus two types of errors must be considered. First, experimental errors arise in the collection of data since no instruments can provide completely accurate measurements. Second, computers will introduce round-off errors. One might intuitively feel that small relative changes in the coefficients in the system will cause small relative errors in the solution. A system that has this property is called *well-conditioned*; otherwise, the system is called *ill-conditioned*.

We now consider several examples of these types of errors, concentrating primarily on changes in b rather than in changes in the entries of A. In addition, we assume that A is a square, complex (or real), invertible matrix, since this is the case most frequently encountered in applications.

Example 1

Consider the system

$$\begin{cases} x_1 + x_2 = 5 \\ x_1 - x_2 = 1. \end{cases}$$

The solution of this system is

$$\begin{pmatrix} 3 \\ 2 \end{pmatrix}.$$

Now suppose that we change the system somewhat and consider the new system

$$\begin{cases} x_1 + x_2 = 5 \\ x_1 - x_2 = 1.0001. \end{cases}$$

This modified system has the solution

$$\begin{pmatrix} 3.00005 \\ 1.99995 \end{pmatrix}.$$

We see that a change of 10^{-4} in one coefficient has caused a change of less than 10^{-4} in each coordinate of the new solution. More generally, the system

$$\begin{cases} x_1 + x_2 = 5 \\ x_1 - x_2 = 1 + \delta \end{cases}$$

has the solution

$$\begin{pmatrix} 3 + \dfrac{\delta}{2} \\ \\ 2 - \dfrac{\delta}{2} \end{pmatrix}.$$

Hence small changes in b introduce small changes in the solution. Of course, we are really interested in "relative changes" since a change in the solution of, say, 10 is considered large if the original solution is of the order of 10^{-2} but small if the original solution is of the order of 10^6.

We will use the notation δb to denote the vector $b' - b$, where b is the vector in the original system and b' is the vector in the modified system. Thus we have

$$\delta b = \begin{pmatrix} 5 \\ 1 + \delta \end{pmatrix} - \begin{pmatrix} 5 \\ 1 \end{pmatrix} = \begin{pmatrix} 0 \\ \delta \end{pmatrix}.$$

We now define the *relative change* in b to be the scalar $\|\delta b\|/\|b\|$, where $\|\cdot\|$ denotes the standard norm in C^n (or R^n); i.e. $\|b\| = \sqrt{\langle b, b\rangle}$. Most of what follows, however, is true for any norm. Similar definitions hold for the *relative change* in x.

$$\frac{\|\delta b\|}{\|b\|} = \frac{|\delta|}{\sqrt{26}} \quad \text{and} \quad \frac{\|\delta x\|}{\|x\|} = \frac{\left\| \begin{pmatrix} 3 + (\delta/2) \\ 2 - (\delta/2) \end{pmatrix} - \begin{pmatrix} 3 \\ 2 \end{pmatrix} \right\|}{\left\| \begin{pmatrix} 3 \\ 2 \end{pmatrix} \right\|} = \frac{|\delta|}{\sqrt{26}}.$$

Thus the relative change in x equals, coincidentally, the relative change in b, so the system is well-conditioned. ∎

Example 2

Consider the system

$$\begin{cases} x_1 + x_2 = 3 \\ x_1 + 1.00001x_2 = 3.00001, \end{cases}$$

which has

$$\begin{pmatrix} 2 \\ 1 \end{pmatrix}$$

as its solution. The solution for the related system

$$\begin{cases} x_1 + x_2 = 3 \\ x_1 + 1.00001x_2 = 3.00001 + \delta \end{cases}$$

is

$$\begin{pmatrix} 2 - (10^5)\delta \\ 1 + (10^5)\delta \end{pmatrix}.$$

Hence

$$\frac{\|\delta x\|}{\|x\|} = 10^5\sqrt{\tfrac{2}{5}}\,|\delta| \geqslant 10^4|\delta|,$$

while

$$\frac{\|\delta b\|}{\|b\|} \approx \frac{|\delta|}{3\sqrt{2}}.$$

Thus the relative change in x is at least 10^4 times the relative change in b! This system is very ill-conditioned. Observe that the lines defined by the two equations in this system are nearly coincident. So a small change in either line could greatly alter the point of intersection, that is, the solution of the system. ∎

To apply the full strength of the theory of self-adjoint matrices to the study of conditioning, we need the notion of the norm of a matrix (see Exercise 22 of Section 6.1 for further results about norms).

Definition. *Let* A *be a complex (or real)* n × n *matrix. Define the* (Euclidean) norm *of* A *by*

$$\|A\| = \max_{x \neq 0} \frac{\|Ax\|}{\|x\|},$$

where $x \in \mathsf{C}^n$ or $x \in \mathsf{R}^n$.

We see intuitively that $\|A\|$ represents the maximum "magnification" of a vector by the matrix A.

The question of whether or not this maximum exists, as well as the problem of how to compute it, can be answered by use of the so-called "Rayleigh quotient."

Definition. *Let* B *an* $n \times n$ *self-adjoint matrix. The* Rayleigh quotient *for* $x \neq 0$ *is defined to be the scalar* $R(x) = \langle Bx, x \rangle / \|x\|^2$.

Theorem 6.36. *For a self-adjoint matrix* B, $\max_{x \neq 0} R(x)$ *is the largest eigenvalue of* B *and* $\min_{x \neq 0} R(x)$ *is the smallest eigenvalue of* B.

Proof. By Theorems 6.19 and 6.20 we may choose an orthonormal basis $\{x_1, \ldots, x_n\}$ of eigenvectors of B such that $Bx_i = \lambda_i x_i$, $1 \leqslant i \leqslant n$, where $\lambda_1 \geqslant \lambda_2 \geqslant \cdots \geqslant \lambda_n$. (Recall that by the lemma to Theorem 6.17 the eigenvalues of B are real.) Now for $x \in \mathsf{C}^n$ (or R^n) there exist scalars $a_1, \ldots, a_n$ such that

$$x = \sum_{i=1}^{n} a_i x_i;$$

hence

$$R(x) = \frac{\langle Bx, x \rangle}{\|x\|^2} = \frac{\left\langle \sum_{i=1}^{n} a_i \lambda_i x_i, \sum_{j=1}^{n} a_j x_j \right\rangle}{\|x\|^2}$$

$$= \frac{\sum_{i=1}^{n} \lambda_i |a_i|^2}{\|x\|^2} \leqslant \frac{\lambda_1 \sum_{i=1}^{n} |a_i|^2}{\|x\|^2} = \lambda_1.$$

It is easy to see that $R(x_1) = \lambda_1$, so we have demonstrated the first half of the theorem. The second half is proved similarly. ∎

Corollary 1. *For any square matrix* A, $\|A\|$ *is finite and, in fact, equals* $\sqrt{\lambda}$, *where* λ *is the largest eigenvalue of* A^*A.

Proof. Let B be the self-adjoint matrix A^*A, and let λ be the largest eigenvalue of B. Since, for $x \neq 0$,

$$0 \leqslant \frac{\|Ax\|^2}{\|x\|^2} = \frac{\langle Ax, Ax \rangle}{\|x\|^2} = \frac{\langle A^*Ax, x \rangle}{\|x\|^2} = \frac{\langle Bx, x \rangle}{\|x\|^2} = R(x),$$

we have from Theorem 6.36 that $\|A\|^2 = \lambda$. ∎

Observe that the proof of Corollary 1 shows that all the eigenvalues of A^*A are nonnegative. For the next corollary we need the following lemma.

Lemma. *For any square matrix* A, *λ is an eigenvalue of* A^*A *if and only if λ is an eigenvalue of* AA^*.

Proof. Let λ be an eigenvalue of A^*A. If $\lambda = 0$, then A^*A is not invertible. Hence A (and A^*) is not invertible, so that λ is also an eigenvalue of AA^*. The proof of the converse is similar.

Suppose now that $\lambda \neq 0$. Then there exists $x \neq 0$ such that $A^*Ax = \lambda x$. Apply A to both sides to obtain $(AA^*)(Ax) = \lambda(Ax)$. Since $Ax \neq 0$ (lest $\lambda x = 0$), we have that λ is an eigenvalue of AA^*. The converse is left as an exercise. ∎

Corollary 2. *Let* A *be an invertible matrix. Then* $\|A^{-1}\| = 1/\sqrt{\lambda}$, *where λ is the smallest eigenvalue of* A^*A.

Proof. Recall that λ is an eigenvalue of an invertible matrix if and only if λ^{-1} is an eigenvalue of its inverse.

Now let $\lambda_1 \geqslant \lambda_2 \geqslant \cdots \geqslant \lambda_n$ be the eigenvalues of A^*A, which by the lemma are the eigenvalues of AA^*. Then $\|A^{-1}\|^2$ equals the largest eigenvalue of $(A^{-1})^*A^{-1} = (AA^*)^{-1}$, which equals $1/\lambda_n$. ∎

For many applications it is only the largest and smallest eigenvalues that are of interest. For example, in the case of vibration problems, the smallest eigenvalue represents the lowest frequency at which vibrations can occur.

We will see the role of both these eigenvalues in our study of conditioning.

Example 3

Let

$$A = \begin{pmatrix} 1 & 0 & 1 \\ -1 & 1 & 0 \\ 0 & 1 & 1 \end{pmatrix}.$$

Then

$$B = A^*A = \begin{pmatrix} 2 & -1 & 1 \\ -1 & 2 & 1 \\ 1 & 1 & 2 \end{pmatrix}.$$

The eigenvalues of B are 3, 3, and 0. Therefore, $\|A\| = \sqrt{3}$. For any

$$x = \begin{pmatrix} a \\ b \\ c \end{pmatrix} \neq 0,$$

we may compute $R(x)$ for the matrix B as

$$3 \geqslant R(x) = \frac{\langle Bx, x\rangle}{\|x\|^2} = \frac{2(a^2 + b^2 + c^2 - ab + ac + bc)}{a^2 + b^2 + c^2}$$

for all $a, b, c \in R$. ∎

Now that we know $\|A\|$ exists for every square matrix, we can make use of the inequality $\|Ax\| \leqslant \|A\| \cdot \|x\|$, which holds for every x.

Assume in what follows that A is invertible, $b \neq 0$, and $Ax = b$. For a given δb, let δx be the vector that satisfies $A(x + \delta x) = b + \delta b$. Then $A(\delta x) = \delta b$, and so $\delta x = A^{-1}(\delta b)$. Hence

$$\|b\| = \|Ax\| \leqslant \|A\| \cdot \|x\| \quad \text{and} \quad \|\delta x\| = \|A^{-1}(\delta b)\| \leqslant \|A^{-1}\| \cdot \|\delta b\|.$$

Thus

$$\frac{\|\delta x\|}{\|x\|} \leqslant \frac{\|A^{-1}\| \cdot \|\delta b\| \cdot \|A\|}{\|b\|} = \|A\| \cdot \|A^{-1}\| \cdot \left(\frac{\|\delta b\|}{\|b\|}\right).$$

Similarly,

$$\frac{1}{\|A\| \cdot \|A^{-1}\|}\left(\frac{\|\delta b\|}{\|b\|}\right) \leqslant \frac{\|\delta x\|}{\|x\|}.$$

The number $\|A\| \cdot \|A^{-1}\|$ is called the *condition number* of A and is denoted *cond*(A). It should be noted that the definition of cond(A) depends on how we define the norm of A. There are many reasonable ways of defining the norm of a matrix. In fact, the only property we used to establish the inequalities above was that $\|Ax\| \leqslant \|A\| \cdot \|x\|$ for all x. We summarize these results in the following theorem.

Theorem 6.37. *For the system* AX = b, *where* A *is invertible and* b ≠ 0, *we have the following two results:*

(a) $\dfrac{1}{\text{cond}(A)}\dfrac{\|\delta b\|}{\|b\|} \leqslant \dfrac{\|\delta x\|}{\|x\|} \leqslant \text{cond}(A)\dfrac{\|\delta b\|}{\|b\|}$ *(for any norm* $\|\cdot\|$*).*

(b) $\text{cond}(A) = \sqrt{\dfrac{\lambda_1}{\lambda_n}}$, *where* λ_1 *and* λ_n *are the largest and smallest eigenvalues, respectively, of* A*A. *(In this part, we assume that* $\|\cdot\|$ *is the Euclidean norm defined in this section.)*

Proof. Statement (a) follows from the previous inequalities, and (b) follows from Corollaries 1 and 2 to Theorem 6.36. ∎

It is clear from Theorem 6.37 that cond(A) $\geqslant 1$. It is left as an exercise to prove that cond(A) $= 1$ if and only if A is a scalar multiple of a unitary or orthogonal matrix as defined in Section 6.5. Moreover, it can be shown with some work that equality can be obtained in (a) by an appropriate choice of b and δb.

We can see immediately from (a) that if cond(A) is close to 1, then we are sure that a small relative error in b forces a small relative error in x. If cond(A) is large, however, then the relative error in x may be small even though the relative error in b is large, or the relative error in x may be large even though the relative error in b is small! In short, cond(A) merely indicates the potential for large relative errors.

We have so far considered only errors in the vector b. If there is an error δA in the coefficient matrix of the system $AX = b$, the situation is more complicated. For example, $A + \delta A$ may fail to be invertible. But under appropriate assumptions it can be shown that a bound for the relative error in x can be given in terms of cond(A). For example, if $A + \delta A$ is invertible, then Forsythe and Moler (G. Forsythe and C. B. Moler, *Computer Solution of Linear Algebraic Systems*, Prentice-Hall, Inc., Englewood Cliffs, N.J., 1976, p. 23) show that

$$\frac{\|\delta x\|}{\|x + \delta x\|} \leqslant \operatorname{cond}(A) \frac{\|\delta A\|}{\|A\|}.$$

It should be mentioned that, in practice, one almost never knows cond(A), for it would be an unnecessary waste of time to compute A^{-1} merely to determine its norm. In fact, if a computer is used to find A^{-1}, the computed inverse of A will in all likelihood only approximate A^{-1}, and the error in the computed inverse will be affected by the size of cond(A). So we are caught in a vicious circle! There are, however, some situations in which a usable approximation of cond(A) can be found. Thus, in most cases, the estimate of the relative error in x is based on an estimate of cond(A).

EXERCISES

1. Label the following statements as being true or false.
(a) If $AX = b$ is well-conditioned, then cond(A) is small.
(b) If cond(A) is large, then $AX = b$ is ill-conditioned.
(c) If cond(A) is small, then $AX = b$ is well-conditioned.
(d) The norm of A equals the Rayleigh quotient.
(e) The norm of A is always equal to the largest eigenvalue of A.

2. Compute the norms of the following matrices.

(a) $\begin{pmatrix} 4 & 0 \\ 1 & 3 \end{pmatrix}$ **(b)** $\begin{pmatrix} 5 & 3 \\ -3 & 3 \end{pmatrix}$

(c) $\begin{pmatrix} 1 & \dfrac{-2}{\sqrt{3}} & 0 \\ 0 & \dfrac{-2}{\sqrt{3}} & 1 \\ 0 & \dfrac{2}{\sqrt{3}} & 1 \end{pmatrix}$

3. Prove that if B is symmetric, then $\|B\|$ is the largest eigenvalue of B.

4. Let A and A^{-1} be as follows:

$$A = \begin{pmatrix} 6 & 13 & -17 \\ 13 & 29 & -38 \\ -17 & -38 & 50 \end{pmatrix} \quad \text{and} \quad A^{-1} = \begin{pmatrix} 6 & -4 & 1 \\ -4 & 11 & 7 \\ -1 & 7 & 5 \end{pmatrix}.$$

The eigenvalues of A are approximately 84.74, 0.2007, and 0.0588.

(a) Approximate $\|A\|$, $\|A^{-1}\|$, and cond(A). (Note Exercise 3.)

(b) Suppose that we have vectors x and $\tilde{x}$ such that $Ax = b$ and $\|b - A\tilde{x}\| \leqslant 0.001$. Use part (a) to determine upper bounds for $\|\tilde{x} - A^{-1}b\|$ (the absolute error) and $\|\tilde{x} - A^{-1}b\|/\|A^{-1}b\|$ (the relative error).

5. Suppose that x is the actual solution of $AX = b$ and that a computer arrives at an approximate solution $\tilde{x}$. If cond$(A) = 100$, $\|b\| = 1$, and $\|b - A\tilde{x}\| = 0.1$, obtain upper and lower bounds for $\|x - \tilde{x}\|/\|x\|$.

6. Let

$$B = \begin{pmatrix} 2 & 1 & 1 \\ 1 & 2 & 1 \\ 1 & 1 & 2 \end{pmatrix}.$$

Compute

$$R\begin{pmatrix} 1 \\ -2 \\ 3 \end{pmatrix}, \qquad \|B\|, \quad \text{and} \quad \text{cond}(B).$$

7. Let B be a symmetric matrix. Prove that $\min_{x \neq 0} R(x)$ equals the smallest eigenvalue of B.

8. Prove that if λ is an eigenvalue of AA^*, then λ is an eigenvalue of A^*A. This completes the proof of the lemma to Corollary 2 to Theorem 6.36.

9. Prove the left inequality of (a) in Theorem 6.37.

10. Prove that $\text{cond}(A) = 1$ if and only if A is a scalar multiple of a unitary or orthogonal matrix as defined in Section 6.5.

11. **(a)** Let A and B be square matrices that are unitarily equivalent as defined in Section 6.5. Prove that $\|A\| = \|B\|$.

(b) Let V be a finite-dimensional inner product space, and let T be a linear operator on T. Define

$$\|\mathsf{T}\| = \max_{x \neq 0} \frac{\|\mathsf{T}(x)\|}{\|x\|}.$$

Prove that $\|\mathsf{T}\| = \|[\mathsf{T}]_\beta\|$, where β is any orthonormal basis of V.

(c) Let V be an infinite-dimensional inner product space with an orthonormal basis $\{x_1, x_2, \ldots\}$. Let T be the linear operator on V such that $\mathsf{T}(x_k) = kx_k$. Prove that $\|\mathsf{T}\|$ [defined in part (b)] does not exist.

6.10* THE GEOMETRY OF ORTHOGONAL OPERATORS

Theorem 6.22 establishes that any rigid motion on a real inner product space is the composition of an orthogonal operator followed by a translation. Thus to understand the geometry of rigid motions thoroughly, we must analyze the structure of orthogonal operators. Such is the aim of this section. As we will discover, an orthogonal operator on a finite-dimensional real inner product space is the composition of rotations and reflections. This material assumes that the reader is familiar with the results about direct sums developed at the end of Section 5.2.

Definitions. *Let* T *be a linear operator on a finite-dimensional real inner product space* V. *The operator* T *is called a* rotation *if* T *is the identity on* V *or if there exists a two-dimensional subspace* W *of* V, *an orthonormal basis* $\beta = \{x_1, x_2\}$ *for* W, *and a real number* θ *such that*

$$\mathsf{T}(x_1) = x_1 \cos\theta + x_2 \sin\theta, \qquad \mathsf{T}(x_2) = -x_2 \sin\theta + x_2 \cos\theta,$$

and $\mathsf{T}(y) = y$ *for all* $y \in \mathsf{W}^\perp$. *In this context* T *is called a* rotation of W about $\mathsf{W}^\perp$. *The subspace* $\mathsf{W}^\perp$ *is called the* axis of rotation.

Rotations were defined in Section 2.1 for the special case that $\mathsf{V} = \mathsf{R}^2$.

Definitions. *Let* T *be a linear operator on a finite-dimensional real inner product space* V. *The operator* T *is called a* reflection *if there exists a one-dimensional subspace* W *of* V *such that* $\mathsf{T}(x) = -x$ *for all* $x \in \mathsf{W}$ *and* $\mathsf{T}(y) = y$ *for all* $y \in \mathsf{W}^\perp$. *In this context* T *is called a* reflection of V about $\mathsf{W}^\perp$.

It should be noted that rotations and reflections (or compositions thereof) are orthogonal operators (see Exercise 2). The principal aim of this section is to establish that the converse is also true, that is, that any orthogonal operator on a finite-dimensional real inner product space is the composition of rotations and reflections.

Example 1

A Characterization of Orthogonal Operators on a One-Dimensional Real Inner Product Space

Let T be an orthogonal operator on a one-dimensional inner product space V. Choose any nonzero vector x in V. Then $V = \text{span}(\{x\})$, and so $T(x) = \lambda x$ for some $\lambda \in R$. Since T is orthogonal and λ is an eigenvalue of T, $\lambda = \pm 1$. If $\lambda = 1$, then T is the identity on V, and hence T is a rotation. If $\lambda = -1$, then $T(x) = -x$ for all $x \in V$, and hence T is a reflection of V about $V^\perp = \{0\}$. Thus T is either a rotation or a reflection. Note that in the first case $\det(T) = 1$, and in the second case $\det(T) = -1$. ∎

Example 2

Some Typical Reflections

(a) Let $T: R^2 \to R^2$ be defined by $T(a, b) = (-a, b)$. If $W = \text{span}(\{e_1\})$, then $T(x) = -x$ for all $x \in W$ and $T(y) = y$ for all $y \in W^\perp$. Thus T is a reflection of R^2 about $W^\perp = \text{span}(\{e_2\})$, the y-axis.

(b) Let $T: R^3 \to R^3$ be defined by $T(a, b, c) = (a, b, -c)$. If $W = \text{span}(\{e_3\})$, then $T(x) = -x$ for all $x \in W$ and $T(y) = y$ for all $y \in W^\perp = \text{span}(\{e_1, e_2\})$, the xy-plane. Hence T is a reflection of R^3 about $W^\perp$. ∎

Example 1 characterizes all orthogonal operators on a one-dimensional real inner product space. The following theorem characterizes all orthogonal operators on a two-dimensional real inner product space. The proof of this result follows easily from Theorem 6.22 since a reflection about the x-axis followed by a rotation by θ is a reflection about the line through the origin with slope $\tan \frac{1}{2}\theta$.

Theorem 6.38. *Let T be an orthogonal operator on a two-dimensional real inner product space V. Then T is either a rotation or a reflection. Furthermore, T is a rotation if and only if* $\det(T) = 1$, *and T is a reflection if and only if* $\det(T) = -1$.

It is immediate from the definition that any reflection on R^2 has eigenvalues of 1 and -1 and that any two eigenvectors corresponding to these eigenvalues are orthogonal. Moreover, the eigenspace of T corresponding to $\lambda = 1$ is one-dimensional and hence can be described as a line passing through the origin. Geometrically, T reflects points in R^2 about this line (see Figure 6.8).

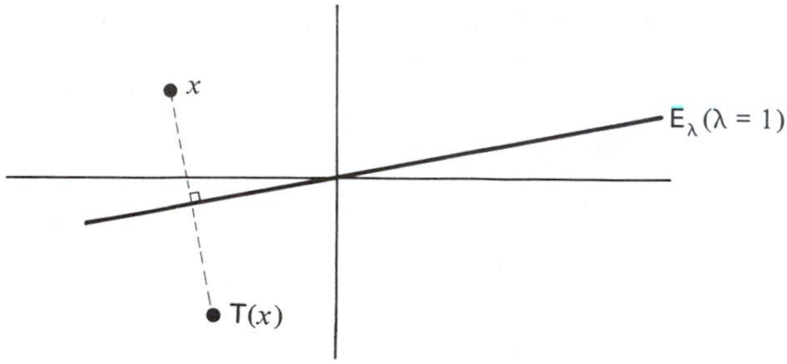

Figure 6.8

For example, if

$$A = \begin{pmatrix} \frac{1}{\sqrt{5}} & \frac{2}{\sqrt{5}} \\ \frac{2}{\sqrt{5}} & \frac{-1}{\sqrt{5}} \end{pmatrix},$$

it is clear that L_A is an orthogonal operator on R^2 and the $\det(\mathsf{L}_A) = \det(A) = -1$. Hence L_A is a reflection by Theorem 6.38. To find the subspace about which L_A reflects, it suffices to find an eigenvector of L_A corresponding to eigenvalue $\lambda = 1$. One such eigenvector is

$$x = \begin{pmatrix} \frac{1}{\sqrt{5}} + 1 \\ \frac{2}{\sqrt{5}} \end{pmatrix}.$$

Consequently, the subspace about which L_A reflects is the line

$$\left\{ t\begin{pmatrix} \frac{1}{\sqrt{5}} + 1 \\ \frac{2}{\sqrt{5}} \end{pmatrix} : \quad t \in R \right\}.$$

Corollary. *Let* V *be a two-dimensional real inner product space. The composition of a reflection and a rotation on* V *is a reflection on* V.

Proof. If T_1 is a reflection on V and T_2 is a rotation on V, then by Theorem 6.38 $\det(\mathsf{T}_1) = 1$ and $\det(\mathsf{T}_2) = -1$. Let $\mathsf{T} = \mathsf{T}_2\mathsf{T}_1$ be the composition. Since T_2 and T_1 are orthogonal, so is T. Moreover, $\det(\mathsf{T}) = \det(\mathsf{T}_2)\cdot\det(\mathsf{T}_1) = -1$. Thus, by Theorem 6.38, T is a reflection. The proof for $\mathsf{T}_1\mathsf{T}_2$ is similar. ∎

We now study orthogonal operators on spaces of higher dimension.

Lemma. *If* T *is a linear operator on a nonzero finite-dimensional real inner product space* V, *then there exists a* T*-invariant subspace* W *of* V *such that* $1 \leqslant \dim(\mathsf{W}) \leqslant 2$.

Proof. Fix an ordered basis $\beta = \{y_1, y_2, \ldots, y_n\}$ for V, and let $A = [\mathsf{T}]_\beta$. Let $\phi_\beta: \mathsf{V} \to \mathsf{R}^n$ be the linear transformation defined by $\phi_\beta(y_i) = e_i$ for $i = 1, 2, \ldots, n$. Then ϕ_β is an isomorphism, and as we have seen in Section 2.4, the diagram in Figure 6.9 is commutative, that is, that $\mathsf{L}_A\phi_\beta = \phi_\beta\mathsf{T}$. As a consequence it suffices to show that there exists an L_A-invariant subspace Z of R^n such that $1 \leqslant \dim(\mathsf{Z}) \leqslant 2$. If we then define $\mathsf{W} = \phi_\beta^{-1}(\mathsf{Z})$, it will follow that W satisfies the conclusion of the theorem (see Exercise 12).

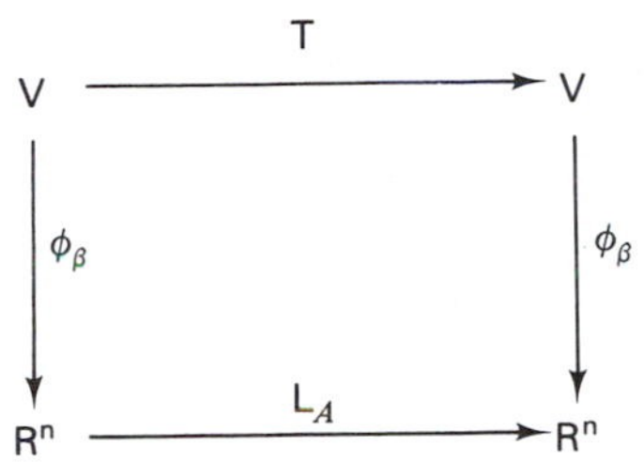

Figure 6.9

The matrix A can be considered as an $n \times n$ matrix over C and as such can be used to define a linear operator U on C^n by $\mathsf{U}(x) = Ax$ for all column vectors x in C^n. Since U is a linear operator on a finite-dimensional vector space over C, it has an eigenvalue $\lambda \in C$. Let $x \in \mathsf{C}^n$ be an eigenvector corresponding to λ. We may write $\lambda = \lambda_1 + i\lambda_2$, where λ_1 and λ_2 are real, and

$$x = \begin{pmatrix} a_1 + ib_1 \\ a_2 + ib_2 \\ \vdots \\ a_n + ib_n \end{pmatrix},$$

where the a_i's and b_i's are real. Thus, setting

$$x_1 = \begin{pmatrix} a_1 \\ a_2 \\ \vdots \\ a_n \end{pmatrix} \quad \text{and} \quad x_2 = \begin{pmatrix} b_1 \\ b_2 \\ \vdots \\ b_n \end{pmatrix},$$

we have $x = x_1 + ix_2$, where x_1 and x_2 are n-tuples with real entries. Note that at least one of x_1 or x_2 is nonzero since $x \neq 0$. Hence

$$\begin{aligned} \mathsf{U}(x) = \lambda x &= (\lambda_1 + i\lambda_2)(x_1 + ix_2) \\ &= (\lambda_1 x_1 - \lambda_2 x_2) + i(\lambda_1 x_2 + \lambda_2 x_1). \end{aligned}$$

Similarly,

$$\mathsf{U}(x) = A(x_1 + ix_2) = Ax_1 + iA_2.$$

Comparing the real and imaginary parts of these two expressions for $\mathsf{U}(x)$, we conclude that

$$Ax_1 = \lambda_1 x_1 - \lambda_2 x_2 \quad \text{and} \quad Ax_2 = \lambda_1 x_2 + \lambda_2 x_1.$$

Finally, let $\mathsf{Z} = \text{span}(\{x_1, x_2\})$, the span being taken as a subspace of R^n. Since $x_1 \neq 0$ or $x_2 \neq 0$, Z is nonzero. Thus $1 \leqslant \dim(\mathsf{Z}) \leqslant 2$, and the preceding pair of equations shows that Z is L_A-invariant. ∎

Theorem 6.39. *Let T be an orthogonal operator on a nonzero finite-dimensional real inner product space V. Then there exists a collection of pairwise orthogonal T-invariant subspaces $\{\mathsf{W}_1, \mathsf{W}_2, \ldots, \mathsf{W}_m\}$ of V such that*

(a) $1 \leqslant \dim(\mathsf{W}_i) \leqslant 2$ *for* $i = 1, 2, \ldots, m$.

(b) $\mathsf{V} = \mathsf{W}_1 \oplus \mathsf{W}_2 \oplus \cdots \oplus \mathsf{W}_m$.

Proof. The proof is by induction on $\dim(\mathsf{V})$. If $\dim(\mathsf{V}) = 1$, the result is obvious. So assume that the result is true whenever $\dim(\mathsf{V}) < n$ for some fixed integer $n > 1$.

Suppose $\dim(\mathsf{V}) = n$. By the lemma there is a T-invariant subspace W_1 of V such that $1 \leqslant \dim(\mathsf{W}_1) \leqslant 2$. If $\mathsf{W}_1 = \mathsf{V}$, the result is established. Otherwise, $\mathsf{W}_1^\perp \neq \{0\}$. By Exercise 13 $\mathsf{W}_1^\perp$ is T-invariant and the restriction of T to $\mathsf{W}_1^\perp$ is orthogonal. Since $\dim(\mathsf{W}_1^\perp) < n$, we may apply the induction hypothesis to $\mathsf{T}_{\mathsf{W}_1^\perp}$ and conclude that there exists a collection of pairwise orthogonal T-invariant subspaces $\{\mathsf{W}_2, \mathsf{W}_3, \ldots, \mathsf{W}_m\}$ of $\mathsf{W}_1^\perp$ such that $1 \leqslant \dim(\mathsf{W}_i) \leqslant 2$ for $i = 2, 3, \ldots, m$ and $\mathsf{W}_1^\perp = \mathsf{W}_2 \oplus \mathsf{W}_3 \oplus \cdots \oplus \mathsf{W}_m$. Thus $\{\mathsf{W}_1, \mathsf{W}_2, \ldots, \mathsf{W}_m\}$ is pairwise orthogonal and by Exercise 12(d) of Section 6.2

$$\mathsf{V} = \mathsf{W}_1 \oplus \mathsf{W}_1^\perp = \mathsf{W}_1 \oplus \mathsf{W}_2 \oplus \cdots \oplus \mathsf{W}_m. \quad ∎$$

Applying Example 1 and Theorem 6.38 in the context of Theorem 6.39, we can conclude that the restriction of T to W_i is either a rotation or a reflection for each $i = 1, 2, \ldots, m$. Thus in some sense T is made up of rotations and reflections. Unfortunately, very little can be said about the uniqueness of the decomposition of V in Theorem 6.39. For example, the W_i's, the number m of W_i's, and the number of W_i's for which $\mathsf{T}_{\mathsf{W}_i}$ is a reflection are not unique. Although the number of W_i's for which $\mathsf{T}_{\mathsf{W}_i}$ is a reflection is not unique, whether this number is even or odd is an intrinsic property of T. Moreover, we can always decompose V so that $\mathsf{T}_{\mathsf{W}_i}$ is a reflection for at most one W_i. These facts are established in the following result.

Theorem 6.40. *Let $\mathsf{T}, \mathsf{V}, \mathsf{W}_1, \ldots, \mathsf{W}_m$ be as in Theorem 6.39.*

(a) *The number of* i's *for which $\mathsf{T}_{\mathsf{W}_i}$ is a reflection is even or odd according to whether* $\det(\mathsf{T}) = 1$ *or* $\det(\mathsf{T}) = -1$.

(b) *It is always possible to decompose* V *as in Theorem 6.39 so that the number of* i's *for which* T_{W_i} *is a reflection is zero or one according to whether det*(T) = 1 *or det*(T) = −1. *Furthermore, if* T_{W_i} *is a reflection, then* $\dim(W_i) = 1$.

Proof. (a) Let r denote the number of W_i's in the decomposition for which T_{W_i} is a reflection. Then by Exercise 14

$$\det(T) = \det(T_{W_1}) \cdot \det(T_{W_2}) \cdot \cdots \cdot \det(T_{W_m}) = (-1)^r,$$

proving (a).

(b) Let $E = \{x \in V : T(x) = -x\}$; then E is a T-invariant subspace of V. If $W = E^\perp$, then W is T-invariant. So by applying Theorem 6.39 to T_W, we obtain a collection of pairwise orthogonal T-invariant subspaces $\{W_1, W_2, \ldots, W_k\}$ of W such that $1 \leqslant \dim(W_i) \leqslant 2$ for $1 \leqslant i \leqslant k$ and $W = W_1 \oplus W_2 \oplus \cdots \oplus W_k$. Observe that, for each $i = 1, 2, \ldots, k$, T_{W_i} is a rotation. For otherwise, if T_{W_i} is a reflection, there exists a nonzero $x \in W_i$ for which $T(x) = -x$. But then $x \in W_i \cap E \subseteq E^\perp \cap E = \{0\}$, a contradiction. If $E = \{0\}$, the result follows. Otherwise, choose an orthonormal basis β for E containing p elements ($p > 0$). It is possible to decompose β into a pairwise disjoint union $\beta = \beta_1 \cup \beta_2 \cup \cdots \cup \beta_r$ such that each β_i contains exactly two elements for $i < r$, and β_r contains two elements if p is even and one element if p is odd. For each $i = 1, 2, \ldots, r$, let $W_{k+i} = \text{span}(\beta_i)$. Then clearly, $\{W_1, W_2, \ldots, W_k, \ldots, W_{k+r}\}$ is pairwise orthogonal and

$$V = W_1 \oplus W_2 \oplus \cdots \oplus W_k \oplus \cdots \oplus W_{k+r}. \tag{27}$$

Moreover, if any β_i contains two elements, then

$$\det(T_{W_{k+i}}) = \det([T_{W_{k+i}}]_{\beta_i}) = \det\begin{pmatrix} -1 & 0 \\ 0 & -1 \end{pmatrix} = 1.$$

So $T_{W_{k+i}}$ is a rotation, and hence T_{W_j} is a rotation for $j < k + r$. If β_r consists of one element, then $\dim(W_{k+r}) = 1$ and

$$\det(T_{W_{k+r}}) = \det([T_{W_{k+r}}]_{\beta_r}) = \det(-1) = -1.$$

Thus $T_{W_{k+r}}$ is a reflection by Theorem 6.39, and we conclude that the decomposition in (27) satisfies the condition of part (b). ∎

As a consequence of the preceding theorem, an orthogonal operator can be factored as a product of rotations and reflections.

Corollary. *Let* T *be an orthogonal operator on a finite-dimensional real inner product space* V. *Then there exists a collection* $\{T_1, T_2, \ldots, T_m\}$ *of orthogonal operators on* V *such that*

(a) *For each* i, T_i *is either a reflection or a rotation.*

(b) *For at most one* i, T_i *is a reflection.*
(c) $\mathsf{T}_i\mathsf{T}_j = \mathsf{T}_j\mathsf{T}_i$ *for all* i *and* j.
(d) $\mathsf{T} = \mathsf{T}_1\mathsf{T}_2 \cdots \mathsf{T}_m$.
(e) $det(\mathsf{T}) = \begin{cases} 1 & \text{if } \mathsf{T}_i \text{ is a rotation for each } i \\ -1 & \text{otherwise.} \end{cases}$

Proof. As in the proof of part (b) of Theorem 6.40 we can write

$$\mathsf{V} = \mathsf{W}_1 \oplus \mathsf{W}_2 \oplus \cdots \oplus \mathsf{W}_m,$$

where $\mathsf{T}_{\mathsf{W}_i}$ is a rotation for $i < m$. For each $i = 1, 2, \ldots, m$, define $\mathsf{T}_i\colon \mathsf{V} \to \mathsf{V}$ by

$$\mathsf{T}_i(x_1 + \cdots + x_m) = x_1 + \cdots + x_{i-1} + \mathsf{T}(x_i) + x_{i+1} + \cdots + x_m,$$

where $x_j \in \mathsf{W}_j$ for all j. It is easily shown that each T_i is an orthogonal operator on V. In fact, T_i is a rotation or a reflection according to whether $\mathsf{T}_{\mathsf{W}_i}$ is a rotation or a reflection. This establishes (a) and (b). The proofs of (c), (d), and (e) are left as exercises (see Exercise 15). ∎

Example 3

Orthogonal Operators on a Three-Dimensional Real Inner Product Space

Let T be an orthogonal operator on a three-dimensional real inner product space V. We will show that T can be decomposed into the composition of a rotation and at most one reflection. Let

$$\mathsf{V} = \mathsf{W}_1 \oplus \mathsf{W}_2 \oplus \cdots \oplus \mathsf{W}_m$$

be a decomposition as in Theorem 6.40(b). Clearly, $m = 2$ or $m = 3$.

If $m = 2$, then $\mathsf{V} = \mathsf{W}_1 \oplus \mathsf{W}_2$. Without loss of generality, suppose that $\dim(\mathsf{W}_1) = 1$ and $\dim(\mathsf{W}_2) = 2$. Thus $\mathsf{T}_{\mathsf{W}_1}$ is a reflection or the identity on W_1, and $\mathsf{T}_{\mathsf{W}_2}$ is a rotation. Defining T_1 and T_2 as in the proof of the corollary to Theorem 6.40, we have that $\mathsf{T} = \mathsf{T}_1\mathsf{T}_2$ is the composition of a rotation and at most one reflection. (Note that if $\mathsf{T}_{\mathsf{W}_1}$ is not a reflection, then T_1 is the identity on V and $\mathsf{T} = \mathsf{T}_2$.)

If $m = 3$, then $\mathsf{V} = \mathsf{W}_1 \oplus \mathsf{W}_2 \oplus \mathsf{W}_3$ and $\dim(\mathsf{W}_i) = 1$ for all i. For each i, let T_i be as in the proof of the corollary to Theorem 6.40. If $\mathsf{T}_{\mathsf{W}_i}$ is not a reflection, then T_i is the identity on W_i. Otherwise, T_i is a reflection. Since $\mathsf{T}_{\mathsf{W}_i}$ is a reflection for at most one i, we conclude that T is either a single reflection or the identity (a rotation). ∎

EXERCISES

1. Label the following statements as being true or false. Assume for the following that the underlying vector spaces are finite-dimensional real inner product spaces.

(a) Any orthogonal operator is either a rotation or a reflection.
(b) The composition of any two rotations on a two-dimensional space is a rotation.
(c) The composition of any two rotations on a three-dimensional space is a rotation.
(d) The composition of any two rotations on a four-dimensional space is a rotation.
(e) The identity operator is a rotation.
(f) The composition of two reflections is a reflection.
(g) Any orthogonal operator is a composition of rotations.
(h) For any orthogonal operator T, if $\det(\mathsf{T}) = -1$, then T is a reflection.
(i) Reflections always have eigenvalues.
(j) Rotations always have eigenvalues.

2. Prove that rotations, reflections, and compositions of rotations and reflections are orthogonal operators.

3. Let

$$A = \begin{pmatrix} \frac{1}{2} & \frac{\sqrt{3}}{2} \\ \frac{\sqrt{3}}{2} & -\frac{1}{2} \end{pmatrix} \quad \text{and} \quad B = \begin{pmatrix} 1 & 0 \\ 0 & -1 \end{pmatrix}.$$

(a) Prove that $\mathsf{L}_A: \mathsf{R}^2 \to \mathsf{R}^2$ is a reflection.
(b) Find the axis in R^2 about which L_A reflects, i.e., the subspace of R^2 on which L_A acts as the identity.
(c) Prove that L_{AB} and L_{BA} are rotations.

4. For any real number ϕ, let

$$A = \begin{pmatrix} \cos\phi & \sin\phi \\ \sin\phi & -\cos\phi \end{pmatrix}.$$

(a) Prove that L_A is a reflection.
(b) Find the axis in R^2 about which L_A reflects.

5. For any real number ϕ, define $\mathsf{T}_\phi = \mathsf{L}_A$, where

$$A = \begin{pmatrix} \cos\phi & -\sin\phi \\ \sin\phi & \cos\phi \end{pmatrix}.$$

(a) Prove that any rotation on R^2 is of the form T_ϕ for some ϕ.
(b) Prove that $\mathsf{T}_\phi \mathsf{T}_\psi = \mathsf{T}_{(\phi+\psi)}$ for any $\phi, \psi \in R$.
(c) Deduce that any two rotations on R^2 commute.

6. Prove that the composition of any two rotations on R^3 is a rotation on R^3.

7. Given real numbers ϕ and ψ, define matrices

$$A = \begin{pmatrix} 1 & 0 & 0 \\ 0 & \cos\phi & -\sin\phi \\ 0 & \sin\phi & \cos\phi \end{pmatrix} \quad \text{and} \quad B = \begin{pmatrix} \cos\psi & -\sin\psi & 0 \\ \sin\psi & \cos\psi & 0 \\ 0 & 0 & 1 \end{pmatrix}.$$

(a) Prove that L_A and L_B are rotations.
(b) Prove that L_{AB} is a rotation.
(c) Find the axis of rotation for L_{AB}.

8. Prove that no orthogonal operator can be both a rotation and a reflection.

9. Prove that if V is a two- or three-dimensional real inner product space, then the composition of two reflections on V is a rotation of V.

10. Give an example of an orthogonal operator that is neither a reflection nor a rotation.

11. Let V be a finite-dimensional real inner product space. Define $\mathsf{T}\colon \mathsf{V} \to \mathsf{V}$ by $\mathsf{T}(x) = -x$. Prove that T is a product of rotations if and only if dim(V) is even.

12. Complete the proof of the lemma to Theorem 6.39 by showing that $\mathsf{W} = \phi_\beta^{-1}(\mathsf{Z})$ satisfies the required conditions.

13. Let T be an orthogonal [unitary] operator on a finite-dimensional real [complex] inner product space V. If W is a T-invariant subspace of V, prove that

(a) T_W is an orthogonal [unitary] operator on W.
(b) $\mathsf{W}^\perp$ is a T-invariant subspace of V. *Hint:* Use the fact that T_W is one-to-one and onto to conclude that, for any $y \in \mathsf{W}$, $\mathsf{T}^*(y) = \mathsf{T}^{-1}(y) \in \mathsf{W}$.
(c) $\mathsf{T}_{\mathsf{W}^\perp}$ is an orthogonal [unitary] operator on W.

14. Let T be a linear operator on a finite-dimensional vector space V. Suppose that V is a direct sum of T-invariant subspaces, say $\mathsf{V} = \mathsf{W}_1 \oplus \mathsf{W}_2 \oplus \cdots \oplus \mathsf{W}_k$. Prove that $\det(\mathsf{T}) = \det(\mathsf{T}_{\mathsf{W}_1})\cdot\det(\mathsf{T}_{\mathsf{W}_2}) \cdot \cdots \cdot \det(\mathsf{T}_{\mathsf{W}_k})$.

15. Complete the proof of the corollary to Theorem 6.40.

16. Let T be an orthogonal operator on an n-dimensional real inner product space V. Suppose that T is not the identity. Prove that

(a) If n is odd, then T can be expressed as the composition of at most one reflection and at most $\frac{1}{2}(n-1)$ rotations.
(b) If n is even, then T can be expressed as the composition of at most $\frac{1}{2}n$ rotations or as the composition of one reflection and at most $\frac{1}{2}(n-2)$ rotations.

17. Let V be a real inner product space of dimension 2. For any $x, y \in \mathsf{V}$ such that $x \neq y$ and $\|x\| = \|y\| = 1$, show that there exists a unique rotation T on V such that $\mathsf{T}(x) = y$.

INDEX OF DEFINITIONS FOR CHAPTER 6

7

Canonical Forms

As we learned in Chapter 5, the advantage of a diagonalizable linear operator lies in the simplicity of its description. Such an operator has a diagonal matrix representation or equivalently, there is a basis for the underlying vector space consisting of eigenvectors of the operator. However, not every linear operator is diagonalizable even if its characteristic polynomial splits. Example 3 of Section 5.2 describes such a linear operator.

It is the purpose of this chapter to consider alternative matrix representations for nondiagonalizable operators. These representations are called *canonical forms.* There are different kinds of canonical forms, and their advantages and disadvantages depend on how they are applied. The choice of a canonical form is determined by the appropriate choice of an ordered basis. Naturally, the canonical forms of a linear operator are not diagonal matrices if the linear operator is not diagonalizable.

In this chapter we treat the two most popular canonical forms. The first of these, the *Jordan canonical form,* requires that the characteristic polynomial of the operator splits. This form is always available if the underlying field is algebraically closed, that is, if every polynomial with coefficients from the field splits. The first two sections deal with this form. The *rational canonical form,* treated in Section 7.4, does not require such factorization.

7.1 GENERALIZED EIGENVECTORS

In the first two sections of this chapter we consider linear operators on finite-dimensional vector spaces for which the characteristic polynomials split. Such operators have at least one eigenvalue. If $\lambda_1, \lambda_2, \ldots, \lambda_n$ (not necessarily distinct) are all the eigenvalues of $\mathsf{T}: \mathsf{V} \to \mathsf{V}$, recall from Theorem 5.4 that T is diagonalizable if and only if there is an ordered basis for V consisting of eigenvectors of T. If $\beta = \{x_1, x_2, \ldots, x_n\}$ is such a basis in which x_j is an eigenvector corresponding to the eigenvalue λ_j, then

$$[\mathsf{T}]_\beta = \begin{pmatrix} \lambda_1 & 0 & \cdots & 0 \\ 0 & \lambda_2 & \cdots & 0 \\ \vdots & \vdots & & \vdots \\ 0 & 0 & \cdots & \lambda_n \end{pmatrix}.$$

Although not every linear operator T on V is diagonalizable, we will prove that for any linear operator whose characteristic polynomial splits, there exists an ordered basis β for V such that

$$[\mathsf{T}]_\beta = \begin{pmatrix} J_1 & O & \cdots & O \\ O & J_2 & \cdots & O \\ \vdots & \vdots & & \vdots \\ O & O & \cdots & J_k \end{pmatrix},$$

where J_i is a square matrix of the form (λ_j) or the form

$$\begin{pmatrix} \lambda_j & 1 & 0 & \cdots & 0 & 0 \\ 0 & \lambda_j & 1 & \cdots & 0 & 0 \\ \vdots & \vdots & \vdots & & \vdots & \vdots \\ 0 & 0 & 0 & \cdots & \lambda_j & 1 \\ 0 & 0 & 0 & \cdots & 0 & \lambda_j \end{pmatrix}$$

for some eigenvalue λ_j of T. Such a matrix J_i is called a *Jordan block* corresponding to λ_j, and the matrix $[\mathsf{T}]_\beta$ is called a *Jordan canonical form* of T. We also say that the ordered basis β is a *Jordan canonical basis* for T. Observe that each Jordan block J_i is "almost" a diagonal matrix—in fact, $[\mathsf{T}]_\beta$ is a diagonal matrix if and only if each J_i is of the form (λ_j).

Example 1

The 8×8 matrix

$$J = \left(\begin{array}{ccc|c|cc|cc} 2 & 1 & 0 & 0 & 0 & 0 & 0 & 0 \\ 0 & 2 & 1 & 0 & 0 & 0 & 0 & 0 \\ 0 & 0 & 2 & 0 & 0 & 0 & 0 & 0 \\ \hline 0 & 0 & 0 & 2 & 0 & 0 & 0 & 0 \\ \hline 0 & 0 & 0 & 0 & 3 & 1 & 0 & 0 \\ 0 & 0 & 0 & 0 & 0 & 3 & 0 & 0 \\ \hline 0 & 0 & 0 & 0 & 0 & 0 & 0 & 1 \\ 0 & 0 & 0 & 0 & 0 & 0 & 0 & 0 \end{array}\right)$$

is a Jordan canonical form of a linear operator $\mathsf{T}: \mathsf{C}^8 \to \mathsf{C}^8$; that is, there exists a basis $\beta = \{x_1, x_2, \ldots, x_8\}$ for C^8 such that $[\mathsf{T}]_\beta = J$. Notice that the characteristic polynomial for T and J is $\det(J - tI) = (t-2)^4(t-3)^2t^2$, and so the multiplicity of each eigenvalue is the number of times that eigenvalue appears on the diagonal of J. Also observe that of the vectors $x_1, x_2, \ldots, x_8$, only x_1, x_4, x_5, and x_7 (the basis vectors corresponding to the first column of each of the Jordan blocks) are eigenvectors of T. ∎

It will be proved that every operator whose characteristic polynomial splits has a unique Jordan canonical form (up to the order of the Jordan blocks). Nevertheless it is not the case that the Jordan canonical form is completely determined by the characteristic polynomial of the transformation. For example, the characteristic polynomial of

$$J' = \begin{pmatrix} 2 & 1 & 0 & 0 & 0 & 0 & 0 & 0 \\ 0 & 2 & 0 & 0 & 0 & 0 & 0 & 0 \\ 0 & 0 & 2 & 1 & 0 & 0 & 0 & 0 \\ 0 & 0 & 0 & 2 & 0 & 0 & 0 & 0 \\ 0 & 0 & 0 & 0 & 3 & 0 & 0 & 0 \\ 0 & 0 & 0 & 0 & 0 & 3 & 0 & 0 \\ 0 & 0 & 0 & 0 & 0 & 0 & 0 & 1 \\ 0 & 0 & 0 & 0 & 0 & 0 & 0 & 0 \end{pmatrix}$$

also is $(t-2)^4(t-3)^2t^2$.

Consider again the matrix J and the basis β of Example 1. Notice that $\mathsf{T}(x_2) = x_1 + 2x_2$ and therefore, $(\mathsf{T} - 2\mathsf{I})(x_2) = x_1$. Similarly, $(\mathsf{T} - 2\mathsf{I})(x_3) = x_2$. Therefore, the first three vectors of β can be written as

$$\{x_1, x_2, x_3\} = \{(\mathsf{T} - 2\mathsf{I})^2(x_3), (\mathsf{T} - 2\mathsf{I})(x_3), x_3\},$$

and of these, only $x_1 = (\mathsf{T} - 2\mathsf{I})^2(x_3)$ is an eigenvector. This pattern is repeated for the other Jordan blocks. For example, x_4 is the only vector of β corresponding to the second block that is an eigenvector. Also note that

$$\{x_5, x_6\} = \{(\mathsf{T} - 3\mathsf{I})(x_6), x_6\}.$$

Because of the very nature of a Jordan canonical form, this pattern must occur for any Jordan canonical basis of a linear operator. This fact gives us some information about the vectors of a Jordan canonical basis. If x lies in a Jordan canonical basis of a linear operator T and corresponds to a Jordan block with diagonal entry λ, then $(\mathsf{T} - \lambda\mathsf{I})^p(x) = 0$ for a sufficiently large p. For example, $(\mathsf{T} - 2\mathsf{I})^3(x_3) = 0$ and $(\mathsf{T} - 2\mathsf{I})^2(x_2) = 0$ for the vectors x_2 and x_3 in the basis β above. Eigenvectors always satisfy this condition for the case $p = 1$. It seems appropriate to identify the nonzero vectors that satisfy this more general condition.

Definition. *Let* T *be a linear operator on a finite-dimensional vector space* V. *A nonzero vector* x *in* V *is called a* generalized eigenvector *of* T *if there exists a* *scalar* λ *such that* $(\mathsf{T} - \lambda\mathsf{I})^p(x) = 0$ *for some positive integer* p. *We say that* x *is a* generalized eigenvector corresponding to λ.

If x is a generalized eigenvector of T corresponding to λ and p is the smallest integer for which $(\mathsf{T} - \lambda\mathsf{I})^p(x) = 0$, then $(\mathsf{T} - \lambda\mathsf{I})^{p-1}(x)$ is an eigenvector of T corresponding to λ. Therefore, λ is an eigenvalue of T.

It is important to identify the sequences of vectors from a Jordan canonical basis that correspond to a Jordan block. From the remarks made above, the following definitions will be useful.

Definitions. *Let* T *be a linear operator on a vector space* V, *and let* x *be a generalized eigenvector of* T *corresponding to the eigenvalue* λ. *If* p *denotes the smallest positive integer such that* $(\mathsf{T} - \lambda\mathsf{I})^p(x) = 0$, *then the ordered set*

$$\{(\mathsf{T} - \lambda\mathsf{I})^{p-1}(x), (\mathsf{T} - \lambda\mathsf{I})^{p-2}(x), \ldots, (\mathsf{T} - \lambda\mathsf{I})(x), x\}$$

is called a cycle *of generalized eigenvectors of* T *corresponding to* λ. *The elements* $(\mathsf{T} - \lambda\mathsf{I})^{p-1}(x)$ *and* x *are called the* initial vector *and the* end vector *of the cycle, respectively. We say that the* length *of the cycle is* p.

For example, recalling the matrix J and the basis β that we have been considering, $\beta_1 = \{x_1, x_2, x_3\}$, $\beta_2 = \{x_4\}$, $\beta_3 = \{x_5, x_6\}$, and $\beta_4 = \{x_7, x_8\}$ are the cycles of generalized eigenvectors of T that occur in β. Notice that β is a disjoint union of these cycles. Theorem 7.1 summarizes these observations and asserts, among other things, that cycles of generalized eigenvectors are always linearly independent.

Theorem 7.1. *Let* T *be a linear operator on* V, *and let* γ *be a cycle of generalized eigenvectors of* T *corresponding to the eigenvalue* λ.

(a) *The initial vector of* γ *is an eigenvector of* T *corresponding to the eigenvalue* λ, *and no other member of* γ *is an eigenvector of* T.
(b) γ *is linearly independent.*
(c) *Let* β *be an ordered basis for* V. *Then* β *is a Jordan canonical basis for* V *if and only if* β *is a disjoint union of cycles of generalized eigenvectors of* T.

Proof. We prove only (b); the proofs of (a) and (c) are left as exercises. The proof is by induction on the length of the cycle γ. If γ has length 1, then $\gamma = \{x_1\}$ is linearly independent since x_1, a generalized eigenvector, is a nonzero vector. Now assume that cycles of length $k - 1$ are linearly independent for some integer $k - 1 \geqslant 1$. Suppose that $\gamma = \{x_1, x_2, \ldots, x_k\}$ is a cycle of generalized eigenvectors corresponding to the eigenvalue λ and that

$$\sum_{i=1}^{k} a_i x_i = 0$$

for some scalars $a_1, a_2, \ldots, a_k$. Applying $\mathsf{T} - \lambda\mathsf{I}$ to the equation above gives

$$\sum_{i=2}^{k} a_i x_{i-1} = 0.$$

But the sum in the preceding equality is a linear combination of elements from a cycle $\{x_1, x_2, \ldots, x_{k-1}\}$ of length $k - 1$. Hence $a_i = 0$ for $i = 2, 3, \ldots, k$. Thus

$$\sum_{i=1}^{k} a_i x_i = 0$$

reduces to $a_1 x_1 = 0$. But since $x_1 \neq 0$, it follows that $a_1 = 0$. So $a_1 = a_2 = \cdots = a_k = 0$, proving that γ is linearly independent. This completes the induction. ∎

We have seen that eigenspaces play a key role in the study of the diagonalization problem. Recall that such a space is spanned by the set of all eigenvectors corresponding to a fixed eigenvalue. For our purposes it is useful to consider the space spanned by the set of all generalized eigenvectors corresponding to a single eigenvalue. Jordan canonical bases are obtained by selecting bases from these spaces in appropriate ways. The following is one way of defining such a space.

Definition. *Let λ be an eigenvalue of a linear operator T on a vector space V. The* generalized eigenspace *of T corresponding to λ, denoted by K_λ, is the set*

$$\mathsf{K}_\lambda = \{x \in \mathsf{V} \colon (\mathsf{T} - \lambda\mathsf{I})^p(x) = 0 \textit{ for some positive integer } p\}.$$

Note that K_λ consists of the zero vector and all the generalized eigenvectors corresponding to λ.

Recall that a subspace W of V is T-invariant if $\mathsf{T}(\mathsf{W}) \subseteq \mathsf{W}$.

Theorem 7.2. *Let λ be an eigenvalue of a linear operator T on a vector space V. Then K_λ is a T-invariant subspace of V containing E_λ (the eigenspace of T corresponding to λ).*

Proof. Clearly, $0 \in \mathsf{K}_\lambda$. Suppose that x and y are in K_λ. Then there exist positive integers p and q such that

$$(\mathsf{T} - \lambda\mathsf{I})^p(x) = 0 \quad \text{and} \quad (\mathsf{T} - \lambda\mathsf{I})^q(y) = 0.$$

Therefore,

$$\begin{aligned}(\mathsf{T} - \lambda\mathsf{I})^{p+q}(x + y) &= (\mathsf{T} - \lambda\mathsf{I})^{p+q}(x) + (\mathsf{T} - \lambda\mathsf{I})^{p+q}(y)\\ &= (\mathsf{T} - \lambda\mathsf{I})^q(0) + (\mathsf{T} - \lambda\mathsf{I})^p(0)\\ &= 0.\end{aligned}$$

By means of a simple calculation it can be shown that $cx \in \mathsf{K}_\lambda$ for any scalar c. Hence K_λ is a subspace of V.

To show that K_λ is T-invariant, consider any $x \in \mathsf{K}_\lambda$. Choose a positive integer p such that $(\mathsf{T} - \lambda\mathsf{I})^p(x) = 0$. Then

$$\begin{aligned}(\mathsf{T} - \lambda\mathsf{I})^p\mathsf{T}(x) &= \mathsf{T}(\mathsf{T} - \lambda\mathsf{I})^p(x)\\ &= \mathsf{T}(0)\\ &= 0.\end{aligned}$$

Therefore, $\mathsf{T}(x) \in \mathsf{K}_\lambda$.

Finally, it is a simple observation that E_λ is contained in K_λ. ∎

We will produce Jordan canonical bases by collecting cycles of generalized eigenvectors from the generalized eigenspaces, but we must take care that the resulting collections are linearly independent. The next lemma and the accompanying Theorem 7.3 localize the problem of guaranteeing linear independence to the individual generalized eigenspaces. Compare these results to Theorem 5.13 and the lemma directly preceding it in Section 5.2.

Lemma. *Let T be a linear operator on a vector space V and let $\lambda_1, \lambda_2, \ldots, \lambda_k$ be distinct eigenvalues of T. For each $i = 1, 2, \ldots, k$, let $x_i \in \mathsf{K}_{\lambda_i}$, the generalized eigenspace corresponding to λ_i. If*

$$x_1 + x_2 + \cdots + x_k = 0, \tag{1}$$

then $x_i = 0$ for all i.

Proof. We prove the lemma by mathematical induction on k, the number of distinct eigenvalues. The result is trivial for $k = 1$. Assume that the lemma holds for any $k - 1$ distinct eigenvalues, where $k - 1 \geqslant 1$, and suppose that we have k distinct eigenvalues $\lambda_1, \lambda_2, \ldots, \lambda_k$ with vectors $x_i \in \mathsf{K}_{\lambda_i}$ satisfying (1). First, suppose that $x_1 \neq 0$. For each i, $1 \leqslant i \leqslant k$, let p_i be the smallest positive integer for which $(\mathsf{T} - \lambda_i\mathsf{I})^{p_i}(x_i) = 0$. Let $z = (\mathsf{T} - \lambda_1\mathsf{I})^{p_1 - 1}(x_1)$. Then z is an eigenvector of T corresponding to λ_1. Let $f(t)$ be the polynomial defined by

$$f(t) = (t - \lambda_2)^{p_2} \cdots (t - \lambda_k)^{p_k}.$$

Then $f(\mathsf{T})$ is the linear operator given by

$$f(\mathsf{T}) = (\mathsf{T} - \lambda_2\mathsf{I})^{p_2} \cdots (\mathsf{T} - \lambda_k\mathsf{I})^{p_k}.$$

Clearly, $f(\mathsf{T})(x_i) = 0$ for all $i > 1$. With the aid of Exercise 22 of Section 5.1, we have

$$\begin{aligned}f(\mathsf{T})(z) &= f(\lambda_1)z\\ &= [(\lambda_1 - \lambda_2)^{p_2} \cdots (\lambda_1 - \lambda_k)^{p_k}]z\\ &\neq 0.\end{aligned}$$

Furthermore,

$$\begin{aligned} f(\mathsf{T})(z) &= f(\mathsf{T})(\mathsf{T} - \lambda_1 \mathsf{I})^{p_1 - 1}(x_1) \\ &= (\mathsf{T} - \lambda_1 \mathsf{I})^{p_1 - 1} f(\mathsf{T})(x_1) \\ &= -(\mathsf{T} - \lambda_1 \mathsf{I})^{p_1 - 1}[f(\mathsf{T})(x_2 + \cdots + x_k)] \qquad \text{by (1)} \\ &= 0. \end{aligned}$$

This is a contraction. So it must be that $x_1 = 0$, and (1) becomes

$$x_2 + \cdots + x_k = 0.$$

Therefore, by the induction hypothesis we have that $x_i = 0$ for $2 \leqslant i \leqslant k$, and this establishes the lemma for k distinct eigenvalues. The lemma now follows. ∎

Theorem 7.3. *Let* T *be a linear operator on a finite-dimensional vector space* V *with distinct eigenvalues* $\lambda_1, \lambda_2, \ldots, \lambda_k$. *For each* $i = 1, 2, \ldots, k$, *let* S_i *be a linearly independent subset of* K_{λ_i}. *Then* $S_i \cap S_j = \varnothing$ *for* $i \neq j$, *and* $S = S_1 \cup S_2 \cup \cdots \cup S_k$ *is a linearly independent subset of* V.

Proof. Suppose that $x \in S_i \cap S_j$ for some $i \neq j$. Let $y = -x$. Then $x \in K_{\lambda_i}$, $y \in K_{\lambda_j}$, and $x + y = 0$. By the lemma $x = 0$, contrary to the fact that x lies in a linearly independent set. Thus $S_i \cap S_j = \varnothing$.

Now suppose that for each i,

$$S_i = \{x_{i1}, x_{i2}, \ldots, x_{in_i}\}.$$

Then $S = \{x_{ij}\colon 1 \leqslant j \leqslant n_i,\ 1 \leqslant i \leqslant k\}$. Consider any scalars $\{a_{ij}\}$ such that

$$\sum_{i=1}^{k} \sum_{j=1}^{n_i} a_{ij} x_{ij} = 0.$$

For each i let

$$y_i = \sum_{j=1}^{n_i} a_{ij} x_{ij}.$$

Then $y_i \in K_{\lambda_i}$ for each i and $y_1 + y_2 + \cdots + y_k = 0$. Therefore, by the lemma, $y_i = 0$ for all i. But S_i is linearly independent for all i. Thus, for each i, it follows that $a_{ij} = 0$ for all j. We conclude that S is linearly independent. ∎

The next step in our program to produce Jordan canonical bases is to describe a method for choosing cycles of generalized eigenvectors from a generalized eigenspace so that the union of the cycles is linearly independent. Notice that the initial vector of each cycle is an eigenvector. Trivially, if the union of the cycles is linearly independent, then the eigenvectors form a linearly independent subset. This observation provides us with a key to the solution.

Theorem 7.4. *Let* T *be a linear operator on a finite-dimensional vector space* V, *and for each* i $(1 \leqslant i \leqslant q)$ *let* Z_i *be a cycle of generalized eigenvectors of* T *corresponding to the eigenvalue λ and having initial vector* y_i. *If the* y_i*'s are distinct and the set* $\{y_1, y_2, \ldots, y_q\}$ *is linearly independent, then the* Z_i*'s are disjoint (no two distinct cycles have any elements in common) and*

$$Z = \bigcup_{i=1}^{q} Z_i$$

is linearly independent.

Proof. That the Z_i's are disjoint follows by Exercise 5 and the fact that the y_i's are distinct.

The proof that Z is linearly independent is by mathematical induction on the number of vectors in Z. If this number is 1, then the result is trivial. Assume that for some integer $n > 1$ the result is valid whenever Z has fewer than n vectors. Now suppose that Z has exactly n vectors. Let W be the subspace of V generated by Z. Clearly, W is $(\mathsf{T} - \lambda\mathsf{I})$-invariant, and $\dim(\mathsf{W}) \leqslant n$. Let U denote the restriction of $\mathsf{T} - \lambda\mathsf{I}$ to W.

For each i let Z_i' denote the cycle obtained from Z_i by deleting the end vector. Each vector of Z_i' is the image under U of a vector of Z_i, and conversely every nonzero image under U of a vector of Z_i is contained in Z_i'. Let

$$Z' = \bigcup_{i=1}^{q} Z_i'.$$

Then by the last statement, Z' generates the range R(U). Furthermore, Z' consists of $n - q$ vectors, and the set of initial vectors of the Z_i''s are also initial vectors of the Z_i's. Thus we may apply the induction hypothesis to conclude that Z' is linearly independent. Therefore, Z' is a basis for R(U). Hence $\dim(\mathsf{R}(\mathsf{U})) = n - q$. Since $\{y_1, y_2, \ldots, y_q\}$ is a linearly independent subset of N(U), we have that $\dim(\mathsf{N}(\mathsf{U})) \geqslant q$. From these inequalities and the dimension theorem we obtain

$$\begin{aligned} n &\geqslant \dim(\mathsf{W}) \\ &= \dim(\mathsf{R}(\mathsf{U})) + \dim(\mathsf{N}(\mathsf{U})) \\ &\geqslant (n - q) + q \\ &= n. \end{aligned}$$

We conclude that $\dim(\mathsf{W}) = n$. Since Z generates W and consists of n vectors, it must be a basis for W, and hence Z is linearly independent. ∎

We are now prepared for the principal theorem of this section.

Theorem 7.5. *Let* T *be a linear operator on an* n*-dimensional vector space* V *such that the characteristic polynomial of* T *splits. Then there exists a Jordan canonical basis for* T*; that is, there exists an ordered basis* β *for* V *that is a disjoint union of cycles of generalized eigenvectors of* T.

Proof. The proof is by induction on n. Clearly, the result is true for $n = 1$. Assume that the conclusion is valid for any vector space of dimension less than a fixed integer $n > 1$, and suppose that $\dim(\mathsf{V}) = n$.

Choose an eigenvalue λ_1 of T, and let $r = \operatorname{rank}(\mathsf{T} - \lambda_1\mathsf{I})$. Then $\mathsf{R}(\mathsf{T} - \lambda_1\mathsf{I})$ is an r-dimensional T-invariant subspace of V, and $r < n$. Let U be the restriction of T to $\mathsf{R}(\mathsf{T} - \lambda_1\mathsf{I})$. The characteristic polynomial of U divides the characteristic polynomial of T (Theorem 5.26) and hence also splits. Thus we may apply the induction hypothesis to obtain a Jordan canonical basis α for U. We wish to extend α to a Jordan canonical basis for T. For each i let S_i consist of the generalized eigenvectors in α corresponding to λ_i. Then S_i is a linearly independent set that is a disjoint union of cycles of generalized eigenvectors of λ_i. Let $Z_1, Z_2, \ldots, Z_p$ be the disjoint cycles whose union is S_1. (It is possible that $p = 0$.) For each cycle Z_i, let $Z_i' = \{y_i\} \cup Z_i$, where y_i is a vector in V such that $(\mathsf{T} - \lambda_1\mathsf{I})(y_i)$ is the end vector of Z_i. Such a y_i always exists because $Z_i \subseteq \mathsf{R}(\mathsf{T} - \lambda_1\mathsf{I})$. Then Z_i' is also a cycle of generalized eigenvectors of T corresponding to λ_1. For each i let z_i denote the initial vector of Z_i. Then $\{z_1, z_2, \ldots, z_p\}$ is a linearly independent subset of $\mathsf{N}(\mathsf{T} - \lambda_1\mathsf{I})$, and this set can be extended to a basis

$$\{z_1, z_2, \ldots, z_p, z_{p+1}, \ldots, z_{n-r}\}$$

for $\mathsf{N}(\mathsf{T} - \lambda_1\mathsf{I})$. If $p < n - r$, then let $Z_i' = \{z_i\}$ for $p < i \leqslant n - r$. Then Z_1', $Z_2', \ldots, Z_{n-r}'$ is a collection of disjoint cycles of generalized eigenvectors corresponding to λ_1. Let S_1' denote the union of this collection. Since the initial vectors of these cycles form a linearly independent set, we have that S_1' is linearly independent by Theorem 7.4. Notice that S_1' is obtained from S_1 by adjoining $n - r$ vectors. Let β be defined by

$$\beta = S_1' \cup S_2 \cup \cdots \cup S_k.$$

Then, β is obtained from α by adjoining $n - r$ vectors. Since α consists of r vectors, we have that β consists of n vectors. Furthermore, β is linearly independent by Theorem 7.3. We conclude that β is a Jordan canonical basis for T. ∎

In the proof of Theorem 7.5 the cycles $\{Z_i'\}$ were constructed so that the set of their initial vectors

$$\{z_1, z_2, \ldots, z_p, z_{p+1}, \ldots, z_{n-r}\}$$

is a basis for $\mathsf{N}(\mathsf{T} - \lambda_1\mathsf{I})$. Thus, in the context of the construction in the proof of Theorem 7.5, the initial vectors of the cycles in a Jordan canonical basis that correspond to an eigenvalue λ form a basis for E_λ, and therefore the number of cycles corresponding to λ equals $\dim(\mathsf{E}_\lambda)$. In the next section we will show (Theorem 7.8) that these relations are true for any Jordan canonical basis.

Having established the existence of a Jordan canonical form, we can now investigate the connection between the generalized eigenspaces and the characteristic polynomial of an operator. It is useful to compare this next theorem to Theorem 5.14 in Section 5.2.

Theorem 7.6. *Let* T *be a linear operator on a finite-dimensional vector space* V *such that the characteristic polynomial of* T *splits. Suppose that* $\lambda_1, \lambda_2, \ldots, \lambda_k$ *are the distinct eigenvalues of* T *with corresponding multiplicities* $m_1, m_2, \ldots, m_k$. *Then:*

(a) $\dim(K_{\lambda_i}) = m_i$ *for all* i.

(b) *If for each* i, S_i *is a basis for* K_{λ_i}, *then the union* $S = S_1 \cup S_2 \cup \cdots \cup S_k$ *is a basis for* V.

(c) *If* β *is a Jordan canonical basis for* T, *then for each* i, $\beta_i = \beta \cap K_{\lambda_i}$ *is a basis for* K_{λ_i}.

(d) $K_{\lambda_i} = N((T - \lambda_i I)^{m_i})$ *for all* i.

(e) T *is diagonalizable if and only if* $E_{\lambda_i} = K_{\lambda_i}$ *for all* i.

Proof. (a), (b), and (c) We prove the first three parts simultaneously. Let β be a Jordan canonical basis for T, and let $J = [T]_\beta$. For each i let S_i and β_i be as in (b) and (c), $d_i = \dim(K_{\lambda_i})$, and $n = \dim(V)$. For each i, the vectors in β_i are in one-to-one correspondence with the columns of J that contain λ_i as the diagonal entry. Since J is an upper triangular matrix, the number of occurrences of λ_i on the diagonal is m_i. Therefore, β_i consists of m_i vectors. Since β_i is a linearly independent subset of K_{λ_i}, we have that $m_i \leqslant d_i$ for all i. By Theorem 7.3 the sets S_i are pairwise disjoint, and their union S is linearly independent. We summarize all of this by

$$n = \sum_{i=1}^{k} m_i \leqslant \sum_{i=1}^{k} d_i \leqslant n,$$

from which it follows that $m_i = d_i$ for all i and

$$\sum_{i=1}^{k} d_i = n.$$

The last equality tells us that S contains n vectors. Therefore, since S is linearly independent, it is a basis for V. Because $m_i = d_i$, we have that each β_i is a basis for K_{λ_i}. Thus we have established (a), (b), and (c).

(d) Clearly, $N((T - \lambda_i I)^{m_i}) \subseteq K_{\lambda_i}$. Suppose that $x \in K_{\lambda_i}$. Then the cycle Z with end vector x is a linearly independent subset of K_{λ_i}, by Theorem 7.1. Since $\dim(K_{\lambda_i}) = m_i$ by (a), it follows that the length of Z cannot exceed m_i; that is, $(T - \lambda_i I)^p(x) = 0$ for $p \geqslant m_i$. Therefore, $x \in N((T - \lambda_i I)^{m_i})$, proving that $N((T - \lambda_i I)^{m_i}) = K_{\lambda_i}$.

(e) If T is diagonalizable, then by Theorem 5.14 we have that $m_i = \dim(E_{\lambda_i})$ for all i. Since E_{λ_i} is a subspace of K_{λ_i} for all i, we have that the two spaces are equal by part (a). Conversely, if $K_{\lambda_i} = E_{\lambda_i}$ for all i, then $\dim(E_{\lambda_i}) = m_i$ for all i by part (a), and T is diagonalizable by Theorem 5.14. ∎

Example 2

Let $\mathsf{T}: \mathsf{C}^3 \to \mathsf{C}^3$ be defined by $\mathsf{T} = \mathsf{L}_A$, where

$$A = \begin{pmatrix} 3 & 1 & -2 \\ -1 & 0 & 5 \\ -1 & -1 & 4 \end{pmatrix}.$$

We will find a basis for each eigenspace and each generalized eigenspace of T.

The characteristic polynomial of T is

$$f(t) = \det(A - tI) = -(t-3)(t-2)^2.$$

Hence $\lambda_1 = 3$ and $\lambda_2 = 2$ are the eigenvalues of T having multiplicities 1 and 2, respectively. By Theorem 7.6, K_{λ_1} has dimension 1, K_{λ_2} has dimension 2, $\mathsf{K}_{\lambda_1} = \mathsf{N}(\mathsf{T} - 3\mathsf{I})$, and $\mathsf{K}_{\lambda_2} = \mathsf{N}((\mathsf{T} - 2\mathsf{I})^2)$. Now $\mathsf{E}_{\lambda_1} = \mathsf{N}(\mathsf{T} - 3\mathsf{I})$ and $\mathsf{E}_{\lambda_2} = \mathsf{N}(\mathsf{T} - 2\mathsf{I})$. Hence $\mathsf{E}_{\lambda_1} = \mathsf{K}_{\lambda_1}$. Since

$$\left\{ \begin{pmatrix} -1 \\ 2 \\ 1 \end{pmatrix} \right\}$$

is a basis for the eigenspace for $\lambda_1 = 3$, it is also a basis for the generalized eigenspace K_{λ_1}.

Now K_{λ_2} has dimension 2 and has a basis consisting of a union of cycles. Hence this basis either is a union of two cycles of length one or consists of one cycle of length 2. The former is impossible because the vector of a cycle of length one is an eigenvector, and hence the resulting basis would be a basis of eigenvectors—contradicting the fact that $\dim(\mathsf{E}_{\lambda_2}) = 1$, which can be easily verified. Therefore, the desired basis is a single cycle of length 2. A vector v is the end vector of such a cycle if and only if $(A - 2I)v \neq 0$, but $(A - 2I)^2 v = 0$. It can be shown that

$$\left\{ \begin{pmatrix} 1 \\ -3 \\ -1 \end{pmatrix} \right\}$$

is a basis for the solution space of the homogeneous system $(A - 2I)X = 0$, and

$$\left\{ \begin{pmatrix} 1 \\ -3 \\ -1 \end{pmatrix}, \begin{pmatrix} -1 \\ 2 \\ 0 \end{pmatrix} \right\}$$

is a basis for the solution space of the homogeneous system $(A - 2I)^2 X = 0$.

Thus the vector

$$\begin{pmatrix} -1 \\ 2 \\ 0 \end{pmatrix}$$

is an acceptable candidate for v. Notice that

$$(A - 2I)v = \begin{pmatrix} 1 \\ -3 \\ -1 \end{pmatrix},$$

and therefore the cycle we have chosen coincides with the basis for the homogeneous system $(A - 2I)^2 X = 0$ given above. It follows that

$$\beta = \left\{ \begin{pmatrix} -1 \\ 2 \\ 1 \end{pmatrix}, \begin{pmatrix} 1 \\ -3 \\ -1 \end{pmatrix}, \begin{pmatrix} -1 \\ 2 \\ 0 \end{pmatrix} \right\}$$

is a basis for C^3, and

$$[\mathsf{T}]_\beta = \begin{pmatrix} 3 & 0 & 0 \\ 0 & 2 & 1 \\ 0 & 0 & 2 \end{pmatrix}$$

is a Jordan canonical form for T. ∎

Example 3

Let $\mathsf{T}\colon \mathsf{P}_2(C) \to \mathsf{P}_2(C)$ be defined by $\mathsf{T}(f) = -f - f'$. We will find a basis for each eigenspace and generalized eigenspace of T. If $\beta = \{1, z, z^2\}$, then β is an ordered basis for $\mathsf{P}_2(C)$ and

$$A = [\mathsf{T}]_\beta = \begin{pmatrix} -1 & -1 & 0 \\ 0 & -1 & -2 \\ 0 & 0 & -1 \end{pmatrix}.$$

Hence the characteristic polynomial of T is $f(t) = \det(A - tI) = -(t+1)^3$. Thus $\lambda = -1$ is the only eigenvalue of T, and hence $\mathsf{K}_\lambda = \mathsf{P}_2(C)$ by Theorem 7.6. So any basis for $\mathsf{P}_2(C)$, for example, β, is a basis for K_λ.

Now $\mathsf{E}_\lambda = \mathsf{N}(\mathsf{T} - \lambda\mathsf{I}) = \mathsf{N}(\mathsf{T} + \mathsf{I})$. Thus, if f is a polynomial in $\mathsf{P}_2(C)$, then $f \in \mathsf{E}_\lambda$ if and only if

$$\begin{aligned} 0 &= \mathsf{T}(f) + f \\ &= -f - f' + f \\ &= -f'. \end{aligned}$$

Therefore, $f \in \mathsf{E}_\lambda$ if and only if f is a constant. Consequently, $\{1\}$ is a basis for E_λ. Since $\mathsf{K}_\lambda = \mathsf{P}_2(C)$ and this space is three-dimensional, a Jordan canonical basis for K_λ must consist of 3 vectors. The initial vector of any cycle corresponding to λ is an eigenvector of T corresponding to λ. Therefore, such a basis must consist of a single cycle since $\dim(\mathsf{E}_\lambda) = 1$. If γ is such a cycle, then

$$[\mathsf{T}]_\gamma = \begin{pmatrix} -1 & 1 & 0 \\ 0 & -1 & 1 \\ 0 & 0 & -1 \end{pmatrix}$$

and this matrix is a Jordan canonical form for T. The set $\gamma = \{2, -2z, z^2\}$ is an example of such a cycle. ∎

In the next section we will develop a more direct approach for finding the Jordan canonical form and a Jordan canonical basis for a linear operator.

Direct Sums*

Let T be a linear operator on a finite-dimensional vector space V, and suppose that the characteristic polynomial of T splits. By Theorem 5.16, T is diagonalizable if and only if V is a direct sum of the eigenspaces of T. If T is diagonalizable, the eigenspaces and the generalized eigenspaces coincide. The following result generalizes Theorem 5.16 to the nondiagonalizable case.

Theorem 7.7. *Let* T *be a linear operator on a finite-dimensional vector space* V *for which the characteristic polynomial of* T *splits. Then* V *is a direct sum of the generalized eigenspaces of* T.

Proof. Exercise. ∎

EXERCISES

1. Label the following statements as being true or false.
   - **(a)** Eigenvectors of a linear operator T are also generalized eigenvectors of T.
   - **(b)** It is possible for a generalized eigenvector of a linear operator T to be associated with a scalar that is not an eigenvalue of T.
   - **(c)** Any linear operator on a finite-dimensional vector space has a Jordan canonical form.
   - **(d)** Cycles of generalized eigenvectors are linearly independent.
   - **(e)** There exists exactly one cycle of generalized eigenvectors corresponding to each eigenvalue of a linear operator on a finite-dimensional vector space.
   - **(f)** Let T be a linear operator on a finite-dimensional vector space whose

characteristic polynomial factors into polynomials of degree 1, and let $\lambda_1, \lambda_2, \ldots, \lambda_k$ be the distinct eigenvalues of T. If, for each i, β_i is any basis for K_{λ_i}, then $\beta_1 \cup \beta_2 \cup \cdots \cup \beta_k$ is a Jordan canonical basis for T.

(g) For any Jordan block J, L_J has Jordan canonical form J.

(h) Let T be a linear operator on an n-dimensional vector space whose characteristic polynomial splits. Then for any eigenvalue λ of T, $\mathsf{K}_\lambda = \mathsf{N}((\mathsf{T} - \lambda\mathsf{I})^n)$.

2. For each of the following linear operators T, find a basis for each eigenspace and each generalized eigenspace.

(a) $\mathsf{T} = \mathsf{L}_A$, where

$$A = \begin{pmatrix} 1 & 1 \\ -1 & 3 \end{pmatrix}$$

(b) $\mathsf{T} = \mathsf{L}_A$, where

$$A = \begin{pmatrix} 11 & -4 & -5 \\ 21 & -8 & -11 \\ 3 & -1 & 0 \end{pmatrix}$$

(c) $\mathsf{T}\colon \mathsf{P}_2(C) \to \mathsf{P}_2(C)$ defined by $\mathsf{T}(f) = 2f - f'$

3. Find the Jordan canonical form of each linear operator in Exercise 2.

4.† Let Z be a cycle of generalized eigenvectors of a linear operator T on V that corresponds to the eigenvalue λ. Prove that span(Z) is a T-invariant subspace of V.

5. Let $Z_1, Z_2, \ldots, Z_p$ be cycles of generalized eigenvectors of a linear operator T corresponding to an eigenvalue λ. Prove that if the initial eigenvectors are distinct, then the cycles are disjoint.

6. Let $\mathsf{T}\colon \mathsf{V} \to \mathsf{W}$ be a linear transformation. Prove the following:

(a) $\mathsf{N}(\mathsf{T}) = \mathsf{N}(-\mathsf{T})$.

(b) $\mathsf{N}(\mathsf{T}^k) = \mathsf{N}((-\mathsf{T})^k)$ for any positive integer k.

(c) If W = V (so that T is a linear operator on V) and λ is an eigenvalue of T, then for any positive integer k

$$\mathsf{N}((\mathsf{T} - \lambda\mathsf{I}_\mathsf{V})^k) = \mathsf{N}((\lambda\mathsf{I}_\mathsf{V} - \mathsf{T})^k).$$

7. Let U be a linear operator on a finite-dimensional vector space V. Prove the following:

(a) $\mathsf{N}(\mathsf{U}) \subseteq \mathsf{N}(\mathsf{U}^2) \subseteq \cdots \subseteq \mathsf{N}(\mathsf{U}^k) \subseteq \mathsf{N}(\mathsf{U}^{k+1}) \subseteq \cdots$.

(b) If $\text{rank}(\mathsf{U}^m) = \text{rank}(\mathsf{U}^{m+1})$ for some positive integer m, then $\text{rank}(\mathsf{U}^m) = \text{rank}(\mathsf{U}^k)$ for any positive integer $k \geqslant m$.

(c) If $\text{rank}(\mathsf{U}^m) = \text{rank}(\mathsf{U}^{m+1})$ for some positive integer m, then $\mathsf{N}(\mathsf{U}^m) = \mathsf{N}(\mathsf{U}^k)$ for any positive integer $k \geqslant m$.

(d) Let T be a linear operator, and let λ be an eigenvalue of T. Prove that if $\text{rank}((\mathsf{T} - \lambda\mathsf{I})^m) = \text{rank}((\mathsf{T} - \lambda\mathsf{I})^{m+1})$ for some integer m, then $\mathsf{K}_\lambda = \mathsf{N}((\mathsf{T} - \lambda\mathsf{I})^m)$.

(e) *Second Test for Diagonalizability.* Let T be a linear operator whose characteristic polynomial splits. Suppose that $\lambda_1, \lambda_2, \ldots, \lambda_k$ are the distinct eigenvalues of T. Then T is diagonalizable if and only if $\text{rank}(\mathsf{T} - \lambda_i \mathsf{I}) = \text{rank}((\mathsf{T} - \lambda_i \mathsf{I})^2)$ for $1 \leqslant i \leqslant k$.

(f) Use part (e) to obtain a simpler proof of Exercise 24 of Section 5.4: If T is a diagonalizable linear operator on a finite-dimensional vector space V and W is any T-invariant subspace of V, then T_{W} is diagonalizable.

8. Let T be a linear operator on a finite-dimensional vector space V such that the characteristic polynomial $f(t)$ of T splits. Prove that $f(\mathsf{T}) = \mathsf{T}_0$; i.e., prove that T satisfies its characteristic polynomial. (This is a special case of the Cayley–Hamilton theorem.) *Hint:* Show that if β is a Jordan canonical basis for T, then $f(\mathsf{T})(x) = 0$ for each $x \in \beta$.

9. Let β be a Jordan canonical basis for a linear operator T on a finite-dimensional vector space V, and let $J = [\mathsf{T}]_\beta$. Fix an eigenvalue λ of T, and let m denote the number of Jordan blocks having λ in the diagonal positions. Prove that $1 \leqslant m \leqslant \dim(\mathsf{E}_\lambda)$. [We will see in the next section that $m = \dim(\mathsf{E}_\lambda)$.]

10. Prove parts (a) and (c) of Theorem 7.1.

Exercises 11 and 12 will be concerned with direct sums.

11. Prove Theorem 7.7.

12. Let T be a linear operator on a finite-dimensional vector space V such that the characteristic polynomial of T splits, and let $\lambda_1, \lambda_2, \ldots, \lambda_k$ be the distinct eigenvalues of T. Prove that

$$J = J_1 \oplus J_2 \oplus \cdots \oplus J_k$$

is the Jordan canonical form for T if and only if each J_i is the Jordan canonical form for the restriction of T to K_{λ_i} for each i.

7.2 JORDAN CANONICAL FORM

For the purposes of this section we fix a linear operator T on an n-dimensional vector space V such that the characteristic polynomial of T splits. Let $\lambda_1, \lambda_2, \ldots, \lambda_k$ denote the distinct eigenvalues of T. Theorem 7.5 assures the existence of a Jordan canonical basis β for T. By Theorem 7.6(c) the cycles of β that correspond to λ_i form a basis for K_{λ_i}. We can reverse this process. Since each generalized eigenspace is T-invariant, there exists a Jordan canonical basis β_i for T_i, the restriction of T to K_{λ_i}. Theorem 7.6(b) now applies, and the union

$$\beta = \bigcup_{i=1}^{k} \beta_i$$

is a Jordan canonical basis for V.

For any Jordan canonical basis β of T, the Jordan canonical forms for the restrictions T_i can be combined to obtain a Jordan canonical form J for T. That is, if $[\mathsf{T}_i]_{\beta_i} = A_i$ for all i, then

$$[\mathsf{T}]_\beta = J = \begin{pmatrix} A_1 & O & \cdots & O \\ O & A_2 & \cdots & O \\ \vdots & \vdots & & \vdots \\ O & O & \cdots & A_k \end{pmatrix},$$

where each O is a zero matrix of the appropriate size.

In this section we compute the matrices A_i and the bases β_i, thereby computing J and β also. While developing a method for finding J, it will become evident that in some sense the matrices A_i are unique. What we mean by "in some sense" will become clear as we proceed.

To aid in formulating a uniqueness theorem for J, we adopt the following convention: The basis β_i for K_{λ_i} will henceforth be ordered in such a way that the cycles appear in order of decreasing length. That is, if β_i is a disjoint union of cycles $Z_1, Z_2, \ldots, Z_{k_i}$, and if the length of the cycle Z_j is p_j, we index the cycles so that $p_1 \geqslant p_2 \geqslant \cdots \geqslant p_{k_i}$. This ordering of the cycles determines an ordering for β_i and hence determines the matrix A_i. It is in this sense that A_i is unique. It then follows that the Jordan canonical form for T is unique up to an ordering of the eigenvalues of T. As we will also see, there is no comparable uniqueness theorem for the bases β_i or for β. Specifically, what will be shown is that the number k_i of cycles that form β_i and the length p_j $(j = 1, 2, \ldots, k_i)$ of each cycle is completely determined by T.

Example 1

To illustrate that the matrix A_i is entirely determined by the numbers k_i, p_1, $p_2, \ldots, p_{k_i}$, suppose that $k_i = 4$ (i.e., there are four cycles), $p_1 = 3$, $p_2 = 3$, $p_3 = 2$, and $p_4 = 1$. Then

$$A_i = \left(\begin{array}{ccc|ccc|cc|c} \lambda_i & 1 & 0 & 0 & 0 & 0 & 0 & 0 & 0 \\ 0 & \lambda_i & 1 & 0 & 0 & 0 & 0 & 0 & 0 \\ 0 & 0 & \lambda_i & 0 & 0 & 0 & 0 & 0 & 0 \\ \hline 0 & 0 & 0 & \lambda_i & 1 & 0 & 0 & 0 & 0 \\ 0 & 0 & 0 & 0 & \lambda_i & 1 & 0 & 0 & 0 \\ 0 & 0 & 0 & 0 & 0 & \lambda_i & 0 & 0 & 0 \\ \hline 0 & 0 & 0 & 0 & 0 & 0 & \lambda_i & 1 & 0 \\ 0 & 0 & 0 & 0 & 0 & 0 & 0 & \lambda_i & 0 \\ \hline 0 & 0 & 0 & 0 & 0 & 0 & 0 & 0 & \lambda_i \end{array}\right). \quad \blacksquare$$

As an aid in computing A_i and β_i, we introduce an array of dots, called a *dot diagram*, to help us visualize the form of the matrix A_i and the basis β_i. Suppose as above that β_i is a disjoint union of cycles $Z_1, Z_2, \ldots, Z_{k_i}$ with lengths $p_1 \geqslant p_2 \geqslant \cdots \geqslant p_{k_i}$, respectively. The dot diagram contains one dot for each member of β_i and is constructed according to the following rules.

1. The array consists of k_i columns (one column for each cycle).
2. Counting from left to right, the jth column consists of p_j dots that correspond to the members of Z_j in the following manner: If x_j is the end vector of Z_j, then the top dot corresponds to $(\mathsf{T} - \lambda_i \mathsf{I})^{p_j - 1}(x_j)$; the second dot corresponds to $(\mathsf{T} - \lambda_i \mathsf{I})^{p_j - 2}(x_j)$; etc. Hence the final (lowermost) dot of the column corresponds to x_j.

Thus the dot diagram associated with β_i may be depicted as

$$
\begin{array}{llll}
\cdot(\mathsf{T} - \lambda_i\mathsf{I})^{p_1-1}(x_1) & \cdot(\mathsf{T} - \lambda_i\mathsf{I})^{p_2-1}(x_2) & \cdots & \cdot(\mathsf{T} - \lambda_i\mathsf{I})^{p_{k_i}-1}(x_{k_i}) \\
\cdot(\mathsf{T} - \lambda_i\mathsf{I})^{p_1-2}(x_1) & \cdot(\mathsf{T} - \lambda_i\mathsf{I})^{p_2-2}(x_2) & & \cdot(\mathsf{T} - \lambda_i\mathsf{I})^{p_{k_i}-2}(x_{k_i}) \\
 & \qquad\vdots & & \qquad\vdots \\
\qquad\vdots & & & \cdot(\mathsf{T} - \lambda_i\mathsf{I})(x_{k_i}) \\
 & \cdot(\mathsf{T} - \lambda_i\mathsf{I})(x_2) & & \cdot x_{k_i} \\
\cdot(\mathsf{T} - \lambda_i\mathsf{I})(x_1) & \cdot x_2 & & \\
\cdot x_1 & & &
\end{array}
$$

In the diagram above we have labeled each dot with the member of β_i to which it corresponds.

Notice that the dot diagram for β_i has k_i columns (one for each cycle) and p_1 rows. Observe also that since $p_1 \geqslant p_2 \geqslant \cdots \geqslant p_{k_i}$, the columns of the dot diagram become shorter (or at least not longer) as we move from left to right.

You might also observe that if r_j denotes the number of dots in the jth row of the array, then $r_1 \geqslant r_2 \geqslant \cdots \geqslant r_{p_1}$. Since the proof of this fact is combinatorial in nature, it will be left to the exercises (see Exercise 7).

Returning to Example 1, where $k_i = 4$, $p_1 = 3$, $p_2 = 3$, $p_3 = 2$, and $p_4 = 1$, we see that the dot diagram for β_i is

$$
\begin{array}{cccc}
\bullet & \bullet & \bullet & \bullet \\
\bullet & \bullet & \bullet & \\
\bullet & \bullet & &
\end{array}
$$

We will devise a method for computing the dot diagram for β_i in terms of T alone. Hence the dot diagram is uniquely determined by T. It is important to understand, however, that when we say that the dot diagram is uniquely determined by T, we are making no assertions about the uniqueness of β_i. Indeed, as we will see, the basis β_i is not unique. By the uniqueness of the dot diagram we mean that if β_i and β_i' are two Jordan canonical bases for K_{λ_i}, then the dot diagrams for β_i and β_i' are identical. Thus, if β_i' is a disjoint union of k_i' cycles of lengths $p_1' \geqslant p_2' \geqslant \cdots \geqslant p_{k_i}'$, then $k_i' = k_i$ and $p_1' = p_1$, $p_2' = p_2, \ldots$, $p_{k_i}' = p_{k_i}$.

To establish this uniqueness result, we will use the following combinatorial fact: Any dot diagram is completely determined by the number of its rows and the number of dots in each row (see Exercise 7). Thus, if these numbers could be computed from properties intrinsic to the transformation T (for example, as the ranks of $(\mathsf{T} - \lambda_i \mathsf{I})^j$ for various values of j), the dot diagram could be reconstructed and the uniqueness of the numbers $k_i, p_1, p_2, \ldots, p_{k_i}$ would be proved. The following results provide the desired method for computing these numbers.

Theorem 7.8. *For any positive integer* r *the basis vectors in* β_i *that are associated with the dots in the first* r *rows of a dot diagram for* β_i *form a basis for* $\mathsf{N}((\mathsf{T} - \lambda_i \mathsf{I})^r)$. *Hence the number of dots in the first* r *rows of a dot diagram for* β_i *equals* $\text{nullity}((\mathsf{T} - \lambda_i \mathsf{I})^r)$.

Proof. By definition of a generalized eigenspace,

$$\mathsf{N}((\mathsf{T} - \lambda_i \mathsf{I})^r) \subseteq \mathsf{K}_{\lambda_i}.$$

Furthermore, K_{λ_i} is invariant under $(\mathsf{T} - \lambda_i \mathsf{I})^r$. Let U denote the restriction of $(\mathsf{T} - \lambda_i \mathsf{I})^r$ to K_{λ_i}. As a consequence of these remarks, $\mathsf{N}((\mathsf{T} - \lambda_i \mathsf{I})^r) = \mathsf{N}(\mathsf{U})$. For this reason it suffices to consider U. Let

$$S_1 = \{x \in \beta_i \colon \mathsf{U}(x) = 0\} \quad \text{and} \quad S_2 = \{x \in \beta_i \colon \mathsf{U}(x) \neq 0\}.$$

Let a and b denote the number of vectors in S_1 and S_2, respectively. Then $a + b = m_i$, where $m_i = \dim(\mathsf{K}_{\lambda_i})$. For any x in β_i, $x \in S_1$ if and only if x is one of the first r vectors of a cycle corresponding to λ_i, and this is true if and only if x corresponds to a dot among the first r rows of the dot diagram of β_i. For any x in S_2, the effect of applying U to x is to move the dot corresponding to x exactly r places up its column of the dot diagram. It follows that U maps S_2 in a one-to-one fashion into β_i. Thus, $\{\mathsf{U}(x) \colon x \in S_2\}$ is a basis for $\mathsf{R}(\mathsf{U})$ consisting of b vectors, and hence $\text{rank}(\mathsf{U}) = b$. Therefore, $\text{nullity}(\mathsf{U}) = m_i - b = a$. But S_1 is a linearly independent subset of $\mathsf{N}(\mathsf{U})$ consisting of a vectors, and is therefore a basis for $\mathsf{N}(\mathsf{U})$. ∎

In the case that $r = 1$, Theorem 7.8 yields the following corollary.

Corollary. *Let* β_i *be a Jordan canonical basis for the restriction of* T *to* K_{λ_i}, *and suppose that* β_i *is the disjoint union of* k_i *cycles of generalized eigenvectors corresponding to* λ_i. *Then the dimension of* E_{λ_i} *equals* k_i. *Hence in a Jordan canonical form for* T *the number of Jordan blocks corresponding to the eigenvalue* λ_i *equals the dimension of* E_{λ_i}.

We are now able to formulate a procedure for computing the dot diagram for β_i directly from T.

Theorem 7.9. *Let* r_j *denote the number of dots in the* j*th row of a dot diagram for* β_i. *Then*

(a) $r_1 = \dim(\mathsf{V}) - \operatorname{rank}(\mathsf{T} - \lambda_i \mathsf{I})$.

(b) $r_j = \operatorname{rank}((\mathsf{T} - \lambda_i \mathsf{I})^{j-1}) - \operatorname{rank}((\mathsf{T} - \lambda_i \mathsf{I})^j)$ *if* $j > 1$.

Proof. By Theorem 7.8,

$$\begin{aligned} r_1 + r_2 + \cdots + r_j &= \operatorname{nullity}((\mathsf{T} - \lambda_i \mathsf{I})^j) \\ &= \dim(\mathsf{V}) - \operatorname{rank}((\mathsf{T} - \lambda_i \mathsf{I})^j) \qquad \text{for any } j \geqslant 1. \end{aligned}$$

Hence

$$r_1 = \dim(\mathsf{V}) - \operatorname{rank}((\mathsf{T} - \lambda_i \mathsf{I})^1)$$

and

$$\begin{aligned} r_j &= (r_1 + r_2 + \cdots + r_j) - (r_1 + r_2 + \cdots + r_{j-1}) \\ &= (\dim(\mathsf{V}) - \operatorname{rank}((\mathsf{T} - \lambda_i \mathsf{I})^j)) - (\dim(\mathsf{V}) - \operatorname{rank}((\mathsf{T} - \lambda_i \mathsf{I})^{j-1})) \\ &= \operatorname{rank}((\mathsf{T} - \lambda_i \mathsf{I})^{j-1}) - \operatorname{rank}((\mathsf{T} - \lambda_i \mathsf{I})^j) \qquad \text{for } j > 1. \quad \blacksquare \end{aligned}$$

This theorem shows that a dot diagram for β_i is completely determined by T. Hence we have proved the following uniqueness result.

Corollary. *For any eigenvalue* λ_i *of* T *the dot diagram for* β_i *is unique. Thus, subject to the convention that cycles are listed in order of decreasing length, the Jordan canonical form of a linear operator is unique up to the ordering of its eigenvalues.*

Before giving some examples of the use of Theorem 7.9, we define the Jordan canonical form of a matrix in the obvious manner.

Definition. *Let* A *be an* $n \times n$ *matrix with entries from* F *such that the characteristic polynomial of* A *(and hence* L_A*)* splits. *Then the* Jordan canonical form *of* A *is defined to be the Jordan canonical form of the linear operator* L_A *on* F^n.

Observe that if J is the Jordan canonical form of a matrix A, then J and A are similar. In fact, if $\beta = \{z_1, z_2, \ldots, z_n\}$ is a Jordan canonical basis for L_A and Q is the $n \times n$ matrix having z_j as its jth column, then $J = Q^{-1}AQ$ by Theorem 5.1.

In the three examples that follow we compute the Jordan canonical forms of two matrices and a linear operator.

Example 2

Let

$$A = \begin{pmatrix} 2 & -1 & 0 & 1 \\ 0 & 3 & -1 & 0 \\ 0 & 1 & 1 & 0 \\ 0 & -1 & 0 & 3 \end{pmatrix}.$$

We will find the Jordan canonical form of A and a Jordan canonical basis for the linear transformation L_A. The characteristic polynomial of A is

$$\det(A - tI) = (t-2)^3(t-3).$$

Thus A has two distinct eigenvalues, $\lambda_1 = 2$ and $\lambda_2 = 3$ with multiplicities 3 and 1, respectively.

Let β_1 be a Jordan canonical basis for the restriction of L_A to K_{λ_1}. Since λ_1 has multiplicity 3, $\dim(\mathsf{K}_{\lambda_1}) = 3$ by Theorem 7.6. Thus the dot diagram for β_1 contains 3 dots. As above, let r_j denote the number of dots in the jth row of this dot diagram. Applying Theorem 7.9, we have

$$r_1 = 4 - \operatorname{rank}(A - 2I) = 4 - \operatorname{rank}\begin{pmatrix} 0 & -1 & 0 & 1 \\ 0 & 1 & -1 & 0 \\ 0 & 1 & -1 & 0 \\ 0 & -1 & 0 & 1 \end{pmatrix} = 4 - 2 = 2$$

and

$$r_2 = \operatorname{rank}(A - 2I) - \operatorname{rank}((A-2I)^2) = 2 - 1 = 1.$$

(Actually, the computation of r_2 is unnecessary in this case. We could deduce that $r_2 = 1$ from the facts that $r_1 = 2$ and that the dot diagram consists of three dots.) Hence the dot diagram associated with β_1 is

$$\begin{matrix} \bullet & \bullet \\ \bullet & \end{matrix}$$

So if T_i denotes the restriction of L_A to K_{λ_i} $(i = 1, 2)$, we must have

$$A_1 = [\mathsf{T}_1]_{\beta_1} = \begin{pmatrix} 2 & 1 & 0 \\ 0 & 2 & 0 \\ 0 & 0 & 2 \end{pmatrix}.$$

Since $\dim(\mathsf{K}_{\lambda_2}) = 1$, any basis β_2 for K_{λ_2} will consist of a single eigenvector corresponding to $\lambda_2 = 3$. Thus

$$A_2 = [\mathsf{T}_2]_{\beta_2} = (3).$$

Setting $\beta = \beta_1 \cup \beta_2$, we have

$$J = [\mathsf{L}_A]_\beta = \left(\begin{array}{ccc:c} 2 & 1 & 0 & 0 \\ 0 & 2 & 0 & 0 \\ 0 & 0 & 2 & 0 \\ \hdashline 0 & 0 & 0 & 3 \end{array}\right),$$

so J is the Jordan canonical form of A.

We now seek a Jordan canonical basis for $\mathsf{T} = \mathsf{L}_A$. First we must find a Jordan canonical basis β_1 for T_1. We know from the preceding computations that the dot diagram corresponding to β_1 must be

$$\begin{array}{ll} \bullet (\mathsf{T} - \lambda_i \mathsf{I})(x_1) & \bullet x_2 \\ \bullet x_1 & \end{array}$$

From this diagram we see that we must choose x_1 so that $x_1 \in \mathsf{N}((\mathsf{T} - \lambda_1 \mathsf{I})^2)$ but $x_1 \notin \mathsf{N}((\mathsf{T} - \lambda_1 \mathsf{I})^1)$. Since

$$A - 2I = \begin{pmatrix} 0 & -1 & 0 & 1 \\ 0 & 1 & -1 & 0 \\ 0 & 1 & -1 & 0 \\ 0 & -1 & 0 & 1 \end{pmatrix}, \qquad (A - 2I)^2 = \begin{pmatrix} 0 & -2 & 1 & 1 \\ 0 & 0 & 0 & 0 \\ 0 & 0 & 0 & 0 \\ 0 & -2 & 1 & 1 \end{pmatrix}.$$

It is now easily seen that

$$\left\{ \begin{pmatrix} 1 \\ 0 \\ 0 \\ 0 \end{pmatrix}, \begin{pmatrix} 0 \\ 1 \\ 2 \\ 0 \end{pmatrix}, \begin{pmatrix} 0 \\ 1 \\ 0 \\ 2 \end{pmatrix} \right\}$$

is a basis for $\mathsf{N}((\mathsf{T} - \lambda_1 \mathsf{I})^2) = \mathsf{K}_{\lambda_1}$. Of these basis vectors,

$$\begin{pmatrix} 0 \\ 1 \\ 2 \\ 0 \end{pmatrix} \quad \text{and} \quad \begin{pmatrix} 0 \\ 1 \\ 0 \\ 2 \end{pmatrix}$$

satisfy the condition of not belonging to $\mathsf{N}((\mathsf{T} - \lambda_1 \mathsf{I})^1)$. Hence we may select x_1 to be either of these vectors, say

$$x_1 = \begin{pmatrix} 0 \\ 1 \\ 2 \\ 0 \end{pmatrix}.$$

Then

$$(\mathsf{T}-\lambda_1\mathsf{I})(x_1)=(A-2I)(x_1)=\begin{pmatrix}0&-1&0&1\\0&1&-1&0\\0&1&-1&0\\0&-1&0&1\end{pmatrix}\begin{pmatrix}0\\1\\2\\0\end{pmatrix}=\begin{pmatrix}-1\\-1\\-1\\-1\end{pmatrix}.$$

Now simply choose x_2 to be an element of E_{λ_1} that is linearly independent of

$$(\mathsf{T}-\lambda_1\mathsf{I})(x_1)=\begin{pmatrix}-1\\-1\\-1\\-1\end{pmatrix};$$

for example, select

$$x_2=\begin{pmatrix}1\\0\\0\\0\end{pmatrix}.$$

Thus we have associated the Jordan canonical basis

$$\beta_1=\left\{\begin{pmatrix}-1\\-1\\-1\\-1\end{pmatrix},\begin{pmatrix}0\\1\\2\\0\end{pmatrix},\begin{pmatrix}1\\0\\0\\0\end{pmatrix}\right\}$$

with the dot diagram in the following manner:

$$\cdot\begin{pmatrix}-1\\-1\\-1\\-1\end{pmatrix}\qquad\cdot\begin{pmatrix}1\\0\\0\\0\end{pmatrix}$$

$$\cdot\begin{pmatrix}0\\1\\2\\0\end{pmatrix}$$

The reader might be concerned that the linear independence of β_1 was not verified. Be assured, however, that this verification is not necessary because of Theorem 7.4. Since x_2 was chosen to be linearly independent of the initial vector

$(\mathsf{T} - \lambda_1 \mathsf{I})(x_1)$ of the cycle

$$\{(\mathsf{T} - \lambda_1 \mathsf{I})(x_1),\ x_1\},$$

it follows from this theorem that β_1 is linearly independent.

Any eigenvector of L_A corresponding to the eigenvalue $\lambda_2 = 3$ will form the desired basis β_2 for K_{λ_2}—for example,

$$\beta_2 = \left\{ \begin{pmatrix} 1 \\ 0 \\ 0 \\ 1 \end{pmatrix} \right\}.$$

Thus

$$\beta = \beta_1 \cup \beta_2 = \left\{ \begin{pmatrix} -1 \\ -1 \\ -1 \\ -1 \end{pmatrix}, \begin{pmatrix} 0 \\ 1 \\ 2 \\ 0 \end{pmatrix}, \begin{pmatrix} 1 \\ 0 \\ 0 \\ 0 \end{pmatrix}, \begin{pmatrix} 1 \\ 0 \\ 0 \\ 1 \end{pmatrix} \right\}$$

is a Jordan canonical basis for L_A.

Notice that if

$$Q = \begin{pmatrix} -1 & 0 & 1 & 1 \\ -1 & 1 & 0 & 0 \\ -1 & 2 & 0 & 0 \\ -1 & 0 & 0 & 1 \end{pmatrix},$$

then $J = Q^{-1}AQ$. ∎

Example 3

Let

$$A = \begin{pmatrix} 2 & -4 & 2 & 2 \\ -2 & 0 & 1 & 3 \\ -2 & -2 & 3 & 3 \\ -2 & -6 & 3 & 7 \end{pmatrix}.$$

Again we will find a Jordan canonical form J for A and a matrix Q such that $J = Q^{-1}AQ$.

The characteristic polynomial of A is $\det(A - tI) = (t-2)^2(t-4)^2$. Let $\lambda_1 = 2$, $\lambda_2 = 4$, and β_i be the Jordan canonical basis for T_i, the restriction of L_A to K_{λ_i}, for $i = 1, 2$.

We begin by computing the dot diagram for β_1. Let r_1 denote the number of dots in the first row of this diagram; then

$$r_1 = 4 - \text{rank}(A - 2I) = 4 - 2 = 2.$$

So the dot diagram for β_1 is

$$\bullet \quad \bullet$$

Thus

$$A_1 = [\mathsf{T}_1]_{\beta_1} = \begin{pmatrix} 2 & 0 \\ 0 & 2 \end{pmatrix}.$$

Next we compute the dot diagram for β_2. Since rank$(A - 4I) = 3$, there is only $4 - 3 = 1$ dot in the first row of the diagram. Since K_{λ_2} has dimension 2 (Theorem 7.6), the dot diagram for β_2 must be

$$\begin{matrix} \bullet \\ \bullet \end{matrix}$$

Thus

$$A_2 = [\mathsf{T}_2]_{\beta_2} = \begin{pmatrix} 4 & 1 \\ 0 & 4 \end{pmatrix}.$$

So if $\beta = \beta_1 \cup \beta_2$, then the Jordan canonical form of L_A is

$$J = [\mathsf{L}_A]_\beta = \left(\begin{array}{cc:cc} 2 & 0 & 0 & 0 \\ 0 & 2 & 0 & 0 \\ \hdashline 0 & 0 & 4 & 1 \\ 0 & 0 & 0 & 4 \end{array}\right).$$

In order to find a matrix Q such that $Q^{-1}AQ = J$, we must first find a Jordan canonical basis β for L_A. The dot diagram for β_1 indicates that β_1 can be chosen to be any linearly independent set of eigenvectors of A corresponding to $\lambda_1 = 2$. For example,

$$\beta_1 = \left\{ \begin{pmatrix} 2 \\ 1 \\ 0 \\ 2 \end{pmatrix}, \begin{pmatrix} 0 \\ 1 \\ 2 \\ 0 \end{pmatrix} \right\}$$

will suffice. For β_2 we must find an element $x_1 \in \mathsf{K}_{\lambda_2} = \mathsf{N}((\mathsf{L}_A - \lambda_2 \mathsf{I})^2)$ such that $x_1 \notin \mathsf{N}((\mathsf{L}_A - \lambda_2 \mathsf{I})^1)$. One way of finding such an element was used to select the vector x_1 in Example 2. In this example we illustrate another method for obtaining such a vector. A simple calculation shows that a basis for the null space of $\mathsf{L}_A - \lambda_2\mathsf{I}$ is

$$\left\{\begin{pmatrix}0\\1\\1\\1\end{pmatrix}\right\}.$$

Let

$$(A-4I)(x_1)=\begin{pmatrix}0\\1\\1\\1\end{pmatrix},$$

and choose x_1 to be any preimage of

$$\begin{pmatrix}0\\1\\1\\1\end{pmatrix}.$$

To do this, we must find a solution to the matrix equation

$$(A-4I)\begin{pmatrix}a\\b\\c\\d\end{pmatrix}=\begin{pmatrix}0\\1\\1\\1\end{pmatrix};$$

i.e.,

$$\begin{pmatrix}-2 & -4 & 2 & 2\\-2 & -4 & 1 & 3\\-2 & -2 & -1 & 3\\-2 & -6 & 3 & 3\end{pmatrix}\begin{pmatrix}a\\b\\c\\d\end{pmatrix}=\begin{pmatrix}0\\1\\1\\1\end{pmatrix}.$$

It is easily verified that

$$\begin{pmatrix}a\\b\\c\\d\end{pmatrix}=\begin{pmatrix}1\\-1\\-1\\0\end{pmatrix}$$

is a solution; so select

$$x_1 = \begin{pmatrix} 1 \\ -1 \\ -1 \\ 0 \end{pmatrix}.$$

Thus

$$\beta_2 = \{(\mathsf{L}_A - \lambda_2 \mathsf{I})(x_1), x_1\} = \left\{ \begin{pmatrix} 0 \\ 1 \\ 1 \\ 1 \end{pmatrix}, \begin{pmatrix} 1 \\ -1 \\ -1 \\ 0 \end{pmatrix} \right\}.$$

Hence

$$\beta = \beta_1 \cup \beta_2 = \left\{ \begin{pmatrix} 2 \\ 1 \\ 0 \\ 2 \end{pmatrix}, \begin{pmatrix} 0 \\ 1 \\ 2 \\ 0 \end{pmatrix}, \begin{pmatrix} 0 \\ 1 \\ 1 \\ 1 \end{pmatrix}, \begin{pmatrix} 1 \\ -1 \\ -1 \\ 0 \end{pmatrix} \right\}$$

is a Jordan canonical basis for L_A.

So if

$$Q = \begin{pmatrix} 2 & 0 & 0 & 1 \\ 1 & 1 & 1 & -1 \\ 0 & 2 & 1 & -1 \\ 2 & 0 & 1 & 0 \end{pmatrix},$$

then $J = Q^{-1}AQ$. ∎

Example 4

Let V denote the vector space of polynomial functions over R in two variables x and y of degree at most 2. (A basis for V is $\alpha = \{1, x, y, x^2, y^2, xy\}$.) Consider the mapping $\mathsf{T}\colon \mathsf{V} \to \mathsf{V}$ defined by

$$\mathsf{T}(f) = \frac{\partial}{\partial x} f.$$

For example, if $f(x, y) = x + 2x^2 - 3xy + y$, then

$$\mathsf{T}(f) = \frac{\partial}{\partial x} f(x, y) = 1 + 4x - 3y.$$

We will find a Jordan canonical basis for T.

First, observe that if $A = [\mathsf{T}]_\alpha$, then

$$A = \begin{pmatrix} 0 & 1 & 0 & 0 & 0 & 0 \\ 0 & 0 & 0 & 2 & 0 & 0 \\ 0 & 0 & 0 & 0 & 0 & 1 \\ 0 & 0 & 0 & 0 & 0 & 0 \\ 0 & 0 & 0 & 0 & 0 & 0 \\ 0 & 0 & 0 & 0 & 0 & 0 \end{pmatrix}.$$

Thus the characteristic polynomial of T is

$$\det(A - tI) = \det\begin{pmatrix} -t & 1 & 0 & 0 & 0 & 0 \\ 0 & -t & 0 & 2 & 0 & 0 \\ 0 & 0 & -t & 0 & 0 & 1 \\ 0 & 0 & 0 & -t & 0 & 0 \\ 0 & 0 & 0 & 0 & -t & 0 \\ 0 & 0 & 0 & 0 & 0 & -t \end{pmatrix} = t^6.$$

Hence T has only one eigenvalue ($\lambda = 0$), and $\mathsf{K}_\lambda = \mathsf{V}$. Let β denote any Jordan canonical basis for T. If r_j denotes the number of dots in the jth row of the dot diagram for β, then $r_1 = 6 - \text{rank}(A) = 6 - 3 = 3$. Since

$$A^2 = \begin{pmatrix} 0 & 0 & 0 & 2 & 0 & 0 \\ 0 & 0 & 0 & 0 & 0 & 0 \\ 0 & 0 & 0 & 0 & 0 & 0 \\ 0 & 0 & 0 & 0 & 0 & 0 \\ 0 & 0 & 0 & 0 & 0 & 0 \\ 0 & 0 & 0 & 0 & 0 & 0 \end{pmatrix},$$

$r_2 = \text{rank}(A) - \text{rank}(A^2) = 3 - 1 = 2$. Thus because $r_1 = 3$, $r_2 = 2$, and there are six dots in the dot diagram, it follows that $r_3 = 1$. So the dot diagram for β is

$$\begin{array}{ccc} \bullet & \bullet & \bullet \\ \bullet & \bullet & \\ \bullet & & \end{array}$$

We conclude that the Jordan canonical form J of T is

$$J = \left(\begin{array}{ccc:cc:c} 0 & 1 & 0 & 0 & 0 & 0 \\ 0 & 0 & 1 & 0 & 0 & 0 \\ 0 & 0 & 0 & 0 & 0 & 0 \\ \hdashline 0 & 0 & 0 & 0 & 1 & 0 \\ 0 & 0 & 0 & 0 & 0 & 0 \\ \hdashline 0 & 0 & 0 & 0 & 0 & 0 \end{array}\right).$$

We now seek a Jordan canonical basis for T. Since the first column of the dot diagram for β consists of three dots, we must find a vector x_1 such that

$$\frac{\partial^2}{\partial x^2}(x_1) \neq 0.$$

Examining the basis $\alpha = \{1, x, y, x^2, y^2, xy\}$ for K_λ, we see that x^2 is a candidate for x_1. Letting $x_1 = x^2$, we find that

$$(\mathsf{T} - \lambda\mathsf{I})(x_1) = \mathsf{T}(x_1) = \frac{\partial}{\partial x}(x_1) = 2x$$

and

$$(\mathsf{T} - \lambda\mathsf{I})^2(x_1) = \mathsf{T}^2(x_1) = \frac{\partial^2}{\partial x^2}(x_1) = 2.$$

Likewise, since the second column of the dot diagram for β consists of two dots, we must find a vector x_2 such that

$$\frac{\partial}{\partial x}(x_2) \neq 0.$$

Examining α with 1, x, and x^2 eliminated from consideration (because they lie in the span of the cycle $\{2, 2x, x^2\}$), we see that we may select $x_2 = xy$. Thus

$$(\mathsf{T} - \lambda\mathsf{I})(x_2) = \mathsf{T}(x_2) = \frac{\partial}{\partial x}(xy) = y.$$

Finally, choose $x_3 = y^2$. Then we have identified the following basis with the dot diagram:

$\cdot 2$	$\cdot y$	$\cdot y^2$
$\cdot 2x$	$\cdot xy$	
$\cdot x^2$		

Thus $\beta = \{2, 2x, x^2, y, xy, y^2\}$ is a Jordan canonical basis for T. ∎

In the three preceding examples we relied upon our ingenuity and the context of the problem to find a Jordan canonical basis. The reader will be able to do the same in the exercises. We are successful in these cases because the dimensions of the generalized eigenspaces under consideration are small. We do not attempt, however, to develop a general algorithm for computing a Jordan canonical basis although one could be formulated by following the steps in the proof of the existence of such a basis (Theorem 7.5).

The following result may be thought of as a corollary to Theorem 7.9.

Theorem 7.10. *Let* A *and* B *be two square matrices of the same size, each having Jordan canonical forms computed according to the conventions of this section. Then* A *and* B *are similar if and only if they have (up to a permutation of their eigenvalues) the same Jordan canonical form.*

Proof. If A and B have the same Jordan canonical form J, then A and B are each similar to J and hence are similar to each other.

Conversely, suppose that A and B are similar. Then A and B must have the same eigenvalues with the same multiplicities. Let J_A and J_B denote the Jordan canonical forms of A and B, respectively, for some fixed ordering of their eigenvalues. Then since A is similar to J_A and B is similar to J_B, the hypothesis implies that J_A and J_B are similar. Hence by Exercise 19 of Section 5.1 there exists a linear operator T on a finite-dimensional vector space V and bases β and γ for V such that $[\mathsf{T}]_\beta = J_A$ and $[\mathsf{T}]_\gamma = J_B$. Thus J_A and J_B are Jordan canonical forms of the same linear operator. Hence, since the eigenvalues of A and B are ordered in the same way, the corollary to Theorem 7.9 implies that $J_A = J_B$. ∎

Example 5

We will determine which of the matrices

$$A = \begin{pmatrix} -3 & 3 & -2 \\ -7 & 6 & -3 \\ 1 & -1 & 2 \end{pmatrix}, \qquad B = \begin{pmatrix} 0 & 1 & -1 \\ -4 & 4 & -2 \\ -2 & 1 & 1 \end{pmatrix},$$

$$C = \begin{pmatrix} 0 & -1 & -1 \\ -3 & -1 & -2 \\ 7 & 5 & 6 \end{pmatrix}, \quad \text{and} \quad D = \begin{pmatrix} 0 & 1 & 2 \\ 0 & 1 & 1 \\ 0 & 0 & 2 \end{pmatrix}$$

are similar. Observe that A, B, and C have the same characteristic polynomial $-(t-1)(t-2)^2$, whereas D has $-t(t-1)(t-2)$ as its characteristic polynomial. Thus, because similar matrices have the same characteristic polynomials, D cannot be similar to A, B, or C. Now each of the matrices A, B, and C has the same eigenvalues $\lambda_1 = 1$ and $\lambda_2 = 2$ with multiplicities 1 and 2, respectively. If J_A, J_B, and J_C denote the Jordan canonical forms of A, B, and C, respectively, with respect to this ordering of their eigenvalues, then

$$J_A = \begin{pmatrix} 1 & 0 & 0 \\ 0 & 2 & 1 \\ 0 & 0 & 2 \end{pmatrix}, \qquad J_B = \begin{pmatrix} 1 & 0 & 0 \\ 0 & 2 & 0 \\ 0 & 0 & 2 \end{pmatrix}, \quad \text{and} \quad J_C = \begin{pmatrix} 1 & 0 & 0 \\ 0 & 2 & 1 \\ 0 & 0 & 2 \end{pmatrix}.$$

Since $J_A = J_C$, A is similar to C, while B is similar to neither A nor C. ∎

The reader should observe that any diagonal matrix is a Jordan canonical form. Thus T *is diagonalizable if and only if its Jordan canonical form is a diagonal matrix. Hence if* T *is a diagonalizable operator on* V, *any Jordan canonical basis for* T *is a basis for* V *consisting of eigenvectors of* T.

EXERCISES

1. Label the following statements as being true or false.

(a) The Jordan canonical form of a diagonal matrix is the matrix itself.

(b) Let T be a linear operator on a finite-dimensional vector space V that has a Jordan canonical form J. If β is any basis for V, then the Jordan canonical form of $[\mathsf{T}]_\beta$ is J.

(c) Linear operators having the same characteristic polynomial are similar.

(d) Matrices having the same Jordan canonical form are similar.

(e) Every matrix is similar to its Jordan canonical form.

(f) Let T be a linear operator on a finite-dimensional vector space with characteristic polynomial $(-1)^n(t-\lambda)^n$. Subject to the convention that the Jordan blocks are ordered by decreasing size, T has a unique Jordan canonical form.

(g) If an operator has a Jordan canonical form, then there is a unique Jordan canonical basis for that operator.

(h) The dot diagram of any linear operator having a Jordan canonical form is unique.

2. Let T be a linear operator on a finite-dimensional vector space V such that the characteristic polynomial of T splits. Let $\lambda_1 = 2$, $\lambda_2 = 4$, and $\lambda_3 = -3$ be the distinct eigenvalues of T, and suppose that the dot diagrams for the restriction of $\mathsf{T} - \lambda_i\mathsf{I}$ to K_{λ_i} $(i = 1, 2, 3)$ are as follows:

$\lambda_1 = 2$			$\lambda_2 = 4$		$\lambda_3 = -3$	
•	•	•	•	•	•	•
•	•		•			
•			•			

Find the Jordan canonical form of T.

3. Let T be a linear operator on a finite-dimensional vector space V such that the Jordan canonical form of T is

$$\left(\begin{array}{ccc|cc|cc}
2 & 1 & 0 & 0 & 0 & 0 & 0\\
0 & 2 & 1 & 0 & 0 & 0 & 0\\
0 & 0 & 2 & 0 & 0 & 0 & 0\\
\hline
0 & 0 & 0 & 2 & 1 & 0 & 0\\
0 & 0 & 0 & 0 & 2 & 0 & 0\\
\hline
0 & 0 & 0 & 0 & 0 & 3 & 0\\
0 & 0 & 0 & 0 & 0 & 0 & 3
\end{array}\right).$$

(a) Find the characteristic polynomial of T.

(b) Find the dot diagram corresponding to each eigenvalue of T.

(c) For which eigenvalues λ_i, if any, does $\mathsf{E}_{\lambda_i} = \mathsf{K}_{\lambda_i}$?

(d) For each eigenvalue λ_i, find the smallest positive integer p_i for which $\mathsf{K}_{\lambda_i} = \mathsf{N}((\mathsf{T} - \lambda_i \mathsf{I})^{p_i})$.

(e) Let U_i denote the restriction of $\mathsf{T} - \lambda_i\mathsf{I}$ to K_{λ_i} for each i. Compute the following for each i.

(1) $\text{rank}(\mathsf{U}_i)$
(2) $\text{rank}(\mathsf{U}_i^2)$
(3) $\text{nullity}(\mathsf{U}_i)$
(4) $\text{nullity}(\mathsf{U}_i^2)$

4. For each of the following matrices A, find a Jordan canonical form J and a matrix Q such that $J = Q^{-1}AQ$. Notice that the matrices in parts (a), (b), and (c) are matrices used in Example 5.

(a) $$A = \begin{pmatrix} -3 & 3 & -2 \\ -7 & 6 & -3 \\ 1 & -1 & 2 \end{pmatrix}$$

(b) $$A = \begin{pmatrix} 0 & 1 & -1 \\ -4 & 4 & -2 \\ -2 & 1 & 1 \end{pmatrix}$$

(c) $$A = \begin{pmatrix} 0 & -1 & -1 \\ -3 & -1 & -2 \\ 7 & 5 & 6 \end{pmatrix}$$

(d) $$A = \begin{pmatrix} 0 & -3 & 1 & 2 \\ -2 & 1 & -1 & 2 \\ -2 & 1 & -1 & 2 \\ -2 & -3 & 1 & 4 \end{pmatrix}$$

5. Let A be an $n \times n$ matrix whose characteristic polynomial splits. Prove that A and A^t have the same Jordan canonical form, and conclude that A and A^t are similar. *Hint:* For any eigenvalue λ of A and A^t and any positive integer r, show that $\text{rank}((A - \lambda I)^r) = \text{rank}((A^t - \lambda I)^r)$.

6. Let V denote the vector space of functions that are linear combinations of e^x, xe^x, x^2e^x, and e^{2x}. Define $\mathsf{T}\colon \mathsf{V} \to \mathsf{V}$ by $\mathsf{T}(f) = f'$ (the derivative of f). Find both a Jordan canonical form and a Jordan canonical basis for T.

7. Suppose that an array of dots (such as a dot diagram) has k columns and m rows and that the jth column of the array contains p_j dots and the ith row of the array contains r_i dots. If $p_1 \geqslant p_2 \geqslant \cdots \geqslant p_k$, prove the following:

(a) $m = p_1$ and $k = r_1$.

(b) $p_j = \max\{i\colon r_i \geqslant j\}$ for $1 \leqslant j \leqslant k$ and $r_i = \max\{j\colon p_j \geqslant i\}$ for $1 \leqslant i \leqslant m$. *Hint:* Use induction on m.

(c) $r_1 \geqslant r_2 \geqslant \cdots \geqslant r_m$.

(d) Conclude that the number of dots in each column of a dot diagram is completely determined if the number of dots in each row is known.

Definition. *A linear operator* T *on* V *is called* nilpotent *if* $\mathsf{T}^p = \mathsf{T}_0$ *for some positive integer* p.

8. Prove that if T is a nilpotent operator on an n-dimensional vector space V, then the characteristic polynomial of T is $(-1)^n t^n$. Hence the characteristic polynomial of T splits, and T has only one eigenvalue (zero) of multiplicity n. *Hint:* Use induction on n. In the general step, assume that the conclusion is true for all vector spaces of dimension less than n, and follow the steps below.

(a) Prove that T has at least one eigenvector corresponding to $\lambda = 0$. Thus $\dim(\mathsf{R}(\mathsf{T})) < \dim(\mathsf{V}) = n$.

(b) Apply the induction hypothesis to the T-invariant subspace R(T).

(c) Extend a basis $\{x_1, x_2, \ldots, x_k\}$ for R(T) to a basis $\beta = \{x_1, x_2, \ldots, x_k, x_{k+1}, \ldots, x_n\}$ for V.

(d) Show that

$$[\mathsf{T}]_\beta = \begin{pmatrix} B_1 & B_2 \\ O & O' \end{pmatrix},$$

where O and O' are $(n-k) \times k$ and $(n-k) \times (n-k)$ zero matrices, respectively.

(e) Deduce that $\det(\mathsf{T} - t\mathsf{I}) = (-1)^n t^n$.

9. Prove the converse of Exercise 8: If T is a linear operator on V having characteristic polynomial $(-1)^n t^n$, then T is nilpotent.

10. Give an example of a linear operator T such that T is not nilpotent but zero is the only eigenvalue of T. Characterize all such transformations.

Definition. *An* n × n *matrix* A *is called* nilpotent *if* A^p *equals the* n × n *zero matrix for some positive integer* p.

11. Let $A \in \mathsf{M}_{n\times n}(F)$. Prove that $A^p = O$, where O denotes the $n \times n$ zero matrix, if and only if $(\mathsf{L}_A)^p = \mathsf{T}_0$. Conclude that A is nilpotent if and only if L_A is nilpotent.

12. Prove that any square upper triangular matrix with each diagonal entry equal to zero is nilpotent.

13. Let T be a linear operator on a finite-dimensional vector space V such that the characteristic polynomial of T splits, and suppose that $\lambda_1, \lambda_2, \ldots, \lambda_k$ are the distinct eigenvalues of T.

(a) Use Theorem 7.6 to prove that for any $x \in \mathsf{V}$ there exist unique vectors $x_1, x_2, \ldots, x_k$ such that $x_i \in \mathsf{K}_{\lambda_i}$ and

$$x = x_1 + x_2 + \cdots + x_k.$$

(b) Let S: V → V be defined as follows: For $x \in \mathsf{V}$, represent x as in part (a) and define

$$\mathsf{S}(x) = \lambda_1 x_1 + \cdots + \lambda_k x_k.$$

Next, define U: V → V by U = T − S. Prove that S is diagonalizable, U is nilpotent, and SU = US.

(c) Prove the converse of part (b). That is, prove that if T is a linear operator on a finite-dimensional vector space V, S and U are linear operators on

V such that $T = S + U$ where S is diagonalizable, U is nilpotent, and $SU = US$, then T and S have the same characteristic polynomial, and S is defined as in part (b). *Hint:* First prove that if x is an eigenvector of S with corresponding eigenvalue λ, then x is also a generalized eigenvector of T corresponding to λ.

14. Let T and U be as in Exercise 13. Suppose that β_i is a Jordan canonical basis for the restriction of T to K_{λ_i}, and let J_i denote the Jordan canonical form of this restriction. Then $\beta = \beta_1 \cup \beta_2 \cup \cdots \cup \beta_k$ is a Jordan canonical basis for T. Let $J = [T]_\beta$ and $S = T - U$. Prove the following:

(a) $[S]_\beta$ is a diagonal matrix whose diagonal entries are identical to the diagonal entries of J; that is, if $D = [S]_\beta$, then

$$D_{ij} = \begin{cases} J_{ij} & \text{if } i = j \\ 0 & \text{otherwise.} \end{cases}$$

(b) If $M = [U]_\beta$, then

$$M_{ij} = \begin{cases} J_{ij} & \text{if } j = i + 1 \\ 0 & \text{otherwise.} \end{cases}$$

(c) $J = D + M$.

(d) $MD = DM$.

(e) As a consequence of parts (c) and (d) there is a binomial expansion for J. Let p be the smallest positive integer for which M^p equals the zero matrix. Then

$$J^r = D^r + rD^{r-1}M + \frac{r(r-1)}{2!}D^{r-2}M^2 + \cdots + rDM^{r-1} + M^r \qquad \text{if } r < p,$$

and

$$J^r = D^r + rD^{r-1}M + \frac{r(r-1)}{2!}D^{r-2}M^2 + \cdots + \frac{r!}{(r-p+1)!(p-1)!}D^{r-p+1}M^{p-1} \qquad \text{if } r \geqslant p.$$

(f) If $T = L_A$, then there exists a matrix Q such that $A = QJQ^{-1}$.

(g) For the matrix Q above and any positive integer r, $A^r = QJ^rQ^{-1}$.

15. Let T be a nilpotent linear operator on a finite-dimensional vector space V. Recall from Exercise 8 that $\lambda = 0$ is the only eigenvalue of T; hence $V = K_\lambda$. Let β be a Jordan canonical basis for T. Prove that for any positive integer i if we delete from β the vectors corresponding to the last i dots in each column of a dot diagram for β, the resulting set is a basis for $R(T^i)$. (If a column of the dot diagram contains fewer than i dots, all the vectors associated with that column are removed from β.)

16. Find a linear operator on a finite-dimensional vector space having two distinct Jordan canonical bases.

17. Let T be a linear operator whose characteristic polynomial splits, and let λ be an eigenvalue of T.

(a) Prove that $\dim(\mathsf{K}_\lambda)$ is the sum of the lengths of all the blocks corresponding to λ in the Jordan canonical form of T.

(b) Deduce that $\mathsf{E}_\lambda = \mathsf{K}_\lambda$ if and only if all the Jordan blocks corresponding to λ are 1×1 matrices.

18. **(a)** Let

$$J = \begin{pmatrix} \lambda & 1 & 0 & \cdots & 0 \\ 0 & \lambda & 1 & \cdots & 0 \\ 0 & 0 & \lambda & \cdots & 0 \\ \vdots & \vdots & \vdots & & \vdots \\ 0 & 0 & 0 & \cdots & 1 \\ 0 & 0 & 0 & \cdots & \lambda \end{pmatrix}.$$

Suppose that J is $m \times m$, and let $N = J - \lambda I_m$. Prove that N^m is the zero matrix.

(b) Observe as in Exercise 14 that for any $r \geqslant m$

$$J^r = \lambda^r I_m + r\lambda^{r-1}N + \frac{r(r-1)}{2!}\lambda^{r-2}N^2 + \cdots + \frac{r(r-1)\cdots(r-m+2)}{(m-1)!}\lambda^{r-m+1}N^{m-1}$$

$$= \begin{pmatrix} \lambda^r & r\lambda^{r-1} & \frac{r(r-1)}{2!}\lambda^{r-2} & \cdots & \frac{r(r-1)\cdots(r-m+2)}{(m-1)!}\lambda^{r-m+1} \\ 0 & \lambda^r & r\lambda^{r-1} & \cdots & \frac{r(r-1)\cdots(r-m+3)}{(m-2)!}\lambda^{r-m+2} \\ \vdots & \vdots & \vdots & & \vdots \\ 0 & 0 & 0 & \cdots & \lambda^r \end{pmatrix}.$$

Prove that $\lim_{r\to\infty} J^r$ exists if and only if one of the following holds:

(1) $|\lambda| < 1$.
(2) $\lambda = 1$ and $m = 1$.

Furthermore, show that $\lim_{r\to\infty} J^r$ is the zero matrix if condition 1 holds and is the matrix (1) if condition 2 holds.

(c) Prove Theorem 5.18.

19. For any $A \in \mathsf{M}_{n\times n}(C)$, define $\|A\| = \max\{|A_{ij}|: 1 \leqslant i, j \leqslant n\}$. Prove the following results for arbitrary $A, B \in \mathsf{M}_{n \times n}(C)$ and $c \in C$.

(a) $\|A\| \geqslant 0$, and $\|A\| = 0$ if and only if A is the zero matrix.

(b) $\|cA\| = |c| \cdot \|A\|$.

(c) $\|A + B\| \leqslant \|A\| + \|B\|$.

(d) $\|AB\| \leqslant n\|A\| \cdot \|B\|$.

20. Let $A \in \mathsf{M}_{n \times n}(R)$ be a transition matrix (see Section 5.3), and $P^{-1}AP = J$ be the Jordan canonical form of A. Let $\|\cdot\|$ be as defined in Exercise 19.

(a) Show that for every positive integer m, $\|A^m\| \leqslant 1$.

(b) Deduce that $\{\|J^m\| : m = 1, 2, \ldots\}$ is bounded.

(c) Using part (b) and Exercise 18(b), prove that each Jordan block corresponding to the eigenvalue $\lambda = 1$ of A is 1×1.

(d) Use part (c), Theorem 5.18, and Exercise 18(b) to show that $\lim_{m \to \infty} A^m$ exists if and only if A has the property that whenever λ is an eigenvalue of A with $|\lambda| = 1$, then $\lambda = 1$.

(e) Prove Theorem 5.25(a) using part (c) and Theorem 5.24.

21. (This exercise requires knowledge of absolutely convergent series.) Recall from page 280 that if $A \in \mathsf{M}_{n \times n}(C)$, then e^A is defined as $\lim_{m \to \infty} B_m$, where

$$B_m = I + A + \frac{A^2}{2!} + \cdots + \frac{A^m}{m!}.$$

Use Exercise 19(d) to show that e^A exists for every $A \in \mathsf{M}_{n \times n}(C)$.

22. Let $X' = AX$ be a system of n linear differential equations where X is an n-tuple of differentiable functions $x_1(t), x_2(t), \ldots, x_n(t)$ of the real variable t, and A is an $n \times n$ coefficient matrix as in Exercise 14 of Section 5.2. In contrast to that exercise, however, suppose that A is not diagonalizable, but that the characteristic polynomial of A splits. Let $\lambda_1, \lambda_2, \ldots, \lambda_k$ be the distinct eigenvalues of A.

(a) Prove that if u is the end vector of a cycle of generalized eigenvectors of L_A of length p and u corresponds to the eigenvalue λ_i, then for any polynomial $f(t)$ of degree less than p the function

$$e^{\lambda_i t}[f(t)(A - \lambda_i \mathsf{I})^{p-1} + f'(t)(A - \lambda_i \mathsf{I})^{p-2} + \cdots + f^{(p-1)}(t)]u$$

is a solution to the system $X' = AX$.

(b) Prove that the general solution to $X' = AX$ is a sum of functions of the form given in part (a), where the vectors u are the end vectors of the distinct cycles that constitute a fixed Jordan canonical basis for L_A.

23. Use Exercise 22(b) to find the general solution to each system of differential equations:

(a) $\begin{cases} x' = 2x + y \\ y' = 2y - z \\ z' = 3z \end{cases}$ **(b)** $\begin{cases} x' = 2x + y \\ y' = 2y + z \\ z' = 2z \end{cases}$

where x, y, and z are unknown real-valued differentiable functions of the real variable t.

7.3 THE MINIMAL POLYNOMIAL

For a given operator T on a finite-dimensional vector space V the Cayley–Hamilton theorem shows that there is a polynomial $f(t)$ for which $f(\mathsf{T}) = \mathsf{T}_0$, namely the characteristic polynomial of T. There are many other polynomials having this property. One of the most important of these, the minimal polynomial, provides another means for studying linear operators.

Definition. *Let* T *be a linear operator on a vector space* V. *A polynomial* p(t) *is called a* minimal polynomial *for* T *if* p(t) *is a monic polynomial of least positive degree for which* $\mathrm{p}(\mathsf{T}) = \mathsf{T}_0$. *(Recall from Appendix E that a monic polynomial is one in which the leading coefficient is* 1.)

It is easy to see that any linear operator T on an n-dimensional vector space has a minimal polynomial of degree at most n. By the Cayley–Hamilton theorem the characteristic polynomial $f(t)$ of T, which is of degree n, satisfies the equation $f(\mathsf{T}) = \mathsf{T}_0$. Choose a polynomial $g(t)$ of least positive degree for which $g(\mathsf{T}) = \mathsf{T}_0$, and let $p(t)$ be the result of dividing $g(t)$ by its leading coefficient. Then $p(t)$ is a minimal polynomial of T and the degree of $p(t)$ is at most n. The next result shows that the requirement that a minimal polynomial be monic guarantees that it is unique.

Theorem 7.11. *Let* p(t) *be a minimal polynomial for a linear operator* T *on a finite-dimensional vector space* V.

(a) *If* g(t) *is any polynomial for which* $\mathrm{g}(\mathsf{T}) = \mathsf{T}_0$, *then* p(t) *divides* g(t). *In particular,* p(t) *divides the characteristic polynomial of* T.

(b) *There is only one minimal polynomial for* T*; i.e.,* p(t) *is unique.*

Proof. (a) Let $g(t)$ be any polynomial for which $g(\mathsf{T}) = \mathsf{T}_0$. The division algorithm for polynomials (see Appendix E) implies that there exist polynomials $q(t)$ and $r(t)$ such that

$$g(t) = q(t)p(t) + r(t), \tag{2}$$

where $r(t)$ has degree less than that of $p(t)$. Substituting T in (2) and using that $g(\mathsf{T}) = p(\mathsf{T}) = \mathsf{T}_0$, we have $r(\mathsf{T}) = \mathsf{T}_0$. Since $r(t)$ has degree less than $p(t)$ and $p(t)$ is a minimal polynomial, $r(t)$ must be the zero polynomial. Thus (2) simplifies to $g(t) = q(t)p(t)$, proving (a).

(b) Suppose that $p_1(t)$ and $p_2(t)$ are each minimal polynomials for T. Then $p_1(t)$ divides $p_2(t)$ by part (a). But since $p_1(t)$ and $p_2(t)$ have the same nonnegative degree, we must have $p_1(t) = cp_2(t)$ for some nonzero scalar c. Moreover, since $p_1(t)$ and $p_2(t)$ are monic, $c = 1$. Thus $p_1(t) = p_2(t)$. ∎

Before continuing our study of the minimal polynomial for an operator, we introduce the minimal polynomial for a matrix.

Definition. *The* minimal polynomial p(t) *for* $A \in \mathsf{M}_{n\times n}(F)$ *is the monic polynomial of least positive degree for which* p(A) *equals the zero matrix.*

Throughout this book, statements about linear transformations have been translated into statements about matrices and vice versa. The following theorem and its corollary are of this type.

Theorem 7.12. *Let* T *be a linear operator on a finite-dimensional vector space* V, *and let* β *be a basis for* V. *Then the minimal polynomial for* T *is the same as the minimal polynomial for* $[\mathsf{T}]_\beta$.

Proof. Exercise. ∎

Corollary. *For any* $A \in M_{n \times n}(F)$, *the minimal polynomial for* A *is the same as the minimal polynomial for* L_A.

Proof. Exercise. ∎

As a consequence of the preceding theorem and corollary, subsequent theorems of this section that are stated for operators are also true for matrices.

In the remainder of this section we study primarily minimal polynomials for operators whose characteristic polynomials split. A more general treatment of minimal polynomials will be given in Section 7.4.

Theorem 7.13. *Let* T *be a linear operator on a finite-dimensional vector space* V, *and let* p(t) *be the minimal polynomial for* T. *A scalar* λ *is an eigenvalue of* T *if and only if* $p(\lambda) = 0$. *Hence the characteristic polynomial and the minimal polynomial for* T *have the same zeros.*

Proof. Let $f(t)$ be the characteristic polynomial of T. Since $p(t)$ divides $f(t)$, $f(t) = q(t)p(t)$ for some polynomial $q(t)$. Let λ be a zero of $p(t)$. Then

$$f(\lambda) = q(\lambda)p(\lambda) = q(\lambda)\cdot 0 = 0.$$

So λ is also a zero for $f(t)$; that is, λ is an eigenvalue of T.

Conversely, suppose that λ is an eigenvalue of T, and let $x \in V$ be an eigenvector corresponding to λ. Then by Exercise 22 of Section 5.1 we have

$$0 = \mathsf{T}_0(x) = p(\mathsf{T})(x) = p(\lambda)x.$$

Since $x \neq 0$, $p(\lambda) = 0$, and so λ is a zero of $p(t)$. ∎

As an immediate consequence of the preceding result we have the following corollary.

Corollary. *Let* T *be a linear operator on a finite-dimensional vector space* V *with minimal polynomial* p(t) *and characteristic polynomial* f(t). *Suppose that* f(t) *factors as*

$$f(t) = (\lambda_1 - t)^{n_1}(\lambda_2 - t)^{n_2}\cdots(\lambda_k - t)^{n_k},$$

where $\lambda_1, \lambda_2, \ldots, \lambda_k$ *are the distinct eigenvalues of* T. *Then there exist integers* m_1,

$m_2, \ldots, m_k$ *such that* $1 \leqslant m_i \leqslant n_i$ *for all* i *and*

$$p(t) = (t - \lambda_1)^{m_1}(t - \lambda_2)^{m_2} \cdots (t - \lambda_k)^{m_k}.$$

Example 1

We compute the minimal polynomial for the matrix

$$A = \begin{pmatrix} 3 & -1 & 0 \\ 0 & 2 & 0 \\ 1 & -1 & 2 \end{pmatrix}.$$

Since A has characteristic polynomial

$$f(t) = \det \begin{pmatrix} 3-t & -1 & 0 \\ 0 & 2-t & 0 \\ 1 & -1 & 2-t \end{pmatrix} = -(t-2)^2(t-3),$$

the minimal polynomial for A must be either $(t-2)(t-3)$ or $(t-2)^2(t-3)$ by the corollary to Theorem 7.13. Substituting A into $p(t) = (t-2)(t-3)$ shows that $p(A)$ is the zero matrix. Thus $p(t)$ is the minimal polynomial for A. ∎

Example 2

Let $\mathsf{T}\colon \mathsf{R}^2 \to \mathsf{R}^2$ be defined by

$$\mathsf{T}(a, b) = (2a + 5b, 6a + b).$$

If β is the standard basis for R^3, then

$$[\mathsf{T}]_\beta = \begin{pmatrix} 2 & 5 \\ 6 & 1 \end{pmatrix}.$$

So the characteristic polynomial of $[\mathsf{T}]_\beta$, and hence of T, is

$$f(t) = \det \begin{pmatrix} 2-t & 5 \\ 6 & 1-t \end{pmatrix} = (t-7)(t+4).$$

Thus the minimal polynomial for T must be $(t-7)(t+4)$ also. ∎

Example 3

Let $\mathsf{D}\colon \mathsf{P}_2(R) \to \mathsf{P}_2(R)$ be the differentiation operator defined by $\mathsf{D}(f) = f'$. We compute the minimal polynomial for D. For the standard basis β we have

$$[\mathsf{D}]_\beta = \begin{pmatrix} 0 & 1 & 0 \\ 0 & 0 & 2 \\ 0 & 0 & 0 \end{pmatrix}.$$

Hence the characteristic polynomial of D is $-t^3$. So the corollary to Theorem

7.13 shows that the minimal polynomial for D is t, t^2, or t^3. Since $D^2(t^2) = 2 \neq 0$, $D^2 \neq T_0$. Thus the minimal polynomial for D must be t^3. ∎

In Example 3 it is easy to verify that $P_2(R)$ is a D-cyclic subspace (of itself). In this example we saw that the minimal and characteristic polynomials are of the same degree. This is no coincidence.

Theorem 7.14. *Let* T *be a linear operator on an* n*-dimensional vector space* V. *If* V *is a* T*-cyclic subspace of itself, then the characteristic polynomial* f(t) *and the minimal polynomial* p(t) *for* T *are of the same degree. Hence* $f(t) = (-1)^n p(t)$.

Proof. If V is a T-cyclic subspace, then there exists an element $x \in V$ such that

$$\beta = \{x, T(x), \ldots, T^{n-1}(x)\}$$

is a basis for V (Theorem 5.27). Let

$$g(t) = a_0 + a_1 t + \cdots + a_k t^k,$$

where $a_k \neq 0$ and $0 \leqslant k < n$. Then

$$g(T)(x) = a_0 x + a_1 T(x) + \cdots + a_k T^k(x)$$

is a linear combination of elements of β having at least one nonzero coefficient, namely a_k. Since β is linearly independent, $g(T)(x) \neq 0$, and hence $g(T) \neq T_0$. Therefore the minimal polynomial for T is of degree n, which is also the degree of the characteristic polynomial of T. ∎

Theorem 7.14 states a condition under which the degree of the minimal polynomial for an operator is as large as possible. We now investigate when the degree of the minimal polynomial is as small as possible. It follows from Theorem 7.13 that if the characteristic polynomial of an operator with k distinct eigenvalues splits, then the minimal polynomial must be of degree at least k. The next theorem shows that the operators for which the degree of the minimal polynomial is as small as possible are precisely the diagonalizable operators.

Theorem 7.15. *Let* T *be a linear operator on a finite-dimensional vector space* V. *Then* T *is diagonalizable if and only if the minimal polynomial for* T *is of the form*

$$p(t) = (t - \lambda_1)(t - \lambda_2) \cdots (t - \lambda_k)$$

where $\lambda_1, \lambda_2, \ldots, \lambda_k$ *are distinct scalars. (Note that in this case the* λ_i*'s are necessarily the distinct eigenvalues of* T.)

Proof. Suppose that T is diagonalizable. Let $\lambda_1, \lambda_2, \ldots, \lambda_k$ be the distinct eigenvalues of T, and define

$$p(t) = (t - \lambda_1)(t - \lambda_2) \cdots (t - \lambda_k).$$

By Theorem 7.13, $p(t)$ divides the minimal polynomial of T. Let $\beta = \{x_1, x_2, \ldots, x_n\}$ be a basis for V consisting of eigenvectors of T, and consider any x_i in β. Then $(\mathsf{T} - \lambda_j \mathsf{I})(x_i) = 0$ for the eigenvalue λ_j to which x_i corresponds. Since $(t - \lambda_j)$ divides $p(t)$, there is a polynomial $q_j(t)$ such that $p(t) = q_j(t)(t - \lambda_j)$ and hence,

$$p(\mathsf{T})(x_i) = q_j(\mathsf{T})(\mathsf{T} - \lambda_j \mathsf{I})(x_i) = 0.$$

It follows that $p(\mathsf{T}) = \mathsf{T}_0$ since $p(\mathsf{T})$ takes each member of a basis of V into the zero vector. Therefore, $p(t)$ is the minimal polynomial for T.

Conversely, suppose that there are distinct scalars $\lambda_1, \lambda_2, \ldots, \lambda_k$ (necessarily eigenvalues of T) such that the minimal polynomial $p(t)$ for T factors as

$$p(t) = (t - \lambda_1)(t - \lambda_2)\cdots(t - \lambda_k).$$

We apply mathematical induction to n, the dimension of V. If $n = 1$, then T is diagonalizable. Next, suppose that T is diagonalizable whenever $\dim(\mathsf{V}) < n$ for some $n > 1$, and suppose that $\dim(\mathsf{V}) = n$. Let $\mathsf{W} = \mathsf{R}(\mathsf{T} - \lambda_k \mathsf{I})$. Clearly, $\mathsf{W} \neq \mathsf{V}$ because λ_k is an eigenvalue of T. If $\mathsf{W} = \{0\}$, then $\mathsf{T} = \lambda_k \mathsf{I}$, which is clearly diagonalizable. Now suppose that $0 < \dim(\mathsf{W}) < n$. Then W is T-invariant, and for any $x \in \mathsf{W}$,

$$(\mathsf{T} - \lambda_1 \mathsf{I})\cdots(\mathsf{T} - \lambda_{k-1}\mathsf{I})(x) = 0.$$

It follows that the minimal polynomial of T_W divides the polynomial $(t - \lambda_1)\cdots(t - \lambda_{k-1})$. Thus we may apply the induction hypothesis to deduce that T_W is diagonalizable. Furthermore, λ_k is not an eigenvalue of T_W by Theorem 7.13. Let $\{x_1, x_2, \ldots, x_m\}$ be a basis for W consisting of eigenvectors of T_W, and hence of T, and let $\{x_{m+1}, \ldots, x_n\}$ be a basis for $\mathsf{N}(\mathsf{T} - \lambda_k \mathsf{I})$, the eigenspace of T corresponding to λ_k. The two sets are disjoint because otherwise λ_k would be an eigenvalue of T_W, contrary to a remark made above. We show that the union of the two sets is linearly independent. Consider scalars $a_1, \ldots, a_n$ such that

$$\sum_{i=1}^{m} a_i x_i + \sum_{i=m+1}^{n} a_i x_i = 0. \tag{3}$$

For each $i \leqslant m$, x_i is an eigenvector of T corresponding to an eigenvalue distinct from λ_k, and therefore there is a scalar $c_i \neq 0$ such that $(\mathsf{T} - \lambda_k \mathsf{I})(x_i) = c_i x_i$. For $i > m$, $(\mathsf{T} - \lambda_k \mathsf{I})(x_i) = 0$. Therefore, if we apply $\mathsf{T} - \lambda_k \mathsf{I}$ to both sides of (3), we obtain

$$\sum_{i=1}^{m} a_i c_i x_i = 0,$$

from which it follows that $a_i c_i = 0$ for $1 \leqslant i \leqslant m$. Since $c_i \neq 0$ for all such i, we have that $a_i = 0$ for $1 \leqslant i \leqslant m$. Thus (3) reduces to

$$\sum_{i=m+1}^{n} a_i x_i = 0,$$

and hence $a_i = 0$ for $m < i \leqslant n$. Thus $\beta = \{x_1, \ldots, x_m, x_{m+1}, \ldots, x_n\}$ is a linearly independent subset of V consisting of n vectors. It follows that β is a basis for V consisting of eigenvectors of T, and therefore, T is diagonalizable. ∎

Example 4

We will determine all matrices $A \in \mathsf{M}_{2\times 2}(R)$ for which $A^2 - 3A + 2I = O$, where O is the 2×2 zero matrix. Define $g(t) = t^2 - 3t + 2 = (t-1)(t-2)$. Since $g(A) = O$, the minimal polynomial $p(t)$ for A divides $g(t)$. Hence the only possible candidates for $p(t)$ are $t - 1$, $t - 2$, or $(t-1)(t-2)$. Note that in all of these cases A is diagonalizable by Theorem 7.15. If $p(t) = t - 1$ or $p(t) = t - 2$, then $A = I$ or $A = 2I$. If $p(t) = (t-1)(t-2)$, then A is similar to

$$\begin{pmatrix} 1 & 0 \\ 0 & 2 \end{pmatrix}. \quad ∎$$

Example 5

We will prove that if A is a real $n \times n$ matrix such that $A^3 = A$, then A is diagonalizable. Note that if $g(t) = t^3 - t = t(t+1)(t-1)$, then $g(A) = O$, where O is the $n \times n$ zero matrix. Hence the minimal polynomial $p(t)$ for A divides $g(t)$. Since $g(t)$ has no repeated factors, neither does $p(t)$. Thus A is diagonalizable by Theorem 7.15. ∎

Example 6

In Example 3 we saw that the minimal polynomial for the differentiation operator $\mathsf{D}\colon \mathsf{P}_2(R) \to \mathsf{P}_2(R)$ is t^3. Hence D is not diagonalizable (Theorem 7.15). ∎

EXERCISES

1. Label the following statements as being true or false. Assume in what follows that all vector spaces are finite-dimensional.
   (a) Every linear operator T has a polynomial $p(t)$ of largest degree for which $p(\mathsf{T}) = \mathsf{T}_0$.
   (b) Every linear operator has a unique minimal polynomial.
   (c) The characteristic polynomial of a linear operator divides the minimal polynomial for that operator.
   (d) The minimal and characteristic polynomials of any diagonalizable operator are identical.
   (e) Let T be a linear operator on an n-dimensional vector space V, $p(t)$ be the minimal polynomial for T, and $f(t)$ be the characteristic polynomial of T. If $f(t)$ splits, then $f(t)$ divides $[p(t)]^n$.

(f) The minimal polynomial for a linear operator always has the same degree as the characteristic polynomial of the operator.
(g) A linear operator is diagonalizable if its minimal polynomial splits.
(h) Let T be a linear operator on V. If V is a T-cyclic subspace, then the degree of the minimal polynomial for T equals dim(V).

2. Compute the minimal polynomials for each of the following matrices.

(a) $\begin{pmatrix} 2 & 1 \\ 1 & 2 \end{pmatrix}$ **(b)** $\begin{pmatrix} 1 & 1 \\ 0 & 1 \end{pmatrix}$

(c) $\begin{pmatrix} 4 & -14 & 5 \\ 1 & -4 & 2 \\ 1 & -6 & 4 \end{pmatrix}$ **(d)** $\begin{pmatrix} 3 & 0 & 1 \\ 2 & 2 & 2 \\ -1 & 0 & 1 \end{pmatrix}$

3. Compute the minimal polynomial for each of the following linear operators.
(a) $\mathsf{T}\colon \mathsf{R}^2 \to \mathsf{R}^2$, where $\mathsf{T}(a, b) = (a + b, a - b)$.
(b) $\mathsf{T}\colon \mathsf{P}_2(R) \to \mathsf{P}_2(R)$, where $\mathsf{T}(f) = f' + 2f$.
(c) $\mathsf{T}\colon \mathsf{P}_2(R) \to \mathsf{P}_2(R)$, where $\mathsf{T}(f) = -xf'' + f' + 2f$.
(d) $\mathsf{T}\colon \mathsf{M}_{n\times n}(R) \to \mathsf{M}_{n\times n}(R)$, where $\mathsf{T}(A) = A^t$. *Hint:* Note that $\mathsf{T}^2 = \mathsf{I}$.

4. Determine which of the matrices and operators in Exercises 2 and 3 are diagonalizable.

5. Describe all linear operators T on R^2 such that T is diagonalizable and $\mathsf{T}^3 - 2\mathsf{T}^2 + \mathsf{T} = \mathsf{T}_0$.

6. Prove Theorem 7.12 and its corollary.

7. Prove the corollary to Theorem 7.13.

8. Let T be a linear operator on a finite-dimensional vector space. Prove that if $g(t)$ is the minimal polynomial of T, then
(a) T is invertible if and only if $g(0) \neq 0$.
(b) If T is invertible and $g(t) = t^n + a_{n-1}t^{n-1} + \cdots + a_1 t + a_0$, then

$$\mathsf{T}^{-1} = -\left(\frac{1}{a_0}\mathsf{T}^{n-1} + \frac{a_{n-1}}{a_0}\mathsf{T}^{n-2} + \cdots + \frac{a_1}{a_0}\mathsf{I}\right).$$

9. Let T be a diagonalizable linear operator on a finite-dimensional vector space V. Prove that V is a T-cyclic subspace if and only if each of the eigenspaces of T is one-dimensional.

10. Let $g(t)$ be the auxiliary polynomial of a homogeneous linear differential equation with constant coefficients (as defined in Section 2.7), and let V denote the solution space of the differential equation. Show that
(a) V is a D-invariant subspace, where $\mathsf{D}\colon \mathsf{C}^\infty \to \mathsf{C}^\infty$ is the differentiation operator.
(b) The minimal polynomial for D_V (the restriction of D to V) is $g(t)$.

(c) If the degree of $g(t)$ is n, then the characteristic polynomial of $\mathsf{D}\colon \mathsf{V} \to \mathsf{V}$ is $(-1)^n g(t)$.

Hint: For parts (b) and (c), use Theorem 2.33.

11. Let $\mathsf{D}\colon \mathsf{P}(R) \to \mathsf{P}(R)$ be the differentiation operator on the space of all polynomials over R. Prove that there exists no polynomial $g(t)$ for which $g(\mathsf{D}) = \mathsf{T}_0$. Hence $\mathsf{D}\colon \mathsf{P}(R) \to \mathsf{P}(R)$ has no minimal polynomial.

12. Let T be a linear operator on a finite-dimensional vector space, and suppose that the characteristic polynomial of T splits. Let $\lambda_1, \lambda_2, \ldots, \lambda_k$ be the distinct eigenvalues of T, and for each i let p_i be the order of the largest Jordan block corresponding to λ_i in a Jordan canonical form of T. Prove that the minimal polynomial of T is

$$(t - \lambda_1)^{p_1}(t - \lambda_2)^{p_2} \cdots (t - \lambda_k)^{p_k}.$$

13. Let T be a linear operator on a finite-dimensional vector space V and suppose that W is a T-invariant subspace of V. Prove that the minimal polynomial of T_W divides the minimal polynomial of T.

The following exercise requires knowledge of direct sums (see Section 5.2).

14. Let V be a finite-dimensional vector space and T a linear operator on V. Let W_1 and W_2 be T-invariant subspaces of V such that $\mathsf{V} = \mathsf{W}_1 \oplus \mathsf{W}_2$, and let $p_1(t)$ and $p_2(t)$ be the minimal polynomials for $\mathsf{T}_{\mathsf{W}_1}$ and $\mathsf{T}_{\mathsf{W}_2}$, respectively. Prove or disprove that $p_1(t)p_2(t)$ is the minimal polynomial for T.

Definition. *Let* T *be a linear operator on a finite-dimensional vector space* V, *and let* x *be a nonzero vector in* V. *The polynomial* p(t) *is called a* T-annihilator *of* x *if* p(t) *is a monic polynomial of least degree for which* $\mathsf{p}(\mathsf{T})(\mathsf{x}) = 0$.

15.† Let T be a linear operator on a finite-dimensional vector space V, and let x be a nonzero vector of V. Prove:

(a) The vector x has a unique T-annihilator.

(b) The T-annihilator of x divides any polynomial $f(t)$ for which $f(\mathsf{T}) = \mathsf{T}_0$.

(c) If $p(t)$ is the T-annihilator of x and W is the T-cyclic subspace generated by x then $p(t)$ is the minimal polynomial for T_W, and dim(W) equals the degree of $p(t)$.

(d) The degree of the T-annihilator of x is 1 if and only if x is an eigenvector of T.

16. Let T be a linear operator on a finite-dimensional vector space V, and let W_1 be a T-invariant subspace of V. If $x \in \mathsf{V}$ and $x \notin \mathsf{W}_1$, prove the following.

(a) There exists a unique monic polynomial $g_1(t)$ of least positive degree such that $g_1(\mathsf{T})(x) \in \mathsf{W}_1$.

(b) If $h(t)$ is a polynomial for which $h(\mathsf{T})(x) \in \mathsf{W}_1$, then $g_1(t)$ divides $h(t)$.

(c) Let W_2 be a T-invariant subspace of V such that $\mathsf{W}_2 \subseteq \mathsf{W}_1$. Prove that if $g_2(t)$ is the unique monic polynomial of least positive degree such that $g_2(\mathsf{T})(x) \in \mathsf{W}_2$, then $g_1(t)$ divides $g_2(t)$. Deduce that $g_1(t)$ divides the minimal and characteristic polynomials of T.

7.4* RATIONAL CANONICAL FORM

Until now we have used eigenvalues, eigenvectors, and generalized eigenvectors in our analysis of linear operators with characteristic polynomials that split. In general, characteristic polynomials need not split, and, indeed, operators need not have eigenvalues! However, the unique factorization theorem for polynomials (Appendix E) guarantees that the characteristic polynomial $f(t)$ of any linear operator T on an n-dimensional vector space factors uniquely as

$$f(t) = (-1)^n(\phi_1(t))^{n_1}(\phi_2(t))^{n_2} \cdots (\phi_k(t))^{n_k},$$

where the $\phi_i(t)$'s $(1 \leqslant i \leqslant k)$ are distinct irreducible monic polynomials, and the n_i's are positive integers. In the case that $f(t)$ splits, each irreducible monic polynomial factor is of the form $\phi_i(t) = t - \lambda_i$, where λ_i is an eigenvalue of T, and there is a one-to-one correspondence between the eigenvalues of T and the irreducible monic factors of the characteristic polynomial. In general, eigenvalues need not exist, but the irreducible monic factors always do exist. In this section we establish structure theorems based on the irreducible monic factors of the characteristic polynomial instead of eigenvalues.

In this context, the following definition is the appropriate replacement for eigenspace and generalized eigenspace.

Definition. *Let* T *be a linear operator on a finite-dimensional vector space* V *with characteristic polynomial*

$$f(t) = (-1)^n(\phi_1(t))^{n_1}(\phi_2(t))^{n_2} \cdots (\phi_k(t))^{n_k},$$

where the $\phi_i(t)$'s $(1 \leqslant i \leqslant k)$ *are distinct irreducible monic polynomials, and the* n_i's *are positive integers. For each* i $(1 \leqslant i \leqslant k)$ *we define the set* K_{ϕ_i} *by*

$$\mathsf{K}_{\phi_i} = \{x \in \mathsf{V} \colon (\phi_i(\mathsf{T}))^p(x) = 0 \text{ for some positive integer } p\}.$$

We will see that each K_{ϕ_i} is a nontrivial T-invariant subspace of V. If $\phi_i(t) = t - \lambda$ is of degree one, then K_{ϕ_i} is the generalized eigenspace of T corresponding to the eigenvalue λ.

Having chosen suitable extensions of the concepts of eigenvalue and eigenspace, the next task is to describe a canonical form of a linear operator suitable to this context. The one that we will study is called the *rational canonical form.* Since a canonical form is a description of a matrix representation of a linear operator, it can be specified by describing the kinds of ordered bases used for the representations.

These bases naturally arise from the generators of certain cyclic subspaces. For this reason the reader should recall the definition of a T-cyclic subspace generated by a vector and Theorem 5.27 of Section 5.4. We briefly review this concept and introduce some new notation and terminology.

Let T be a linear operator on a finite-dimensional vector space V, and let x be a nonzero vector in V. We use the notation $\mathsf{C}_x(\mathsf{T})$ for the T-cyclic subspace generated by x. Recall (Theorem 5.27) that if $\dim(\mathsf{C}_x(\mathsf{T})) = k$, then the set

$$\{x, \mathsf{T}(x), \mathsf{T}^2(x), \ldots, \mathsf{T}^{k-1}(x)\}$$

is an ordered basis for $\mathsf{C}_x(\mathsf{T})$. To distinguish this basis from all other ordered bases for $\mathsf{C}_x(\mathsf{T})$, we call it the *T-cyclic basis generated by x* and denote it by $\mathsf{B}_x(\mathsf{T})$. Let A be the matrix representation of the restriction of T to $\mathsf{C}_x(\mathsf{T})$ relative to the ordered basis $\mathsf{B}_x(\mathsf{T})$. Recall from the proof of Theorem 5.27 that

$$A = \begin{pmatrix} 0 & 0 & \cdots & 0 & -a_0 \\ 1 & 0 & \cdots & 0 & -a_1 \\ 0 & 1 & \cdots & 0 & -a_2 \\ \vdots & \vdots & & \vdots & \vdots \\ 0 & 0 & \cdots & 1 & -a_{k-1} \end{pmatrix},$$

where

$$\mathsf{T}^k(x) = -(a_0 x + a_1 \mathsf{T}(x) + \cdots + a_{k-1}\mathsf{T}^{k-1}(x)).$$

Furthermore, the characteristic polynomial of A is given by

$$\det(A - tI) = (-1)^k(a_0 + a_1 t + \cdots + a_{k-1}t^{k-1} + t^k).$$

The matrix A is called the *companion matrix* of the monic polynomial $h(t) = a_0 + a_1 t + \cdots + a_{k-1}t^{k-1} + t^k$. Every monic polynomial has a companion matrix, and the characteristic polynomial of the companion matrix of any monic polynomial $g(t)$ of degree k is equal to $(-1)^k g(t)$ (see Exercise 19 of Section 5.4). By Theorem 7.14, the monic polynomial $h(t)$ is also the minimal polynomial for A. Since A is the matrix representation of the restriction of T to $\mathsf{C}_x(\mathsf{T})$, $h(t)$ is also the minimal polynomial for this restriction. By Exercise 15 of Section 7.3, $h(t)$ is also the T-annihilator of x.

It is the object of this section to prove that for every linear operator T on a finite-dimensional vector space V, there exists an ordered basis β for V such that the matrix representation $[\mathsf{T}]_\beta$ is of the form

$$\begin{pmatrix} C_1 & O & \cdots & O \\ O & C_2 & \cdots & O \\ \vdots & \vdots & & \vdots \\ O & O & \cdots & C_r \end{pmatrix},$$

where each C_i is the companion matrix of some polynomial $(\phi(t))^m$, where $\phi(t)$ is a monic irreducible divisor of the characteristic polynomial of T, and m is a positive integer. Such a matrix representation is called a *rational canonical form* of T. We will call the accompanying basis a *rational canonical basis* of T. The following theorem is a simple consequence of the accompanying lemma.

Lemma. *Let* T *be a linear operator on a finite-dimensional vector space* V, *let* x *be a nonzero vector in* V, *and suppose that the* T*-annihilator of* x *is of the form* $(\phi(t))^p$ *for some irreducible monic polynomial* $\phi(t)$. *Then* $\phi(t)$ *divides the minimal polynomial of* T, *and so* $x \in K_\phi$.

Proof. Suppose that $(\phi(t))^p$ is the T-annihilator of x. Then $(\phi(t))^p$, and hence $\phi(t)$, divides the minimal polynomial of T by Exercise 15 of Section 7.3; and by definition, $x \in K_\phi$. ∎

Theorem 7.16. *Let* T *be a linear operator on a finite-dimensional vector space* V, *and let* β *be an ordered basis for* V. *Then* β *is a rational canonical basis of* T *if and only if* β *is the disjoint union of* T*-cyclic bases* $B_{x_i}(T)$, *where each* x_i *lies in* K_ϕ *for some irreducible monic divisor* $\phi(t)$ *of the characteristic polynomial of* T.

Proof. Exercise. ∎

Example 1

The 8×8 matrix C given by

$$C = \left(\begin{array}{cc|cccc|cc} 0 & -3 & 0 & 0 & 0 & 0 & 0 & 0 \\ 1 & 1 & 0 & 0 & 0 & 0 & 0 & 0 \\ \hline 0 & 0 & 0 & 0 & 0 & -1 & 0 & 0 \\ 0 & 0 & 1 & 0 & 0 & 0 & 0 & 0 \\ 0 & 0 & 0 & 1 & 0 & -2 & 0 & 0 \\ 0 & 0 & 0 & 0 & 1 & 0 & 0 & 0 \\ \hline 0 & 0 & 0 & 0 & 0 & 0 & 0 & -1 \\ 0 & 0 & 0 & 0 & 0 & 0 & 1 & 0 \end{array}\right)$$

is a rational canonical form of the linear operator $T: R^8 \to R^8$ defined by $T = L_C$, and the standard ordered basis for R^8 is the accompanying rational canonical basis for T. In this case the submatrices C_1, C_2, and C_3 are the companion matrices of the polynomials $\phi_1(t)$, $(\phi_2(t))^2$, and $\phi_2(t)$, respectively, where

$$\phi_1(t) = t^2 - t + 3 \quad \text{and} \quad \phi_2(t) = t^2 + 1.$$

In the context of Theorem 7.16, β is the disjoint union of the T-cyclic bases

$$\beta = \mathsf{B}_{e_1}(\mathsf{T}) \cup \mathsf{B}_{e_3}(\mathsf{T}) \cup \mathsf{B}_{e_7}(\mathsf{T})$$
$$= \{e_1, e_2\} \cup \{e_3, e_4, e_5, e_6\} \cup \{e_7, e_8\}.$$

By Exercise 38 of Section 5.4 the characteristic polynomial $f(t)$ of T is given by the product of the characteristic polynomials of the companion matrices:

$$f(t) = \phi_1(t)(\phi_2(t))^2\phi_2(t) = \phi_1(t)(\phi_2(t))^3. \quad \blacksquare$$

This example suggests that if a linear operator has a rational canonical form, then the companion matrix of some power of each monic irreducible divisor $\phi(t)$ of its characteristic polynomial appears at least once as a submatrix of the canonical form. This follows easily from Exercise 38 of Section 5.4. In this case, K_ϕ is not trivial. However, the nontriviality of K_ϕ rests upon the existence of the rational canonical form, and the proof of this existence is the main goal of this section.

Since the minimal polynomial of a linear operator divides the characteristic polynomial (Theorem 7.11), the irreducible monic divisors of the minimal polynomial are also divisors of the characteristic polynomial, and it is easy to show that K_ϕ is nontrivial for any such divisor ϕ. Consequently, we use the minimal polynomial in place of the characteristic polynomial.

We begin with a result that lists several properties of irreducible divisors of the minimal polynomial. The reader is advised to review the definition of T-annihilator and the accompanying Exercise 15 of Section 7.3.

Theorem 7.17. *Let* T *be a linear operator on a finite-dimensional vector space* V, *and let*

$$\mathrm{p(t)} = (\phi_1(\mathrm{t}))^{\mathrm{m}_1}(\phi_2(\mathrm{t}))^{\mathrm{m}_2}\cdots(\phi_\mathrm{k}(\mathrm{t}))^{\mathrm{m}_\mathrm{k}}$$

be the minimal polynomial of T, *where the* $\phi_\mathrm{i}(\mathrm{t})$*'s* $(1 \leqslant \mathrm{i} \leqslant \mathrm{k})$ *are the distinct irreducible monic factors of* p(t), *and the* $\mathrm{m_i}$*'s are positive integers. Then*

(a) *For each* i, $\mathsf{K}_{\phi_\mathrm{i}}$ *is a nontrivial* T*-invariant subspace of* V.
(b) *For* $\mathrm{i} \neq \mathrm{j}$, $\mathsf{K}_{\phi_\mathrm{i}} \cap \mathsf{K}_{\phi_\mathrm{j}} = \{0\}$.
(c) *For* $\mathrm{i} \neq \mathrm{j}$, $\mathsf{K}_{\phi_\mathrm{i}}$ *is invariant under* $\phi_\mathrm{j}(\mathsf{T})$, *and the restriction of* $\phi_\mathrm{j}(\mathsf{T})$ *to* $\mathsf{K}_{\phi_\mathrm{i}}$ *is one-to-one.*
(d) *For each* i, $\mathsf{K}_{\phi_\mathrm{i}} = \mathsf{N}((\phi_\mathrm{i}(\mathrm{t}))^{\mathrm{m_i}})$.

Proof. (a) The proof that K_{ϕ_i} is a T-invariant subspace of V is an exercise. Let $f_i(t)$ be the polynomial obtained from $p(t)$ by omitting the factor $(\phi_i(t))^{m_i}$. To prove that K_{ϕ_i} is nontrivial, first observe that $f_i(t)$ is a proper divisor of $p(t)$, and therefore, there is a vector z in V for which $x = f_i(\mathsf{T})(z) \neq 0$. However, $x \in \mathsf{K}_{\phi_i}$ because

$$(\phi_i(\mathsf{T}))^{m_i}(x) = (\phi_i(\mathsf{T}))^{m_i} f_i(\mathsf{T})(z) = p(\mathsf{T})(z) = 0.$$

(b) Let $x \in \mathsf{K}_{\phi_i} \cap \mathsf{K}_{\phi_j}$, and suppose that $x \neq 0$. Then there exist positive integers p_i and p_j such that $(\phi_i(t))^{p_i}(x) = (\phi_j(t))^{p_j}(x) = 0$. Let $g(t)$ be the T-annihilator of x. Then $g(t)$ divides both $(\phi_i(t))^{p_i}$ and $(\phi_j(t))^{p_j}$. But this is impossible because these two polynomials are relatively prime (see Appendix E). We conclude that $x = 0$.

(c) Since K_{ϕ_i} is T-invariant, it is also $\phi_j(\mathsf{T})$-invariant. Suppose that $\phi_j(\mathsf{T})(x) = 0$ for some $x \in \mathsf{K}_{\phi_i}$. Then $x \in \mathsf{K}_{\phi_i} \cap \mathsf{K}_{\phi_j}$, and hence $x = 0$ by (b). We conclude that the restriction of $\phi_j(\mathsf{T})$ to K_{ϕ_i} is one-to-one.

(d) Consider any i. Clearly, $\mathsf{N}((\phi_i(t))^{m_i}) \subseteq \mathsf{K}_{\phi_i}$. Let $f_i(t)$ be as in (a). Since $f_i(t)$ is a product of polynomials of the form $\phi_j(t)$, for $j \neq i$ we have by (c) that K_{ϕ_i} is invariant under $f_i(\mathsf{T})$, and the restriction of $f_i(\mathsf{T})$ to K_{ϕ_i} is one-to-one. Hence this restriction is also onto. Let $x \in \mathsf{K}_{\phi_i}$. Then there exists $y \in \mathsf{K}_{\phi_i}$ such that $f_i(\mathsf{T})(y) = x$. Therefore,

$$(\phi_i(\mathsf{T}))^{m_i}(x) = (\phi_i(\mathsf{T}))^{m_i} f_i(\mathsf{T})(y) = p(\mathsf{T})(y) = 0,$$

and hence $x \in \mathsf{N}((\phi_i(\mathsf{T}))^{m_i})$. Thus $\mathsf{K}_{\phi_i} = \mathsf{N}((\phi_i(\mathsf{T}))^{m_i})$. ∎

Since rational canonical bases of an operator T are formed by taking unions of T-cyclic bases, we must take care that the resulting unions are linearly independent. Theorem 7.18 reduces this problem to the linear independence of subsets of K_ϕ, where ϕ is an irreducible monic divisor of the minimal polynomial of T.

Lemma. *Let T be a linear operator on a finite-dimensional vector space V, and let $\phi_1, \phi_2, \ldots \phi_k$ be the distinct irreducible monic divisors of the minimal polynomial of T. For each i $(1 \leqslant i \leqslant k)$ let $x_i \in \mathsf{K}_{\phi_i}$ be such that*

$$x_1 + x_2 + \cdots + x_k = 0. \tag{4}$$

Then $x_i = 0$ for all i.

Proof. The proof is by mathematical induction on k, the number of distinct divisors. The result is trivial if $k = 1$. Assume that the lemma holds for any $k - 1$ distinct divisors, where $k - 1 \geqslant 1$, and suppose that there are k distinct divisors $\phi_1, \phi_2, \ldots, \phi_k$ and vectors $x_i \in \mathsf{K}_{\phi_i}$ $(1 \leqslant i \leqslant k)$ satisfying (4). Since $x_k \in \mathsf{K}_{\phi_k}$, there is a positive integer p such that $(\phi_k(\mathsf{T}))^p(x_k) = 0$. We apply the operator $(\phi_k(\mathsf{T}))^p$ to both sides of (4) to obtain

$$(\phi_k(\mathsf{T}))^p(x_1) + (\phi_k(\mathsf{T}))^p(x_2) + \cdots + (\phi_k(\mathsf{T}))^p(x_{k-1}) = 0.$$

For each $i < k$, $(\phi_k(\mathsf{T}))^p(x_i) \in \mathsf{K}_{\phi_i}$ and hence, by the induction hypothesis, $(\phi_k(\mathsf{T}))^p(x_i) = 0$ for all i. Therefore, by part (c) of Theorem 7.17, $x_i = 0$ for $i < k$. Thus, (4) simplifies to the equation $x_k = 0$. ∎

Theorem 7.18. *Suppose that T is a linear operator on a finite-dimensional vector space V. Let $\phi_1, \phi_2, \ldots, \phi_k$ be distinct irreducible monic divisors of the* *minimal polynomial of* T, *and for each* i $(1 \leqslant i \leqslant k)$ *let* S_i *be a linearly independent subset of* K_{ϕ_i}. *Then* $S_i \cap S_j = \varnothing$ *for* $i \neq j$, *and* $S = S_1 \cup S_2 \cup \cdots \cup S_k$ *is a linearly independent subset of* V.

Proof. That $S_i \cap S_j = \varnothing$ for $i \neq j$ is a trivial consequence of part (b) of Theorem 7.17. The rest of the proof is formally identical to the proof of Theorem 7.3 of Section 7.1, except that subspaces of the form K_{ϕ_i} are used in place of the generalized eigenspaces. ∎

We now focus our attention on linearly independent subsets of K_ϕ that are unions of T-cyclic bases. These subsets are used to form rational canonical bases. The next several results give us ways of building up such sets. They serve the dual purposes of leading to the existence theorem for the rational canonical form and of providing methods for constructing rational canonical bases.

For Theorems 7.19 and 7.20 we fix a linear operator T on a finite-dimensional vector space V and an irreducible monic divisor $\phi(t)$ of the minimal polynomial of T.

Theorem 7.19. *Let* $x_1, x_2, \ldots, x_k$ *be distinct vectors in* K_ϕ *such that*

$$S_1 = B_{x_1}(T) \cup \cdots \cup B_{x_k}(T)$$

is linearly independent. For each i *let* z_i *be a vector in* V *such that* $\phi(T)(z_i) = x_i$. *Then*

$$S_2 = B_{z_1}(T) \cup \cdots \cup B_{z_k}(T)$$

is also linearly independent.

Proof. Consider any linear combination of the vectors of S_2 that sums to zero, say,

$$\sum_{i=1}^{k} \sum_{j=0}^{n_i} a_{ij} T^j(z_i) = 0. \tag{5}$$

For each i let $f_i(t)$ be the polynomial defined by

$$f_i(t) = \sum_{j=0}^{n_i} a_{ij} t^j.$$

Then (5) can be rewritten as

$$\sum_{i=1}^{k} f_i(T)(z_i) = 0. \tag{6}$$

Now apply $\phi(T)$ to both sides of (6) to obtain

$$\sum_{i=1}^{k} f_i(T)\phi(T)(z_i) = \sum_{i=1}^{k} f_i(T)(x_i) = 0.$$

The preceding sum can be rewritten as a linear combination of vectors in S_1.

Since S_1 is linearly independent, it follows that

$$f_i(\mathsf{T})(x_i) = 0 \qquad \text{for all } i.$$

Therefore, the T-annihilator of x_i divides $f_i(t)$ for all i (see Exercise 15 of Section 7.3). Since $\phi(t)$ divides the T-annihilator of x_i, it follows that $\phi(t)$ divides $f_i(t)$ for all i. Thus for each i there exists a polynomial $g_i(t)$ such that $f_i(t) = g_i(t)\phi(t)$. Hence (6) becomes

$$\sum_{i=1}^{k} g_i(\mathsf{T})\phi(\mathsf{T})(z_i) = \sum_{i=1}^{k} g_i(\mathsf{T})(x_i) = 0.$$

Again, linear independence of S_1 requires that

$$f_i(\mathsf{T})(z_i) = g_i(\mathsf{T})(x_i) = 0 \qquad \text{for all } i.$$

But $f_i(\mathsf{T})(z_i)$ is the result of grouping the terms of the linear combination in (5) that arise from the linearly independent set $\mathsf{B}_{z_i}(\mathsf{T})$. We conclude that for each i, $a_{ij} = 0$ for all j. Therefore, S_2 is linearly independent. ∎

We now show that K_ϕ has a basis consisting of a union of T-cyclic bases.

Lemma. *Let* W *be a* T*-invariant subspace of* K_ϕ*, and let* β *be a basis for* W*. Then*

(a) *For any* $x \in \mathsf{N}(\phi(\mathsf{T}))$*, if* $x \notin \mathsf{W}$*, then* $\beta \cup \mathsf{B}_x(\mathsf{T})$ *is linearly independent.*

(b) *For some* $z_1, z_2, \ldots, z_s$ *in* $\mathsf{N}(\phi(\mathsf{T}))$*,* β *can be extended to a linearly independent set*

$$\beta' = \beta \cup \mathsf{B}_{z_1}(\mathsf{T}) \cup \cdots \cup \mathsf{B}_{z_s}(\mathsf{T})$$

whose span contains $\mathsf{N}(\phi(\mathsf{T}))$.

Proof. (a) Let $\beta = \{x_1, x_2, \ldots, x_k\}$, and suppose that

$$\sum a_i x_i + z = 0,$$

where $z = \sum b_j \mathsf{T}^j(x)$. Then $z \in \mathsf{C}_x(\mathsf{T}) \cap \mathsf{W}$. Suppose that $z \neq 0$. Then z has T-annihilator $\phi(t)$, and therefore,

$$d = \dim(\mathsf{C}_z(\mathsf{T})) \leqslant \dim(\mathsf{C}_x(\mathsf{T}) \cap \mathsf{W}) \leqslant \dim(\mathsf{C}_x(\mathsf{T})) = d,$$

where d is the degree of $\phi(t)$. It follows that $\mathsf{C}_x(\mathsf{T}) \cap \mathsf{W} = \mathsf{C}_x(\mathsf{T})$, and hence $x \in \mathsf{W}$, contrary to the hypothesis. Therefore, $z = \sum b_j \mathsf{T}^j(x) = 0$, from which it follows that $b_j = 0$ for all j. Similarly, $a_i = 0$ for all i, and hence $\beta \cup \mathsf{B}_x(\mathsf{T})$ is linearly independent.

(b) If W does not contain $\mathsf{N}(\phi(\mathsf{T}))$, choose a vector z_1 in $\mathsf{N}(\phi(\mathsf{T}))$ not in W. Now apply (a) to conclude that $\beta_1 = \beta \cup \mathsf{B}_{z_1}(\mathsf{T})$ is linearly independent. Let W_1 denote the span of β_1. If W_1 does not contain $\mathsf{N}(\phi(\mathsf{T}))$, proceed as above to choose a vector z_2 in $\mathsf{N}(\phi(\mathsf{T}))$, not in W_1, and such that the set $\beta_2 = \beta \cup \mathsf{B}_{z_1}(\mathsf{T}) \cup \mathsf{B}_{z_2}(\mathsf{T})$ is linearly independent. Continue this process eventually to obtain vectors $z_1, z_2, \ldots, z_s$ in $\mathsf{N}(\phi(\mathsf{T}))$ such that the union

$$\beta' = \beta \cup \mathsf{B}_{z_1}(\mathsf{T}) \cup \cdots \cup \mathsf{B}_{z_s}(\mathsf{T})$$

is a linearly independent set whose span contains $\mathsf{N}(\phi(\mathsf{T}))$. ∎

Theorem 7.20. *If the minimal polynomial of* T *is of the form* $p(t) = (\phi(t))^m$, *then* T *has a rational canonical basis.*

Proof. The proof is by mathematical induction on m. Suppose that $m = 1$. Apply (b) of the lemma to $\mathsf{W} = \{0\}$ to obtain a linearly independent subset of V of the form $\mathsf{B}_{x_1}(\mathsf{T}) \cup \cdots \cup \mathsf{B}_{x_s}(\mathsf{T})$ whose span contains $\mathsf{N}(\phi(\mathsf{T}))$. Since $\mathsf{V} = \mathsf{N}(\phi(\mathsf{T}))$, this set is a rational canonical basis for V.

Now suppose that for some integer $m > 1$ the result is valid whenever the minimal polynomial for T is of the form $(\phi(t))^k$ where $k < m$, and assume that the minimal polynomial for T is $p(t) = (\phi(t))^m$. Let $r = \operatorname{rank}(\phi(\mathsf{T}))$. Then $\mathsf{R}(\phi(\mathsf{T}))$ is a T-invariant subspace of V, and the restriction of T to this subspace has minimal polynomial $(\phi(t))^{m-1}$. Therefore, we may apply the induction hypothesis to obtain a rational canonical basis for the restriction of T to $\mathsf{R}(\phi(\mathsf{T}))$. Suppose that $x_1, x_2, \ldots, x_k$ are the generating vectors of the T-cyclic bases that constitute this rational canonical basis. For each i choose z_i in V such that $x_i = \phi(\mathsf{T})(z_i)$. By Theorem 7.19 the union of the sets $\mathsf{B}_{z_i}(\mathsf{T})$ is linearly independent. Let β denote this union, and let W denote the space generated by β. Then W contains $\mathsf{R}(\phi(\mathsf{T}))$. Apply (b) of the lemma and adjoin additional T-cyclic bases (if necessary) $\mathsf{B}_{z_{k+1}}(\mathsf{T}), \ldots, \mathsf{B}_{z_s}(\mathsf{T})$ to β, where z_i is in $\mathsf{N}(\phi(\mathsf{T}))$ for $i \geqslant k$, to obtain a linearly independent set

$$\beta' = \mathsf{B}_{z_1}(\mathsf{T}) \cup \cdots \cup \mathsf{B}_{z_k}(\mathsf{T}) \cup \cdots \cup \mathsf{B}_{z_s}(\mathsf{T})$$

whose span W′ contains both W and $\mathsf{N}(\phi(\mathsf{T}))$.

We argue that $\mathsf{W}' = \mathsf{V}$. Since W′ is T-invariant, let U denote the restriction of $\phi(\mathsf{T})$ to W′. By the way in which W′ was obtained from $\mathsf{R}(\phi(\mathsf{T}))$, it follows that $\mathsf{R}(\mathsf{U}) = \mathsf{R}(\phi(\mathsf{T}))$ and $\mathsf{N}(\mathsf{U}) = \mathsf{N}(\phi(\mathsf{T}))$. Therefore,

$$\begin{aligned}\dim(\mathsf{W}') &= \operatorname{rank}(\mathsf{U}) + \operatorname{nullity}(\mathsf{U})\\ &= \operatorname{rank}(\phi(\mathsf{T})) + \operatorname{nullity}(\phi(\mathsf{T}))\\ &= \dim(\mathsf{V}).\end{aligned}$$

Thus $\mathsf{W}' = \mathsf{V}$, and β' is a rational canonical basis of T. ∎

Corollary. *In general,* K_ϕ *has a basis consisting of a union of* T*-cyclic bases.*

Proof. Apply Theorem 7.20 to the restriction of T to K_ϕ. ∎

We now extend Theorem 7.20 to the general case.

Theorem 7.21. *Every linear operator on a finite-dimensional vector space has a rational canonical basis and hence, a rational canonical form.*

Proof. Let T be a linear operator on the finite-dimensional vector space V, and let $p(t) = (\phi_1(t))^{m_1}(\phi_2(t))^{m_2} \cdots (\phi_k(t))^{m_k}$ be the minimal polynomial of T, where the $\phi_i(t)$'s are the distinct irreducible monic factors of $p(t)$, and $m_i > 0$ for all i. The proof is by mathematical induction on k. The case $k = 1$ is proved in Theorem 7.20.

Now suppose that the result is valid whenever the minimal polynomial contains fewer than k distinct irreducible factors for some $k > 1$, and suppose that $p(t)$ contains k distinct factors. Let U be the restriction of T to the T-invariant subspace $W = R((\phi_k(T))^{m_k})$, and let $q(t)$ be the minimal polynomial of U. Then $q(t)$ divides $p(t)$ by Exercise 13 of Section 7.3. Furthermore, $\phi_k(t)$ does not divide $q(t)$. For otherwise, there would exist a nonzero vector x in W such that $\phi_k(U)(x) = 0$, and a vector y in V such that $x = (\phi_k(T))^{m_k}(y)$. It follows that $(\phi_k(T))^{m_k+1}(y) = 0$, and hence $y \in K_{\phi_k}$ and $x = (\phi_k(T)^{m_k}(y) = 0$ by Theorem 7.17(d), a contradiction. Thus, $q(t)$ contains fewer than k distinct irreducible divisors. Therefore, by the induction hypothesis U has a rational canonical basis S_1 consisting of a union of U-cyclic bases (and hence of T-cyclic bases) of vectors from some of the subspaces K_{ϕ_i}, $1 \leqslant i \leqslant k-1$. By the corollary to Theorem 7.20, K_{ϕ_k} has a basis S_2 consisting of a union of T-cyclic bases. By Theorem 7.18, S_1 and S_2 are disjoint, and $S = S_1 \cup S_2$ is linearly independent. Let s denote the number of elements in S. Then

$$\begin{aligned} s &= \dim(R((\phi_k(T))^{m_k})) + \dim(K_{\phi_k}) \\ &= \text{rank}((\phi_k(T))^{m_k}) + \text{nullity}((\phi_k(T))^{m_k}) \\ &= n. \end{aligned}$$

We conclude that S is a basis for V. Therefore, S is a rational canonical basis, and T has a rational canonical form. ∎

The following theorem relates the rational canonical form to the characteristic polynomial.

Theorem 7.22. *Let T be a linear operator on an n-dimensional vector space V with characteristic polynomial*

$$f(t) = (-1)^n(\phi_1(t))^{n_1}(\phi_2(t))^{n_2} \cdots (\phi_k(t))^{n_k},$$

where the $\phi_i(t)$'s ($1 \leqslant i \leqslant k$) are distinct irreducible monic polynomials and the n_i's are positive integers. Then

(a) *For each i, $\phi_i(t)$ divides the minimal polynomial of T.*
(b) *For each i, $\dim(K_{\phi_i}) = d_i n_i$, where d_i denotes the degree of $\phi_i(t)$.*
(c) *If β is a rational canonical basis for V, then for each i, $\beta \cap K_{\phi_i}$ is a basis for K_{ϕ_i}.*

(d) *If* S_i *is a basis for* K_{ϕ_i} *for each* i, *then* $S = S_1 \cup \cdots \cup S_k$ *is a basis for* V. *In particular, if each* S_i *is a disjoint union of* T*-cyclic bases, then* S *is a rational canonical basis for* T.

Proof. (a) By Theorem 7.21, T has a rational canonical form C. By Exercise 38 of Section 5.4 the characteristic polynomial of C, and hence of T, is the product of the characteristic polynomials of the companion matrices that compose C. Therefore, each irreducible monic divisor $\phi(t)$ of $f(t)$ divides the characteristic polynomial of at least one of the companion matrices, and hence for some positive integer p, $(\phi(t))^p$ is the T-annihilator of a nonzero vector of V. We conclude that $(\phi(t))^p$, and hence $\phi(t)$, divides the minimal polynomial of T.

(b), (c), and (d) Let $C = [\mathsf{T}]_\beta$. For each i, $d_i n_i$ is the sum of the orders of the companion matrices of C that arise from T-cyclic bases in K_{ϕ_i}, and this number is equal to the number of elements in $\beta_i = \beta \cap K_{\phi_i}$, which is a linearly independent subset of K_{ϕ_i}. It follows that $n_i d_i \leqslant \dim(K_{\phi_i})$ for all i. Furthermore, the sum of the $d_i n_i$'s is the degree of $f(t)$. For each i, let S_i be any basis of K_{ϕ_i}, and let $S = S_1 \cup \cdots \cup S_k$. By Theorem 7.18, S is linearly independent and the bases are disjoint. Therefore,

$$n = \sum_{i=1}^{k} d_i n_i \leqslant \sum_{i=1}^{k} \dim(K_{\phi_i}) \leqslant n,$$

and it follows that $\dim(K_{\phi_i}) = d_i n_i$ for all i. Therefore, β_i is a basis for K_{ϕ_i}. Moreover, S contains n elements, and hence is a basis for V. ∎

Uniqueness of the Rational Canonical Form

Having shown that a rational canonical form exists, we are now in a position to ask about the extent to which it is unique. Certainly, the rational canonical form of an operator T can be modified by permuting the T-cyclic bases that constitute the corresponding rational canonical basis. This has the effect of permuting the companion matrices that make up the rational canonical form. As in the case of the Jordan canonical form, we will show that except for these permutations, the rational canonical form is unique, although rational canonical bases are not.

To simplify this task we adopt the convention of ordering every rational canonical basis so that all the T-cyclic bases associated with the same irreducible monic divisor of the characteristic polynomial are grouped together. Furthermore, within each such grouping we will always arrange the T-cyclic bases in order of decreasing size. Our task is to show that subject to this order, the rational canonical form of a linear operator is unique up to the arrangement of the irreducible monic divisors.

As in the case of the Jordan canonical form of Section 7.2, for each irreducible monic divisor ϕ of the characteristic polynomial of T we introduce a diagram of dots to describe the order of that part of the rational canonical basis contained in K_ϕ. We prove that the diagram is completely determined by a property intrinsic to operators, namely rank.

For what follows T is a linear operator with a rational canonical basis β, $\phi(t)$ is an irreducible monic divisor of its characteristic polynomial of degree d, and $\mathsf{B}_{x_1}(\mathsf{T}), \mathsf{B}_{x_2}(\mathsf{T}), \ldots, \mathsf{B}_{x_k}(\mathsf{T})$ are the T-cyclic bases of β that are contained in K_ϕ. For each j let $(\phi(t))^{p_j}$ be the T-annihilator of x_j. This polynomial has degree dp_j, and therefore, by Exercise 15 of Section 7.3, $\mathsf{B}_{x_j}(\mathsf{T})$ contains dp_j elements. Furthermore, $p_1 \geqslant p_2 \geqslant \cdots \geqslant p_k$ since the T-cyclic bases are arranged in decreasing order of size. The *dot diagram* associated with this configuration is defined to be the array of dots consisting of k columns with p_j dots in the jth column and arranged so that the jth column begins at the top and terminates after p_j dots. For example, if $k = 3$, $p_1 = 4$, $p_2 = 2$, and $p_3 = 2$, the dot diagram is given below.

In contrast to the Jordan canonical form, the dots of a dot diagram for a rational canonical form do not correspond to individual members of the rational canonical basis.

For each i, let r_i be the number of dots in the ith row of a dot diagram. In the diagram above, $r_1 = r_2 = 3$, and $r_3 = r_4 = 1$. According to Exercise 7 of Section 7.2, the r_i's determine the p_j's, and vice versa.

Example 2

Recall the rational canonical form C of Example 1. Since there are two irreducible monic divisors of the characteristic polynomial of C, there are two dot diagrams to consider, one for $\phi_1(t) = t^2 - t + 3$ and the other for $\phi_2(t) = t^2 + 1$. Since $\phi_1(t)$ is the T-annihilator of e_1 and $\mathsf{B}_{e_1}(\mathsf{T})$ is a basis for K_{ϕ_1}, the dot diagram for $\phi_1(t)$ consists of a single dot. The other two T-cyclic bases, $\mathsf{B}_{e_3}(\mathsf{T})$ and $\mathsf{B}_{e_7}(\mathsf{T})$, lie in K_{ϕ_2}. Since e_3 has T-annihilator $(\phi_2(t))^2$ and e_7 has T-annihilator $\phi_2(t)$, $p_1 = 2$ and $p_2 = 1$ for the dot diagram of $\phi_2(t)$. These diagrams are displayed below.

Dot diagram for $\phi_1(t)$ Dot diagram for $\phi_2(t)$ ∎

The next theorem tells us that the r_i's are expressible in terms of the ranks of powers of $\phi(\mathsf{T})$, and thus are independent of the choice of a rational canonical basis. The p_j's are therefore independent of the choice of basis. Hence, subject to the conventions we have discussed, the rational canonical form of an operator is unique.

Theorem 7.23. *Let T be a linear operator on a finite-dimensional vector space V, let $\phi(t)$ be an irreducible monic divisor of the characteristic polynomial of T of degree d, and let r_i and p_j be as above for the dot diagram for $\phi(t)$ with respect to a rational canonical basis of T. Then*

$$r_1 = \frac{1}{d}[\dim(\mathsf{V}) - \operatorname{rank}(\phi(\mathsf{T}))]$$

and

$$r_i = \frac{1}{d}[\operatorname{rank}((\phi(\mathsf{T}))^{i-1}) - \operatorname{rank}((\phi(\mathsf{T}))^i)] \qquad \textit{for } i > 1.$$

The key to the proof of this theorem rests on a lemma which we now state. An outline of the proof of the lemma is provided with many of the details left to the exercises.

Lemma. *Let a be the total number of dots in the dot diagram for $\phi(t)$. Then $da = \dim(\mathsf{K}_\phi)$. Furthermore, for any $i \leqslant p_1$,*

$$\operatorname{nullity}((\phi(\mathsf{T}))^i) = d(r_1 + \cdots + r_i).$$

Outline of Proof. Let $\mathsf{B}_{x_1}(\mathsf{T}), \ldots, \mathsf{B}_{x_k}(\mathsf{T})$ be the T-cyclic bases in that part of the rational canonical basis lying in K_ϕ. By Theorem 7.22, the union is a basis for K_ϕ. For each j, the number of vectors in $\mathsf{B}_{x_j}(\mathsf{T})$ is dp_j. Therefore,

$$\dim(\mathsf{K}_\phi) = \sum dp_j = da.$$

Let U be the restriction of T to K_ϕ. Then $\mathsf{N}((\phi(\mathsf{T}))^i) = \mathsf{N}((\phi(\mathsf{U}))^i)$, and therefore it suffices to prove the second part of the lemma for $\operatorname{nullity}((\phi(\mathsf{U}))^i)$. The range $\mathsf{R}((\phi(\mathsf{U}))^i)$ is generated by the images under $(\phi(\mathsf{U}))^i$ of the vectors in $\mathsf{B}_{x_1}(\mathsf{T}) \cup \cdots \cup \mathsf{B}_{x_k}(\mathsf{T})$. For each j such that $p_j > i$, let $y_j = (\phi(\mathsf{U}))^i(x_j)$. Then the range of $(\phi(\mathsf{U}))^i$ is generated by $\bigcup_{p_j > i} \mathsf{B}_{y_j}(\mathsf{T})$, and this set is linearly independent. Furthermore, each y_j has $(\phi(\mathsf{T}))^{p_j - i}$ as its annihilator, and hence $\mathsf{B}_{y_j}(\mathsf{T})$ consists of $d(p_j - i)$ elements. Therefore,

$$\operatorname{rank}((\phi(\mathsf{U}))^i) = \sum_{p_j > i} d(p_j - i).$$

Since $p_j - i$ is the number of dots in the jth column of the diagram after the first i rows have been removed, we have that

$$a - \sum_{p_j > i} (p_j - i) = r_1 + \cdots + r_i.$$

Since $\dim(\mathsf{K}_\phi) = da$,

$$\begin{aligned}\text{nullity}((\phi(\mathsf{T}))^i) &= \text{nullity}((\phi(\mathsf{U}))^i)\\ &= da - \text{rank}((\phi(\mathsf{U}))^i)\\ &= da - \sum_{p_j > i} d(p_j - i)\\ &= d\left[a - \sum_{p_j > i}(p_j - i)\right]\\ &= d(r_1 + \cdots + r_i). \quad \blacksquare\end{aligned}$$

Proof of Theorem 7.23. For $i = 1$

$$dr_1 = \text{nullity}(\phi(\mathsf{T})) = \dim(\mathsf{V}) - \text{rank}(\phi(\mathsf{T})).$$

For $i > 1$,

$$\begin{aligned}dr_i &= d(r_1 + \cdots + r_i) - d(r_1 + \cdots + r_{i-1})\\ &= \text{nullity}((\phi(\mathsf{T}))^i) - \text{nullity}((\phi(\mathsf{T}))^{i-1})\\ &= [\dim(\mathsf{V}) - \text{rank}((\phi(\mathsf{T}))^i)] - [\dim(\mathsf{V}) - \text{rank}((\phi(\mathsf{T}))^{i-1})]\\ &= [\text{rank}((\phi(\mathsf{T}))^{i-1}) - \text{rank}((\phi(\mathsf{T}))^i)],\end{aligned}$$

and the result follows. $\blacksquare$

Corollary. *Under the convention described earlier, the rational canonical form of a linear operator is unique up to the arrangement of the irreducible monic divisors of the characteristic polynomial.*

Example 3

Let $\beta = \{e^x \cos(2x), e^x \sin(2x), xe^x \cos(2x), xe^x \sin(2x)\}$, and let W be the span of β in C^∞ (see Section 2.7). Then W is a four-dimensional subspace of C^∞, and β is an ordered basis for W. Let D: W → W be the mapping defined by $\mathsf{D}(f) = f'$, the derivative of f. Then D is a linear operator on W. We will find the rational canonical form and a rational canonical basis of D. Let $A = [\mathsf{D}]_\beta$. Then

$$A = \begin{pmatrix} 1 & 2 & 1 & 0\\ -2 & 1 & 0 & 1\\ 0 & 0 & 1 & 2\\ 0 & 0 & -2 & 1\end{pmatrix},$$

and the characteristic polynomial of D, and hence of A, is

$$f(t) = (t^2 - 2t + 5)^2.$$

Thus $\phi(t) = t^2 - 2t + 5$ is the only irreducible monic divisor of $f(t)$. Since $\phi(t)$ has degree 2 and W is four-dimensional, the dot diagram for $\phi(t)$ contains only two dots. Therefore, the diagram is determined by r_1, the number of dots in the first row. Because ranks are preserved under matrix representations, we use A in place of D in the formula given in Theorem 7.23. Since

$$\phi(A) = \begin{pmatrix} 0 & 0 & 0 & 4 \\ 0 & 0 & -4 & 0 \\ 0 & 0 & 0 & 0 \\ 0 & 0 & 0 & 0 \end{pmatrix},$$

we have that

$$r_1 = \tfrac{1}{2}[4 - \operatorname{rank}(\phi(A))] = \tfrac{1}{2}[4 - 2] = 1.$$

It follows that the second dot lies in the second row, and the dot diagram is given by

$$\begin{array}{c} \bullet \\ \bullet \end{array}$$

Therefore, W is a cyclic space, and it is generated by a single function f with D-annihilator $(\phi(t))^2$. Furthermore, its rational canonical form is given by the companion matrix of $(\phi(t))^2 = t^4 - 4t^3 + 14t^2 - 20t + 25$,

$$\begin{pmatrix} 0 & 0 & 0 & -25 \\ 1 & 0 & 0 & 20 \\ 0 & 1 & 0 & -14 \\ 0 & 0 & 1 & 4 \end{pmatrix}.$$

For the cyclic generator it suffices to find a function g in W for which $\phi(\mathsf{D})(g) \neq 0$. Since $\phi(A)e_3 \neq 0$, it follows that $\phi(\mathsf{D})(xe^x \cos(2x)) \neq 0$, and therefore, $g(x) = xe^x \cos(2x)$ can be chosen as the cyclic generator. Therefore,

$$\mathsf{B}_g(\mathsf{D}) = \{xe^x \cos(2x),\ \mathsf{D}(xe^x \cos(2x)),\ \mathsf{D}^2(xe^x \cos(2x),\ \mathsf{D}^3(xe^x \cos(2x))\}$$

is a rational canonical basis of D. Notice that the function h defined by $h(x) = xe^x \sin(2x)$ can be chosen in place of g. This shows that the rational canonical basis is not unique. ∎

We now define the rational canonical form of a matrix in the natural way.

Definition. *The* rational canonical form *of a matrix* A *in* $\mathsf{M}_{n \times n}(F)$ *is defined to be the rational canonical form of the linear operator* $\mathsf{L}_A\colon \mathsf{F}^n \to \mathsf{F}^n$. *The accompanying basis will be called a* rational canonical basis *of* A.

Example 4

Find the rational canonical form and a rational canonical basis of the real matrix

$$A = \begin{pmatrix} 0 & 2 & 0 & -6 & 2 \\ 1 & -2 & 0 & 0 & 2 \\ 1 & 0 & 1 & -3 & 2 \\ 1 & -2 & 1 & -1 & 2 \\ 1 & -4 & 3 & -3 & 4 \end{pmatrix}.$$

The characteristic polynomial of A is

$$f(t) = \det(A - tI) = -(t^2 + 1)^2(t - 2),$$

and therefore $\phi_1(t) = t^2 + 1$ and $\phi_2(t) = t - 2$ are the distinct irreducible monic divisors of $f(t)$. By Theorem 7.22, $\dim(\mathsf{K}_{\phi_1}) = 4$ and $\dim(\mathsf{K}_{\phi_2}) = 1$. Since the degree of $\phi_1(t)$ is 2, the total number of dots in the dot diagram for $\phi_1(t)$ is $4/2 = 2$, and the number of dots in the first row r_1 is given by

$$\begin{aligned} r_1 &= \tfrac{1}{2}[\dim(\mathsf{R}^5) - \operatorname{rank}(\phi_1(A))] \\ &= \tfrac{1}{2}[5 - \operatorname{rank}(A^2 + 2I)] \\ &= \tfrac{1}{2}(5 - 1) = 2. \end{aligned}$$

Thus the dot diagram for $\phi_1(t)$ is given by

$$\bullet \quad \bullet$$

and each column of this diagram contributes the companion matrix for $\phi_1(t) = t^2 + 2$

$$\begin{pmatrix} 0 & -2 \\ 1 & 0 \end{pmatrix}$$

to the rational canonical form. Since $\dim(\mathsf{K}_{\phi_2}) = 1$, the dot diagram of $\phi_2(t) = t - 2$ consists of a single dot, which contributes the 1×1 matrix (2). Therefore, the rational canonical form of A is

$$\left(\begin{array}{cc:cc:c} 0 & -2 & 0 & 0 & 0 \\ 1 & 0 & 0 & 0 & 0 \\ \hdashline 0 & 0 & 0 & -2 & 0 \\ 0 & 0 & 1 & 0 & 0 \\ \hdashline 0 & 0 & 0 & 0 & 2 \end{array}\right).$$

If β is a rational canonical basis of A, then $\beta \cap \mathsf{K}_{\phi_1}$ is the union of two cyclic bases $\mathsf{B}_{x_1}(\mathsf{L}_A)$ and $\mathsf{B}_{x_2}(\mathsf{L}_A)$, where x_1 and x_2 each have annihilator $\phi_1(t)$. It follows that each x_i lies in $\mathsf{N}(\phi(\mathsf{L}_A))$. It can be verified that

$$\left\{\begin{pmatrix}1\\0\\0\\0\\0\end{pmatrix}, \begin{pmatrix}0\\1\\0\\0\\0\end{pmatrix}, \begin{pmatrix}0\\0\\2\\1\\0\end{pmatrix}, \begin{pmatrix}0\\0\\-1\\0\\1\end{pmatrix}\right\}$$

is a basis for $\mathsf{N}(\phi(\mathsf{L}_A))$. Setting

$$x_1 = e_1 = \begin{pmatrix}1\\0\\0\\0\\0\end{pmatrix},$$

we see that

$$Ax_1 = \begin{pmatrix}0\\1\\1\\1\\1\end{pmatrix}.$$

Next, choose x_2 to be a vector in K_{ϕ_1} that it is linearly independent of $\mathsf{B}_{x_1}(\mathsf{L}_A) = \{x_1, Ax_1\}$, for example,

$$x_2 = e_2 = \begin{pmatrix}0\\1\\0\\0\\0\end{pmatrix}.$$

Then it can be seen that

$$Ax_2 = \begin{pmatrix}2\\-2\\0\\-2\\-4\end{pmatrix},$$

and $\mathsf{B}_{x_1}(\mathsf{L}_A) \cup \mathsf{B}_{x_2}(\mathsf{L}_A)$ is a basis for K_{ϕ_1}.

Since the dot diagram of $\phi_2(t) = t - 2$ consists of a single dot, any nonzero vector in K_{ϕ_2} is an eigenvector of A corresponding to eigenvalue $\lambda = 2$. For example, choose

$$x_3 = \begin{pmatrix} 0 \\ 1 \\ 1 \\ 1 \\ 2 \end{pmatrix}.$$

By Theorem 7.22, $\beta = \{x_1, Ax_1, x_2, Ax_2, x_3\}$ is a rational canonical basis of A. ∎

Example 5

For the matrix A of Example 4, we find an invertible matrix Q such that $Q^{-1}AQ$ is the rational canonical form of A. Let Q be the matrix whose columns are the vectors of the rational canonical basis β of Example 4. That is,

$$Q = \begin{pmatrix} 1 & 0 & 0 & 2 & 0 \\ 0 & 1 & 1 & -2 & 1 \\ 0 & 1 & 0 & 0 & 1 \\ 0 & 1 & 0 & -2 & 1 \\ 0 & 1 & 0 & -4 & 2 \end{pmatrix}.$$

Then $Q^{-1}AQ$ is the rational canonical form of A by Theorem 5.1. ∎

Example 6

Find the rational canonical form and a rational canonical basis of the real matrix

$$A = \begin{pmatrix} 2 & 1 & 0 & 0 \\ 0 & 2 & 1 & 0 \\ 0 & 0 & 2 & 0 \\ 0 & 0 & 0 & 2 \end{pmatrix}.$$

The characteristic polynomial of A is $f(t) = (t - 2)^4$. Therefore, $\phi(t) = t - 2$ is the only irreducible monic divisor of $f(t)$, and $\mathsf{K}_\phi = \mathsf{R}^4$. In this case $\phi(t)$ has degree 1. If we apply Theorem 7.23 to compute the dot diagram for $\phi(t)$, the result is

$$r_1 = 4 - \operatorname{rank}(\phi(A)) = 4 - 2 = 2,$$

$$r_2 = \operatorname{rank}(\phi(A)) - \operatorname{rank}((\phi(A))^2) = 2 - 1 = 1,$$

and

$$r_3 = \operatorname{rank}((\phi(A))^2) - \operatorname{rank}((\phi(A))^3) = 1 - 0 = 1.$$

Since there are $4 = \dim(\mathsf{R}^4)/1$ dots in the diagram, we may terminate the computation with r_3. Thus the dot diagram for A is

$$\begin{array}{cc} \bullet & \bullet \\ \bullet & \\ \bullet & \end{array}$$

Since $(t-2)^3$ has the companion matrix

$$\begin{pmatrix} 0 & 0 & 8 \\ 1 & 0 & -12 \\ 0 & 1 & 6 \end{pmatrix}$$

and $(t-2)$ has the companion matrix

$$(2),$$

the rational canonical form of A is given by

$$\left(\begin{array}{ccc:c} 0 & 0 & 8 & 0 \\ 1 & 0 & -12 & 0 \\ 0 & 1 & 6 & 0 \\ \hdashline 0 & 0 & 0 & 2 \end{array}\right).$$

We now find a rational canonical basis of A. The dot diagram above indicates that there are vectors x_1 and x_2 in R^4 with annihilators $(\phi(t))^3$ and $\phi(t)$, respectively, and such that

$$\beta = \{\mathsf{B}_{x_1}(\mathsf{L}_A) \cup \mathsf{B}_{x_2}(\mathsf{L}_A)\} = \{x_1, Ax_1, A^2x_1, x_2\}$$

is a rational canonical basis of A. Moreover, $x_1 \notin \mathsf{N}((\mathsf{L}_A - 2\mathsf{I})^2)$, and $x_2 \in \mathsf{N}(\mathsf{L}_A - 2\mathsf{I})$. It can easily be shown that

$$\mathsf{N}(\mathsf{L}_A - 2\mathsf{I}) = \operatorname{span}(\{e_1, e_4\})$$

and

$$\mathsf{N}((\mathsf{L}_A - 2\mathsf{I})^2) = \operatorname{span}(\{e_1, e_2, e_4\}).$$

The standard vector e_3 meets the criteria for x_1, so we set $x_1 = e_3$. It follows that

$$Ax_1 = \begin{pmatrix} 0 \\ 1 \\ 2 \\ 0 \end{pmatrix} \quad \text{and} \quad A^2x_1 = \begin{pmatrix} 1 \\ 4 \\ 4 \\ 0 \end{pmatrix}.$$

Next, we choose a vector x_2 not in the span of $\mathsf{B}_{x_1}(\mathsf{T})$ but in $\mathsf{N}(\mathsf{L}_A - 2\mathsf{I})$. Clearly, e_4 satisfies this condition. Thus

$$\beta = \left\{ \begin{pmatrix} 0 \\ 0 \\ 1 \\ 0 \end{pmatrix}, \begin{pmatrix} 0 \\ 1 \\ 2 \\ 0 \end{pmatrix}, \begin{pmatrix} 1 \\ 4 \\ 4 \\ 0 \end{pmatrix}, \begin{pmatrix} 0 \\ 0 \\ 0 \\ 1 \end{pmatrix} \right\}$$

is a rational canonical basis of A. ∎

Direct Sums*

The following theorem is a simple consequence of Theorem 7.22.

Theorem 7.24. *Let* T *be a linear operator on an* n*-dimensional vector space* V *with characteristic polynomial*

$$f(t) = (-1)^n(\phi_1(t))^{n_1}(\phi_2(t))^{n_2} \cdots (\phi_k(t))^{n_k},$$

where the $\phi_i(t)$*'s* $(1 \leqslant i \leqslant k)$ *are distinct irreducible monic polynomials and the* n_i*'s are positive integers. Then*

(a) $\mathsf{V} = \mathsf{K}_{\phi_1} \oplus \mathsf{K}_{\phi_2} \oplus \cdots \oplus \mathsf{K}_{\phi_k}$.

(b) *If* T_i *is the restriction of* T *to* K_{ϕ_i} *and* C_i *is the rational canonical form of* T_i*, then*

$$C_1 \oplus C_2 \oplus \cdots \oplus C_k$$

is the rational canonical form of T.

Proof. Exercise. ∎

The following theorem is a simple consequence of Theorem 7.16.

Theorem 7.25. *Let* T *be a linear operator on a finite-dimensional vector space* V*. Then* V *is a direct sum of* T*-cyclic subspaces* $\mathsf{C}_{x_i}(\mathsf{T})$*, where each* x_i *lies in* K_ϕ *for some irreducible monic divisor* $\phi(t)$ *of the characteristic polynomial of* T.

Proof. Exercise. ∎

EXERCISES

1. Label the following statements as being true or false.

(a) Every rational canonical basis of a linear operator T is the union of T-cyclic bases.

(b) If a basis β is the union of T-cyclic bases of a linear operator T, then β is a rational canonical basis of T.

(c) There exist square matrices having no rational canonical form.
(d) A square matrix is similar to its rational canonical form.
(e) The Jordan canonical form and rational canonical form of any linear operator are the same.
(f) For any linear operator T on a finite-dimensional vector space, any irreducible factor of the characteristic polynomial of T divides the minimal polynomial of T.
(g) Let $\phi(t)$ be an irreducible monic divisor of the characteristic polynomial of a linear operator T. The dots in the dot diagram used to compute the rational canonical form of $\mathsf{T}_{\mathsf{K}_\phi}$ are in one-to-one correspondence with the vectors in a basis for K_ϕ.

2. For each of the following, find the rational canonical form and an accompanying rational canonical basis.

(a) The real matrix

$$A = \begin{pmatrix} 3 & 1 & 0 \\ 0 & 3 & 1 \\ 0 & 0 & 3 \end{pmatrix}$$

(b) The real matrix

$$A = \begin{pmatrix} 0 & -1 \\ 1 & -1 \end{pmatrix}$$

(c) The complex matrix

$$A = \begin{pmatrix} 0 & -1 \\ 1 & -1 \end{pmatrix}$$

(d) The real matrix

$$A = \begin{pmatrix} 0 & -7 & 14 & -6 \\ 1 & -4 & 6 & -3 \\ 0 & -4 & 9 & -4 \\ 0 & -4 & 11 & -5 \end{pmatrix}$$

(e) The real matrix

$$A = \begin{pmatrix} 0 & -4 & 12 & -7 \\ 1 & -1 & 3 & -3 \\ 0 & -1 & 6 & -4 \\ 0 & -1 & 8 & -5 \end{pmatrix}$$

3. Prove that if T is a linear operator on a finite-dimensional vector space V with minimal polynomial $(\phi(t))^m$ for some positive integer m, then $\mathsf{N}((\phi(\mathsf{T}))^{m-1})$ is a proper T-invariant subspace of V.

4. Let T be a linear operator on a finite-dimensional vector space with minimal polynomial $(\phi(t))^m$ for some irreducible monic polynomial $\phi(t)$ and some positive integer m. Prove that the restriction of T to $\mathsf{R}(\phi(\mathsf{T}))$ has minimal polynomial $(\phi(t))^{m-1}$.

5. Let T be a linear operator on a finite-dimensional vector space. Prove that the rational canonical form of T is a diagonal matrix if and only if T is diagonalizable.

6. Let T be a linear operator on a finite-dimensional vector space V with characteristic polynomial $f(t) = (-1)^n \phi_1(t)\phi_2(t)$, where $\phi_1(t)$ and $\phi_2(t)$ are distinct irreducible monic polynomials and $n = \dim(\mathsf{V})$.

(a) Prove that there exist elements x_1 and x_2 in V such that x_1 has T-annihilator $\phi_1(t)$, x_2 has T-annihilator $\phi_2(t)$, and $\mathsf{B}_{x_1}(\mathsf{T}) \cup \mathsf{B}_{x_2}(\mathsf{T})$ is a basis for V.

(b) Prove that there exists a vector x_3 in V with T-annihilator $\phi_1(t)\phi_2(t)$ for which $\mathsf{V} = \mathsf{C}_{x_3}(\mathsf{T})$.

(c) Describe the difference between the matrix representation of T with respect to $\mathsf{B}_{x_1}(\mathsf{T}) \cup \mathsf{B}_{x_2}(\mathsf{T})$ and the matrix representation of T with respect to $\mathsf{B}_{x_3}(\mathsf{T})$.

Thus to assure uniqueness of the rational canonical form, we require that the generators of these T-cyclic bases that constitute a rational canonical basis have T-annihilators equal to powers of irreducible monic factors of the characteristic polynomial of T.

7. Let T be a linear operator with minimal polynomial

$$f(t) = (\phi_1(t))^{m_1}(\phi_2(t))^{m_2} \cdots (\phi_k(t))^{m_k}$$

where the $\phi_i(t)$'s are the distinct irreducible monic factors of $f(t)$. Prove that for each i, m_i is the number of entries in the first column of the dot diagram for $\phi_i(t)$.

8. Let T be a linear operator on a finite-dimensional vector space V. Prove that for any irreducible polynomial $\phi(t)$, if $\phi(\mathsf{T})$ is not one-to-one on V, then $\phi(t)$ divides the characteristic polynomial of T. *Hint:* Apply Exercise 15 of Section 7.3.

9. Fill in the details of the proof of the lemma within the proof of Theorem 7.23.

10. Let T be a linear operator on a finite-dimensional vector space, and suppose that $\phi(t)$ is an irreducible monic factor of the characteristic polynomial of T. Prove that if $\phi(t)$ is the T-annihilator of vectors x and y, then $x \in \mathsf{C}_y(\mathsf{T})$ if and only if $\mathsf{C}_x(\mathsf{T}) = \mathsf{C}_y(\mathsf{T})$.

Exercises 11 and 12 are concerned with direct sums.

11. Prove Theorem 7.24.

12. Prove Theorem 7.25.

INDEX OF DEFINITIONS FOR CHAPTER 7

Appendices

APPENDIX A SETS

A *set* is a collection of objects, called *elements* or *members* of the set. If x is an element of the set A, then we write $x \in A$; if x is not an element of A, then we write $x \notin A$. For example, if Z is the set of integers, then $3 \in Z$ and $\frac{1}{2} \notin Z$.

Two sets A and B are called *equal*, denoted $A = B$, if they contain exactly the same elements. Sets may be described in one of two ways:

1. By listing the elements of the set between set braces $\{\ \}$
2. By describing the elements of the set in terms of some characteristic property

For example, the set consisting of the elements 1, 2, 3, and 4 can be written as $\{1, 2, 3, 4\}$ or as

$$\{x\colon x \text{ is a positive integer less than } 5\}.$$

Note that the order in which the elements of a set are listed is immaterial; hence

$$\{1, 2, 3, 4\} = \{3, 1, 2, 4\} = \{1, 3, 1, 4, 2\}.$$

Example 1

Let A denote the set of real numbers between 1 and 2. Then A may be written as

$$A = \{x\colon x \text{ is a real number and } 1 < x < 2\}$$

or, if R is the set of real numbers, as

$$A = \{x \in R\colon 1 < x < 2\}. \quad \blacksquare$$

A set B is said to be a *subset* of a set A, written $B \subseteq A$ or $A \supseteq B$, if every element of B is an element of A. For example, $\{1, 2, 6\} \subseteq \{2, 8, 7, 6, 1\}$. Observe that $A = B$ if and only if $A \subseteq B$ and $B \subseteq A$, a fact that is often used to prove that two sets are equal.

The *empty set*, denoted by $\varnothing$, is the set containing no elements. The empty set is a subset of every set.

Sets may be combined to form other sets in two basic ways. The *union* of two sets A and B, denoted $A \cup B$, is the set of elements that are in A, or B, or both; that is,

$$A \cup B = \{x: x \in A \text{ or } x \in B\}.$$

The *intersection* of two sets A and B, denoted $A \cap B$, is the set of elements that are in both A and B; that is,

$$A \cap B = \{x: x \in A \text{ and } x \in B\}.$$

Two sets are called *disjoint* if their intersection is the empty set.

Example 2

Let $A = \{1, 3, 5\}$ and $B = \{1, 5, 7, 8\}$. Then

$$A \cup B = \{1, 3, 5, 7, 8\} \quad \text{and} \quad A \cap B = \{1, 5\}.$$

Likewise, if $X = \{1, 2, 8\}$ and $Y = \{3, 4, 5\}$, then

$$X \cup Y = \{1, 2, 3, 4, 5, 8\} \quad \text{and} \quad X \cap Y = \varnothing.$$

Thus X and Y are disjoint sets. ∎

The union and intersection of more than two sets can be defined analogously. Specifically, if $A_1, A_2, \ldots, A_n$ are sets, then the union and intersection of these sets are defined as

$$\bigcup_{i=1}^{n} A_i = \{x: x \in A_i \text{ for some } i = 1, 2, \ldots, n\}$$

and

$$\bigcap_{i=1}^{n} A_i = \{x: x \in A_i \text{ for all } i = 1, 2, \ldots, n\}.$$

Similarly, if Λ is an index set and $\{A_\alpha: \alpha \in \Lambda\}$ is a collection of sets, the union and intersection of these sets are defined by

$$\bigcup_{\alpha \in \Lambda} A_\alpha = \{x: x \in A_\alpha \text{ for some } \alpha \in \Lambda\}$$

and

$$\bigcap_{\alpha \in \Lambda} A_\alpha = \{x: x \in A_\alpha \text{ for all } \alpha \in \Lambda\}.$$

Example 3

Let $\Lambda = \{\alpha \in R: \alpha > 1\}$, and let

$$A_\alpha = \left\{x \in R: \frac{-1}{\alpha} \leqslant x \leqslant 1 + \alpha\right\}$$

for each $\alpha \in \Lambda$, where R denotes the set of real numbers. Then

$$\bigcup_{\alpha \in \Lambda} A_\alpha = \{x \in R: x > -1\} \quad \text{and} \quad \bigcap_{\alpha \in \Lambda} A_\alpha = \{x \in R: 0 \leqslant x \leqslant 2\}. \quad \blacksquare$$

By a relation on a set A we mean a rule for determining whether or not, for any elements x and y in A, x stands in a given relationship to y. More precisely, a *relation* on A is a set S of ordered pairs of elements of A such that $(x, y) \in S$ if and only if x stands in the given relationship to y. On the set of real numbers, for instance, "is equal to," "is less than," and "is greater than or equal to" are familiar relations. A relation S on a set A is called an *equivalence relation* on A if these three conditions hold.

1. For each $x \in A$, $(x, x) \in S$ (*reflexivity*).
2. If $(x, y) \in S$, then $(y, x) \in S$ (*symmetry*).
3. If $(x, y) \in S$ and $(y, z) \in S$, then $(x, z) \in S$ (*transitivity*).

If S is an equivalence relation on a set A, we usually write $x \sim y$ in place of $(x, y) \in S$. For example, if we define $x \sim y$ to mean that $x - y$ is divisible by a fixed integer n, then $\sim$ is an equivalence relation on the set of integers.

APPENDIX B FUNCTIONS

If A and B are sets, then a *function* f from A into B, written $f: A \to B$, is a rule that associates to each element x in A a unique element denoted $f(x)$ in B. The element $f(x)$ is called the *image of* x (*under* f) and x is called a *preimage of* f(x) (*under* f). If $f: A \to B$, then A is called the *domain* of f, and the set $\{f(x): x \in A\}$ of all images of elements in A is called the *range* of f. Note that the range of f is a subset of B. If $S \subseteq A$, we denote by $f(S)$ the set $\{f(x): x \in S\}$ of all images of elements of S. Likewise, if $T \subseteq B$, we denote by $f^{-1}(T)$ the set $\{x \in A: f(x) \in T\}$ of all preimages of elements in T. Finally, two functions $f: A \to B$ and $g: A \to B$ are *equal* if $f(x) = g(x)$ for all $x \in A$.

Example 1

Suppose that $A = [-10, 10]$ and $B = R$, the set of real numbers. Let $f: A \to B$ be the function that assigns to each element x in A the element $x^2 + 1$ in B; that is, f is defined by $f(x) = x^2 + 1$. Then A is the domain of f and $[1, 101]$ is the range of f. Since $f(2) = 5$, the image of 2 is 5 and 2 is a preimage of 5. Notice that -2 is another preimage of 5. Moreover, if $S = [1, 2]$ and $T = [82, 101]$, then $f(S) = [2, 5]$ and $f^{-1}(T) = [-10, -9] \cup [9, 10]$. $\blacksquare$

As Example 1 shows, the preimage of an element in the range need not be unique. Functions such that each element of the range has a unique preimage are called *one-to-one*; that is, $f: A \to B$ is one-to-one if $f(x) = f(y)$ implies $x = y$ or, equivalently, if $x \neq y$ implies $f(x) \neq f(y)$.

If $f: A \to B$ is a function with range B, i.e., if $f(A) = B$, then f is called *onto*.

Suppose that $f: A \to B$ is a function and $S \subseteq A$. Then a function $f_S: S \to B$, called the *restriction of* f *to* S, can be formed by defining $f_S(x) = f(x)$ for each $x \in S$.

The following example illustrates these concepts.

Example 2

Let $f: [-1, 1] \to [0, 1]$ be defined by $f(x) = x^2$. This function is onto but not one-to-one since $f(-1) = f(1) = 1$. Note that if $S = [0, 1]$, then f_S is both onto and one-to-one. Finally, if $T = [\frac{1}{2}, 1]$, then f_T is one-to-one but not onto. ∎

Let A, B, and C be sets and $f: A \to B$ and $g: B \to C$ be functions. By following f with g we obtain a function $g \circ f: A \to C$ called the *composite* of g and f. Thus $(g \circ f)(x) = g(f(x))$ for all $x \in A$. For example, let $A = B = C = R$ (the set of real numbers), $f(x) = \sin x$, and $g(x) = x^2 + 3$. Then $(g \circ f)(x) = g(f(x)) = \sin^2 x + 3$, whereas $(f \circ g)(x) = f(g(x)) = \sin(x^2 + 3)$. Hence $g \circ f \neq f \circ g$. Functional composition is associative, however; that is, if $h: C \to D$, then $h \circ (g \circ f) = (h \circ g) \circ f$.

A function $f: A \to B$ is said to be *invertible* if there exists a function $g: B \to A$ such that $(f \circ g)(y) = y$ for all $y \in B$ and $(g \circ f)(x) = x$ for all $x \in A$. If such a function g exists, then it is unique and is called the *inverse* of f. We denote the inverse of f (when it exists) by f^{-1}. It can be shown that f is invertible if and only if f is both one-to-one and onto.

Example 3

The function $f: R \to R$ defined by $f(x) = 3x + 1$ is one-to-one and onto; hence f is invertible. The inverse of f is the function $f^{-1}: R \to R$ defined by $f^{-1}(x) = (x - 1)/3$. ∎

The following facts about invertible functions are easily proved:

1. If $f: A \to B$ is invertible, then f^{-1} is invertible and $(f^{-1})^{-1} = f$.
2. If $f: A \to B$ and $g: B \to C$ are invertible, then $g \circ f$ is invertible and $(g \circ f)^{-1} = f^{-1} \circ g^{-1}$.

APPENDIX C FIELDS

The set of real numbers is an example of an algebraic structure called a "field." Basically, a field is a set in which four operations (called addition, multiplication, subtraction, and division) can be defined so that, with the exception of division by zero, the sum, product, difference, and quotient of any two elements in the set is an element of the set. More precisely, a field is defined as follows.

Definitions. *A* field F *is a set on which two operations* + *and* · *(called* addition *and* multiplication, *respectively) are defined so that for each pair of elements* x, y *in* F *there are unique elements* x + y *and* x · y *in* F *for which the following conditions hold for all elements* a, b, c *in* F:

(F 1) $a + b = b + a$ *and* $a \cdot b = b \cdot a$
(commutativity of addition and multiplication).

(F 2) $(a + b) + c = a + (b + c)$ *and* $(a \cdot b) \cdot c = a \cdot (b \cdot c)$
(associativity of addition and multiplication).

(F 3) *There exist distinct elements* 0 *and* 1 *in* F *such that*

$$0 + a = a \quad \textit{and} \quad 1 \cdot a = a$$

(existence of identity elements for addition and multiplication).

(F 4) *For each element* a *in* F *and each nonzero element* b *in* F *there exist elements* c *and* d *in* F *such that*

$$a + c = 0 \quad \textit{and} \quad b \cdot d = 1$$

(existence of inverses for addition and multiplication).

(F 5) $a \cdot (b + c) = a \cdot b + a \cdot c$
(distributivity of multiplication over addition).

The elements x + y *and* x · y *are called the* sum *and* product, *respectively, of* x *and* y. *The elements* 0 *(read* "zero"*) and* 1 *(read* "one"*) mentioned in* (F 3) *are called* identity elements *for addition and multiplication, respectively, and the elements* c *and* d *referred to in* (F 4) *are called an* additive inverse *for* a *and a* multiplicative inverse *for* b, *respectively.*

Example 1

The set of real numbers with the usual definitions of addition and multiplication is a field, which will be denoted by *R*. ■

Example 2

The set of rational numbers with the usual definitions of addition and multiplication is a field. ■

Example 3

The set of all real numbers of the form $a + b\sqrt{2}$, where a and b are rational numbers, with addition and multiplication as in *R* is a field. ■

Example 4

The field Z_2 consists of two elements 0 and 1 with the operations of addition and multiplication defined by the equations

$$0+0=0, \qquad 0+1=1+0=1, \qquad 1+1=0,$$

$$0\cdot 0=0, \qquad 0\cdot 1=1\cdot 0=0, \quad \text{and} \quad 1\cdot 1=1. \quad \blacksquare$$

Example 5

Neither the set of positive integers nor the set of integers with the usual definitions of addition and multiplication is a field, for in either case (F 4) does not hold. ▮

The identity and inverse elements whose existence is guaranteed by (F 3) and (F 4) are unique; this is a consequence of the following theorem.

Theorem C.1 (Cancellation Laws). *Let* a, b, *and* c *be arbitrary elements of a field* F.

(a) *If* $a+b=c+b$, *then* $a=c$.
(b) *If* $a\cdot b=c\cdot b$ *and* $b\neq 0$, *then* $a=c$.

Proof. The proofs of (a) and (b) are similar; so only (b) will be proved.

If $b\neq 0$, then (F 4) guarantees the existence of an element d in F such that $b\cdot d=1$. Multiply both sides of the equality $a\cdot b=c\cdot b$ by d to obtain $(a\cdot b)\cdot d=(c\cdot b)\cdot d$. Consider the left side of this equality: by (F 2) and (F 3) we have

$$(a\cdot b)\cdot d=a\cdot(b\cdot d)=a\cdot 1=a.$$

Similarly, the right side of the equality reduces to c. Thus

$$a=(a\cdot b)\cdot d=(c\cdot b)\cdot d=c. \quad \blacksquare$$

Corollary. *The elements* 0 *and* 1 *mentioned in* (F 3) *and the elements* c *and* d *mentioned in* (F 4) *are unique.*

Proof. Suppose that $0'\in F$ satisfies $0'+a=a$ for each $a\in F$. Since $0+a=a$ for each $a\in F$, we have $0'+a=0+a$ for each $a\in F$. Thus $0'=0$ by Theorem C.1.

The proofs of the remaining parts are similar. ▮

Thus each element b in a field has a unique additive inverse and, if $b\neq 0$, a unique multiplicative inverse. (It will be shown in the corollary to Theorem C.2 that 0 has no multiplicative inverse.) The additive inverse and the multiplicative inverse of b are denoted by $-b$ and b^{-1}, respectively. Notice that $-(-b)=b$ and that $(b^{-1})^{-1}=b$.

Subtraction and division can be defined in terms of addition and multiplication by using the additive and multiplicative inverses. Specifically, subtraction of b is defined to be addition of $-b$ and division by $b \neq 0$ is defined to be multiplication by b^{-1}; that is,

$$a - b = a + (-b) \quad \text{and} \quad a/b = a \cdot b^{-1}.$$

Division by zero is undefined, but with this exception the sum, product, difference, and quotient of any two elements of a field are defined.

Many of the familiar properties of multiplication of real numbers are true in any field, as the following theorem shows.

Theorem C.2. *Let* a *and* b *be arbitrary elements of a field. Then each of the following are true.*

(a) $a \cdot 0 = 0$.

(b) $(-a) \cdot b = a \cdot (-b) = -(a \cdot b)$.

(c) $(-a) \cdot (-b) = a \cdot b$.

Proof. (a) Since $0 + 0 = 0$, (F 5) shows that

$$a \cdot 0 = a \cdot (0 + 0) = a \cdot 0 + a \cdot 0.$$

Thus $0 + a \cdot 0 = a \cdot 0 + a \cdot 0$, and cancellation of $a \cdot 0$ by Theorem C.1 gives $0 = a \cdot 0$.

(b) By definition $-(a \cdot b)$ is the unique element of F such that $a \cdot b + [-(a \cdot b)] = 0$. So in order to prove that $(-a) \cdot b = -(a \cdot b)$ it suffices to show that $a \cdot b + (-a) \cdot b = 0$. But $-a$ is the element of F such that $a + (-a) = 0$; so

$$a \cdot b + (-a) \cdot b = [a + (-a)] \cdot b = 0 \cdot b = b \cdot 0 = 0$$

by (F 5) and part (a). Thus $(-a) \cdot b = -(a \cdot b)$. The proof that $a \cdot (-b) = -(a \cdot b)$ is similar.

(c) By twice applying part (b), we find

$$(-a) \cdot (-b) = -[a \cdot (-b)] = -[-(a \cdot b)] = a \cdot b. \quad \blacksquare$$

Corollary. *The additive identity of a field has no multiplicative inverse.*

In an arbitrary field F it may happen that a sum $1 + 1 + \cdots + 1$ (p summands) equals 0 for some positive integer p. For example, in the field Z_2 (defined in Example 4), $1 + 1 = 0$. In this case the smallest positive integer p for which a sum of p 1's equals 0 is called the *characteristic* of F; if no such positive integer exists, then F is said to have *characteristic zero*. Thus Z_2 has characteristic two, and R has characteristic zero. Observe that if F is a field of characteristic $p \neq 0$, then $x + x + \cdots + x$ (p summands) equals 0 for all $x \in F$. In a field having finite characteristic (especially characteristic two), many unnatural problems arise. For this reason some of the results about vector spaces stated in this book require that the field over which the vector space is defined be of characteristic zero (or, at least, of some characteristic other than two).

Finally, note that in other sections of this book the product of two elements a and b in a field is denoted ab rather than $a \cdot b$.

APPENDIX D COMPLEX NUMBERS

For the purposes of algebra the field of real numbers is not sufficient, for there are polynomials of nonzero degree with real number coefficients that have no zeros in the field of real numbers (for example, $x^2 + 1$). It is often desirable to have a field in which any polynomial of nonzero degree with coefficients from that field has a zero in the field. For this reason we shall "enlarge" the field of real numbers to obtain such a field.

Definitions. *A* complex number *is an expression of the form* $\mathrm{z} = \mathrm{a} + \mathrm{bi}$, *where* a *and* b *are real numbers called the* real part *and the* imaginary part *of* z, *respectively.*

The sum *and* product *of two complex numbers* $\mathrm{z} = \mathrm{a} + \mathrm{bi}$ *and* $\mathrm{w} = \mathrm{c} + \mathrm{di}$ (*where* a, b, c, *and* d *are real numbers*) *are defined as follows:*

$$\mathrm{z} + \mathrm{w} = (\mathrm{a} + \mathrm{bi}) + (\mathrm{c} + \mathrm{di}) = (\mathrm{a} + \mathrm{c}) + (\mathrm{b} + \mathrm{d})\mathrm{i}$$

and

$$\mathrm{zw} = (\mathrm{a} + \mathrm{bi})(\mathrm{c} + \mathrm{di}) = (\mathrm{ac} - \mathrm{bd}) + (\mathrm{bc} + \mathrm{ad})\mathrm{i}.$$

Example 1

The sum and product of $z = 3 - 5i$ and $w = 9 + 7i$ are

$$z + w = (3 - 5i) + (9 + 7i) = (3 + 9) + [(-5) + 7]i = 12 + 2i$$

and

$$\begin{aligned} zw &= (3 - 5i)(9 + 7i) = [3 \cdot 9 - (-5) \cdot 7] + [(-5) \cdot 9 + 3 \cdot 7]i \\ &= 62 - 24i. \quad \blacksquare \end{aligned}$$

Any real number c may be regarded as a complex number by associating c with the complex number $c + 0i$. Observe that this correspondence preserves sums and products; that is,

$$(c + 0i) + (d + 0i) = (c + d) + 0i, \quad \text{and} \quad (c + 0i)(d + 0i) = cd + 0i.$$

Any complex number of the form $bi = 0 + bi$, where b is a nonzero real number, is called *imaginary*. The product of two imaginary numbers is real since

$$\begin{aligned} (bi)(di) &= (0 + bi)(0 + di) = (0 - bd) + (b \cdot 0 + 0 \cdot d)i \\ &= -bd. \end{aligned}$$

In particular, for $i = 0 + 1i$, we have $i \cdot i = -1$.

The observation that $i^2 = i \cdot i = -1$ provides an easy way to remember the definition of multiplication of complex numbers: simply multiply two complex numbers as you would any two algebraic expressions and replace i^2 by -1. Example 2 illustrates this technique.

Example 2

The product of $-5 + 2i$ and $1 - 3i$ is

$$\begin{aligned}(-5 + 2i)(1 - 3i) &= -5(1 - 3i) + 2i(1 - 3i)\\ &= -5 + 15i + 2i - 6i^2\\ &= -5 + 15i + 2i - 6(-1)\\ &= 1 + 17i. \quad \blacksquare\end{aligned}$$

The real number 0, regarded as a complex number, is an additive identity element for the set of complex numbers since

$$\begin{aligned}(a + bi) + 0 = (a + bi) + (0 + 0i) &= (a + 0) + (b + 0)i\\ &= a + bi.\end{aligned}$$

Likewise the real number 1, regarded as a complex number, is a multiplicative identity element for the set of complex numbers since

$$\begin{aligned}(a + bi) \cdot 1 = (a + bi)(1 + 0i) &= (a \cdot 1 - b \cdot 0) + (b \cdot 1 - a \cdot 0)i\\ &= a + bi.\end{aligned}$$

Clearly each complex number $a + bi$ has an additive inverse, namely $(-a) + (-b)i$. But also each complex number except 0 has a multiplicative inverse. In fact,

$$(a + bi)^{-1} = \left(\frac{a}{a^2 + b^2}\right) - \left(\frac{b}{a^2 + b^2}\right)i.$$

In view of the preceding statements the following result is not surprising.

Theorem D.1. *The set of complex numbers with the operations of addition and multiplication defined above is a field.*

We denote the field of complex numbers by C.

Definition. *The* (complex) conjugate *of a complex number* $a + bi$ *is the complex number* $a - bi$. *We denote the conjugate of the complex number* z *by* $\bar{z}$.

Example 3

The conjugates of $-3 + 2i$, $4 - 7i$, and 6 are as follows:

$$\overline{-3 + 2i} = -3 - 2i, \qquad \overline{4 - 7i} = 4 + 7i,$$

and

$$\overline{6} = \overline{6 + 0i} = 6 - 0i = 6. \quad \blacksquare$$

The following result is an easy consequence of the definition of the complex conjugate.

Theorem D.2. *A complex number* z *is a real number if and only if* $z = \bar{z}$.

For any complex number $z = a + bi$, $z\bar{z}$ is real and nonnegative, for

$$z\bar{z} = (a + bi)(a - bi) = a^2 + b^2.$$

This fact can be used to define the absolute value of a complex number.

Definition. *The* absolute value (*or* modulus) *of a complex number* $z = a + bi$ *is the real number* $\sqrt{a^2 + b^2}$. *We denote the absolute value of* z *by* $|z|$. *Observe that* $z\bar{z} = |z|^2$.

The fact that the product of a complex number and its conjugate is real provides an easy method for determining the quotient of two complex numbers; for if $c + di \neq 0$, then

$$\frac{a + bi}{c + di} = \frac{a + bi}{c + di} \cdot \frac{c - di}{c - di} = \frac{(ac + bd) + (bc - ad)i}{c^2 + d^2}$$

$$= \frac{ac + bd}{c^2 + d^2} + \frac{bc - ad}{c^2 + d^2} i.$$

Example 4

We will illustrate the procedure described above by computing the quotient $(1 + 4i)/(3 - 2i)$:

$$\frac{1 + 4i}{3 - 2i} = \frac{1 + 4i}{3 - 2i} \cdot \frac{3 + 2i}{3 + 2i} = \frac{-5 + 14i}{9 + 4} = -\frac{5}{13} + \frac{14}{13} i. \quad \blacksquare$$

The absolute value of a complex number has the familiar properties of the absolute value of a real number, as the following result shows.

Theorem D.3. *Let* z *and* w *denote any two complex numbers. Then*
(a) $|z + w| \leqslant |z| + |w|$.
(b) $|zw| = |z| \cdot |w|$.
(c) $|z| - |w| \leqslant |z + w|$.

Proof. Let $z = a + bi$ and $w = c + di$, where a, b, c, and d are real numbers.

(a) Observe first that

$$0 \leqslant (ad - bc)^2 = a^2d^2 - 2abcd + b^2c^2,$$

so $2abcd \leqslant a^2d^2 + b^2c^2$. Adding $a^2c^2 + b^2d^2$ to both sides of the inequality gives

$$\begin{aligned}(ac + bd)^2 &= a^2c^2 + 2abcd + b^2d^2\\ &\leqslant a^2c^2 + a^2d^2 + b^2c^2 + b^2d^2 = (a^2 + b^2)(c^2 + d^2).\end{aligned}$$

By taking square roots, we obtain

$$ac + bd \leqslant \sqrt{a^2 + b^2}\sqrt{c^2 + d^2}.$$

Now

$$\begin{aligned}|z + w|^2 &= |(a + c) + (b + d)i|^2\\ &= (a + c)^2 + (b + d)^2\\ &= a^2 + c^2 + b^2 + d^2 + 2(ac + bd)\\ &\leqslant a^2 + c^2 + b^2 + d^2 + 2\sqrt{a^2 + b^2}\sqrt{c^2 + d^2}\\ &= (\sqrt{a^2 + b^2} + \sqrt{c^2 + d^2})^2\\ &= (|z| + |w|)^2.\end{aligned}$$

By taking square roots, we obtain (a).

(b) From the definition of absolute value we see that

$$\begin{aligned}|zw| &= |(a + bi)(c + di)| = |(ac - bd) + (bc + ad)i|\\ &= \sqrt{(ac - bd)^2 + (bc + ad)^2} = \sqrt{a^2c^2 + b^2d^2 + b^2c^2 + a^2d^2}\\ &= \sqrt{a^2 + b^2}\sqrt{c^2 + d^2} = |a + bi| \cdot |c + di| = |z| \cdot |w|.\end{aligned}$$

(c) From (a) and (b) it follows that

$$|z| = |(z + w) - w| \leqslant |z + w| + |-w| = |z + w| + |w|.$$

So

$$|z| - |w| \leqslant |z + w|. \quad \blacksquare$$

Our motivation for enlarging the set of real numbers to the set of complex numbers was to obtain a field such that every polynomial with nonzero degree having coefficients in that field has a zero. Our next result guarantees that the field of complex numbers has this property.

Theorem D.4 (The Fundamental Theorem of Algebra). *Let* $a_0, \ldots, a_n$ $(n \geqslant 1)$ *be complex numbers such that* $a_n \neq 0$. *Then*

$$a_nz^n + a_{n-1}z^{n-1} + \cdots + a_1z + a_0$$

has a zero in the field of complex numbers.

For a proof, see Walter Rudin's *Principles of Mathematical Analysis* (McGraw-Hill, New York, 1964).

The following important corollary follows from Theorem D.4 and the division algorithm for polynomials (Theorem E.1).

Corollary. *If* $p(z) = a_n z^n + \cdots + a_1 z + a_0$ *is a polynomial of degree* $n \geqslant 1$ *with complex coefficients, then there exist complex numbers* $c_1, \ldots, c_n$ *(not necessarily distinct) such that*

$$p(z) = a_n(z - c_1) \cdots (z - c_n).$$

A field is called *algebraically closed* if it has the property that every polynomial with coefficients from that field factors as a product of polynomials of degree 1. Thus the corollary above shows that the field of complex numbers is algebraically closed.

APPENDIX E POLYNOMIALS

In this appendix we discuss some properties of polynomials with coefficients from a field. For the definition of a polynomial, refer to Section 1.2.

Definition. *A polynomial* $f(x)$ divides *a polynomial* $g(x)$ *if there exists a polynomial* $q(x)$ *such that* $g(x) = f(x)q(x)$.

Our first result shows that the familiar long division process for polynomials with real coefficients is valid for polynomials with coefficients from an arbitrary field.

Theorem E.1 (The Division Algorithm for Polynomials). *Let* $f_1(x)$ *be a polynomial of degree* n, *and let* $f_2(x)$ *be a polynomial of degree* $m \geqslant 0$. *Then there exist polynomials* $q(x)$ *and* $r(x)$ *such that*

(a) *The degree of* $r(x)$ *is less than* m.

(b) $f_1(x) = q(x)f_2(x) + r(x)$.

(c) $q(x)$ *and* $r(x)$ *are unique with respect to conditions* (a) *and* (b).

Proof. We will begin by establishing the existence of $q(x)$ and $r(x)$ that satisfy conditions (a) and (b). If $n < m$, we can take $q(x) = 0$ and $r(x) = f_1(x)$ to satisfy (a) and (b).

Assume, therefore, that $m \leqslant n$. In this case we will establish the existence of $q(x)$ and $r(x)$ by induction on n. Suppose first that $n = 0$; then $m \leqslant n$ implies that $m = 0$. Thus $f_1(x)$ and $f_2(x)$ are nonzero constants. Hence we may take $q(x) = f_1(x)(f_2(x))^{-1}$ and $r(x) = 0$ to satisfy (a) and (b).

Now suppose that the theorem is true whenever $f_1(x)$ has degree less than $n > 0$. Let

$$f_1(x) = a_n x^n + a_{n-1} x^{n-1} + \cdots + a_1 x + a_0$$

and

$$f_2(x) = b_m x^m + b_{m-1}x^{m-1} + \cdots + b_1 x + b_0,$$

where $m \leqslant n$. Define a polynomial $h(x)$ by

$$\begin{aligned} h(x) &= f_1(x) - a_n b_m^{-1} x^{n-m} f_2(x) \\ &= (a_{n-1} - a_n b_m^{-1} b_{m-1})x^{n-1} + (a_{n-2} - a_n b_m^{-1} b_{m-2})x^{n-2} \\ &\qquad + \cdots + (a_0 - a_n b_m^{-1} b_0). \end{aligned} \tag{1}$$

Then $h(x)$ is a polynomial of degree less than n.

CASE 1. $h(x)$ is of degree less than m. In this case, let $q(x) = a_n b_m^{-1} x^{n-m}$ and $r(x) = h(x)$. Then by (1) we obtain

$$f_1(x) = q(x)f_2(x) + r(x),$$

and $r(x)$ has degree less than m.

CASE 2. $h(x)$ has degree at least m. Since $h(x)$ has degree less than n, we may apply the induction hypothesis to obtain polynomials $q_1(x)$ and $r(x)$ such that $r(x)$ has degree less than m, and

$$h(x) = q_1(x)f_2(x) + r(x). \tag{2}$$

Combining (1) and (2) and solving for $f_1(x)$, we have

$$f_1(x) = [a_n b_m^{-1} x^{n-m} + q_1(x)]f_2(x) + r(x).$$

In this case, let $q(x) = a_n b_m^{-1} x^{n-m} + q_1(x)$, so that $f_1(x) = q(x)f_2(x) + r(x)$, where $r(x)$ has degree less than m. This proves the existence of $q(x)$ and $r(x)$.

We now show the uniqueness of q and r. Suppose that $q_1(x)$, $q_2(x)$, $r_1(x)$, and $r_2(x)$ exist such that $r_1(x)$ and $r_2(x)$ each has degree less than m and

$$f_1(x) = q_1(x)f_2(x) + r_1(x) = q_2(x)f_2(x) + r_2(x).$$

Then

$$[q_1(x) - q_2(x)]f_2(x) = r_2(x) - r_1(x). \tag{3}$$

The right side of (3) is a polynomial of degree less than m. Since $f_2(x)$ has degree m, it must follow that $q_1(x) - q_2(x)$ is the zero polynomial. Hence $q_1(x) = q_2(x)$; thus $r_1(x) = r_2(x)$ by (3). ∎

In the context of Theorem E.1 we call $q(x)$ and $r(x)$ the *quotient* and *remainder*, respectively, for the division of $f_1(x)$ by $f_2(x)$. For example, the quotient and remainder for the division of the complex polynomial

$$f_1(x) = (3 + i)x^5 - (1 - i)x^4 + 6x^3 + (-6 + 2i)x^2 + (2 + i)x + 1$$

by the complex polynomial

$$f_2(x) = (3 + i)x^2 - 2ix + 4$$

are

$$q(x) = x^3 + ix^2 - 2 \quad \text{and} \quad r(x) = (2 - 3i)x + 9.$$

Corollary 1. *Let* f(x) *be a polynomial of degree at least* 1, *and let* $a \in F$. *Then* $f(a) = 0$ *if and only if* $x - a$ *divides* f(x).

Proof. Suppose that $x - a$ divides $f(x)$. Then there exists a polynomial $q(x)$ such that $f(x) = (x - a)q(x)$. Thus $f(a) = (a - a)q(a) = 0 \cdot q(a) = 0$.

Conversely, suppose that $f(a) = 0$. By the division algorithm there exist polynomials $q(x)$ and $r(x)$ such that $r(x)$ has degree less than one and

$$f(x) = q(x)(x - a) + r(x).$$

Substituting a for x in the above we obtain $r(a) = 0$. Since $r(x)$ has degree less than 1, it must be the constant polynomial $r(x) = 0$. Thus $f(x) = q(x)(x - a)$. ∎

For any polynomial $f(x)$ with coefficients from a field F, an element $a \in F$ is called a *zero* of $f(x)$ if $f(a) = 0$. With this terminology the corollary above states that a is a zero of $f(x)$ if and only if $x - a$ divides $f(x)$.

Corollary 2. *Any polynomial of degree* $n \geq 1$ *has at most* n *distinct zeros.*

Proof. The proof is by induction on n. The result is obvious if $n = 1$. Suppose therefore that the result is true for some positive integer n, and let $f(x)$ be a polynomial of degree $n + 1$. If $f(x)$ has no zeros, then there is nothing to prove. Otherwise, if a is a zero of $f(x)$, then by Corollary 1 we may write $f(x) = (x - a)q(x)$ for some polynomial $q(x)$. Note that $q(x)$ must be of degree n; therefore by the induction hypothesis $q(x)$ can have at most n distinct zeros. Thus, since any zero of $f(x)$ distinct from a is also a zero of $q(x)$, $f(x)$ can have at most $n + 1$ distinct zeros. ∎

Polynomials having no common divisors arise naturally in the study of canonical forms.

Definition. *Two nonzero polynomials are called* relatively prime *if no polynomial of positive degree divides each of them.*

For example, the polynomials with real coefficients $f(x) = x^2(x - 1)$ and $h(x) = (x - 1)(x - 2)$ are not relatively prime because $x - 1$ divides each of them. On the other hand, $f(x)$ and $g(x) = (x - 2)(x - 3)$ are relatively prime because they have no common factors of positive degree.

Proposition E.2. *If* $f_1(x)$ *and* $f_2(x)$ *are relatively prime polynomials, there exist polynomials* $q_1(x)$ *and* $q_2(x)$ *such that*

$$q_1(x)f_1(x) + q_2(x)f_2(x) = 1,$$

where 1 *denotes the constant polynomial with value* 1.

Proof. Without loss of generality assume that the degree of $f_1(x)$ is greater than or equal to the degree of $f_2(x)$. The proof will be by mathematical induction on the degree of $f_2(x)$. If $f_2(x)$ has degree 0, then $f_2(x)$ is a nonzero constant c. So in this case we can take

$$q_1(x) = 0 \quad \text{and} \quad q_2(x) = c^{-1},$$

which clearly satisfy the condition

$$q_1(x)f_1(x) + q_2(x)f_2(x) = 1.$$

Now suppose that the theorem holds whenever $f_2(x)$ has degree less than n for some integer $n \geq 1$, and suppose that $f_2(x)$ has degree n. By the division algorithm there exist polynomials $q(x)$ and $r(x)$ such that $r(x)$ has degree less than n and

$$f_1(x) = q(x)f_2(x) + r(x). \tag{4}$$

Since $f_1(x)$ and $f_2(x)$ are relatively prime, $r(x)$ is not the zero polynomial. If $r(x)$ has degree 0, then $r(x)$ is a nonzero constant, c, and we obtain the conclusion as before. Suppose then that $r(x)$ has degree greater than zero. Since $r(x)$ has degree less than n, we may apply the induction hypothesis to $f_2(x)$ and $r(x)$ provided that we can show these polynomials to be relatively prime. Suppose otherwise; then there exists a polynomial $g(x)$ of positive degree that divides both $f_2(x)$ and $r(x)$. So there exist polynomials $h_1(x)$ and $h_2(x)$ such that

$$r(x) = g(x)h_1(x) \quad \text{and} \quad f_2(x) = g(x)h_2(x). \tag{5}$$

Combining (4) and (5), we obtain

$$f_1(x) = [q(x)h_2(x) + h_1(x)]g(x),$$

and so $g(x)$ divides $f_1(x)$. But $g(x)$ divides $f_2(x)$, contradicting the fact that $f_1(x)$ and $f_2(x)$ are relatively prime. Thus $r(x)$ and $f_2(x)$ are relatively prime. Hence by the induction hypothesis there exist $g_1(x)$ and $g_2(x)$ such that

$$g_1(x)f_2(x) + g_2(x)r(x) = 1. \tag{6}$$

Combining (4) and (6), we obtain

$$g_1(x)f_2(x) + g_2(x)[f_1(x) - q(x)f_2(x)] = 1.$$

Thus

$$g_2(x)f_1(x) + [g_1(x) - g_2(x)q(x)]f_2(x) = 1.$$

Setting $q_1(x) = g_2(x)$ and $q_2(x) = g_1(x) - g_2(x)q(x)$, we obtain the desired conclusion. ▮

Example 1

Let $f_1(x) = x^3 - x^2 + 1$ and $f_2(x) = (x - 1)^2$. As polynomials with real coefficients, $f_1(x)$ and $f_2(x)$ are relatively prime. It is easily verified that the polynomials $q_1(x) = -x + 2$ and $q_2(x) = x^2 - x - 1$ satisfy

$$q_1(x)f_1(x) + q_2(x)f_2(x) = 1,$$

and hence these polynomials satisfy the conclusion of Proposition E.2. ∎

Throughout Chapters 5, 6, and 7 we consider linear operators that are polynomials in a particular operator T and matrices that are polynomials in a particular matrix A. For these operators and matrices the following notation is convenient.

Definitions. *Let*

$$f(x) = a_0 + a_1x + \cdots + a_nx^n$$

be a polynomial with coefficients from a field F. *If* T *is a linear operator on a vector space* V *over* F, *we define* f(T) *by*

$$f(T) = a_0I + a_1T + \cdots + a_nT^n.$$

Similarly, if A *is an* $n \times n$ *matrix with entries from* F, *we define* f(A) *by*

$$f(A) = a_0I + a_1A + \cdots + a_nA^n.$$

Example 2

Let T be the linear operator on R^2 defined by $T(a, b) = (2a + b, a - b)$, and let $f(x) = x^2 + 2x - 3$. It is easily checked that $T^2(a, b) = (5a + b, a + 2b)$; so

$$\begin{aligned} f(T)(a, b) &= (T^2 + 2T - 3I)(a, b) \\ &= (5a + b, a + 2b) + (4a + 2b, 2a - 2b) - 3(a, b) \\ &= (6a + 3b, 3a - 3b). \end{aligned}$$

Similarly, if

$$A = \begin{pmatrix} 2 & 1 \\ 1 & -1 \end{pmatrix},$$

then

$$f(A) = A^2 + 2A - 3I = \begin{pmatrix} 5 & 1 \\ 1 & 2 \end{pmatrix} + 2\begin{pmatrix} 2 & 1 \\ 1 & -1 \end{pmatrix} - 3\begin{pmatrix} 1 & 0 \\ 0 & 1 \end{pmatrix} = \begin{pmatrix} 6 & 3 \\ 3 & -3 \end{pmatrix}.$$ ∎

The next three results utilize this notation.

Proposition E.3. *Let* f(x) *be a polynomial with coefficients from a field* F, *and let* T *be a linear operator on a vector space* V *over* F. *Then*

(a) f(T) *is a linear operator on* V.

(b) *If* β *is a finite ordered basis for* V *and* $A = [T]_\beta$, *then* $[f(T)]_\beta = f(A)$.

Proof. Exercise. ∎

Proposition E.4. *Let* T *be a linear operator on a vector space* V *over* F, *and let* A *be a square matrix with entries from* F. *Then for any polynomials* $f_1(x)$ *and* $f_2(x)$ *with coefficients from* F

(a) $f_1(T)f_2(T) = f_2(T)f_1(T)$.
(b) $f_1(A)f_2(A) = f_2(A)f_1(A)$.

Proof. Exercise. ∎

Proposition E.5. *Let* T *be a linear operator on a vector space* V *over a field* F, *and let* A *be an* $n \times n$ *matrix with entries from* F. *If* $f_1(x)$ *and* $f_2(x)$ *are relatively prime polynomials with entries from* F, *then there exist polynomials* $q_1(x)$ *and* $q_2(x)$ *with entries from* F *such that*

(a) $q_1(T)f_1(T) + q_2(T)f_2(T) = I$.
(b) $q_1(A)f_1(A) + q_2(A)f_2(A) = I$.

Proof. Exercise. ∎

In Chapters 5 and 7 we are concerned with determining when a linear operator T on a finite-dimensional vector space can be "diagonalized" and with finding a simple (canonical) representation of T. Both of these problems are affected by the factorization of a certain polynomial determined by T (the "characteristic polynomial" of T). In this setting certain types of polynomials play an important role.

Definitions. *A polynomial* f(x) *with coefficients from a field* F *is called* monic *if its leading coefficient is* 1. *If* f(x) *has positive degree and cannot be expressed as a product of polynomials with coefficients from* F *each having positive degree, then* f(x) *is called* irreducible.

Observe that whether or not a polynomial is irreducible depends on the field from which its coefficients come. For example, $f(x) = x^2 + 1$ is irreducible over the field of real numbers but not irreducible over the field of complex numbers since $x^2 + 1 = (x + i)(x - i)$.

Clearly, any polynomial of degree 1 is irreducible. Moreover, for polynomials with coefficients from an algebraically closed field, the polynomials of degree 1 are the only irreducible polynomials.

The following facts are easily established.

Proposition E.6. *Let* $\phi(x)$ *and* f(x) *be polynomials with coefficients from a field* F. *If* $\phi(x)$ *is irreducible and* $\phi(x)$ *does not divide* f(x), *then* $\phi(x)$ *and* f(x) *are relatively prime.*

Proof. Exercise. ∎

Proposition E.7. *Any two distinct irreducible monic polynomials are relatively prime.*

Proof. Exercise. ∎

Proposition E.8. *Let* f(x), g(x), *and* ϕ(x) *be polynomials with coefficients from the same field. If* ϕ(x) *is irreducible and divides the product* f(x)g(x), *then* ϕ(x) *divides* f(x) *or* ϕ(x) *divides* g(x).

Proof. Suppose that $\phi(x)$ does not divide $f(x)$. Then $\phi(x)$ and $f(x)$ are relatively prime by Proposition E.6, and so there exist polynomials $q_1(x)$ and $q_2(x)$ such that

$$1 = q_1(x)\phi(x) + q_2(x)f(x).$$

Multiplying both sides of this equation by $g(x)$ yields

$$g(x) = q_1(x)\phi(x)g(x) + q_2(x)f(x)g(x). \tag{7}$$

Since $\phi(x)$ divides $f(x)g(x)$, there exists a polynomial $h(x)$ such that $f(x)g(x) = \phi(x)h(x)$. Thus (7) becomes

$$g(x) = q_1(x)\phi(x)g(x) + q_2(x)\phi(x)h(x) = \phi(x)[q_1(x)g(x) + q_2(x)h(x)].$$

So $\phi(x)$ divides $g(x)$. ∎

Corollary. *Let* ϕ(x), ϕ_1(x), ϕ_2(x), ..., ϕ_n(x) *be irreducible monic polynomials with coefficients from the same field. If* ϕ(x) *divides the product* ϕ_1(x)ϕ_2(x) ⋯ ϕ_n(x), *then* ϕ(x) = ϕ_i(x) *for some* i (i = 1, 2, ..., n).

Proof. We will prove the corollary by induction on n. For $n = 1$ the result is an immediate consequence of Proposition E.7. Suppose then that the corollary is true for any $n - 1$ irreducible monic polynomials and that we are given n irreducible monic polynomials $\phi_1(x), \phi_2(x), \ldots, \phi_n(x)$. If $\phi(x)$ divides

$$\phi_1(x)\phi_2(x)\cdots\phi_n(x) = [\phi_1(x)\phi_2(x)\cdots\phi_{n-1}(x)]\phi_n(x),$$

then $\phi(x)$ divides the product $\phi_1(x)\phi_2(x)\cdots\phi_{n-1}(x)$ or $\phi(x)$ divides $\phi_n(x)$ by Proposition E.8. In the first case, $\phi(x) = \phi_i(x)$ for some i ($i = 1, 2, \ldots, n - 1$) by the induction hypothesis; in the second case, $\phi(x) = \phi_n(x)$ by Proposition E.7. ∎

We are now able to establish the unique factorization theorem, which is used throughout Chapters 5 and 7. This result states that every polynomial of positive degree is uniquely expressible as a constant times a product of irreducible monic polynomials.

Theorem E.9 (Unique Factorization Theorem for Polynomials). *For any polynomial* f(x) *of positive degree, there exist a unique constant* c, *unique distinct irreducible monic polynomials* ϕ_1(x), ϕ_2(x), ..., ϕ_k(x), *and unique positive integers* $n_1, n_2, \ldots, n_k$ *such that*

$$f(x) = c[\phi_1(x)]^{n_1}[\phi_2(x)]^{n_2}\cdots[\phi_k(x)]^{n_k}.$$

Proof. We begin by showing the existence of such a factorization using induction on the degree of $f(x)$. If $f(x)$ is of degree 1, then $f(x) = ax + b$ for some constants a and b with $a \neq 0$. Setting $\phi(x) = x + b/a$, we have $f(x) = a\phi(x)$. Since $\phi(x)$ is an irreducible monic polynomial, the result is proved in this case. Now suppose that the conclusion is true for any polynomial with positive degree less than some integer $n > 1$, and let $f(x)$ be a polynomial of degree n. Then

$$f(x) = a_n x^n + \cdots + a_1 x + a_0$$

for some scalars a_i with $a_n \neq 0$. If $f(x)$ is irreducible, then

$$f(x) = a_n\left(x^n + \frac{a_{n-1}}{a_n}x^{n-1} + \cdots + \frac{a_1}{a_n}x + \frac{a_0}{a_n}\right)$$

is a representation of $f(x)$ as a product of a_n and a monic irreducible polynomial. If $f(x)$ is not irreducible, then $f(x) = g(x)h(x)$ for some polynomials $g(x)$ and $h(x)$ each of positive degree less than n. The induction hypothesis guarantees that both $g(x)$ and $h(x)$ factor as products of a constant and powers of distinct irreducible monic polynomials. Consequently $f(x) = g(x)h(x)$ also factors in this way. Thus in either case $f(x)$ can be factored as a product of a constant and powers of distinct irreducible monic polynomials.

It remains to establish the uniqueness of such a factorization. Suppose that

$$\begin{aligned} f(x) &= c[\phi_1(x)]^{n_1}[\phi_2(x)]^{n_2}\cdots[\phi_k(x)]^{n_k} \\ &= d[\psi_1(x)]^{m_1}[\psi_2(x)]^{m_2}\cdots[\psi_r(x)]^{m_r}, \end{aligned} \tag{8}$$

where c and d are constants, $\phi_i(x)$ and $\psi_j(x)$ are irreducible monic polynomials, and n_i and m_j are positive integers for $i = 1, 2, \ldots, k$ and $j = 1, 2, \ldots, r$. Clearly both c and d must be the leading coefficient of $f(x)$; hence $c = d$. By dividing, (8) becomes

$$[\phi_1(x)]^{n_1}[\phi_2(x)]^{n_2}\cdots[\phi_k(x)]^{n_k} = [\psi_1(x)]^{m_1}[\psi_2(x)]^{m_2}\cdots[\psi_r(x)]^{m_r}. \tag{9}$$

So $\phi_i(x)$ divides the right side of (9) for $i = 1, 2, \ldots, k$. Consequently, by the corollary to Proposition E.8, for each i ($i = 1, 2, \ldots, k$), $\phi_i(x) = \psi_j(x)$ for some $j = 1, 2, \ldots, r$, and for any j ($j = 1, 2, \ldots, r$), $\psi_j(x) = \phi_i(x)$ for some $i = 1, 2, \ldots, k$. We conclude that $r = k$ and that by renumbering if necessary, $\phi_i(x) = \psi_i(x)$ for $i = 1, 2, \ldots, k$. Suppose that $n_i \neq m_i$ for some i. Without loss of generality we may suppose that $i = 1$ and $n_1 > m_1$. Then by canceling $[\phi_1(x)]^{m_1}$ from both sides of (9), we obtain

$$[\phi_1(x)]^{n_1 - m_1}[\phi_2(x)]^{n_2}\cdots[\phi_k(x)]^{n_k} = [\phi_2(x)]^{m_2}\cdots[\phi_k(x)]^{m_k}. \tag{10}$$

Since $n_1 - m_1 > 0$, $\phi_1(x)$ divides the left side of (10) and hence divides the right side also. So $\phi_1(x) = \phi_i(x)$ for some $i = 2, \ldots, k$ by the corollary to Proposition E.8. But this contradicts that $\phi_1(x), \phi_2(x), \ldots, \phi_k(x)$ are distinct. Hence the factorizations of $f(x)$ in (8) are the same. ∎

It is often useful to regard a polynomial $f(x) = a_n x^n + \cdots + a_1 x + a_0$ with coefficients from a field F as a function $f\colon F \to F$. In this case the value of f at $c \in F$ is $f(c) = a_n c^n + \cdots + a_1 c + a_0$. Unfortunately, for arbitrary fields F, there is not a one-to-one correspondence between polynomials and polynomial functions. For example, if $f(x) = x^2$ and $g(x) = x$ are two polynomials over the field Z_2 (as defined in Example 4 of Appendix C), then $f(x)$ and $g(x)$ have different degrees and hence are not equal as polynomials. But $f(a) = g(a)$ for all $a \in Z_2$, so that f and g are equal polynomial functions. Our final result shows that this anomaly cannot occur if F is an infinite field.

Theorem E.10. *Let* f(x) *and* g(x) *be polynomials with coefficients from an infinite field* F. *If* f(a) = g(a) *for all* a ∈ F, *then* f(x) *and* g(x) *are equal.*

Proof. Suppose that $f(a) = g(a)$ for all $a \in F$. Define $h(x) = f(x) - g(x)$, and suppose that $h(x)$ is of degree $n \geqslant 1$. It follows from the corollary to Theorem E.9 that $h(x)$ can have at most n zeros. But

$$h(a) = f(a) - g(a) = 0$$

for any $a \in F$, contradicting the assumption that $h(x)$ has positive degree. Thus $h(x)$ is a constant polynomial, and since $h(a) = 0$ for each $a \in F$, it follows that $h(x)$ is the zero polynomial. Hence $f(x) = g(x)$. ∎

Answers to Selected Exercises

SECTION 1.1

1. Only the pairs in parts (b) and (c) are parallel.

2. **(a)** $(3, -2, 4) + t(-8, 9, -3) = x$

(c) $(3, 7, 2) + t(0, 0, -10) = x$

3. **(a)** $(2, -5, -1) + t_1(-2, 9, 7) + t_2(-5, 12, 2) = x$

(c) $(-8, 2, 0) + t_1(9, 1, 0) + t_2(14, -7, 0) = x$

SECTION 1.2

1. **(a)** T **(b)** F **(c)** F **(d)** F **(e)** T **(f)** F
(g) F **(h)** F **(i)** T **(j)** T **(k)** T

3. $M_{13} = 3$, $M_{21} = 4$, and $M_{22} = 5$

4. **(a)** $\begin{pmatrix} 6 & 3 & 2 \\ -4 & 3 & 9 \end{pmatrix}$ **(c)** $\begin{pmatrix} 8 & 20 & -12 \\ 4 & 0 & 28 \end{pmatrix}$

(e) $2x^4 + x^3 + 2x^2 - 2x + 10$

(g) $10x^7 - 30x^4 + 40x^2 - 15x$

13. No, (VS 4) fails.

14. Yes.

15. No.

20. 2^{mn}

SECTION 1.3

1. **(a)** F **(b)** F **(c)** T **(d)** F **(e)** T **(f)** F

2. **(a)** $\begin{pmatrix} -4 & 5 \\ 2 & -1 \end{pmatrix}$; the trace is -5.

(c) $\begin{pmatrix} -3 & 0 & 6 \\ 9 & -2 & 1 \end{pmatrix}$ **(e)** $\begin{pmatrix} 1 \\ -1 \\ 3 \\ 5 \end{pmatrix}$

8. **(a)** Yes. **(e)** No. **(f)** No.
11. No, the set is not closed under addition.
14. Yes.

SECTION 1.4

1. **(a)** T **(b)** F **(c)** T **(d)** F **(e)** T **(f)** F
2. **(a)** $\{x_2(1, 1, 0, 0) + x_4(-3, 0, -2, 1) + (5, 0, 4, 0): x_2, x_4 \in R\}$
(c) There are no solutions.
(e) $\{x_3(10, -3, 1, 0, 0) + x_4(-3, 2, 0, 1, 0) + (-4, 3, 0, 0, 5): x_3, x_4 \in R\}$
3. **(a)** Yes. **(c)** No. **(e)** No.
4. **(a)** Yes. **(c)** Yes. **(e)** No.

SECTION 1.5

1. **(a)** F **(b)** T **(c)** F **(d)** F **(e)** T **(f)** T
5. $\left\{\begin{pmatrix} 1 & 0 \\ 0 & 0 \end{pmatrix}, \begin{pmatrix} 0 & 0 \\ 0 & 1 \end{pmatrix}\right\}$
8. 2^n

SECTION 1.6

1. **(a)** F **(b)** T **(c)** F **(d)** F **(e)** T **(f)** F
(g) F **(h)** T **(i)** F **(j)** T **(k)** T
2. **(a)** Yes. **(c)** Yes. **(e)** No.
3. **(a)** No. **(c)** No. **(e)** No.
4. No.
5. No.
8. $\{x_1, x_3, x_5, x_7\}$
9. $(a_1, a_2, a_3, a_4) = a_1x_1 + (a_2 - a_1)x_2 + (a_3 - a_2)x_3 + (a_4 - a_3)x_4$
14. $n^2 - 1$
16. $\frac{1}{2}n(n-1)$
21. $\dim(\mathsf{W}_1) = 3$, $\dim(\mathsf{W}_2) = 2$, $\dim(\mathsf{W}_1 + \mathsf{W}_2) = 4$, and $\dim(\mathsf{W}_1 \cap \mathsf{W}_2) = 1$

SECTION 1.7

1. **(a)** F **(b)** F **(c)** F **(d)** T **(e)** T **(f)** T

SECTION 2.1

1. (a) T (b) F (c) F (d) T (e) F (f) F (g) T (h) F
2. The nullity is 1, and the rank is 2. T is not one-to-one but is onto.
4. The nullity is 4, and the rank is 2. T is neither one-to-one nor onto.
5. The nullity is 0, and the rank is 3. T is one-to-one but not onto.
10. $T(2, 3) = (5, 11)$. T is one-to-one. **12.** No.

SECTION 2.2

1. (a) T (b) T (c) F (d) T (e) T (f) F
2. (a) $\begin{pmatrix} 2 & -1 \\ 3 & 4 \\ 1 & 0 \end{pmatrix}$ (c) $(2 \quad 1 \quad -3)$ (d) $\begin{pmatrix} 0 & 2 & 1 \\ -1 & 4 & 5 \\ 1 & 0 & 1 \end{pmatrix}$

   (f) $\begin{pmatrix} 0 & 0 & \cdots & 0 & 1 \\ 0 & 0 & \cdots & 1 & 0 \\ \vdots & \vdots & & \vdots & \vdots \\ 0 & 1 & \cdots & 0 & 0 \\ 1 & 0 & \cdots & 0 & 0 \end{pmatrix}$ (g) $(1 \quad 0 \quad \cdots \quad 0 \quad 1)$

3. $$[T]_\beta^\gamma = \begin{pmatrix} -\frac{1}{3} & -1 \\ 0 & 1 \\ \frac{2}{3} & 0 \end{pmatrix} \quad \text{and} \quad [T]_\alpha^\gamma = \begin{pmatrix} -\frac{7}{3} & -\frac{11}{3} \\ 2 & 3 \\ \frac{2}{3} & \frac{4}{3} \end{pmatrix}$$

5. (a) $\begin{pmatrix} 1 & 0 & 0 & 0 \\ 0 & 0 & 1 & 0 \\ 0 & 1 & 0 & 0 \\ 0 & 0 & 0 & 1 \end{pmatrix}$ (b) $\begin{pmatrix} 0 & 1 & 0 \\ 2 & 2 & 2 \\ 0 & 0 & 0 \\ 0 & 0 & 2 \end{pmatrix}$ (e) $\begin{pmatrix} 1 \\ -2 \\ 0 \\ 4 \end{pmatrix}$

9. $$\begin{pmatrix} 1 & 1 & 0 & \cdots & 0 \\ 0 & 1 & 1 & \cdots & 0 \\ 0 & 0 & 1 & \cdots & 0 \\ \vdots & \vdots & \vdots & & \vdots \\ 0 & 0 & 0 & \cdots & 1 \\ 0 & 0 & 0 & \cdots & 1 \end{pmatrix}$$

SECTION 2.3

1. (a) F (b) T (c) F (d) T (e) F (f) F
   (g) F (h) F (i) T (j) T
2. (a) $$A(2B + 3C) = \begin{pmatrix} 20 & -9 & 18 \\ 5 & 10 & 8 \end{pmatrix} \quad \text{and} \quad A(BD) = \begin{pmatrix} 29 \\ -26 \end{pmatrix}$$

(b) $A^tB = \begin{pmatrix} 23 & 19 & 0 \\ 26 & -1 & 10 \end{pmatrix}$ and $CB = (27 \quad 7 \quad 9)$

3. (a)
$$[\mathsf{T}]_\beta = \begin{pmatrix} 2 & 3 & 0 \\ 0 & 3 & 6 \\ 0 & 0 & 4 \end{pmatrix}, \quad [\mathsf{U}]_\beta^\gamma = \begin{pmatrix} 1 & 1 & 0 \\ 0 & 0 & 1 \\ 1 & -1 & 0 \end{pmatrix}, \quad \text{and} \quad [\mathsf{UT}]_\beta^\gamma = \begin{pmatrix} 2 & 6 & 6 \\ 0 & 0 & 4 \\ 2 & 0 & -6 \end{pmatrix}$$

4. (a) $\begin{pmatrix} 1 \\ -1 \\ 4 \\ 6 \end{pmatrix}$ **(c)** (5)

11. (a) No. **(b)** No.

SECTION 2.4

1. (a) F **(b)** T **(c)** F **(d)** F **(e)** T **(f)** F
(g) T **(h)** T **(i)** T

17. (b) $\begin{pmatrix} 1 & 0 & 0 & 0 \\ 0 & 0 & 1 & 0 \\ 0 & 1 & 0 & 0 \\ 0 & 0 & 0 & 1 \end{pmatrix}$

SECTION 2.5

1. (a) F **(b)** T **(c)** T **(d)** F **(e)** T

2. (a) $\begin{pmatrix} a_1 & b_1 \\ a_2 & b_2 \end{pmatrix}$ **(c)** $\begin{pmatrix} 3 & -1 \\ 5 & -2 \end{pmatrix}$

3. (a) $\begin{pmatrix} a_2 & b_2 & c_2 \\ a_1 & b_1 & c_1 \\ a_0 & b_0 & c_0 \end{pmatrix}$ **(c)** $\begin{pmatrix} 0 & -1 & 0 \\ 1 & 0 & 0 \\ -3 & 2 & 1 \end{pmatrix}$ **(e)** $\begin{pmatrix} 5 & -6 & 3 \\ 0 & 4 & -1 \\ 3 & -1 & 2 \end{pmatrix}$

4. $[\mathsf{T}]_{\beta'} = \begin{pmatrix} 2 & -1 \\ -1 & 1 \end{pmatrix}\begin{pmatrix} 2 & 1 \\ 1 & -3 \end{pmatrix}\begin{pmatrix} 1 & 1 \\ 1 & 2 \end{pmatrix} = \begin{pmatrix} 8 & 13 \\ -5 & -9 \end{pmatrix}$

5. $[\mathsf{T}]_{\beta'} = \begin{pmatrix} \frac{1}{2} & \frac{1}{2} \\ \frac{1}{2} & -\frac{1}{2} \end{pmatrix}\begin{pmatrix} 0 & 1 \\ 0 & 0 \end{pmatrix}\begin{pmatrix} 1 & 1 \\ 1 & -1 \end{pmatrix} = \begin{pmatrix} \frac{1}{2} & -\frac{1}{2} \\ \frac{1}{2} & -\frac{1}{2} \end{pmatrix}$

6. $\mathsf{T}(x, y) = \dfrac{1}{1 + m^2}((1 - m^2)x + 2my, 2mx + (m^2 - 1)y)$

SECTION 2.6

1. **(a)** F **(b)** T **(c)** T **(d)** T **(e)** F **(f)** T **(g)** T **(h)** F
2. The functions in parts (a), (c), (e), and (f) are linear functionals.
3. **(a)** $\mathsf{f}_1(x, y, z) = x - \frac{1}{2}y$, $\mathsf{f}_2(x, y, z) = \frac{1}{2}y$, and $\mathsf{f}_3(x, y, z) = -x + z$
5. The basis for V is $\{p_1(x), p_2(x)\}$, where $p_1(x) = 2 - 2x$ and $p_2(x) = -\frac{1}{2} + x$.
7. **(a)** $\mathsf{T}^t(\mathsf{f}) = \mathsf{g}$, where $\mathsf{g}(a + bx) = -3a - 4b$

   (b) $[\mathsf{T}^t]_{\gamma^*}^{\beta^*} = \begin{pmatrix} -1 & 1 \\ -2 & 1 \end{pmatrix}$ **(c)** $[\mathsf{T}]_\beta^\gamma = \begin{pmatrix} -1 & -2 \\ 1 & 1 \end{pmatrix}$

SECTION 2.7

1. **(a)** T **(b)** T **(c)** F **(d)** F **(e)** T **(f)** F **(g)** T
2. **(a)** F **(b)** F **(c)** T **(d)** T **(e)** F
3. **(a)** $\{e^{-t}, te^{-t}\}$ **(c)** $\{e^{-t}, te^{-t}, e^t, te^t\}$ **(e)** $\{e^{-t}, e^t \cos 2t, e^t \sin 2t\}$
4. **(a)** $\{e^{(1+\sqrt{5})t/2}, e^{(1-\sqrt{5})t/2}\}$ **(c)** $\{1, e^{-4t}, e^{-2t}\}$

SECTION 3.1

1. **(a)** T **(b)** F **(c)** T **(d)** F **(e)** T **(f)** F
   (g) T **(h)** F **(i)** T
2. Adding -2 times column 1 to column 2 transforms A into B.

SECTION 3.2

1. **(a)** F **(b)** F **(c)** T **(d)** T **(e)** F **(f)** T
   (g) T **(h)** T **(i)** T
2. **(a)** 2 **(c)** 2 **(e)** 3 **(g)** 1
4. **(a)** $D = \begin{pmatrix} 1 & 0 & 0 & 0 \\ 0 & 1 & 0 & 0 \\ 0 & 0 & 0 & 0 \end{pmatrix}$; the rank is 2.
5. **(a)** The rank is 2, and the inverse is $\begin{pmatrix} -1 & 2 \\ 1 & -1 \end{pmatrix}$.

   (c) The rank is 2; so no inverse exists.

   (e) The rank is 3, and the inverse is $\begin{pmatrix} \frac{1}{6} & -\frac{1}{3} & \frac{1}{2} \\ \frac{1}{2} & 0 & -\frac{1}{2} \\ -\frac{1}{6} & \frac{1}{3} & \frac{1}{2} \end{pmatrix}$.

(g) The rank is 4, and the inverse is $\begin{pmatrix} -51 & 15 & 7 & 12 \\ 31 & -9 & -4 & -7 \\ -10 & 3 & 1 & 2 \\ -3 & 1 & 1 & 1 \end{pmatrix}$.

6. (a) $\mathsf{T}^{-1}(ax^2 + bx + c) = -ax^2 - (4a + b)x - (10a + 2b + c)$
(c) $\mathsf{T}^{-1}(a, b, c) = (\frac{1}{6}a - \frac{1}{3}b + \frac{1}{2}c, \frac{1}{2}a - \frac{1}{2}c, -\frac{1}{6}a + \frac{1}{3}b + \frac{1}{2}c)$
(e) $\mathsf{T}^{-1}(a, b, c) = (\frac{1}{2}a - b + \frac{1}{2}c)x^2 + (-\frac{1}{2}a + \frac{1}{2}c)x + b$

7. $\begin{pmatrix} 1 & 0 & 0 \\ 0 & 1 & 0 \\ 1 & 0 & 1 \end{pmatrix}\begin{pmatrix} 1 & 0 & 0 \\ 1 & 1 & 0 \\ 0 & 0 & 1 \end{pmatrix}\begin{pmatrix} 1 & 0 & 0 \\ 0 & -2 & 0 \\ 0 & 0 & 1 \end{pmatrix}\begin{pmatrix} 1 & 2 & 0 \\ 0 & 1 & 0 \\ 0 & 0 & 1 \end{pmatrix}\begin{pmatrix} 1 & 0 & 0 \\ 0 & 1 & 0 \\ 0 & -1 & 1 \end{pmatrix}\begin{pmatrix} 1 & 0 & 1 \\ 0 & 1 & 0 \\ 0 & 0 & 1 \end{pmatrix}$

SECTION 3.3

1. (a) F (b) F (c) T (d) F (e) F (f) F
(g) T (h) F

2. (a) $\left\{\begin{pmatrix} -3 \\ 1 \end{pmatrix}\right\}$ (c) $\left\{\begin{pmatrix} -1 \\ 1 \\ 1 \end{pmatrix}\right\}$

(e) $\left\{\begin{pmatrix} -2 \\ 1 \\ 0 \\ 0 \end{pmatrix}, \begin{pmatrix} 3 \\ 0 \\ 1 \\ 0 \end{pmatrix}, \begin{pmatrix} -1 \\ 0 \\ 0 \\ 1 \end{pmatrix}\right\}$ (g) $\left\{\begin{pmatrix} -3 \\ 1 \\ 1 \\ 0 \end{pmatrix}, \begin{pmatrix} 1 \\ -1 \\ 0 \\ 1 \end{pmatrix}\right\}$

3. (a) $\left\{\begin{pmatrix} 5 \\ 0 \end{pmatrix} + t\begin{pmatrix} -3 \\ 1 \end{pmatrix}: \quad t \in R\right\}$

(c) $\left\{\begin{pmatrix} 2 \\ 1 \\ 1 \end{pmatrix} + t\begin{pmatrix} -1 \\ 1 \\ 1 \end{pmatrix}: \quad t \in R\right\}$

(e) $\left\{\begin{pmatrix} 1 \\ 0 \\ 0 \\ 0 \end{pmatrix} + r\begin{pmatrix} -2 \\ 1 \\ 0 \\ 0 \end{pmatrix} + s\begin{pmatrix} 3 \\ 0 \\ 1 \\ 0 \end{pmatrix} + t\begin{pmatrix} -1 \\ 0 \\ 0 \\ 1 \end{pmatrix}: \quad r, s, t \in R\right\}$

(g) $\left\{\begin{pmatrix} 0 \\ 0 \\ 0 \\ 1 \end{pmatrix} + r\begin{pmatrix} -3 \\ 1 \\ 1 \\ 0 \end{pmatrix} + s\begin{pmatrix} 1 \\ -1 \\ 0 \\ 1 \end{pmatrix}: \quad r, s \in R\right\}$

4. **(b)** (2) $A^{-1} = \begin{pmatrix} \frac{1}{3} & 0 & \frac{1}{3} \\ \frac{1}{9} & \frac{1}{3} & -\frac{2}{9} \\ -\frac{4}{9} & \frac{2}{3} & -\frac{1}{9} \end{pmatrix}$ (3) $\begin{pmatrix} x_1 \\ x_2 \\ x_3 \end{pmatrix} = \begin{pmatrix} 3 \\ 0 \\ -2 \end{pmatrix}$

6. $$\mathsf{T}^{-1}\{(1, 11)\} = \left\{ \begin{pmatrix} \frac{11}{2} \\ -\frac{9}{2} \\ 0 \end{pmatrix} + t \begin{pmatrix} 1 \\ -1 \\ 2 \end{pmatrix} : \quad t \in R \right\}$$

7. The systems in parts (b), (c), and (d) have solutions.

11. The farmer, tailor, and carpenter must have incomes in the proportions 4:3:4.

13. There must be 7.8 units of the first commodity and 9.5 units of the second.

SECTION 3.4

1. **(a)** F **(b)** T **(c)** T **(d)** T **(e)** F **(f)** T **(g)** T

2. **(a)** $\begin{pmatrix} 4 \\ -3 \\ -1 \end{pmatrix}$ **(c)** $\begin{pmatrix} 2 \\ 3 \\ -2 \\ -1 \end{pmatrix}$

(e) $$\left\{ \begin{pmatrix} 4 \\ 0 \\ 1 \\ 0 \end{pmatrix} + r \begin{pmatrix} 4 \\ 1 \\ 0 \\ 0 \end{pmatrix} + s \begin{pmatrix} 1 \\ 0 \\ 2 \\ 1 \end{pmatrix} : \quad r, s \in R \right\}$$

(g) $$\left\{ \begin{pmatrix} -23 \\ 0 \\ 7 \\ 9 \\ 0 \end{pmatrix} + r \begin{pmatrix} 1 \\ 1 \\ 0 \\ 0 \\ 0 \end{pmatrix} + s \begin{pmatrix} -23 \\ 0 \\ 6 \\ 9 \\ 1 \end{pmatrix} : \quad r, s \in R \right\}$$

(i) $$\left\{ \begin{pmatrix} 2 \\ 0 \\ 0 \\ -1 \\ 0 \end{pmatrix} + r \begin{pmatrix} 0 \\ 2 \\ 1 \\ 0 \\ 0 \end{pmatrix} + s \begin{pmatrix} 1 \\ -4 \\ 0 \\ -2 \\ 1 \end{pmatrix} : \quad r, s \in R \right\}$$

4. **(a)** $$\left\{ \begin{pmatrix} \frac{4}{3} \\ \frac{1}{3} \\ 0 \\ 0 \end{pmatrix} + t \begin{pmatrix} 1 \\ -1 \\ 1 \\ 2 \end{pmatrix} : \quad t \in R \right\}$$

(c) There are no solutions.

SECTION 4.1

1. **(a)** T **(b)** F **(c)** T **(d)** F **(e)** T **(f)** F
2. **(a)** 30 **(c)** -8
3. **(a)** $-10+15i$ **(c)** -24
4. **(a)** 19 **(c)** 14

SECTION 4.2

1. **(a)** T **(b)** T **(c)** F **(d)** F **(e)** T **(f)** F
3. **(a)** -34 **(c)** -49
4. The functions in parts (c), (d), and (g) are 3-linear.

SECTION 4.3

1. **(a)** T **(b)** T **(c)** F **(d)** T **(e)** F **(f)** T **(g)** F
   (h) T **(i)** T **(j)** F **(k)** T **(l)** F **(m)** F
2. **(a)** 90 **(c)** 0
3. **(a)** 100 **(c)** 0 **(e)** 86 **(g)** $-102+76i$

16. **(a)** $x_1 = \dfrac{b_1a_{22} - b_2a_{12}}{a_{11}a_{22} - a_{12}a_{21}},\ x_2 = \dfrac{b_2a_{11} - b_1a_{21}}{a_{11}a_{22} - a_{12}a_{21}}$
    (c) $x_1 = -1,\ x_2 = -1.2,\ x_3 = -1.4$
    (e) $x_1 = -43,\ x_2 = -109,\ x_3 = -17$

18. **(a)** $\begin{pmatrix} A_{22} & -A_{12} \\ -A_{21} & A_{11} \end{pmatrix}$ **(c)** $\begin{pmatrix} -3i & 0 & 0 \\ 4 & -1+i & 0 \\ 10+16i & -5-3i & 3+3i \end{pmatrix}$
    (e) $\begin{pmatrix} 6 & 22 & 12 \\ 12 & -2 & 24 \\ 21 & -38 & -27 \end{pmatrix}$ **(g)** $\begin{pmatrix} 18 & 28 & -6 \\ -20 & -21 & 37 \\ 48 & 14 & -16 \end{pmatrix}$

SECTION 4.4

1. **(a)** T **(b)** T **(c)** T **(d)** F **(e)** F **(f)** T
   (g) T **(h)** F **(i)** T **(j)** T **(k)** T
2. **(a)** 22 **(c)** $2-4i$
3. **(a)** -12 **(c)** 22 **(e)** -3
4. **(a)** 88 **(c)** -6 **(e)** $17-3i$ **(g)** $24+24i$

SECTION 5.1

1. **(a)** F **(b)** T **(c)** T **(d)** F **(e)** F **(f)** F
   (g) F **(h)** T **(i)** T **(j)** F **(k)** F

2. **(a)** $Q = \begin{pmatrix} 1 & 1 \\ 1 & 2 \end{pmatrix}$, $[\mathsf{L}_A]_\beta = \begin{pmatrix} 6 & 11 \\ -2 & -4 \end{pmatrix}$

(c) $$Q = \begin{pmatrix} 1 & 1 & 1 \\ 1 & 0 & 1 \\ 1 & 1 & 2 \end{pmatrix}, \quad [\mathsf{L}_A]_\beta = \begin{pmatrix} 2 & 2 & 2 \\ -2 & -3 & -4 \\ 1 & 1 & 2 \end{pmatrix}$$

3. **(a)** The eigenvalues are 4 and -1, a basis of eigenvectors is

$$\left\{\begin{pmatrix} 2 \\ 3 \end{pmatrix}, \begin{pmatrix} 1 \\ -1 \end{pmatrix}\right\}, \quad Q = \begin{pmatrix} 2 & 1 \\ 3 & -1 \end{pmatrix}, \quad \text{and} \quad D = \begin{pmatrix} 4 & 0 \\ 0 & -1 \end{pmatrix}.$$

(c) The eigenvalues are 1 and -1, a basis of eigenvectors is

$$\left\{\begin{pmatrix} 1 \\ 1-i \end{pmatrix}, \begin{pmatrix} 1 \\ -1-i \end{pmatrix}\right\}, \quad Q = \begin{pmatrix} 1 & 1 \\ 1-i & -1-i \end{pmatrix}, \quad \text{and} \quad D = \begin{pmatrix} 1 & 0 \\ 0 & -1 \end{pmatrix}.$$

4. The eigenvalues are 1, 2, and 3, and a basis of eigenvectors is $\{1, x, x^2\}$.

25. 4

SECTION 5.2

1. **(a)** F **(b)** F **(c)** F **(d)** T **(e)** T **(f)** T
(g) T **(h)** T **(i)** F

2. **(a)** Not diagonalizable. **(c)** $Q = \begin{pmatrix} 1 & 4 \\ 1 & -3 \end{pmatrix}$

(e) Not diagonalizable. **(g)** $$Q = \begin{pmatrix} 1 & 1 & 1 \\ 2 & -1 & 0 \\ -1 & 0 & -1 \end{pmatrix}$$

3. **(a)** Not diagonalizable. **(c)** Not diagonalizable.
(d) $\beta = \{x - x^2, 1 - x - x^2, x + x^2\}$

(e) $\beta = \left\{\begin{pmatrix} 1 \\ 1 \end{pmatrix}, \begin{pmatrix} 1 \\ -1 \end{pmatrix}\right\}$

7. $$A^n = \begin{pmatrix} \dfrac{5^n}{3} + \dfrac{2(-1)^n}{3} & \dfrac{2(5^n)}{3} - \dfrac{2(-1)^n}{3} \\[2ex] \dfrac{5^n}{3} - \dfrac{(-1)^n}{3} & \dfrac{2(5^n)}{3} + \dfrac{(-1)^n}{3} \end{pmatrix}$$

13. **(b)** $X(t) = c_1 e^{3t}\begin{pmatrix} -2 \\ 1 \end{pmatrix} + c_2 e^{-2t}\begin{pmatrix} 1 \\ -1 \end{pmatrix}$

(c) $$X(t) = e^t\left[c_1\begin{pmatrix} 1 \\ 0 \\ 0 \end{pmatrix} + c_2\begin{pmatrix} 0 \\ 1 \\ 0 \end{pmatrix}\right] c_3 e^{2t}\begin{pmatrix} 1 \\ 1 \\ 1 \end{pmatrix}$$

SECTION 5.3

1. **(a)** T **(b)** T **(c)** F **(d)** F **(e)** T **(f)** T
(g) T **(h)** F **(i)** F **(j)** T

2. **(a)** $\begin{pmatrix} 0 & 0 \\ 0 & 0 \end{pmatrix}$ **(c)** $\begin{pmatrix} \frac{7}{13} & \frac{7}{13} \\ \frac{6}{13} & \frac{6}{13} \end{pmatrix}$ **(e)** No limit exists.

(g) $\begin{pmatrix} -1 & 0 & -1 \\ -4 & 1 & -2 \\ 2 & 0 & 2 \end{pmatrix}$ **(i)** No limit exists.

6. One month after arrival 25% of the patients have recovered, 20% are ambulatory, 41% are bedridden, and 14% have died; eventually $\frac{59}{90}$ recover and $\frac{31}{90}$ die.

7. $\frac{3}{7}$

8. Only the matrices in parts (a) and (b) are regular transition matrices.

9. **(a)** $\begin{pmatrix} \frac{1}{3} & \frac{1}{3} & \frac{1}{3} \\ \frac{1}{3} & \frac{1}{3} & \frac{1}{3} \\ \frac{1}{3} & \frac{1}{3} & \frac{1}{3} \end{pmatrix}$ **(c)** No limit exists.

(e) $\begin{pmatrix} 0 & 0 & 0 \\ \frac{1}{2} & 1 & 0 \\ \frac{1}{2} & 0 & 1 \end{pmatrix}$ **(g)** $\begin{pmatrix} 0 & 0 & 0 & 0 \\ 0 & 0 & 0 & 0 \\ \frac{1}{2} & \frac{1}{2} & 1 & 0 \\ \frac{1}{2} & \frac{1}{2} & 0 & 1 \end{pmatrix}$

10. **(a)** $\begin{pmatrix} 0.225 \\ 0.441 \\ 0.334 \end{pmatrix}$ after two stages and $\begin{pmatrix} 0.20 \\ 0.60 \\ 0.20 \end{pmatrix}$ eventually.

(c) $\begin{pmatrix} 0.372 \\ 0.225 \\ 0.403 \end{pmatrix}$ after two stages and $\begin{pmatrix} 0.50 \\ 0.20 \\ 0.30 \end{pmatrix}$ eventually.

(e) $\begin{pmatrix} 0.329 \\ 0.334 \\ 0.337 \end{pmatrix}$ after two stages and $\begin{pmatrix} \frac{1}{3} \\ \frac{1}{3} \\ \frac{1}{3} \end{pmatrix}$ eventually.

12. $\frac{9}{19}$ new, $\frac{6}{19}$ once-used, and $\frac{4}{19}$ twice-used.

13. In 1995 24% will own large cars, 34% will own intermediate-sized cars, and 42% will own small cars; the corresponding eventual proportions are 0.10, 0.30, and 0.60.

18. $e^O = I$ and $e^I = eI$.

SECTION 5.4

1. **(a)** F **(b)** T **(c)** F **(d)** F **(e)** T **(f)** T **(g)** T

2. The subspaces in parts (a), (c), and (d) are T-invariant.

6. **(a)** $\left\{\begin{pmatrix}1\\0\\0\\0\end{pmatrix}, \begin{pmatrix}1\\0\\1\\1\end{pmatrix}, \begin{pmatrix}1\\-1\\2\\2\end{pmatrix}\right\}$ **(c)** $\left\{\begin{pmatrix}0 & 1\\1 & 0\end{pmatrix}\right\}$

9. **(a)** $-t(t^2 - 3t + 3)$ **(c)** $1 - t$

12. **(a)** $t(t - 1)(t^2 - 3t + 3)$ **(c)** $(t - 1)^3(t + 1)$

18. **(c)**

$$A^{-1} = \tfrac{1}{2}\begin{pmatrix}2 & -2 & -4\\0 & 1 & 3\\0 & 0 & -2\end{pmatrix}$$

30. **(a)** $6 - 6t + t^2$

SECTION 6.1

1. **(a)** T **(b)** T **(c)** F **(d)** F **(e)** F **(f)** F **(g)** T

2. $\langle x, y\rangle = 8 + 5i$, $\|x\| = \sqrt{7}$, $\|y\| = \sqrt{14}$, and $\|x + y\|^2 = 37$.

3. $\langle f, g\rangle = 1$, $\|f\| = \dfrac{\sqrt{3}}{3}$, $\|g\| = \sqrt{\dfrac{e^2 - 1}{2}}$,

   and $\|f + g\| = \sqrt{\dfrac{11 + 3e^2}{6}}$.

16. **(b)** No.

SECTION 6.2

1. **(a)** F **(b)** T **(c)** T **(d)** F **(e)** T **(f)** F **(g)** T

2. **(b)** The orthonormal basis is

$$\left\{\frac{\sqrt{3}}{3}(1, 1, 1), \frac{\sqrt{6}}{6}(-2, 1, 1), \frac{\sqrt{2}}{2}(0, -1, 1)\right\}.$$

   The Fourier coefficients are $2\sqrt{3}/3$, $-\sqrt{6}/6$, and $\sqrt{2}/2$.

   (c) The orthonormal basis is $\{1, 2\sqrt{3}(x - \tfrac{1}{2}), 6\sqrt{5}(x^2 - x + \tfrac{1}{6})\}$.

   The Fourier coefficients are $3/2$, $\sqrt{3}/6$, and 0.

4. $S^\perp = \text{span}\{(i, -\tfrac{1}{2}(1 + i), 1)\}$

5. $S^\perp$ is the plane through the origin that is perpendicular to x_0; $S_0^\perp$ is the line through the origin that is perpendicular to the plane containing x_1 and x_2.

17. **(a)** $\dfrac{1}{17}\begin{pmatrix}26\\104\end{pmatrix}$ **(b)** $\dfrac{1}{14}\begin{pmatrix}29\\17\\40\end{pmatrix}$

18. **(b)** $1/\sqrt{14}$

SECTION 6.3

1. **(a)** T **(b)** F **(c)** F **(d)** T **(e)** F **(f)** T **(g)** T
2. **(a)** $y = (1, -2, 4)$ **(c)** $y = 210x^2 - 204x + 33$
3. **(a)** $\mathsf{T}^*(x) = (11, -12)$ **(c)** $\mathsf{T}^*(f(x)) = 69x^2 - 9x - 5$
14. $\mathsf{T}^*(x) = \langle x, z \rangle y$
18. The line is $y = -2t + \frac{5}{2}$ with $E = 1$, and the parabola is $y = t^2/3 - 4t/3 + 2$ with $E = 0$.
19. The spring constant is approximately 2.1.
20. $x = \frac{2}{7}, y = \frac{3}{7}, z = \frac{1}{7}$

SECTION 6.4

1. **(a)** T **(b)** F **(c)** F **(d)** T **(e)** T **(f)** T
   (g) F **(h)** T
2. **(a)** T is self-adjoint; the orthonormal basis is $\left\{\frac{1}{\sqrt{5}}(1, -2), \frac{1}{\sqrt{5}}(2, 1)\right\}$.
   (b) T is normal but not self-adjoint.
   (c) T is not normal.

SECTION 6.5

1. **(a)** T **(b)** F **(c)** F **(d)** T **(e)** F **(f)** T
   (g) F **(h)** F **(i)** F
2. **(a)** $P = \frac{1}{\sqrt{2}}\begin{pmatrix} 1 & 1 \\ 1 & -1 \end{pmatrix}$ and $D = \begin{pmatrix} 3 & 0 \\ 0 & -1 \end{pmatrix}$

   (d)
$$P = \begin{pmatrix} \frac{1}{\sqrt{2}} & \frac{1}{\sqrt{6}} & \frac{1}{\sqrt{3}} \\ -\frac{1}{\sqrt{2}} & \frac{1}{\sqrt{6}} & \frac{1}{\sqrt{3}} \\ 0 & -\frac{2}{\sqrt{6}} & \frac{1}{\sqrt{3}} \end{pmatrix} \quad \text{and} \quad D = \begin{pmatrix} -2 & 0 & 0 \\ 0 & -2 & 0 \\ 0 & 0 & 4 \end{pmatrix}$$

4. T_z is normal for all $z \in C$; T_z is self-adjoint if and only if $z \in R$; T_z is unitary if and only if $|z| = 1$.
5. Only the pair of matrices in part (d) is unitarily equivalent.
21. **(a)** $x = \frac{1}{\sqrt{2}}x' + \frac{1}{\sqrt{2}}y'$ and $y = \frac{1}{\sqrt{2}}x' - \frac{1}{\sqrt{2}}y'$
    The new quadratic form is $3(x')^2 - (y')^2$.

    (c) $x = \frac{3}{\sqrt{13}}x' + \frac{2}{\sqrt{13}}y'$ and $y = \frac{-2}{\sqrt{13}}x' + \frac{3}{\sqrt{13}}y'$
    The new quadratic form is $5x^2 - 8y'^2$.

23. **(c)**
$$Q = \begin{pmatrix} \frac{1}{\sqrt{2}} & \frac{1}{\sqrt{3}} & -\frac{\sqrt{6}}{6} \\ \frac{1}{\sqrt{2}} & -\frac{1}{\sqrt{3}} & \frac{\sqrt{6}}{6} \\ 0 & \frac{1}{\sqrt{3}} & \frac{\sqrt{6}}{3} \end{pmatrix} \quad \text{and} \quad R = \begin{pmatrix} \sqrt{2} & \sqrt{2} & 2\sqrt{2} \\ 0 & \sqrt{3} & \frac{\sqrt{3}}{3} \\ 0 & 0 & \frac{\sqrt{6}}{3} \end{pmatrix}$$

(e) $x_1 = 3,\ x_2 = -5,\ x_3 = 4$

SECTION 6.6

1. **(a)** F **(b)** T **(c)** T **(d)** T **(e)** F **(f)** F

2. For $\mathsf{W} = \text{span}(\{(1, 2)\})$, $[\mathsf{T}]_\beta = \begin{pmatrix} \frac{1}{5} & \frac{2}{5} \\ \frac{2}{5} & \frac{4}{5} \end{pmatrix}$.

3. (2) **(a)** $\mathsf{T}_1(a, b) = \frac{1}{2}(a + b, a + b)$ and $\mathsf{T}_2(a, b) = \frac{1}{2}(a - b, -a + b)$
(d) $\mathsf{T}_1(a, b, c) = \frac{1}{3}(2a - b - c, -a + 2b - c, -a - b + 2c)$ and $\mathsf{T}_2(a, b, c) = \frac{1}{3}(a + b + c, a + b + c, a + b + c)$

SECTION 6.7

1. **(a)** F **(b)** F **(c)** T **(d)** F **(e)** T **(f)** F
(g) F **(h)** F **(i)** T **(j)** F

4. **(a)** Yes. **(b)** No. **(c)** No. **(d)** Yes. **(e)** Yes.
(f) No.

5. **(a)** $\begin{pmatrix} 0 & 2 & -2 \\ 2 & 0 & -2 \\ 1 & 1 & 0 \end{pmatrix}$ **(b)** $\begin{pmatrix} 1 & 0 & 0 & 1 \\ 0 & 0 & 0 & 0 \\ 0 & 0 & 0 & 0 \\ 1 & 0 & 0 & 1 \end{pmatrix}$ **(c)** $\begin{pmatrix} 0 & 0 & 0 & 0 \\ -1 & 0 & -4 & 0 \\ 0 & 0 & 0 & 0 \\ -2 & 0 & -8 & 0 \end{pmatrix}$

16. **(a)** $\left\{ \begin{pmatrix} \frac{2}{\sqrt{5}} \\ -\frac{1}{\sqrt{5}} \end{pmatrix}, \begin{pmatrix} \frac{1}{\sqrt{5}} \\ \frac{2}{\sqrt{5}} \end{pmatrix} \right\}$

(b) Same as part (a).

(c) $\left\{ \begin{pmatrix} \frac{1}{\sqrt{2}} \\ 0 \\ \frac{1}{\sqrt{2}} \end{pmatrix}, \begin{pmatrix} 0 \\ 1 \\ 0 \end{pmatrix}, \begin{pmatrix} \frac{1}{\sqrt{2}} \\ 0 \\ -\frac{1}{\sqrt{2}} \end{pmatrix} \right\}$

17. Same as Exercise 16(c).

21. **(a)** $Q = \begin{pmatrix} 1 & -3 \\ 0 & 1 \end{pmatrix}$ and $D = \begin{pmatrix} 1 & 0 \\ 0 & -7 \end{pmatrix}$

(b) $Q = \begin{pmatrix} 1 & -\frac{1}{2} \\ 1 & \frac{1}{2} \end{pmatrix}$ and $D = \begin{pmatrix} 2 & 0 \\ 0 & -\frac{1}{2} \end{pmatrix}$

(c) $Q = \begin{pmatrix} 0 & 0 & 1 \\ 0 & 1 & -0.25 \\ 1 & 0 & 2 \end{pmatrix}$ and $D = \begin{pmatrix} -1 & 0 & 0 \\ 0 & 4 & 0 \\ 0 & 0 & 6.75 \end{pmatrix}$

SECTION 6.9

1. **(a)** F **(b)** T **(c)** T **(d)** F **(e)** F

2. **(a)** $\sqrt{18}$ **(c)** approximately 2.34

4. **(a)** $\|A\| \approx 84.74$, $\|A^{-1}\| \approx 17.01$, and $\text{cond}(A) \approx 1441$

(b) $\|\tilde{x} - A^{-1}b\| \leqslant \|A^{-1}\| \cdot \|A\tilde{x} - b\| \approx 0.17$ and

$$\frac{\|\tilde{x} - A^{-1}b\|}{\|A^{-1}b\|} \leqslant \text{cond}(A)\frac{\|b - A\tilde{x}\|}{\|b\|} \approx \frac{14.41}{\|b\|}$$

5. $0.001 \leqslant \dfrac{\|x - \tilde{x}\|}{\|x\|} \leqslant 10$

6. $R\begin{pmatrix} 1 \\ -2 \\ 3 \end{pmatrix} = \dfrac{9}{7}$, $\|B\| = 2$, and $\text{cond}(B) = 2$

SECTION 6.10

1. **(a)** F **(b)** T **(c)** T **(d)** F **(e)** T **(f)** F
(g) F **(h)** T **(i)** T **(j)** F

3. **(b)** $\left\{t\begin{pmatrix} \sqrt{3} \\ 1 \end{pmatrix} : \quad t \in R\right\}$

4. **(b)** $\left\{t\begin{pmatrix} 1 \\ 0 \end{pmatrix} : \quad t \in R\right\}$ if $\phi = 0$ and $\left\{t\begin{pmatrix} \cos\phi + 1 \\ \sin\phi \end{pmatrix} : \quad t \in R\right\}$ if $\phi \neq 0$

7. **(c)** There are six possibilities:

(1) Any line through the origin if $\phi = \psi = 0$.

(2) $\left\{t\begin{pmatrix} 0 \\ 0 \\ 1 \end{pmatrix} : \quad t \in R\right\}$ if $\phi = 0$ and $\psi = \pi$.

(3) $\left\{t\begin{pmatrix} \cos\psi + 1 \\ -\sin\psi \\ 0 \end{pmatrix} : \quad t \in R\right\}$ if $\phi = \pi$ and $\psi \neq \pi$.

(4) $\left\{ t \begin{pmatrix} 0 \\ \cos\phi - 1 \\ \sin\phi \end{pmatrix} : \quad t \in R \right\}$ if $\psi = \pi$ and $\phi \neq \pi$.

(5) $\left\{ t \begin{pmatrix} 0 \\ 1 \\ 0 \end{pmatrix} : \quad t \in R \right\}$ if $\phi = \psi = \pi$.

(6) $\left\{ t \begin{pmatrix} \sin\phi(\cos\psi + 1) \\ -\sin\phi\sin\psi \\ \sin\psi(\cos\phi + 1) \end{pmatrix} : \quad t \in R \right\}$ otherwise.

SECTION 7.1

1. **(a)** T **(b)** F **(c)** F **(d)** T **(e)** F **(f)** F
   (g) T **(h)** T
2. **(a)** For $\lambda = 2$,

$$\left\{ \begin{pmatrix} 1 \\ 1 \end{pmatrix} \right\}$$

is a basis for E_λ; any basis for R^2 is a basis for K_λ.

(b) For $\lambda = -1$,

$$\left\{ \begin{pmatrix} 1 \\ 3 \\ 0 \end{pmatrix} \right\}$$

is a basis for both E_λ and K_λ.
For $\lambda = 2$,

$$\left\{ \begin{pmatrix} 1 \\ 1 \\ 1 \end{pmatrix} \right\}$$

is a basis for E_λ and

$$\left\{ \begin{pmatrix} 0 \\ 1 \\ -1 \end{pmatrix}, \begin{pmatrix} 1 \\ 1 \\ 1 \end{pmatrix} \right\}$$

is a basis for K_λ.

(c) For $\lambda = 2$,

$$\left\{ \begin{pmatrix} 2 \\ 0 \\ 0 \end{pmatrix} \right\}$$

is a basis for E_λ and

$$\left\{ \begin{pmatrix} 0 \\ 0 \\ 1 \end{pmatrix}, \begin{pmatrix} 0 \\ -2 \\ 0 \end{pmatrix}, \begin{pmatrix} 2 \\ 0 \\ 0 \end{pmatrix} \right\}$$

is a basis for K_λ.

3. (a) $\begin{pmatrix} 2 & 1 \\ 0 & 2 \end{pmatrix}$ **(b)** $\begin{pmatrix} -1 & 0 & 0 \\ 0 & 2 & 1 \\ 0 & 0 & 2 \end{pmatrix}$ **(c)** $\begin{pmatrix} 2 & 1 & 0 \\ 0 & 2 & 1 \\ 0 & 0 & 2 \end{pmatrix}$

SECTION 7.2

1. (a) T **(b)** T **(c)** F **(d)** T **(e)** T **(f)** F
(g) F **(h)** T

2.

$$\left(\begin{array}{ccc|cc|c|ccc|c|c|c}
2 & 1 & 0 & 0 & 0 & 0 & 0 & 0 & 0 & 0 & 0 & 0 \\
0 & 2 & 1 & 0 & 0 & 0 & 0 & 0 & 0 & 0 & 0 & 0 \\
0 & 0 & 2 & 0 & 0 & 0 & 0 & 0 & 0 & 0 & 0 & 0 \\
\hline
0 & 0 & 0 & 2 & 1 & 0 & 0 & 0 & 0 & 0 & 0 & 0 \\
0 & 0 & 0 & 0 & 2 & 0 & 0 & 0 & 0 & 0 & 0 & 0 \\
\hline
0 & 0 & 0 & 0 & 0 & 2 & 0 & 0 & 0 & 0 & 0 & 0 \\
\hline
0 & 0 & 0 & 0 & 0 & 0 & 4 & 1 & 0 & 0 & 0 & 0 \\
0 & 0 & 0 & 0 & 0 & 0 & 0 & 4 & 1 & 0 & 0 & 0 \\
0 & 0 & 0 & 0 & 0 & 0 & 0 & 0 & 4 & 0 & 0 & 0 \\
\hline
0 & 0 & 0 & 0 & 0 & 0 & 0 & 0 & 0 & 4 & 0 & 0 \\
\hline
0 & 0 & 0 & 0 & 0 & 0 & 0 & 0 & 0 & 0 & -3 & 0 \\
\hline
0 & 0 & 0 & 0 & 0 & 0 & 0 & 0 & 0 & 0 & 0 & -3
\end{array}\right)$$

3. (a) $-(t-2)^5(t-3)^2$

(b) For $\lambda_1 = 2$ For $\lambda_2 = 3$

$$\begin{array}{cc} \cdot & \cdot \\ \cdot & \cdot \\ \cdot & \end{array} \qquad\qquad \begin{array}{cc} \cdot & \cdot \end{array}$$

(c) $\lambda_2 = 3$

(d) $p_1 = 3$ and $p_2 = 1$

(e) (1) $\operatorname{rank}(\mathsf{U}_1) = 3$ and $\operatorname{rank}(\mathsf{U}_2) = 0$

(2) $\operatorname{rank}(\mathsf{U}_1^2) = 1$ and $\operatorname{rank}(\mathsf{U}_2^2) = 0$

(3) $\operatorname{nullity}(\mathsf{U}_1) = 2$ and $\operatorname{nullity}(\mathsf{U}_2) = 2$

(4) $\operatorname{nullity}(\mathsf{U}_1^2) = 4$ and $\operatorname{nullity}(\mathsf{U}_2^2) = 2$

4. (a)

$$J = \left(\begin{array}{c|cc} 1 & 0 & 0 \\ \hline 0 & 2 & 1 \\ 0 & 0 & 2 \end{array}\right) \quad \text{and} \quad Q = \begin{pmatrix} 1 & 1 & 1 \\ 2 & 1 & 2 \\ 1 & -1 & 0 \end{pmatrix}$$

(d)
$$J = \left(\begin{array}{cc|cc} 0 & 1 & 0 & 0 \\ 0 & 0 & 0 & 0 \\ \hline 0 & 0 & 2 & 0 \\ 0 & 0 & 0 & 2 \end{array}\right) \quad \text{and} \quad Q = \begin{pmatrix} 1 & 0 & 1 & -1 \\ 1 & -1 & 0 & 1 \\ 1 & -2 & 0 & 1 \\ 1 & 0 & 1 & 0 \end{pmatrix}$$

6. The Jordan canonical form is

$$\left(\begin{array}{ccc|c} 1 & 1 & 0 & 0 \\ 0 & 1 & 1 & 0 \\ 0 & 0 & 1 & 0 \\ \hline 0 & 0 & 0 & 2 \end{array}\right),$$

and a Jordan canonical basis is $\{2e^x, 2xe^x, x^2e^x, e^{2x}\}$.

23. (a)
$$\begin{pmatrix} x \\ y \\ z \end{pmatrix} = e^{2t}\left[(c_1 + c_2t)\begin{pmatrix} 1 \\ 0 \\ 0 \end{pmatrix} + c_2\begin{pmatrix} 0 \\ 1 \\ 0 \end{pmatrix}\right] + c_3e^{3t}\begin{pmatrix} 1 \\ 1 \\ -1 \end{pmatrix}$$

(b)
$$\begin{pmatrix} x \\ y \\ z \end{pmatrix} = e^{2t}\left[(c_1 + c_2t + c_3t^2)\begin{pmatrix} 1 \\ 0 \\ 0 \end{pmatrix} + (c_2 + 2c_3t)\begin{pmatrix} 0 \\ 1 \\ 0 \end{pmatrix} + 2c_3\begin{pmatrix} 0 \\ 0 \\ 1 \end{pmatrix}\right]$$

SECTION 7.3

1. (a) F **(b)** T **(c)** F **(d)** F **(e)** T **(f)** F
(g) F **(h)** T

2. (a) $(t-1)(t-3)$ **(c)** $(t-1)^2(t-2)$
(d) $(t-2)^2$

3. (a) $t^2 - 2$ **(c)** $(t-2)^2$
(d) $(t-1)(t+1)$

4. For (2), (a); For (3), (a) and (d).

5. The operators are T_0, I, and those operators having both 0 and 1 as eigenvalues.

SECTION 7.4

1. (a) T **(b)** F **(c)** F **(d)** T **(e)** F **(f)** T **(g)** F

2. (a)
$$\begin{pmatrix} 0 & 0 & 27 \\ 1 & 0 & -27 \\ 0 & 1 & 9 \end{pmatrix}$$
(b)
$$\begin{pmatrix} 0 & -1 \\ 1 & -1 \end{pmatrix}$$

(c)
$$\begin{pmatrix} \frac{1}{2}(-1 + i\sqrt{3}) & 0 \\ 0 & \frac{1}{2}(-1 - i\sqrt{3}) \end{pmatrix}$$
(e)
$$\left(\begin{array}{cc|cc} 0 & -2 & 0 & 0 \\ 1 & 0 & 0 & 0 \\ \hline 0 & 0 & 0 & -3 \\ 0 & 0 & 1 & 0 \end{array}\right)$$

List of Frequently Used Symbols

Index of Theorems

Index